Trace Elemental Analysis of Metals

Trace Elemental Analysis of Metals

Methods and Techniques

Thomas R. Dulski

Carpenter Technology Corporation
Reading, Pennsylvania

CRC Press is an imprint of the
Taylor & Francis Group, an **informa** business

The methods and techniques described in this book are potentially hazardous. The author and the publisher. Marcel Dekker. Inc.. assume no liability whatsoever for any material. financial. or personal loss or injury incurred from the implementation of the equipment. chemicals. or procedures described herein. The ideas and statements contained in this book are those of the individual author and are not in any way endorsed by. and shall not be construed as. the ideas or statements of Carpenter Technology Corporation.

CRC Press
Taylor & Francis Group
6000 Broken Sound Parkway NW, Suite 300
Boca Raton, FL 33487-2742

First issued in paperback 2019

ISBN-13: 978-0-8247-1985-2 (hbk)
ISBN-13: 978-0-367-39969-6 (pbk)

Visit the Taylor & Francis Web site at
http://www.taylorandfrancis.com

and the CRC Press Web site at
http://www.crcpress.com

Preface

This book grew out of a perception that the analysis of metals has become increasingly focused on the characterization of lower component concentrations. An earlier work of mine, which was an attempt to survey metals analysis broadly, did not exhaustively treat this low-concentration realm now so important in commerce.

This volume provides a detailed but practical resource of information on minor, trace, and ultratrace methods and techniques for metals. The extensive references are intended to guide the working analyst to the exact details needed to solve real-world analytical problems.

The introduction places the subject area in perspective within the contexts of both materials development and analytical chemistry. The first three chapters are devoted to essential prerequisites and refinement tools that are directly applicable to measurement approaches. Each of next five chapters deals with a specific measurement technique. The ninth chapter describes a special topic—inclusions and phases—and how the chemistry laboratory can be used to provide information to supplement metallographic analysis, while safeguarding the accuracy of its own work. A final chapter summarizes the quality features that are particular to working at low analyte concentrations.

In researching and correlating the diverse literature on the subject of this volume, I was impressed by the number of both academic and industrial chemists who have contributed to this field. Many of these people I have known by reputation; a few I am honored to regard as personal friends. The high-quality, often unsung laboratory work that forms the core of this body of knowledge deserves our respect and acknowledgement. I must also here indulge in a personal note by acknowledging the support, understanding, and encouragement of my wife, Grace, and daughter, Brittany, to whom this book is dedicated.

Thomas R. Dulski

A Manual for the Chemical Analysis of Metals, MNL25, American Society for Testing and Materials, West Conshohocken, PA, 1996.

Contents

Introduction

What we understand today by the term *elemental trace analysis* represents a technical arena that is scarcely more than fifty years old. To many specialists the term is most closely associated with environmental, clinical, or nutritional surveys, but the roots of this subject and much of the continuing impetus for its advancement derive from materials research. If one views the Manhattan Project as the somber dawn of this era, then the commercial nuclear and aerospace developments that followed might be regarded as a somewhat brighter midday. From nuclear weapons to commercial power reactors, from manned space vehicles to the Concorde, trace levels of the elements required exact characterization. More recently, the development of the semiconductor industry has shifted the paradigm to a new level of requirements. The production of silicon integrated circuit chips and fiberoptic cable even has changed the term—and we hear for the first time of *ultratrace analysis*. Clean rooms and strange new instruments have entered the picture and academics are talking wistfully about counting atoms.

The analysis of metals is a major pattern inextricably woven into this fabric. When a new methodology is developed it is not long before it is being applied to gas turbine superalloys or surgical implant alloys, to ultrapurity aluminum, or to the new "clean" low alloy steels. Sometimes metals needs have driven developments. Often the application to metals followed closely upon other materials needs. Occasionally, the technology drew out the metals application, such as when a new methodology required an electrically conductive sample.

To specify exactly what is meant by this term "trace" we seem to need to designate both an era and an industry. For example, in the nickel alloy industry in the mid-1950s 0.01% was considered a trace level for the concentration of an impurity [1]. Today fractional part-per-million levels of several elements are routinely measured in nickel alloys under the appellation of trace analysis. In other metals industries today 0.001% is considered a trace level. Each industrial niche seems to retain its own implicit meanings for terms such as "trace" and "ultratrace," as well as for "major" and "minor" component concentration levels. In a work such as this, which purports to deal with these matters in a practical manner, it makes a great deal of sense to define these terms and others at the outset of any substantive discussion. If you prefer your own or your industry's nomenclature, translation should be a simple and straightforward matter.

In this text, while we will use the term "trace analysis" in general discussion to refer to all levels below 0.01% on a weight basis, let us allow the following rigorous concentration definitions to apply when we need to refer to them later in the text:

Residual (or *tramp*): <100 ppm [<0.01%(w/w)]
Trace: 1 – 100 ppm [0.0001 – 0.01%(w/w)]
Ultra-trace: <1 ppm [<0.0001%(w/w)]
Minor: 1.0 – 0.01%(w/w)
Major: >1.0%(w/w)

Let us also be sure that here at the onset we understand these levels to apply to the metal sample itself and not to some solution-based preparation. For it is easy to lay claims to detection limits and sensitivities that sound and *are* too good to be true when we control dilutions and aliquots.

What then does all this mean, and why is it important? It is easier to answer the second part of this question first. It is now well established that the physical properties of metals and alloys can be profoundly affected by extremely minute concentrations of certain elements. In the present we can say that, perhaps 90% of these effects are deleterious—the result of unwanted contamination from furnace charge materials, from materials that contact the product in processing or storage. The remaining 10% of trace elements are known to exert beneficial effects and are intentionally added to alloys either for their own influence or for the synergistic influence they exert on the effects of other alloying additions.

The list of metallurgical properties that can be significantly modified (for good or ill) by residual levels of elements is quite extensive. In steels, for example, hydrogen at very low levels can cause embrittlement by reducing the material's ductility but not its strength. Other impurities, like lead and bismuth, when concentrated at ferrous alloy grain boundaries can similarly lead to temper embrittlement and other failure modes. Some elements effect inclusion shape at trace levels (rare earths, for example) often with dramatic anisotropic effects related to rolling direction. Trace levels of certain elements affect the size and growth of grains during solidification. The cracks that sometimes form during hot work of steels (often called "hot shortness") are often the result of liquid films of traces of low melting impurity compounds at grain boundaries. Trace levels of interstitial nitrogen are responsible for the phenomenon known as "strain aging" (the return of yield point after removal by cold work). Changes in creep strength and creep plasticity have also been ascribed to the presence of residual elements.

And mechanical properties are but the tip of the iceberg. Magnetic properties of ferromagnetic alloys are affected by trace levels of carbon, sulfur, titanium, zirconium, and aluminum. Corrosion resistance in certain stainless steels has been related to traces of certain carbide- and nitride-stabilizing elements. Other oxidation resisting effects in other alloy systems have also been noted. Otherwise innocuous levels of impurities have been related to serious neutron absorption problems with alloys utilized in nuclear applications. Toxic effects are often a concern for traces of impurities in alloys that are used for surgical implants or in the food service industry.

If we are now convinced that these very low concentration levels really do matter, we should return to the larger question of what these tiny quantities, in fact, represent. How does one get some feeling for what we are dealing with? Perhaps the most effective way to achieve this is to examine some examples from everyday life, or at least from something we might be able to visualize, however dimly. Table 1 is an attempt to bring this realm of low concentrations a little closer to subjective reality.

We must also ask ourselves what accuracy means in the world of such minute and elusive quantities. Part of the answer lies in a consideration of sample size. Consider, for example a copper penny. It weighs about 2.5 grams. One part-per-million of silver in it is 2.5 micrograms—about 23 quadrillion silver atoms. One part-per-billion of lead in it is 2.5 nanograms—around 7 billion lead atoms. However, if we choose to analyze for these elements using an optical emission spark excitation technique we are effectively sampling a tiny portion of the penny—a microgram, say. We must also consider that only a fraction (hopefully a reproducible fraction) of each analyte is actually being

Table 1 Examples of Trace and Ultratrace Concentration Levels

Concentration	Example
15.8 ppm	One inch in a mile
11.6 ppm	One second in a day
1.9 ppm	One minute in a year
0.10 ppm	One penny in $100.000.00
31.8 ppb	One second in a year
5.6 ppb	One inch on the road trip from New York City to Los Angeles
0.18 ppb	One person in the present world population
0.07 ppb	One year in the age of the universe

measured. If we accept a value of 10% for the analytical efficiency we are down to 900 million silver atoms and 300 thousand lead atoms. One begins, perhaps, to see that an error in counting 300 thousand atoms has a more serious effect on the result than a similar error in counting 7 billion atoms. Likewise, an error in counting 900 million sounds worse than the same error in counting 23 quadrillion. Clearly, then, within practical limits, sample size plays as important a role in the trace realm as it does in any other concentration range. And we are cautioned not to be sanguine about microsampling techniques—in these cases we are really measuring precious little analyte.

Another part of the meaning of accuracy at these levels relates to sample homogeneity. How uniformly distributed are those silver and lead atoms in that penny? If I spark Abe Lincoln's nose will I find more lead than if I spark his beard? In this particular example the answer may be no, but for other samples—many other samples—the equivalent answer will be yes. Trace elements are notoriously heterogeneous in many metals and alloys. They accumulate at grain boundaries, they are associated with phases and inclusions. They may be concentrated or depleted at metal surfaces. Such effects make microsampling techniques especially perilous. We must look for ways to represent the average trace element concentration of the bulk material. This can take two forms—we can either statistically sample and analyze microregions and average the results, or we can take a large, representative sample and analyze it.

All of this presupposes that we know the level of heterogeneity of the analyte in the sample. Unfortunately, in the modern industrial environment this is seldom the case. It is a rarity in the application of trace analysis techniques to find oneself firmly confident of the relative error contributions of the method itself and of that due to inhomogeneity of the analyte in the test sample.

One important question we must ask ourselves is, what is an acceptable level of precision in the replicate determinations? Many of the statistical guidelines that are useful at higher concentrations become impractical at trace levels. Instrument readings taken near the noise limit do not necessarily represent a normal distribution. In this realm practical exigencies may make us tolerant of errors of 100% or more. In general, we find ourselves captives of the technology with only limited powers to improve precision by modification of the analysis scheme.

Bias is another concern, but here we have more opportunities to exert influence. In trace analysis the negative systematic error is seldom serious unless it stems from sampling or electronic circuitry in the measurement device. Positive systematic error is often a major concern, however. It usually arises from reagent or apparatus contamination. Solving a positive bias problem may be as simple as switching reagent suppliers or modifying some part of the manipulative technique. Most trace methods control bias by means of a blank correction. But finding and using the correct blank is sometimes a major undertaking.

Perhaps the largest concern about accuracy in trace analysis is the means of its assessment. The fact is that for the great majority of work at trace levels appropriate certified reference materials are not available. One can sometimes make do with less than ideal validation standards, but unaccounted matrix effects may be at work. The near ideal approach is to verify a trace result from one technique by comparing it to a trace result from another technique based on different chemical or physical principles. This alternative may not be practical, however. And it always proves costly and time consuming.

It should be evident from the preceding remarks that the trace regime is a rarified region where some of the rules and conventional wisdom of analytical chemistry need to be modified. This volume represents an attempt to present a practical "traveler's guide" to this strange land. It does not purport to exhaustively map all of the terrain. The reader seeking information on the latest state-of-the-art techniques must look elsewhere, for here we walk only the well-traveled paths. Our purpose will be to always keep an eye toward the useful, the practical, the germane. This is intended to serve as a reference for the working metals analyst who must routinely or occasionally tread these paths at someone else's request. The analyst's personal agenda is probably not to write a paper on an obscure surface phenomenon, or to characterize a subtle mechanical property, although the client may have these things in mind. Our analyst's goal is to obtain the most reliable analytical result with the least expense of time and effort.

In analytical chemistry there are nearly always alternative approaches to a given problem. Usually there is a "quick and dirty" answer and a "slow and elegant" one. Often there is a dichotomy between a fast, expensive instrumental approach and a slow, inexpensive wet chemical scheme. However, since time is money in many people's view, this distinction begins to blur somewhat.

A laboratory can spend a great deal of money to operate in the trace realm—on analytical instruments, on contamination control, and on personnel. Nearly always the level of expertise among the analysts must be high, which means educated, experienced chemists employed at a fair living wage.

The most basic approaches to trace analysis and the first that were developed were preconcentration schemes utilizing conventional technology. These are techniques that extend emission and absorption spectroscopy into the trace realm by utilizing chemical manipulations to isolate and concentrate the analyte. They work just fine, even with relatively unsophisticated instruments, but are labor intensive, and rely on a diminishing resource—the skilled "classical" inorganic analyst. Voltammetric techniques, such as polarography in its various forms, are similar in a sense since they require near complete separation of the analyte in most cases. When electrothermal atomic absorption spectrophotometry became commercial in the 1970s it represented a breakthrough since it allowed trace analysis without separation at an equipment cost that was approachable by the ordinary industrial laboratory.

Inorganic mass spectrometry was always the unapproachable and impractical debutante, too expensive and volatile to be bothered with. Yet today, it has re-emerged as the most promising hope for the future. Its recent transformations are now less quirky and more affordable. There are many other trace analysis techniques that function well in their respective niches: the new low limit carbon and sulfur determinators, the latest oxygen, nitrogen, and hydrogen determinators, nuclear methods, and isotope dilution mass spectrometry. It is also possible to isolate inclusions and phases from an alloy and characterize them in the same manner as the alloy matrix. This last subject is closely related to trace work since tramp elements are often associated with compounds and metallic phases in an alloy.

In the chapters that follow we will devote some attention to each of these topics in turn. Our purpose is not to provide an exhaustive theoretical grounding in the principles behind these techniques since that is available in many current authoritative texts (many of which are listed in the Bibliography at the back of this book and in the References at the end of each chapter.) Rather, our intent is to furnish very pragmatic detail of the sort that is needed to apply these approaches to practical problems in metals analysis. The laboratory worker may regard this volume as a practical guide to help solve problems associated with the trace analysis requests he receives. The laboratory manager may use it to assist in the planning of present and future analytical protocols. The student will find it a supplement to a theoretical foundation in these disciplines, and a useful glimpse of how these techniques are utilized in a real-world environment.

There is little more to be said before we embark into this Lilliputian world of minute concentrations except to admit to a few prejudices that may bias this presentation. First, the author's background is largely from the basic and specialty steel industries, and so many examples will be based on iron-, nickel-, or cobalt-based alloys. Second, the author's training and experience have centered around classical wet chemical techniques, and as a result there will be a marked tendency to solve problems by means of chemical separations. Finally, there is a distinctly industrial perspective to this work, reflective of the author's background. Thus, we may view the economics of an analytical problem on an equal footing with its scientific challenge. With those disclaimers in place it is fair to say that this is an honest effort to be *useful*. Only the reader may judge if it has succeeded.

REFERENCE

1. Davis, C.M., Accuracy in Trace Analysis: Sample Handling and Analysis, Proceedings of the 7th IMR Symposium, NIST Special Publication 422, National Institute of Standards and Technology, Gaithersburg, MD, 1976, pp. 1005–1022.

1
The Prelude to Measurement

1. INTRODUCTION

The analytical chemist is keenly aware that the measurement step wherein the analyte is quantified sits at the end of an essential manipulative sequence. That sequence, however, varies widely in complexity based on the nature of the sample's analytical requirements and the approach chosen. In this regard trace analysis is neither more nor less demanding than work at major concentration levels, but it is, fundamentally, different.

All analytical procedures begin with a sampling step designed to represent the concentration of some analyte or set of analytes in a laboratory-sized sample in just the same manner as they occur in some larger, macrosized entity (sometimes termed the "lot"). The next step is to process this *laboratory sample* in some manner so that it becomes practical to select a *test portion* for an analytical determination that still represents the macrosized entity. The degree of further manipulation that this test portion must receive is highly variable. In solids-based spectrometry the test portion is merely the eroded and/or excited portions of the solid test sample. In solution-based work a weighed portion of comminuted and blended material is reacted in some manner to produce a solution—an approach which has the *potential* to compensate for low order inhomogeneity in the material. In some cases it is necessary to perform chemical separations in order to obtain a solution in which the analyte or analytes are free of interferences—a subject which will be taken up in the next chapter.

How, then, is trace analysis unique? First, there are special sampling considerations with even moderately inhomogeneous test materials. Second, intrinsic to the analytical scheme are features designed to enhance measurement response to the analyte. These may involve chemical preconcentration in some form, ranging from coarse group separation to complete isolation of the analyte. They may also involve the selection of special chemical and instrumental conditions to control or minimize interferences and background effects. And, finally, there is the matter of contamination and its control. The reagents, the water, the vessels, the air, the analyst himself, may all contribute to that elusive but essential entity known as the blank.

2. THE SAMPLING IMPERATIVE

It is generally recognized that not much of any value can happen in analytical chemistry unless a proper sample is presented to the analyst. What escapes some otherwise knowledgeable people, however, is that the trace realm presents special challenges that may mandate major modification of the sampling plan.

Any modern discussion of materials sampling must acknowledge the theoretical and semi-empirical treatments that have emerged from the minerals and ores industries. By far the most comprehensive sampling theory is that of Pierre Gy [1–3]. It is a deep and robust formalism that is applicable to any material form. For our purposes here it is only possible to touch upon a few key features.

Gy recognized that sampling representativeness depends upon an understanding of two forms of material heterogeneity—each is a scalar function, approaching at its lower limit a zero state called homogeneity. The first is *constitution heterogeneity*, which is an intrinsic property of the constituent particles or units of material at some defined scale. The second is *distribution heterogeneity*, which describes the aggregate collection of particles or units. Thus, if one were sampling a mixture of sand and soil for the determination of iron, the variation of iron among the sand grains and the variation of iron among the soil particles would constitute the element's constitution heterogeneity, while the effect of the uniformity in the mixture of sand and soil would represent its distribution heterogeneity. Grinding of the proper sort may lower constitution heterogeneity, but, clearly, thorough mixing of the proper sort will lower distribution heterogeneity.

Gy then distinguishes two forms of sampling. Nonprobalistic sampling is the common, but flawed, approach in which some constituent elements have a zero probability of being selected. So-called "grab-sampling" is an example. Here, the easiest and cheapest approach is employed using one motion of a sampling device at the most accessible sampling location. In probalistic sampling, on the other hand, all constituent elements have *some* probability of being selected. Nonprobalistic sampling will never adequately represent the lot unless the material has a very low order of heterogeneity. Probalistic sampling has the *possibility* of representing the lot, but true representation will only be achieved with a properly designed sampling plan in which all constituent elements have an *equal* probability of being selected [4].

Gy's theoretical treatment deals with probalistic sampling and how it can be made to most nearly represent the lot. Clearly, factors such as the shape, density, and size of the constituents, as well as the distribution heterogeneity, enter into this representation process, which can be termed a minimization of the sampling error.

Nonprobalistic sampling for trace and ultratrace elemental determinations is nearly always meaningless. But even correct, probalistic sampling for these components will often display statistics that deviate from a normal distribution. Both the nature of the sampling process and the sample size must be considered in order to minimize the sampling error while remaining within practical economic constraints. As we will see, sampling for trace work almost always involves some degree of compromise.

For metals and metal-related commodities it is best to view sampling as a three-stage process. First, the lot or batch is *sampled.* This nascent sample is then presented to the threshold of the laboratory as the *laboratory sample.* The *sample preparation* function is a laboratory activity designed to reduce the laboratory sample to one or several small units designated as the *test samples.* The final stage involves the *selection* of the *test portion,* the part of the test sample actually employed in an analytical determination. The sampling plan should be designed so that at each stage the lot or batch is faithfully represented.

2.1 Sampling the Lot

Here we begin by examining those features of the sampling plan that take place outside of the laboratory, typically by nonlaboratory, sometimes by nontechnical, personnel. In the sampling of metals we can discern three types of lot or batch: molten metal (at some intermediate or final stage in its processing), solid metal forms, and particulate metals and metal-related materials. Each of these is unique and each raises special problems in regard to the representation of trace levels of analyte.

2.1.1 Molten Metal Sampling

Most metals industries sample and test the in-process molten alloy to confirm that no errors have occurred and that refining is proceeding as planned. Many use one or several in-process or "prelim" analyses to make alloy "trim" additions and to fine-tune the processing parameters. A "final" sample (or samples) is intended to confirm that the cast solidified metal meets compositional requirements. These samples are all prepared and measured with all possible alacrity to allow real-time decisions to be made.

Only in special cases is the trace realm scrutinized during processing because it is extremely difficult and expensive to provide that sort of analytical information "on-the-fly." Most alloy producers who must meet trace and ultratrace limits start with the purest virgin metals, optimize processing variables to refine out residual impurities, then sample at some convenient point *after* the metal is solidified, and wait nervously for the laboratory results, which may take hours or days.

However, one area of trace analyte molten metal sampling that has grown considerably is delineated by those elements determined using thermal evolution techniques. Relatively affordable and rugged instrumentation is available to measure carbon, sulfur, hydrogen, nitrogen, and oxygen at levels below 10 ppm within minutes of receipt of the sample. Much less practical are mass spectrometers and simultaneous multi-element graphite furnace atomic absorption spectrophotometers, which do not appear to meld favorably with a mill process control environment, at least at their current state of development.

The present question then becomes, how does one sample molten metal for trace carbon, sulfur, hydrogen, nitrogen, and oxygen? While we must always bear in mind Gy's admonition against the "opportunistic" grab sample, how else in a practical sense does one sample an intensely hot, vigorously reactive, slag-covered mass that weighs, perhaps, a hundred tons? One advantage is the fact that most molten metal baths are actively stirred by convection currents, and sometimes by gas injection, or applied electromagnetic fields.

Even so, the "dip sample," whether it is a spoon used to fill a mold, or a self-contained immersion probe, must usually penetrate a layer of slag without contaminating the metal sample. Sampling spoons may be fabricated from steel or cast iron for use in the ferrous metals industry, or from ceramic or graphite for use in the nonferrous metals industry. The bowl generally has a capacity of about 500 mL and the handle is 2–5 m long. Sample molds come in a number of different designs. Some produce small vertically or horizontally cast disks, others may produce conical forms. Molds may be fabricated from steel, cast iron, copper, or other metals when rapid cooling is desired, or from sand or ceramic when a slower cooling is necessary to achieve lower hardness in the test piece.

The spoon and the mold may receive special preparation treatments such as the application of special coatings and/or preheating. In steel sampling the spoon is often purposely coated with hot slag before it is plunged into the molten metal below. The filled spoon is then extracted, placed on a spoon rest and skimmed of slag before the metal is poured into the mold. Reactive metal, such as rim grade steel, must be deoxidized in the spoon or the mold by adding metal deoxidants (typically aluminum, but sometimes titanium, zirconium, or silicon metal).

Immersion probes may be simple heat-resistant glass tubes that are plunged into the metal layer in the furnace (or in two-stage sampling into the charged sampling spoon) with suction then applied to draw out a long pin sample. Others are evacuated glass devices of special designs to produce disk and pin samples simultaneously. These are already charged with a preweighed amount of deoxidant. They are plunged into the molten metal using specially designed tongs. Modern practices in many metals industries utilize sophisticated probe designs that fill by hydrostatic pressure once protective material at the entrance port ablates away, thus preventing slag from entering the probe. The probe itself may contain a mixing chamber with deoxidant, a pin and disk mold, and even a thermocouple. In all cases the withdrawn probe is allowed to cool somewhat before it is quenched in water.

Spoons and immersion probes are used for preliminary test samples that are typically taken at one location in the furnace, trough, argon–oxygen decaburization (AOD) vessel, tundish, or other molten metal source. Generally, the exigencies of the refining process do not allow time for multiple samples for each "prelim" test, although such a plan might average out any incorrectness in the sampling. "Final" tests are another matter, however.

"Final" tests (which are often really not final at all, as we will see) may be taken during the teeming of the molten metal from the ladle into ingot molds or casts. As the molten metal is poured a portion is diverted into test molds to produce laboratory samples. Large amounts of metal may be sampled from three to five times during the teeming operation. Slag contamination is less of a problem here since it floats above the tap hole.

Sometimes a series of spoon or immersion samples taken from the tundish of a continuous caster are regarded as the "final" test. However, even here, we cannot always dismiss questions about the representativeness of the samples which are taken.

Can one reasonably expect to accurately represent a hundred tons of metal with a 50-g "lollipop," or even a 5 lb conical-shaped casting, no matter how carefully taken? The answer for trace work is simply that one cannot. Unless the material is extremely homogeneous there is little chance that *all* constituents (to use Gy's terminology) have an equal probability of being selected, no matter how well-stirred the pot.

Why, then, does molten metal sampling persist and what function does it serve? The answer lies in the fact that truth can sometimes be approached by knowledgeable people as an iterative function. The important proviso is that every constituent must have *some* chance of being selected. Thus, with molten metal sampling we are in the realm of the empirical. The lab result might tell us that, for example, the manganese content of the alloy is low. And it really *is* low, but perhaps not to exactly the same degree as the test result suggests. In most instances this kind of information provides sufficient feedback to seasoned process personnel to allow the molten metal to be adjusted into the specification range.

Moreover, there is good data to support the perception that the final ("ladle") test for certain alloy commodities (such as plain carbon and low alloy steel) is a better measure of the average concentrations of major and minor components than any later tests on solid forms of the same material [5]. This odd fact stems from the highly heterogeneous form in which these products are commonly sold.

Tramp elements are more problematic, however. Since these elements must meet a maximum level criterion one is faced with the task of proving a negative—"lead does not exceed 1 ppm," for example—with a sample that imperfectly represents the batch. If all or most of the lead is associated with constituents of the molten metal that have a low probability of being selected, false negatives are likely.

For this reason alone it is imprudent to rely upon molten metal sample test results as the final word on tramp levels in a material. But there are other reasons to resample for tramps at a later stage in the processing. The solidification process itself is often a refining phenomenon in many alloy systems. Impurity compounds may be carried along by the liquid-solid interface and deposited in high concentrations at the center, top, or bottom of an ingot. Many critical end-use alloys are remelted several times by techniques designed to remove impurities. Thus a measurement of tramp analytes at the initial melting stage will bear little relation to final tramp levels after this additional processing. Moreover, these later refining steps, such as electroslag remelting and vacuum arc remelting, do not easily lend themselves to molten metal sampling.

Subsequent thermal treatments, such as tempering and annealing, may alter tramp levels as well. In particular small atomic radius interstitial elements may migrate to the surface of solid forms where they reside as a concentrated layer (often machined off) or react with furnace atmospheres to escape as a volatile species. Tramps can also be picked up from furnace atmospheres or from solid surface contact with the hot metal. Even the inks used in surface markers have been suspected as a potential source of tramps in alloys.

Despite these objections many metallurgists request tramp analysis at the initial melting stage (or shortly thereafter) to help define the task for subsequent processing. Once it has been recognized that a molten metal dip or pour sample imperfectly represents the tramp analytes, it nevertheless still remains the task to perform the sampling operation, transport the cooled, solidified laboratory sample to the lab, prepare the sample, and analyze it, all without contaminating it.

This is not an easy task considering the reactive nature of molten and hot, solidified metal, and the usually quite dirty environment of the melting and refining operations in most metals industries. In addition there is generally a slag layer in furnaces, degassers, and tundishes that must be breached without contaminating the metal withdrawn from below it. Mold powders used in continuous casting operations are another contamination source.

For trace carbon determination the most significant source of contamination may be the sampling device itself. Many immersion samplers utilize paper tubes and paper slag caps designed to burn away once the slag layer is breached. This paper and the adhesives used in the assembly can be a major problem when alloy carbon levels below 100 ppm are being measured. This concern is particularly acute in the production of so-called "ultralow carbon" (ULC) steels. Many steelmakers ship finished metal with a certificate of analysis based on molten metal samples taken at the tundish or the mold at a continuous caster. While this practice may be questioned on some grounds, it becomes vital that special noncontaminating samplers be employed when ULC grades are sampled [6,7].

Sulfur, nitrogen, and oxygen are also frequently measured at trace levels from molten metal samples. Entrained slag or mold powder particles, as well as microscopic void porosity can be a problem. An unsound or suspect pin sample should be discarded and a resampling immediately conducted.

It is hydrogen determination, however, that presents special problems. Because hydrogen is mobile at room temperature in the lattice of some metals, including low alloy steel, such samples begin to lose hydrogen as they solidify and cool. Trace and ultratrace levels of hydrogen can be determined using a molten metal immersion probe which remains sealed until it is inserted into a hydrogen determinator. The probe casing is crushed inside the instrument and so-called "mobile" (or "free" or "noncombined") hydrogen is measured. Then the solid pin sample is subjected to a heating program wherein the "combined" (or "fixed") hydrogen is released. Typically, the total hydrogen sum is the most useful value.

A different (typically higher) value is obtained by measuring hydrogen directly in the molten metal with a closed-loop of recirculating nitrogen carrier gas. In this commercial system hydrogen from the molten metal reaches an equilibrium partial pressure in the carrier gas and is measured after about one minute. An accurate value for hydrogen in molten metal often is a useless predictor of hydrogen levels in finished commodities, but it *can* sometimes be used to predict casting behavior. In some alloys, like plain carbon steels, an excessive hydrogen level in the molten metal means the casting may crack upon cooling.

Molten metal sampling for spectrometric trace and ultratrace analyses is typically confined to high purity metals. So-called book molds (which open like a book to release the solid disk sample) are one common approach. The fill channel (also called the "riser" or the "sprue") is cropped off and the flat surface of the disk is abraded and polished to remove any copper or other contaminant material from the mold itself. Techniques which are able to resolve small spatial areas on the disk are extremely useful in determining the "microheterogeneity" of a given analyte. A surprising degree of heterogeneity at the trace level has sometimes been revealed by such studies [8].

Whether the applied measurement technique is optical emission or mass spectrometry the solidified sample should be presumed segregated with regard to tramps until proved otherwise. Unfortunately, this is not always realistically possible because techniques with appropriate spatial resolution may lack sufficient sensitivity to accurately measure analyte concentrations at extremely low (trace and ultratrace) levels. The answer is again a compromise. If one or two measurable tramps show no significant segregation we can make a perilous and marginally valid conditional assumption that *all* tramps are uniformly distributed. Repeated resurfacing and reanalyzing for tramps by a "whole surface" measurement technique such as X-ray fluorescence, or looking at surface quadrants by an arc source optical emission technique is one approach. Taking a multiple series of samples from the molten bath and analyzing each for all tramp elements is another. These approaches are not employed much because of time constraints and costs, however.

2.1.2 Solid Metal Sampling

Once the metal product has solidified there is now something relatively stable upon which a rational sampling plan can be designed. That is not to say that its composition or its degree of analyte heterogeneity will not change with subsequent processing. But at least at some given subsequent stage in the manufacture of a metal commodity these properties will, for the most part, remain invariant. This condition is inherently more manageable than the dynamically changing conditions of the molten state. And, thus, in general, solid metal sampling is much more likely to lead to information about tramps relevant to the end use of the commodity.

The types of metals and metal alloys that require trace analysis typically undergo a series of refining stages wherein tramp elements are reduced ultimately below some threshold concentration limit. At each stage there is an opportunity to sample the solid material to provide analytical information which can help guide subsequent processing steps. Finished material at the end of this process should always be analyzed as well. This "final analysis" represents the lot as it is sold. For high purity metals and premium alloys the material at this stage should be at a minimal level of heterogeneity and tramp element concentrations should be lower than at any of the previous processing stages.

We must never assume, however, that solid metal sampling is a trivial exercise at any stage. For example, conventional wisdom dictates that large steel cast cylindrical sections be sampled at midradius. But this instruction is usually technically impractical and often destructive to the casting. There are imperfect, but often acceptable alternative locations. Such sampling options are often highly empirical, and, in practice, are frequently the result of a formal agreement between the vendor and the customer.

In casting operations thermal convection currents and the chemical and physical equilibria occurring at the solid/liquid interface determine the distribution pattern of inclusions and second phases in the finished form. Tramp elements are frequently associated with these constituents and thus their distribution heterogeneity must be correctly, probalistically sampled. Most metals and alloys are crystalline solids, and tramp elements are frequently concentrated in the boundaries between crystal grains. This phenomenon has implications for solid optical emission spectrometric techniques because certain types of excitation discharges selectively excite these regions, yielding erroneously high values for tramp elements.

Metal working operations such as blooming and forging are designed to homogenize alloys as well as to impart desired physical properties. Thus, typically a wrought material is less heterogeneous than a cast form of the same alloy. Hot or cold rolling tends to impart a directional element to tramp distribution since formable inclusion compounds tend to elongate as "stringers" in the rolling direction. Heat treating operations, as described above, tend to drive certain elements toward the alloy's outer surface. And furnace atmospheres can contribute to the dynamic removal of susceptible species.

The sampling plan for solid forms must incorporate all of these contingencies as they apply to a specific alloy composition at a specific stage in its processing. The assignment is not as daunting as it sounds, however. Most designs will incorporate full cross-sectional milling at one or more selected locations. Large castings will be drilled (usually with a large, slowly turned drill bit) as close to the midradius as is practical. An alternative is to drill from the side, discarding all chips until the drill bit reaches the midradius region.

Sheet and strip is usually milled or nibbled across the entire transverse width. Special precautions are taken to avoid regions where the tramp composition is known or suspected of being nonrepresentative. Thus torch- or arc-cut edges and the weld seam region of finished shapes are never included. With cast ingots the center shrinkage known as "pipe," as well as seams and cracks, are always avoided. Drillings or millings from scaled regions (i.e., oxidized, heat-affected outer surfaces) are always discarded before chip collection begins.

Surface enhancement or depletion can be appropriately averaged by cross-sectional milling. Alternatively, it is possible to study the phenomenon in bar forms by lathe-turning a defined series of steps and collecting the turnings from each. Sometimes it is feasible to study the "edge-inward" concentration gradient using a cross-sectional solid sample and an instrumental approach with appropriate spatial resolution.

Solid test samples for instrumental work of all kinds sometimes deserve Gy's pejorative appellation, "specimen" (either a nonprobalistic sample, or an incorrect probalistic sample [4]). It *is* possible, however, to select a solid test sample cut out from a region of the lot that yields faithful representation of the lot as a whole. The problem is related to similar sampling problems for various forms of physical testing (tensile test samples,for example, must be selected to represent the average material property). But sampling for solid sample spectrometry is even more demanding. The most accurate work may sometimes require a statistical homogeneity study based on a series of solid test samples. The results of such a study may require that data from some pieces be rejected, with a consensus average calculated from the remainder of the data.

A length of bar, for example, may be the subject of such a study as representative of a large lot of bars. The bar may be cut into a series of slices, each coded so that its original location can be identified. The faces of these disks are then analyzed randomly or exhaustively and the observed analyte variation is compared to analytical method variability on one slice. A highly reproducible instrumental approach such as X-ray fluorescence spectrometry is typically employed. An alternate design in which transverse heterogeneity is added to the study typically utilizes a point-to-plane optical emission technique. Here the surface of each disk is divided into quadrants and each is arced or sparked as a separate region.

The time and effort involved in such exhaustive solid sampling studies is rarely justified in routine commodity analysis. For most work, if the material is known to be reasonably homogeneous, a single, judiciously chosen, sampling location will suffice. Highly heterogeneous commodities are by convention certified by the "final heat analysis" of the molten metal. There *are* occasions, however, where a homogeneity test is absolutely essential. These include the preparation of reference material samples and special vendor/customer agreements in which a detailed plan of sublot testing is specified.

Sampling solid metal for oxygen and hydrogen determination requires that excessive heating be avoided. Surface oxidation from atmospheric oxygen during sample removal may be difficult to eliminate during sample preparation for oxygen determination. For this reason ultralow oxygen metals should be sampled under a sheath of inert gas. On the other hand, hydrogen tends to be mobile in certain matrices, such as iron. So it becomes necessary to take special precautions, such as packing a casting with dry ice prior to removal of a sample by drilling with a trepanning tool and breaking off the center pin with pliers. Such samples are stored in liquid nitrogen prior to hydrogen determination.

Sampling finished or semifinished metal forms will require specially designed protocols. For example, in sampling pig iron, typically, three pigs are selected from the lot. Wire samples may require selecting separate portions from the beginning, middle, and end of a spool. Thick-walled tubing is typically drilled at a point on the center of the end wall. Thin-wall seamless tubing may be flattened and milled across the ends. Thin-wall welded tubing must have the weld-seam cut out and discarded before flattening and milling.

Perhaps the most challenging form of solid metal sampling is for scrap analysis. In many metal industries some form of scrap constitutes a part, often a dominant part, of the total furnace charge. Scrap can consist of large, diverse pieces or finely divided machine shop dross. It may be all of one alloy grade, all of one class of alloys, or a hodgepodge of materials whose only commonalty is the base metal component. In some industries the proper use of scrap is the key to economic viability, and so a great deal of attention should be paid to its sampling and analysis.

The proper sampling of scrap can be such a refined art in some industries that alloy producers may be forced to place all their confidence in the scrap dealer's specialized procedures. These may involve a rotating cross-stream "cutter"—a device that sweeps across a moving conveyor belt to divert a portion of the material to a separate sample bin [3]. This representative fraction of the total lot is then melted into a solid button which is analyzed and then sent to the alloy producer customer along with the analysis certificate.

Sampling scrap for tramp elements is especially critical for those metals industries who use it to produce low tramp alloys. The only practical way to sample for such work is to collect a sizeable amount of properly diverted material and melt it in a vacuum induction furnace under a subatmospheric partial pressure of argon. Under such conditions tramp elements in many matrices tend to be retained and to be uniformly distributed in the cast mini-ingot. The casting is cropped of pipe and scale and then appropriate solid sections can be used for spectrometric analysis. Additional pieces are typically milled across the full cross-section to obtain chips for graphite furnace atomic absorption (GFAA), inductively coupled plasma mass spectrometry (ICP-MS), or other tramp solution-based techniques. Electric arc remelting, which is often used to prepare a spectrometric solid from small pieces of metal sample, is inappropriate for most tramp work since bismuth, lead, zinc, cadmium, magnesium, tellurium, thallium, and other elements are lost by volatilization, and copper may be picked up from the copper mold which is commonly used. In both techniques trace levels of oxygen, nitrogen, hydrogen, carbon, and sulfur must be considered at least partially lost [9].

There are many technical details and much lore associated with both molten metal and solid metal sampling. A few out-of-print older texts can provide valuable information if they can be found [5,10,11]. There are also currently an important group of commodity-specific sampling documents available from the American Society for Testing and Materials (ASTM) [12–18] and the International Organization for Standardization (ISO) [19].

2.1.3 Particulate Metal Sampling

In the previous section we confined our discussion to monolithic batches or lots. Here we move on to the subject of aggregate batches or lots. This could be a carload of small boulders or a vial full of dust-sized particles. The only provision for inclusion here is that by some consensus it is agreed that all the particles, collectively, are to be considered as part of a whole. Sampling that consensus "whole" is directly amenable to the theories and techniques that Gy and others have described.

With the exception of ores and minerals the greatest sampling challenge in the inorganic analysis realm may be ferroalloys. These commodities are produced and sold in large tonnage lots which are comprised of diverse-sized pieces of rather heterogeneous composition. The sampling problem is enhanced by the need for tramp analysis, which derives from the fact that these commodities are the starting materials for producing many critical end-use alloys. The alloy producer must, therefore, be assured that these materials, which sometimes constitute a significant portion of the total furnace charge, will not introduce undesirable levels of tramp elements.

At the other end of the scale are the metal and alloy powders produced by molten metal atomization and by other processes. These materials are generally very homogeneous, although they may exhibit some variation in composition with mesh size, particularly with respect to oxygen concentration, which is often a surface area effect. Powders are used to produce finished forms directly through some form of consolidation technique. This may involve blending with organic or inorganic binder compounds, extrusion, sintering, hot isostatic pressing, or other processes. One of the key parameters of any commercial consolidation process is the "percent density" achieved in the finished form since voids and entrained compounds represent potential problems with the product. In most cases oxygen is the key tramp element.

In between these two examples lie a number of metal particulate products with intermediate levels of heterogeneity. They are all charge materials used in the production of commercial alloys. Metals in a range of purities, sold as lumps, shot, or powder, master alloys and stoichiometric intermetallic compounds all fall into this category. Finally, there are those materials that are not metals at all but need to be considered here since they are essential to the production of metals and alloys—ores, limestone, fluorspar, crushed refractories, and much more. Each of these can and will contribute some burden of tramp elements to the finished product.

Returning to the theoretical treatments is advantageous here because there are many practical applications to be directly derived from them. As we discussed in Section 2, an analyte may be partially or exclusively associated with a constituent of the lot. Earlier we used the example of iron associated with sand grains in a sand/soil mixture. If we begin to grind this mixture we eventually reach a point where the iron is no longer associated with the sand particles but may be regarded as a component of the entire sample. This point is termed the *liberation size*. By comminuting the sample we have minimized the constitution heterogeneity and it is now possible to minimize the distribution heterogeneity by mixing. Moreover, it makes intuitive sense that a trace analyte associated with a constituent phase requires that a large initial sample be taken so that the constituents have an appropriate probability of being selected.

The comminution process must always be examined critically because it can introduce sample contamination. It is also possible to grind a material too finely, resulting in a number of problems. Dusting losses and moisture and oxygen pickup are examples. In some cases a malleable analyte (like gold flecks in silicate ore) may actually coat the inside of the grinding apparatus. In a number of cases an overly ground material will be difficult or impossible to mix properly.

Sampling particulates is, of course, a process highly dependent upon mixing and pouring operations which for powders are strongly affected by the density, size, and shape of the particles. Particle density affects momentum; light particles tend to drop straight down while heavy particles tend to follow a somewhat ballistic line. Particle size affects both momentum and air resistance. Particle shape produces effects due to friction and the angle of repose. Fines tend to work their way into the center of a poured pile. Angular particles tend to accumulate in the center of the pile while rounded particles tend to accumulate at the pile periphery. Pouring or mixing a powder is also affected by local air turbulence, and by magnetic or electrostatic fields [3].

There are numerous designs for sampling particulate systems. The grab sample should sound even more of a cautionary note with particulates than with the typically more homogeneous liquid and solid material types. Its use persists, however, with the use of "thief probes" to remove a sample of finely divided particulates from storage or shipment vessels. Most useful sampling plan designs either sample continuously during a transfer of the commodity or incorporate a composite composed of a random selection from numerous locations in the lot. The two volume book by F.F. Pitard (*Pierre Gy's Sampling Theory and Sampling Practice* [3]) treats correct and incorrect sampling approaches in great detail and will prove an invaluable reference for anyone engaged in designing and implementing a sampling plan.

Very heterogeneous materials like ferroalloys are sampled by dumping every fifth container then selecting samples to create a composite that is composed of approximately 0.5% of the total lot. The composite is then piled into a cone and flattened into a disk which is divided into four equal quadrants. Opposing quadrants are selected and shoveled into another pile which is, in turn, similarly divided. This "coning and quartering" process continues until the sample size has been reduced to the appropriate level. The appropriate level is the minimum size that may be transported to the laboratory which still adequately represents the lot.

The ASTM standard for sampling ferroalloys [20] specifies a minimum sample of 100 lb (46 kg) for material composed of 1/4-in. (6.4 mm) diameter pieces and one ton (0.9 Mg) for 2.5 in. (64 mm) diameter material. The ASTM standard for sampling iron ores [21] uses an empirical formula for calculating the increment size when a fixed number of sampling increments (termed one operation of a sampling device) is specified. The formula is based on the material's average particle size and density. These standards are attempts to provide practical guidelines based on experience with the typical nature and properties of industrial commodities.

Gy's sampling theory is an attempt to place such considerations on a theoretical basis. He defines a *sampling constant* (C) which can be used to calculate the minimum sample weight needed to achieve a given sampling variance. One first estimates C from the following.

$$C = fglm \tag{1}$$

Here f is a *shape factor*, which is a relative measure of how different the average particle shape is from an ideal cube (for which $f = 1$). The next term, g, is a *particle size distribution factor*. The next, l, is the *liberation factor*—it drops to zero at ideal homogeneity. The last term, m, is called the *mineralogical factor*; it is an estimate of the maximum heterogeneity of the analyte. Gy and other workers have shown how to obtain practical estimates for each of these four parameters [22–24].

The next step is to solve for the minimum sample weight (M) in the following equation.

$$M = Cd^3/(s_r)^2 \tag{2}$$

Here d is the largest linear dimension of the largest particles and the quantity $(s_r)^2$ is the relative sampling variance (a user selected quantity). We will return to this concept in a different form in Section 2.3 where it is applied to the analytical test portion size.

2.2 Preparing the Test Sample

What has been described up to this point are operations which take place outside of the laboratory. The next phase involves laboratory processing of the *laboratory sample*, which will reduce it in size and weight to the *test sample* while still representing the original lot. The test sample is that bag of metal chips or that solid disk that is presented to the analyst and from which the *test portion* is selected. Sometimes many manipulative steps are involved in preparing the test sample—pickling in acids, shot or sand blasting, cutting with abrasive or metal or diamond saws, crushing, grinding, polishing, riffling, sieving, electric arc or induction melting, solvent cleaning, and other procedures are all used at one time or another by some laboratories.

In the case of preparing a sample for trace analysis it must be recognized that each one of these procedures has the potential to introduce analyte contamination in the test sample. In addition some of these operations may be responsible for losses of analyte. They may also introduce undesirable nonanalyte elements that act as interferences. The problem is often compounded by the sheer size of the laboratory sample and the difficulties involved in adequate equipment cleaning to prevent cross contamination between samples.

Here we have divided this subject into two categories which treat the preparation of solid test samples and of particulate test samples as the separate tasks that they represent. Note that Section 2.2.1 includes solids from solids and solids from particulates and that Section 2.2.2 includes particulates from solids and particulates from particulates.

2.2.1 The Solid Test Sample

The solid metal laboratory samples that arrive at the lab's receiving area may vary in size from small pins and disks to miniature ingots weighing 25 lb (11.25 kg) or more. Pieces larger than that are inappropriate unless the laboratory is equipped with a crane or other proper handling equipment. Once the identifying codes and associated information have been recorded (or entered into a laboratory information management system) the laboratory sample should undergo a visual inspection to assess the applicability of the sampling plan and to check for scale, "pipe," holes, cracks, and visible inclusions. Defects likely to undermine the analytical work should initiate a retest.

Mini-ingots and other large sample castings are cropped of any "pipe," often using a water-cooled abrasive cut-off wheel. The resulting surface may be further cropped if visual inspection reveals inclusions or porosity. Slices for spectrometry are then removed and stamped or electrolytically etched with an identifying number. Forged or rolled alloys are simply sliced. Nearly all spectrometric work is from a surface perpendicular to the rolling direction. The opposite surface of the piece is then polished with a disk or belt sander, which may be a simple manual unit or an automatic or semi-automatic device. If the polishing equipment lacks an adequate dust removal system, use of an appropriate respirator is essential.

Laboratory sample preparation rooms must be posted as an area requiring safety glasses with wire mesh side shields. Face shields and goggles are needed for acid pickling and other operations where wire mesh side shields offer inadequate protection. Neckties and loose fitting clothing are dangerously inappropriate in a sample preparation area.

Pickling to remove scale and corrosion is common if heavy swing grinding equipment is unavailable, or if the sample piece is small or irregularly shaped. The acid medium chosen must be selected for the alloy system to clean the oxidized surface efficiently without severely undercutting the base metal. It is convenient to use borosilicate glass trays containing the pickling medium gently warmed on a hotplate. Sample pieces are added and removed with tongs and rinsed thoroughly in distilled water, then dried. If the pickling requires hydrofluoric acid a large Teflon TFE beaker, warmed on a sandbath, is needed. Alternatives to pickling include grit-blasting and shot-blasting. All of these operations must be followed by polishing to an appropriate surface finish.

If the material is too hard to be cut it can usually be annealed into a condition in which solid spectrometric test samples are easily removed. It is not necessary to achieve a "dead soft" condition but merely a condition in which the material can be machined to size without excessive heating or tool wear. Expert metallurgical advice is always an advantage, as are access to heat treating facilities. But it *is* possible in some instances to achieve the desired condition by air cooling an alloy after heating it in a laboratory muffle furnace. It must be remembered, however, that each time a metal alloy is heated some compositional change might occur with analytes either lost or redistributed. For certain materials surface depletions and enrichments are a real possibility.

Disks from immersion probe sampling are ground to clean metal on a swing grinder, then polished on a disk or belt sander. Pins from immersion tests may be cut on a twin-bladed abrasive wheel to quickly obtain a test sample of appropriate size. They may be pickled, polished, or hand-filed, depending on the analyte, the technique, and the alloy matrix.

Disks are typically used for optical emission, X-ray fluorescence, and sometimes glow discharge mass spectrometry. As with all solid test samples, the greatest source of contamination is likely to be the final polishing operation. The disk or belt abrasive material must be chosen to avoid obvious problems. Thus, test samples for trace aluminum determination are polished on a silicon carbide abrasive, while test samples for trace silicon are polished on an alumina abrasive. There are other choices as well. Another problem is cross-contamination between test samples of different composition polished on the same belt or disk. Frequent changing of belts and disks is a necessary precaution, especially in tramp work. A third problem is the smearing of soft inclusions and phases during polishing which can produce erroneous data from spectrometric techniques. This difficulty has no facile solution, although approaches that might prove useful involve substituting the abrasive polish with diamond saw cutting, lathe turning, or electropolishing.

When a sealed immersion sampler for hydrogen determination is submitted the laboratory sample *is* the test portion. The sampler is inserted in the hydrogen determinator and the outer casing is mechanically crushed, releasing the "mobile" hydrogen. Then the material is subjected to a heating cycle wherein the remaining lattice-bound hydrogen is released and measured. The sum of these two sets of numbers is reported as total hydrogen. Other hydrogen pin samples are submitted to the laboratory in liquid nitrogen and retrieved by strings bearing identifying tags. Solids for oxygen and nitrogen are often hand-filed in a vise using a clean file, then cleaned in low-residue acetone or methylene chloride in a hood, where they are allowed to air dry, then stored in a desiccator.

Inorganic mass spectrometry based on solids usually has some special test sample preparation requirements. Spark source mass spectrometry generally requires two machined sample electrodes of exact dimensions. Glow discharge mass spectrometry can be configured for either a pin or a disk sample shape, although pins are more common. Acid etching is a common preparation and lengthy "preburns" in the instrument are actually part of the sample preparation task. Spark- and laser ablation-inductively coupled plasma mass spectrometries both require disks or appropriately mounted small pieces.

We have already touched upon the process of making a solid spectrometric sample from chips, drillings, or a handful of small parts. There are several designs for commercial units to produce spectrometric disks from such starting materials. Some utilize the direct current arc produced by an arc welding power supply; many utilize an induction coil to melt the sample. There is generally provision for an argon or nitrogen protective gas sheath to prevent oxidation. The molten metal may be cast in copper or ceramic molds. One design incorporates a centrifuge arm to centrifugally cast the molten metal. Generally, at least 50 g of starting material is required.

These devices have proven extremely useful, but the potential user must be cautioned by a similar warning to that given for heat treating alloys. When an alloy is melted volatile components may be lost. Tramps are a particular concern and, as noted above, generally require a laboratory size vacuum induction furnace operated under a small partial pressure of argon. Materials that are difficult to cast into good spectrometric disks (such as certain ferroalloy compositions) can be accurately diluted with pure iron (or possibly another metal) and melted and cast by one of these techniques. The resultant disk has good spectrometric properties and the analyte concentrations obtained are easily corrected to their concentrations in the original material.

Some laboratories have utilized these disk casting devices to prepare calibration standards for spectrometric work. Such practices cannot be condoned, however, unless homogeneity studies and wet chemical analyses from chips obtained from full cross-sectional milling are conducted. Under no circumstances should the "taken" compositional values be assumed to be free of recovery losses.

2.2.2 The Particulate Test Sample

Particulate test samples are needed for most analytical methodologies that involve preparing a solution from the test portion prior to measurement. If the laboratory sample is a solid it must be drilled, milled, nibbled, or crushed to obtain particles of the desired mesh size. In this process an approach must be taken that does not overheat the sample.

A drill press fitted with a large diameter drill bit is often the method of choice. A typical steel analysis laboratory uses a flat-beaded track bit of 3/4 in. (19 mm) diameter with nicked cutting edges to prevent long, "bedspring" type drillings. A slow speed and manual feed are always employed. The drilling location is marked initially with a center punch and chips are collected on glazed paper or in a customized pan fabricated from a sheet of stainless steel. Small or unusually shaped samples may require smaller diameter drill bits. In general, the harder the sample material, the slower the drill speed that should be used. Cast material should be drilled at several points around the midradius of a cropped casting, and the drill bit should not approach the bottom of the cast. Annealing is sometimes necessary to soften the sample sufficiently for drilling, or other chip-making processes [10]. It should be approached with the same caution as when applied to obtaining solid test samples.

Representation of a full cross section of a solid is best handled with a milling machine. In some equipment designs the sample may be either horizontally or vertically mounted. Sheet can be folded and milled across the folded edges. Thin-wall tubing can be flattened and milled across the end wall. Sheet can also be nibbled with either a manual or a power nibbling tool. In general, no lubricants are used in all of these chip-making operations. Nevertheless, it is usually necessary to degrease the resultant chips in a low residue solvent (methylene chloride or acetone are popular choices), then dry them at 110°C or lower.

Test sample chips should never be sieved unless one is preparing a chip reference material (where midsize fractions are often used for their superior homogeneity). Some labs screen chips of nonmagnetic alloys with a magnet to ensure that no pieces of drill bit have contaminated the test sample. This practice should be cautiously employed, however, since some nonmagnetic alloys contain a minor magnetic phase.

Alloys that are hard but not tough are brittle enough to be crushed. This can be accomplished in a ball or disk mill. The mill is stopped at intervals and the contents sieved. Fines are collected and the fraction retained on the sieve is returned to the mill. This process is repeated until the entire sample has passed through the sieve. The sieve mesh size, of course, varies with the analytical requirements but 100 mesh (150 mm) and 325 mesh (45 mm) are popular choices. Alloys that can be crushed can usually also be annealed into a condition suitable for machining. Thus it becomes a judgment call as to the best way to proceed. Crushing operations offer greater opportunities for sample contamination, but there is always the danger of analyte loss in annealing.

In general all sample chip generation should be approached with cautions about contamination. All equipment should be kept clean with a high pressure air hose and a shop vacuum. Many laboratories believe that for the thermal evolution determination of carbon, sulfur, and nitrogen the best work is done with chips. In these cases it becomes essential that overheating during machining be avoided and that the test samples for ultralow levels be transferred and stored appropriately. Thus paper envelopes are often replaced by glass vials.

Laboratory samples that arrive as particulates and must be prepared as particulate test samples sometimes represent a great challenge. The degree of difficulty is generally related to the heterogeneity and hardness of the material, the level of variability which can be tolerated in the analytical result, and whether or not the analyte is concentrated in a minor phase. Laboratory samples of heterogeneous materials, such as ferroalloys, should arrive as very large volumes of material. Metal alloy powders, on the other hand, may be submitted as a vial of presieved powder. The former requires a great deal of preparation, the latter, none at all.

If the individual pieces of a particulate laboratory sample are 4 in. (100 mm) in diameter or larger some form of manual comminution is required as a first step. This may involve use of a compressed air jack hammer if the material is very hard. Pieces 4 in. (100 mm) in diameter or smaller can be crushed to 1/4 in. (6.4 mm) in a mechanical jaw crusher. At this size the next step is usually mixing, followed by division. Large volume samples are best mixed in an appropriate sized V-blender. Smaller samples may be blended in a laboratory jar mixer or by hand. The time required for mixing and its relative effectiveness are, of course, material specific. It is generally wise to err in the direction of too much mixing. One hour in a properly used V-blender, however, is adequate for most materials.

Large samples are divided by a laboratory version of the "coning and quartering" operation which was described in Section 2.1.3. Smaller samples are best divided by riffling, which requires a set of riffles (also known as Jones dividers). Stationary versions of the most common 50% split riffles are devices with linear arrays of slotted openings connected to chutes that alternate in their connection to two collection vessels. Proper use requires a riffles with opening slots 2–4 times larger than the top-size particles in the sample material. The blended sample is merely poured through the riffles and one of the two collected samples is selected for further processing. Riffling may be repeated any number of times to reduce the sample volume appropriately. Stationary riffles with other ratios, for example, 25% (4 collection vessels) and 12.5% (8 collection vessels) are also available. In addition to stationary riffles there are also rotating designs with a circular array of wedge-shaped collectors. In some versions the collectors rotate beneath a stationary vibrating feeder, in others the collectors are stationary beneath a rotating feeder.

While there are exceptions for specific commodities sample preparation for heterogeneous materials generally consists of a series of comminution, mixing, and dividing steps that are repeated until an adequate amount of test sample of just the right mesh size is achieved. Throughout this cycle of "crush-mix-split" operations correct representation of the original lot must be faithfully maintained.

At 1/4 in. (6.4 mm) top-size the roll crusher or cone crusher are effective devices. These can reduce this starting material to 1–2 mm–sized particles. After mixing and dividing this material can be reduced further in a lab pulverizer, a planetary or centrifugal mill, or a vibrating disk or cup mill. Most of these mills are capable of grinding the 1–2 mm feed much finer than required for analytical work. Grinding a sample too fine will produce dusting problems and often produces a hygroscopic and difficult to weigh test sample. Moreover, if occluded gases or moisture or the partitioning of oxidation states are part of the analytical requirements the test sample will inadequately represent these species.

To prevent over-grinding at this stage the mill should be stopped at regular intervals and the product sieved. Material retained on the selected sieve is returned to the mill for further processing. This procedure is repeated until *all* material has passed the selected sieve size. Even small amounts of abrasive resistant material retained on the sieve must not be discarded since they may be rich in analyte.

Sieving should always utilize a tight fitting collection pan below and cover above the sieve. A vibrating sieve shaker is useful for laboratories that do a great deal of sieving. Sieves, like all the material utilized in the preparation of the test sample, must be thoroughly cleaned between samples. While brushing and a compressed air hose on a trigger release and a shop vacuum will remove major contamination, sieves and grinding mill components must be thoroughly flushed with distilled water and completely dried between uses. The sample preparer and anyone else in the room must wear a suitable respirator during all dust forming operations. These include all grinding, mixing, dividing, and sieving operations as well as all cleanup with an air hose, brush, or vacuum.

2.3 Selecting the Test Portion

Even very experienced analysts tend to underestimate the importance of test portion size to the ultimate accuracy and representativeness of the result. Clearly, there are considerations that are strictly analytical chemistry concerns. Is the test portion large enough to provide sufficient analyte to be measured accurately? Is the test portion small enough so that matrix background and interference effects can be adequately handled by the method?

But there is also a critical relationship between test portion size and the representativeness of the analytical work. To return to Gy's language: do all constituents of the test sample have a 50% chance of being selected in the test portion? It turns out that it is possible to calculate a minimum test portion size necessary to achieve accurate representation of the test sample. Ingamells and Switzer developed the concept of the laboratory sampling constant (K_s) for this purpose [25–27].

To calculate K_s one must select an analytical procedure known to have a variability which is less than 1/3 of the variability due to the test sample. One must then select a test sample weight which is known to produce a reasonable level of accuracy with this procedure when reasonably homogeneous samples are analyzed. One then analyzes replicate test portions of this weight of test sample and calculates the relative standard deviation,

$$R = 100s/x \tag{3}$$

where R is the relative standard deviation, s is the standard deviation (1-σ), and x is the mean. The laboratory sampling constant, K_s is then given by

$$K_s = R^2 w \tag{4}$$

where w is the sample weight employed. K_s is in mass units and represents the weight of sample that will result in a 1% subsampling error at the 68%

Table 1.1 Calculated Values of the Laboratory Sampling Constant

Data Set	Test Portion (g)	Mean Value (%)	s, ± (%)	R, (%)	K_s (g)
A	0.25	25.13	0.18	0.72	0.13
B	1.0	5.29	0.06	1.13	1.5
C	1.0	0.295	0.009	3.05	9.3
D	2.0	0.0072	0.0003	4.17	35.
E	3.0	0.00054	0.00004	7.41	164.

confidence interval. K_s is specific to the test sample and the analyte, but is independent of the analytical method employed. Further comminution and mixing of the sample may sometimes, but not always, reduce the value of K_s.

Analysts who perform this exercise are often surprised by the results. Values of K_s are sometimes very large, especially for low concentration analytes. Once the value for K_s at weight w is in hand, the expected relative standard deviation (expected subsampling error), R_a, at a different test portion weight, w_a, can be obtained from

$$R_a = (K_s / w_a)^{1/2} \tag{5}$$

Tables 1.1 and 1.2 illustrate these relationships with a series of data summaries and calculated quantities. Table 1.1 lists the calculated values of K_s for five data sets. Table 1.2 shows the expected subsampling error if we insist upon taking what are often regarded as "routine" test portion sizes.

In solution-based analytical work the test portion may range between 100 mg and 5 g. But in spectrometric analysis using solid samples the test portion must be regarded as that part of the test sample responsible for the measured analytical signal. In DC arc optical emission the test portion might be the 1 or 10 mg shown in Table 1.2. In glow discharge mass spectrometry the test portion might be measured in nanograms. In X-ray fluorescence only a few atomic layers are measured in situ.

Table 1.2 Calculated Subsampling Error for Selected Test Portions

Data Set	10 G	1 G	100 Mg	10 Mg	1 Mg
A	0.11%	0.36%	1.14%	3.61%	11.40%
B	0.39%	1.22%	3.87%	12.25%	38.73%
C	0.96%	3.05%	9.64%	30.50%	96.44%
D	1.87%	5.92%	18.71%	59.16%	187.08%
E	4.05%	12.81%	40.50%	128.06%	404.97%

Clearly, this information must raise some warning flags about the complaisant routine use of solid sample instrumental techniques for samples of moderate to low homogeneity. And the situation is always most alarming at the near trace, trace, and ultratrace analyte levels where the subsampling error can climb into hundreds of percent.

There are other unique aspects of the relation of test portion size to the analytical data distribution when we operate in the trace realm. Ingamells and Switzer consider the situation in which the analyte is very concentrated in a heterogeneously distributed constituent phase in the sample. A test portion of such a material which contains six or less analyte-rich particles will produce data that is described by a skewed Poisson distribution, rather than a normal Gaussian distribution. For this case these authors describe the value for a single result (x_i) as

$$x_i = H + cz_i \quad (6)$$

where H is the average concentration of analyte in the sample matrix, z_i is the number of analyte-rich particles in the selected test portion, and c is the average concentration of analyte in the analyte-rich particles.

At a test portion size where the value of z_i is between zero and six, only a small number of the total replicates taken will accurately represent the test sample. Such a large number of low results will be obtained that when the true value occurs it may be rejected as an outlier. The erratic behavior of the data at this test portion size may lead the analyst to believe that he is observing instrument noise or some other form of analytical method error. However, if a very large number of values are averaged without rejecting suspicious data the mean can be reasonably accurate.

Increasing the test portion size so that z_i is greater than six leads to much more well-behaved data. The data follows a Gaussian distribution and the mean is both accurate and precise. Unfortunately, lowering the test portion size so that z_i is nearly always zero also leads to well-behaved data. In this case the data also follows a Gaussian distribution and the mean is again precise, but the mean value is completely inaccurate, reflecting only the analyte which is present in the matrix.

We can see in this description a possible explanation for some of those puzzling discrepancies between sets of trace results determined by different analytical techniques. If the test portion size employed for a given analytical method is too small the resultant data will appear fine but only H in Eq. (6) will have been measured. A different method utilizing a larger test portion may generate badly scattered data with a higher mean that may be much closer to the truth. In fact a cautionary note is suggested here: if small test portions lead to precise data at trace levels either the material is extremely homogeneous or the results are wrong.

3. THE CONTROL OF CONTAMINATION

It is generally recognized that the most pervasive impediment to accurate trace analysis is a large or uncontrolled blank. If the blank is too large it places a barrier before the inherent sensitivity of an instrumental measurement. While the barrier is seldom opaque, it is also seldom completely transparent, and thus we must peer at truth through some degree of translucence. If the blank is very erratic our view through the barrier is clouded, and we must only report that truth lies somewhere beyond it. But a controlled, repeatable blank allows us to sight beyond the barrier and estimate a value.

The sources of the blank are diverse but derive largely from failure to perfectly control contamination during the manipulative steps of the analysis. The other sources of the blank are inherent in the sample (positive interference from matrix elements) and the measurement (positive bias in the instrument response). These will be discussed in Section 4. Here we will take up the subject of contamination by analyte as it may occur between sampling and measurement.

3.1 The Laboratory

Sample preparation, dissolution, chemical separation, and preconcentration all take place in some area of a laboratory. In the case of trace analysis special precautions to ensure clean work spaces have often proven necessary. Keeping equipment, hoods, and work surfaces clean begins, of course, with prompt and meticulous housekeeping. Spills are immediately wiped up and great care is taken to prevent cross-contamination between samples. But the key issue with trace and ultratrace work is airborne particulate contamination. The technical and administrative staff of each facility must decide how much effort and expenditure is justified to limit or eliminate such contamination.

A practical first step in an investigation of airborne contamination is to place outfall collection tests in the work environment. These can be circles of filter paper or open beakers. The diameter of the collection area and the time of outfall exposure should be noted. The samples are then digested in acid, diluted to volume, and measured for the analytes of interest. A 9-cm diameter circular collection area (63.6 cm^2) exposed for two working days should yield a reasonable first estimate of the problem (if there *is* a discernible problem). Longer collection times may be needed to take into account the effect of periodic or intermittent activity within the laboratory environment, as well as cyclic activity adjacent to the laboratory, such as automobile traffic patterns.

Seasonal activity, such as fertilizer application near farming regions, or road salt application in winter, will require longer term, more complicated outfall studies. The placement of the outfall collection tests is critical to the validity of the results. In general they should be placed where the work related to trace analysis will be performed, but protected from drafts or human intervention that may inadvertently invalidate the study.

The results of such tests are often surprising. Before the abolition of leaded gasoline lead at the trace level was pervasive in the environment. Iron and silicon are always everywhere. Zinc is common when anything galvanized is used in the room. Other elements vary widely based on the nature of the work conducted in the laboratory, and the local industry, and local air quality controls. Sodium will be common near the sea, calcium will be common near farmlands.

While the containment of particulate airborne contamination elicits visions of clean rooms with capped, gowned, and bootied analysts, in fact, there are stages to this process that allow a limited budget chemistry lab to moderate the effect of the lab environment on the blank. A positive pressure environment with conventional filters on the air outlet ducts is a relatively painless upgrade for most work environments. Special floor and benchtop coatings and cleaning protocols with, perhaps, the installation of a peelable layered "tacky mat" at the room threshold are relatively inexpensive changes as well. A benchtop clean air station with an *integrated high efficiency particulate air* (HEPA)* filter is an economic way to provide a clean area for aliquotting and diluting operations. If samples are digested in sealed high pressure vessels this may be all the airborne contamination control needed.

For open vessel hotplate digestion it may be necessary in certain situations to invest in a class 100 laminar flow hood. This type of hood draws room air through HEPA filters and showers it evenly over the work surface which may include an electric hotplate. Acid fumes are drawn off through louvers at the back, then exhausted up the stack. "Laminar flow" indicates that no air turbulence occurs. "Class 100" means that the distribution of particle sizes in the work environment falls below a limiting curve—a hundred particles at the 0.5 mm size are permitted per cubic foot of air (0.035 particles/l). This class 100 environment is ideal for analytical work since the greatest airborne contribution to an inorganic analysis blank typically comes from particles between 1 and 10 μm in diameter. Particles below 1 μm, even in massive numbers, usually have insufficient mass to contribute significantly to the blank [28,29]. Federal Standard 209a defines class 100 conditions [30].

If the laboratory decides to invest in this technology HEPA filters with plastic separators must be specified. Aluminum separators will be corroded by acid fumes and then contribute significantly to the blank for aluminum

* These filters were developed originally for the Manhattan Project to control radioactive dusts. They consist of pleated sheets of glass fibers.

determination. The replacement of metal with plastic is a good idea throughout an area where trace work is performed. Metal fixtures that have been corroded by acid fumes are a significant source of airborne particulate blank problems.

The question that must be answered before embarking on courses such as these is: what do such investments buy you? Many analytes which are commonly determined at the trace and ultratrace levels are very rarely found in airborne particulates. With other analytes, like chloride, all the HEPA filters of a clean room will not prevent sample contamination from gases (say, if someone should open a bottle of hydrochloric acid in the room). For the majority of routine trace analysis requirements a clean workplace with good analytical and housekeeping procedures in place may be all that is required.

3.2 The Equipment

Obviously, any piece of laboratory apparatus that contacts the test portion has the potential of contaminating it. This potential increases dramatically in the presence of aggressive conditions—heat, acids, oxidants, and molten salts, for example. Thus, it is a great advantage to seek out and employ only equipment manufactured from high purity or analyte-free materials.

3.2.1 Vessels

Borosilicate glass (81% SiO_2) from which most labware is manufactured is far from an ideal material for trace analysis. It contains major amounts of aluminum, boron, and sodium, as well as significant minor amounts of potassium, iron, calcium, magnesium, and manganese. Trace levels of antimony and zinc are also commonly present [31]. Fortunately, many common mineral acids are not particularly corrosive with this composition, allowing higher level trace work in such vessels. Hydrofluoric acid, phosphoric acid, and caustic conditions, however, release massive contamination from borosilicate glass. For this reason the use of borosilicate glass vessels in trace analysis is highly restricted. In ultratrace work such vessels have no place at all.

Vycor glass (a trademark of Corning, Inc.) is 96% SiO_2. Vycor vessels are manufactured by chemical leaching of the sodium and boron from a borosilicate vessel, then firing the vessel at 1200°C to produce a pore-free product. Vycor shows four times the resistance to thermal shock of borosilicate glass. Unfortunately, it retains 0.5% Al_2O_3 and 3% B_2O_3, making it unsuitable for certain very low level work with aluminum and boron. Fused quartz, on the other hand, is 99.9% SiO_2 and in its water-clear version is available as beakers, flasks, evaporating dishes, and other forms. This material is extremely resistant to thermal shock. And, although it is attacked by HF, H_3PO_4, and caustic, for the most part only silicon is added to the sample solution. Part per million level impurities remain, however, if the starting material of manufacture was natural quartz crystal. For the ultimate high purity glass there is synthetic quartz, produced by the vapor phase hydrolysis of $SiCl_4$, but it is not currently available as labware [31].

Beakers, flasks, bottles, and much other laboratory equipment is available manufactured from plastic. Low density polyethylene (LDPE) is tough and flexible, while high density polyethylene (HDPE) is more rigid and somewhat brittle. Both forms are resistant to common acids and solvents, but neither can withstand a temperature above that of a boiling water bath. For trace work LDPE is the preferred form. This is because HDPE shows a higher residual metal content (aluminum, titanium, zinc, chromium, and cobalt have been found) [32].

Two common forms of fluorocarbon polymer differ in trace metal content as well. White, opaque polytetrafluoroethylene—known by the duPont trademark "Teflon TFE"—resists all chemicals except molten alkali metals, fluorine gas, and carbon trifluoride. It can be used at temperatures up to 250°C. Both perchloric acid and sulfuric acid can be heated in TFE vessels in a sandbath on a hotplate until the acid anhydride fumes are evolved. Little or no contamination from TFE vessels has been observed. Fluorinated ethylene propylene—known by the duPont trademark "Teflon FEP"—on the other hand, reacts with perchloric acid to produce hydrofluoric acid. It also reacts with those substances that attack Teflon TFE. FEP has a temperature limit of 205°C and is thus unsuitable for heating "nonvolatile" mineral acids to fumes. Traces of iron, manganese, zinc, aluminum, copper, and nitrogen have been found in particles near the surface of FEP vessels—most likely a residue from the molding process [33]. Isotope dilution mass spectrometry and neutron activation analysis have been applied in the characterization of leachable trace impurities from various forms of plastic materials, including polyethylene, polypropylene, polymethylpentene, polycarbonate, polyvinyl chloride and both TFE and FEP Teflon [34].

In recent years a new fluoropolymer has emerged for labware applications. It is perfluoroalkoxy polymer, known by the duPont trademark, "Teflon PFA." It exceeds the temperature limit of Teflon TFE and is translucent like Teflon FEP. Like FEP it should not be used with perchloric acid. Teflon PFA is available as beakers and bottles. After an acid cleaning and water rinse regimen PFA bottles have been used for the storage of high purity acids.

Crucibles for ignitions, volatilizations, and molten salt fusions are manufactured from a number of materials. In the analytical laboratory porcelain, platinum, Vycor, fused quartz, graphite, and zirconium are the most common crucible materials. Iron, nickel, and silver crucibles are also found at some facilities. Porcelain in either glazed or unglazed form is not ideal for trace work due to the possible pick-up of aluminum, potassium, and silicon from the substrate material. Additionally, the silica glazes may contain copper. Of course, porcelain, Vycor, and fused quartz crucibles are only suitable for ignitions and fusions with acid fluxes like potassium pyrosulfate. They must never be used for basic fluxes such as carbonates, hydroxides, or potassium peroxide, since they will be destroyed by such treatment.

Platinum labware is usually alloyed with rhodium or iridium to impart hardness. Gold is used as an alloying element to impart "nonwetting" properties to platinum crucibles and dishes used for casting fused borate disks for X-ray fluorescence work. Although generally regarded as noncontaminating, most molten salt fusions dissolve some measurable amount of platinum. Platinum crucibles should never be used for sodium peroxide or any hydroxide fusions. *Aqua regia* or any other mixture of hydrochloric and nitric acid will dissolve platinum. Platinum should also never be exposed to chlorine or bromine water. Permanent or semipermanent alloying is also a concern with platinumware. Igniting large amounts of carbonaceous substances in the absence of sufficient oxygen, or heating in a luminous gas flame will produce a carbon alloy reaction. Large amounts of sulfur, selenium, tellurium, phosphorus, arsenic, or antimony compounds will also alloy with platinum. Platinumware should never be heated directly in contact with a metal surface such as a hotplate, and metal samples should never be fused in platinum vessels.

Zirconium, nickel, and iron crucibles are used principally for sodium peroxide fusions, which dissolve massive amounts of the crucible material. Iron crucibles are suitable for only one fusion and should then be discarded. More expensive nickel crucibles may be useable for up to four fusions. In addition they are commonly used for sodium, potassium, or lithium hydroxide fusions where they show a longer service life. Zirconium crucibles are uniquely resistant to molten sodium peroxide, although with each use a significant amount of zirconium dissolves. New zirconium crucibles appear shiny and metallic but darken to dull purple upon first heating. They should only be heated in the reducing flame of a Meker burner. Zirconium will oxidize in an air muffle furnace forming a surface coating of powdery white ZrO_2.

The purest substance from which crucibles are manufactured is graphite. While porous graphite crucibles are available and are used widely for muffle furnace fusions (especially with lithium tetraborate), they contaminate the sample with particles of graphite. Pyrolytic graphite-coated crucibles are produced by chemical vapor deposition on a crystalline carbon substrate involving the "cracking" of toluene vapor in a vacuum furnace. They are harder, more chemically resistant, and less contaminating than porous graphite crucibles [29].

3.2.2 Other Equipment

Besides vessels that contain the sample or the reagents there are numerous pieces of laboratory equipment with a potential to add to the blank. A galvanized splashboard for a coated wire drying rack can be a significant source of zinc contamination. Unshielded chromel heating elements in a drying oven can contribute chromium and nickel blanks [29]. Similarly, the refractory ceiling of a muffle furnace can contribute silica. Even highly alloyed nickel or stainless steel crucible tongs can contaminate a hot platinum crucible. Platinum-tipped tongs should be used instead. Forceps for removing a crucible from a leach solution should be made of plastic or Teflon-coated metal.

Filter papers, filter pulp, and membrane filters all contain detectable levels of trace impurities. Membrane filters are for the most part better in this regard than even acid-washed, ashless filter papers, but phosphorus and nitrogen are frequently significant contaminants from both types of filters.

Plastic ion exchange columns are best constructed from polymethylmethacrylate which is transparent, reasonably acid resistant, and has a very low trace element content. On the other hand, polyvinyl chloride tubing should not be allowed to contact acid solutions of the sample or any reagents used in the methodology. Contamination from iron, zinc, tin, lead, titanium, and magnesium, among other elements, is possible due to the plasticizers and stabilizers added to this polymer. Iron, zinc, and barium are extractable contaminants from silicone rubber [29].

The "lint-free" tissues used to wipe pipets and micropipets have detectable levels of calcium, aluminum, silicon, magnesium, iron, sodium, and copper. Open gas flames outside of a hood will contaminate the laboratory air with sulfur due to the mercaptan odorant added to natural gas. The detergent utilized to clean glassware may contaminate it with carbon, phosphorus, sodium, magnesium, boron, sulfur, or other trace elements [35].

3.2.3 Cleaning Labware

Much has been written concerning the cleaning of new labware prior to its first use in trace analysis. In particular acid soaking regimens of up to a week or more for plastic vessels have been recommended. These are followed by lengthy soaks with high purity water [29,34]. The analyst must remain the final arbiter concerning the need for and efficacy of such pretreatments, keeping always in mind the analytical purpose.

A much larger problem than sample contamination from new equipment is the blank burden from previously used equipment. Every time a piece of laboratory apparatus is utilized it is contaminated, quite often with adherent matrix elements which are difficult to remove completely. How, then, is laboratory equipment restored to very near its pristine state? The answer is complicated because each form of contamination has its own best treatment. The subject is large enough to, perhaps, one day deserve a text of its own.

By far the best individual to decontaminate apparatus is the individual who used it last and now expects to use it again. He or she should know what contaminants are likely to be present and what must be absent for the anticipated work. Other scenarios are hazardous at best. Apparatus utilized for trace work should be kept isolated from general use equipment and stored where it is least likely to be contaminated by the laboratory atmosphere. In a large laboratory engaged in a great deal of trace level work it may be necessary to clean equipment thoroughly before storage and then again just before use to ensure that inadvertent contamination has not occurred.

Among the adherent stains on glassware encountered in metals analysis some of the most tenacious are from titanium, niobium, tantalum, tungsten, molybdenum, tin, and mercury. Titanium dioxide is sometimes deposited as a monolayer which produces a rainbow effect in bright light. Heavier deposits are white, as is titanium phosphate. Both forms of titanium compound, as well as white niobium and tantalum oxides, are best removed by adding sulfuric acid to the vessel and heating to strong fumes of SO_3. When the vessel has cooled to room temperature the excess acid is cautiously disposed of and the vessel is rinsed with copious water, scrubbed with a brush and a nylon abrasive pad, rinsed, and inspected. Stubborn areas may necessitate a second treatment.

Tungsten oxide (WO_3) is bright yellow, molybdenum oxide (MoO_3) appears white. Both are easily removed from a vessel by adding strong caustic and swirling. The vessel is then rinsed in hot water and scrubbed. Tin forms an adherent layer of white SnO_2 (glassware used to dispense stannous chloride solutions typically becomes stained in this manner). It is easily removed with concentrated hydrochloric acid. Black or gray deposits of mercury, or an adherent "mercury mirror" are easily released with concentrated nitric acid. Sometimes a drop of mercury forms in the vessel. In this case the acid must be diluted and decanted until neutral water is present in the vessel; then the mercury drop is transferred to an approved mercury waste container. Naturally, these operations, as with all laboratory operations involving noxious materials, must be conducted in an efficient fume hood, using approved safety equipment and approved laboratory practices.

Occasionally other stubborn stains are encountered on glassware. Silver and lead chlorides succumb to concentrated ammonium hydroxide. Elemental selenium and tellurium require *aqua regia* (i.e., a 3:1 mixture of $HCl:HNO_3$), as does gold. A significant problem for many metal industries is the routine cleaning of volumetric flasks and other glassware where the sample mix is highly varied. It may be necessary to utilize a vessel for trace nickel in an iron matrix, for example, then next use it for trace iron in a nickel matrix. Obviously, it is impossible to keep a separate set of glassware for every conceivable contingency. And so a routine cleaning regimen needs to be foolproof.

One approach applicable to many common transition metals is to rinse the vessels first in tap water, then allow them to soak in concentrated hydrochloric acid for about an hour. This is followed by thorough rinsing and scrubbing with a good noncontaminating laboratory detergent and a final thorough rinse with distilled water. The vessels are then dried in a particle-free environment and stored. When needed for the next use they are removed from storage and rinsed with 18 megohm-cm, filtered water, such as that produced by a recirculating "polisher" system. For work with lead, bismuth, cadmium, selenium, and tellurium a similar regimen, substituting nitric acid soaking for hydrochloric acid soaking is appropriate.

Organic films and other contaminants that "de-wet" glass surfaces causing droplets of liquid to cling to vessel walls are best removed by soaking in concentrated sulfuric acid. Chromic acid (a mixture of $K_2Cr_2O_7$ or CrO_3 and H_2SO_4) is *not* recommended for trace analysis since it deposits an adherent layer of chromium on the glass surfaces.

Crucibles require special cleaning procedures since their contamination is often more adherent than that encountered with glassware. Also the analyst must be aware that certain materials, like platinum, for all their chemical inertness, are easily damaged. Platinum crucibles are best cleaned by fusing with potassium pyrosulfate to red heat, then pouring out the melt into a dry metal waste container. The crucible is allowed to cool, then it is placed in a boiling solution of 1:1 $HCl:H_2O$ for about ten minutes. Then it is rinsed thoroughly with distilled water and allowed to dry. Heavily contaminated crucibles may require two such treatments. Zirconium and nickel crucibles are boiled in 1:1 $HCl:H_2O$, rinsed, and dried. Vycor and fused quartz crucibles can be soaked in a room temperature *aqua regia* solution for 10 min.

3.3 The Reagents

Usually the largest positive contribution to the blank derives from reagents which were added to or exposed to the sample solution. It behooves the trace analyst to seek out the highest purity reagents possible while striking some balance between analytical needs and budgetary constraints because high purity reagents command imposing prices. For example, if ACS reagent grade nitric acid is contaminated with titanium a method for trace titanium requires at least the next level of purity. The ACS grade may be just fine for trace aluminum determinations, however.

There is a complex argot of terms associated with reagent purity that is not completely consistent between manufacturers. "ACS" grade always means that the reagent meets the specifications described in *Reagent Chemicals, American Chemical Society Specifications* [36]. "C.P." and "Technical" grades are below this in quality and are never used in the analytical chemistry lab unless there is no practical alternative. For inorganic chemicals the next higher level of purity above ACS grade may be called "Instrument" grade, "Electronic" grade, or even "Trace Metal" grade. Included with these reagents is an extensive analysis of the lot. But the analyst must be cautious because this lot analysis may omit certain elements present in measurable quantities and which may contribute to the blank. Above this level is the "Ultrapurity" grade (generally, at 2–10 times the cost of ACS grade), which is provided with a very detailed and usually nearly exhaustive lot analysis. These reagents may be delivered in sealed glass ampoules or in Teflon PFA bottles sealed with Teflon TFE tape.

Solvents follow a different progression of terms, partly directed toward an anticipated usage. "HPLC" grade, intended for high-performance liquid chromatographic use, from the viewpoint of inorganic analysis is actually at or slightly below the ACS grade level. "Pesticide" grade represents a level where solvent residue after evaporation is very low. "Spectrophotometric" grade solvents are very pure and are accompanied by a UV absorption curve representing the lot. There are also solvents sold for use in specified EPA procedures.

Gases at 99.6% purity may be called "Extra dry" or "Bone dry." At 99.99% purity they are generally called "High purity." Gases with very low total hydrocarbon content are called "Zero gas." Extreme purity levels begin with "Prepurified," then proceed to "Ultrahigh purity," and end at "Research" grade (at 99.999% purity or higher).

Clearly, the trace analyst needs to use high purity chemicals. But choosing a level of purity far beyond the true analytical needs imposes an irresponsible economic burden on the analytical operation. Thus here, as everywhere in this endeavor, the analytical chemist must become intimately familiar with the tools of the trade and select among them wisely.

3.3.1 Water

Water of high and consistent quality must be conveniently and copiously available to the trace analyst. The American Society for Testing and Materials (ASTM) publishes a standard specifying four types of reagent water [37]. Each type is characterized by the means of preparation from feed water and by the parameter values that the resultant product is expected to possess.

Type IV, the lowest level of purity, is produced by any available technology (distillation, ion exchange, reverse osmosis, electrodialysis, or any combination of these techniques). It may show a minimum resistivity of 0.2 megohm · cm at 25°C and a maximum total solids content of 2.0 ppm. Type IV water is considered useful for washing and rinsing labware and for preparing some synthetic test solutions. Type III water, the next level, is prepared the same way except for a final filter through a 0.45 μm membrane. Type III must show at least 1.0 megohm · cm resistivity and 1.0 ppm total solids. It is recommended for most general laboratory testing. Type II water must be prepared by distillation and must show at least 1.0 megohm•cm resistivity and 0.1 ppm total solids. To achieve this level preliminary treatment of the feedwater or multiple distillations may be required. Type II water also must meet stringent requirements for the absence of oxidizable organic substances and the absence of microbes. Type II water is specified for analytical laboratory procedures and all general procedures where low levels of organic matter are required.

Type I water must be prepared by distillation from feedwater of at least 0.05 megohm · cm resistivity. The distillate is then polished with a mixed bed of ion exchange resins and then filtered through a 0.2 μm membrane. The product must have a resistivity of at least 16.67 megohm · cm and show no more than 0.1 ppm total solids. It must also meet the other requirements of Type II water. Type I water is generally considered suitable for most trace analysis procedures.

A large laboratory can easily justify the installation of a system that provides Type I, II, and III water at convenient locations. An example of such a system might involve an ion exchange system using tap water as its input feed. At a filtered tap on this installation Type III water is available. This water then feeds a central still with distillate reservoirs used to distribute Type II water to taps throughout the building. Such reservoirs should be equipped with UV lights to prevent the growth of microorganisms. At tap locations where Type I water is needed a recirculating ion exchange polisher with a final 0.22 μm filter is installed. A reverse osmosis system followed by an activated charcoal bed, then a mixed bed (strong acid and strong base) ion exchange resin, followed by a 0.22 μm final filter has been shown to, likewise, produce very low trace impurity water [38] that can be tapped at different stages.

It is a good policy to establish a regular program of trace element monitoring of the Type I water. This can be a relatively painless exercise if the laboratory is equipped with a simultaneous inductively-coupled plasma optical emission spectrometer. A selected group of key elements (iron, silicon, calcium, magnesium, sodium, and copper, for example) can even be control charted to monitor trends that might give early warning of problems with the water system.

3.3.2 Acids

All metals and alloys dissolve at least to some degree in some combination of acids. And even in those analytical schemes employing other means to effect dissolution, the aqueous solution ultimately measured is nearly always acidic.

Thus acids play a large role in all metal analysis, and their purity plays a critical role in trace analysis.

Hydrochloric acid (HCl) is probably the most widely useful mineral acid. It is nonoxidizing. Much of its dissolving power derives from the formation of metal–chloride complexes. These complexes also prove to have great utility in solvent extraction and ion exchange separations. ACS reagent grade hydrochloric acid is a 37% (w/w) solution of HCl gas in water (12 M; specific gravity: 1.18). For determinations at analyte levels below about 0.001% the analyst should consider an "Instrument" grade. A perusal of the lot assay will quickly reveal if this intermediate level of reagent will be adequate. If not, it will be necessary to utilize an expensive "Ultrapurity" grade.

Nitric acid (HNO_3) is a strong oxidant that dissolves many metals and forms a passivating oxide layer on many others. It is highly effective in combination with other acids, such as hydrochloric, hydrofluoric, sulfuric, or perchloric. ACS reagent grade is 70% (w/w) (16 M; specific gravity: 1.14). For titanium determinations at levels below 0.1% it is necessary to use an "Ultrapurity" grade.

Hydrofluoric acid (HF) is incompletely ionized in aqueous solution, but it is a highly active complex former. For metals analysis it is used primarily in combination with other acids. Special precautions in its use are necessary, and the analyst must be familiar with procedures for its safe handling, as well as with emergency medical protocols. ACS reagent grade is 49% (w/w) (29M; specific gravity: 1.15). "Ultrapurity" grade may be ten times as expensive as ACS grade, but it is occasionally necessary.

Sulfuric acid (H_2SO_4) exhibits distinct dissolution properties in concentrated and dilute solution, some metals dissolving in one concentration and relatively unreactive in the other. At room temperature it is nonoxidizing, but it becomes slightly oxidizing at its fuming temperature, when white vapors of the anhydride, SO_3, are evolved. ACS reagent grade is 96% (w/w) (18 M; specific gravity: 1.84). "Instrument" grade and "Ultrapurity" grade are available.

Perchloric acid ($HClO_4$) is one of the "strongest" acids (in the hydrogen ion donor sense). For example, it is completely ionized in the presence of sulfuric acid which is also a very strong acid. It is completely nonoxidizing at room temperature but becomes powerfully oxidizing at its fuming temperature. It forms no metal ion complexes—a property used to advantage in some ion exchange schemes. Two common concentrations are sold as ACS reagent grade: 60% (w/w) (10 M; specific gravity: 1.53) and 70% (w/w) (12 M; specific gravity: 1.66). "Instrument" and "Ultrapurity" grades are generally sold at 70% (w/w). Except in the basic steel industry, perchloric acid is seldom used to dissolve metal. It finds use in destroying inclusion compounds and in evolving volatile species from solution. Special hoods with a complete stack wash-down feature are required for its use since explosive compounds tend to accumulate. The spontaneously explosive anhydrous acid is evolved from heated sulfuric/perchloric acid mixtures. Such evolved distillates should never be collected.

Phosphoric acid (H_3PO_4) is generally used in combination with either sulfuric or perchloric acid where use is made of its complexing properties. ACS reagent grade is 86% (w/w) (15 M; specific gravity: 1.69). Both "Instrument" grade and "Ultra-purity" grade are available. Many elements which precipitate as hydrous oxides in acid solution can be held in solution by the presence of phosphoric acid.

Acetic acid (CH_3COOH) finds some limited use in the trace analysis of metals. So-called "glacial" acetic acid is nearly 100% CH_3COOH. Because it finds use in both inorganic and organic chemistry the available grades include not only ACS reagent grade, "Instrument" grade, and "Ultrapurity" grade, but also "Aldehyde-free" grade and "HPLC" grade, among others.

For laboratories that perform a great deal of trace analysis it may prove economical to prepare high-purity versions of the above acids and others in-house. There are three viable means to this end, although not all are practical for every acid. The dissolution of a gaseous compound in high purity water is an expensive and hazardous option. However, some laboratories do find it practical to prepare sulfurous acid (H_2SO_3) by sparging a dilute sulfuric acid solution with SO_2 gas.

Isopiestic distillation (isothermal distillation) relies upon the diffusion of acid vapors at room temperature. In this lengthy process two vessels are placed a short distance apart in a sealed container (such as a desiccator minus the desiccant). One vessel contains an ACS reagent grade volatile acid the other contains high purity water. After several days the water will have turned into a very high purity dilute form of the acid. The process can be accelerated by either changing the reagent grade acid daily or by increasing the amount of reagent grade acid relative to the volume of high purity water [39] (the ratio of exposed surface areas is obviously a critical parameter).

The most widely useful approach to the in-house preparation of high purity acids is subboiling distillation. In this process a reagent grade acid is heated below its boiling point in a quartz or Teflon TFE apparatus. Generally, the heat source is an infrared glow bar or an infrared lamp placed just above the surface of the feedstock. The condensate is collected in a specially cleaned vessel (sometimes a Teflon FEP or PFA bottle). The entire process is conducted in a class 100 laminar flow hood, or at least the collection bottle is shielded in some manner from room air particulate outfall.

Subboiling distillation has been applied to the laboratory production of high purity hydrochloric, nitric, hydrofluoric, sulfuric, and perchloric acids, among others. Of course, an all-Teflon apparatus is required for hydrofluoric acid [40].

A very simple and comparatively inexpensive alternative to the purchase of a subboiling distillation apparatus was described by Mattinson [41]. It consists of two Teflon FEP bottles connected at right angles through a threaded Tefon TFE joint. Reagent grade acid in one bottle is heated by means of a heat lamp and the condensate is collected in the second bottle which is cooled in a water bath. Other investigators have confirmed that this simple arrangement can produce hydrochloric and hydrofluoric acids of exceptional purity [42]. Spark source mass spectrometry has been utilized to monitor the impurity levels in hydrochloric, hydrofluoric, and nitric acids produced by subboiling distillation [43].

3.3.3 Other Reagents

There are hundreds of reagents that find use in the trace analysis of metals. Only a few of them, however, are used frequently enough to bear mentioning here.

Ammonium hydroxide (NH_4OH) is sold commercially as a 56.6% (w/w) solution (14.5 M; specific gravity: 0.90). In addition to ACS grade both an "Instrumental" and an "Ultrapurity" grade are available. High purity solutions are easily prepared by an isopiestic approach. Sparging ammonia gas into high purity water is another route, but it requires an EDTA solution trap to stop contamination by copper and nickel (presumably from the valve fittings) which will otherwise contaminate the product [29].

Hydrogen peroxide (H_2O_2), 30% (w/w), can be obtained in an "Ultrapurity" grade. Unlike ACS reagent grade this product is not stabilized with sodium stannate (Na_2SnO_3) and thus has a more limited shelf life. The container should be kept in an loose-fitting plastic bag in a refrigerator.

Most, but not all, needed molten salt fluxes can be obtained in either an "Instrumental" or an "Ultrapurity" grade or both. Sodium carbonate (Na_2CO_3), available in both grades, as well as ACS reagent grade, is an excellent alkaline flux for siliceous residues and refractory nitrides. The principal acid flux, potassium pyrosulfate ($K_2S_2O_7$), which is most useful for high alumina residues, is sold both as an ACS grade reagent and as an "Instrumental" grade. Various high purity forms of lithium tetraborate ($Li_2B_4O_7$), lithium metaborate ($LiBO_2$), sodium tetraborate ($Na_2B_4O_7$), and boric acid (H_3BO_3) are available. Alone or in mixtures with carbonates these fluxes find utility as molten salt compounds for a broad range of materials.

Unfortunately, one of the potentially most useful flux compounds, sodium peroxide (Na_2O_2) is not currently available in any purity above ACS reagent grade. This may limit its utility in certain ultra-trace work with highly refractory compounds. Sodium hydroxide (NaOH) and potassium hydroxide (KOH) are similarly unavailable in ultra-purity form, limiting some contemplated approaches.

When acid fusion melts need to be leached in a dilute perchloric acid solution potassium pyrosulfate is an impractical choice due to the limited solubility of potassium perchlorate. In this case an anhydrous fused sodium bisulfate ($NaHSO_4$) is substituted. Again, however, nothing purer than ACS reagent grade is commercially available.

Certain oxidizing agents in pure form are considered primary standards for titrimetry and other work. Such compounds are generally sold in a "Primary Standard" grade by commercial laboratory chemical manufacturers. Many are also available as Standard Reference Materials (SRMs) from the National Institute of Standards and Technology (NIST). Examples include potassium dichromate ($K_2Cr_2O_7$), ceric ammonium nitrate [$(NH_4)_2Ce(NO_3)_6$], potassium iodate (KIO_3), and potassium bromate ($KBrO_3$). In addition an "Instrumental" grade of potassium peroxydisulfate (commonly known as potassium persulfate) ($K_2S_2O_8$) is available.

Among reducing agents both sodium oxalate (NaC_2O_4) and oxalic acid ($H_2C_2O_4 \cdot 2H_2O$), as well as arsenious oxide (As_2O_3) and sodium meta-arsenite ($NaAsO_2$) are sold as "Primary Standard" grades. The principal acid/base titration standard, potassium biphthalate [$KH(COO)_2C_6H_4$] is, likewise, sold as a "Primary Standard" grade. Ammonium phosphate [$(NH_4)_3PO_4$] and potassium phosphate (K_3PO_4), both useful in preparing pH buffers, are sold in "Ultrapurity" grades.

Many other compounds are sold in "Instrumental" and "Ultrapurity" grades by laboratory chemical manufacturers. Some of these can be used to good advantage in trace and ultratrace analysis even though that end use was not the target market for the product. And even in the analytical area high-purity can be diverted to meet novel needs. For example, potassium biphthalate makes a good solution standard for carbon determination, and potassium bromide intended for infrared work makes an excellent solution standard for bromine determination.

We have already discussed purity grades for organic solvents (Section 3.3). It should only be further noted here that suitability for an *organic* trace analysis application is no guarantee of suitability for an *inorganic* trace analysis application. Other needed organic reagents are seldom available at high purity although the lot analysis may include some statement purporting the product's suitability for a trace inorganic analysis application. Sometimes trace level work will require the analyst to purify an organic reagent before use. Solvent extraction, recrystallization, or both may be involved.

There are companies specializing in supplying high purity compounds and elements in a range of forms and purities. These materials are used for preparing calibration solutions and synthetic matrix solutions. Several cautionary notes need to be made in this regard. High purity metals in the form of powders are very often contaminated with surface oxygen, an element which is almost never included in the accompanying lot analysis. Wire, rod, or even shot are the preferable forms because of their lower surface area (and thus lower surface oxygen burden). Oxychloride salts (such as $ZrOCl_2 \cdot 8H_2O$) are hygroscopic and thus nearly impossible to use unless the solutions prepared from them are standardized by classical methods (like gravimetry). Oxides may be difficult to dissolve and a few may not be stoichiometric. The "high-fired" ignited oxides of some metals have no stoichiometry problems but are typically the most resistant to dissolution.

A final note needs to be made about commercially prepared calibration solutions. In today's increasingly burdened analytical laboratories with their heavy dependence on solution-based comparative methodology it is tempting to place blind faith in commercial standard solutions. This course is fraught with problems, however, since poor quality in such products can undermine the work of the best intentioned laboratory. The only practical approach to the use of such solutions is to screen each potential supplier by applying each of their products to the analysis of appropriate validation reference materials before it is put into routine use. Of course all routine work must be validated as well. When a problem with a data set occurs the commercial solution employed should be one of the first parameters scrutinized.

3.4 The Analyst

Perhaps the least obvious source of contamination is the analyst himself. The human body and its accouterments (clothing, adornments, and cosmetics) produce a veritable snowstorm of particles. Dried skin particles are sources of zinc and copper, hair contains iron and lead as well as zinc and copper. Zinc is a component of many cosmetics and skin lotions. Zinc and bismuth occur in lipstick. Cosmetics may also contain aluminum, calcium, titanium, iron, potassium, and other elements. Antiperspirants are likely to contain zirconium.

Touching surfaces that come in contact with sample solutions can contribute contamination as well. In addition to massive amounts of sodium and chloride, human perspiration contains potassium, calcium, magnesium, as well as sulfate, phosphate, and ammonium ions. Tobacco smoke was banned from the trace analysis laboratory long before it was banned from the workplace in general. It contains boron, potassium, and iron, among other elements.

The literature contains some discussion about contamination resulting from the wearing of jewelry. Watches, rings, bangles, and bracelets are generally considered possible sources of gold, silver, and platinum group metals, as well as iron, copper, and chromium. It might also be argued as a side issue that such hand and wrist adornments represent a safety hazard in the laboratory [29,31].

While the use of disposable gloves is a good idea, they must never be the type that are prepowdered with talcum or corn starch (for easy slipping on and removal). For reasons of safety as well as contamination control all food, drink, tobacco, and cosmetics must be strictly banned from the trace analysis lab. Individuals who must use medicinal skin preparations should be assigned to other tasks.

Finally, the analyst is, in a sense, the biggest single factor among all the contributions to the blank because it is his/her skill, foresight, and attention to detail that is brought to bear in minimizing all the other contributing factors.

4. THE CONTROL OF OTHER POSITIVE ERRORS

As we have just seen, contamination from the laboratory, the equipment, the reagents, and the analyst all contribute significantly to a positive bias on the measured analytical signal. But these are not the only sources of positive error. Other sources are inherent in the nature of the sample and in the behavior of the measurement device. Still others are ascribable to the design of the sampling and analytical protocol. This subject is very broad and so only a few illustrations must suffice.

Most (but not all) matrix elemental interferences in optical emission spectrometry exert a positive bias on the analytical measurement. There are two ways to correct for them: a mathematical approach and a chemical approach. In the mathematical approach a known concentration of the interferent element is measured as a pure solution. The apparent analyte signal is used to establish an interelement correction (IEC) factor. In the analysis of an unknown sample the interfering element is measured. The result is then multiplied by the IEC and subtracted from the total analyte signal. Isobaric interferences in mass spectrometry are handled in a similar manner.

In the chemical approach (for both optical emission and mass spectrometry) a chemical separation scheme is utilized to remove the interfering element from the sample matrix, or to isolate the analyte from the entire sample matrix. Another chemical approach is to synthesize the sample matrix, including all interferents, in a series of solutions, using various levels of analyte concentration from zero to some level just above the highest level of interest. A calibration curve properly prepared in this manner is immune to this type of positive interference.

In X-ray fluorescence spectrometry interelement enhancement effects are handled mathematically, either through empirically derived coefficients or through a fundamental parameters approach. In graphite furnace atomic absorption spectrophotometry most positive bias derives from memory effects. Here, residual analyte left in the tube from the previous sample enhances the analyte signal. Memory effects are also observed in inductively coupled plasma optical emission and inductively coupled plasma mass spectrometries. Proper clean-out or flushing procedures between samples are usually all that is needed to correct the problem. However, in extreme cases the graphite tube may have to be replaced, or the ICP nebulizer, torch, and other sample inlet devices may have to be removed for special cleaning.

In some situations the sample preparation scheme will generate a positive bias. For example, silicon values might run high if a ferroalloy were incompletely crushed and only the portion that passed a 325 mesh (45 mm) sieve were taken for the test sample. In some situations the analytical scheme might be flawed. For example, aluminum values will run high if hydrofluoric acid is added to a Vycor flask containing the sample.

5. THE CONTROL OF NEGATIVE ERRORS

It is possible that negative biases do not immediately spring to mind in connection with trace analysis, but, in fact, they should, since the problem is very real. Negative errors can occur from the same sources as those discussed in Section 4 above. For example, background shifts and self-absorption effects in optical emission can result in signal depression (the former sometimes necessitating "negative IECs"). Interelement absorption effects occur in X-ray fluorescence and are handled by software corrections. In graphite furnace atomic absorption Zeeman background correction systems are subject to the "rollover" phenomenon in which at some concentration level additional analyte results in a *decrease* in absorbance.

There are, of course, innumerable ways the sampling or the analytical plan can produce negative errors. For example, overgrinding gold ore may coat the grinder with gold, leaving the test sample depleted in the gold analyte. Adding hydrochloric acid to dissolve a steel sample before nitric acid is added could result in the volatile loss of some of the phosphorus analyte. And, of course, physical loss due to some manipulative error is possible at any stage in the analytical process.

However, the source of negative errors that requires special consideration here is vessel surface adsorption. Aqueous ions interact with vessel walls in a complex and incompletely understood manner, especially at very high dilutions. These phenomena often result in significant negative errors if appropriate corrective measures are not taken. The literature is replete with references that document such behavior for a number of different ionic species and container materials. Clearly, the solution matrix, including the hydrogen ion concentration, plays a critical role in these effects. Another key parameter is the preliminary treatment, if any, that the vessel surface may have received.

A certain amount of laboratory lore is associated with this subject, reflective of the uncertainty in theoretical models. For example, many of the preservatives added to water samples are designed to retard or prevent surface adsorption losses. Some empirical studies have been conducted, such as those which demonstrated the efficacy of sodium thiosulfate in preventing adsorption losses of silver ion. Less than 1% adsorption loss of 1.0 and 0.05 ppm silver solutions occurred over a 30 day period in Vycor, Teflon, polypropylene, and polystyrene vessels in the presence of sodium thiosulfate [44]. Without thiosulfate trace level concentrations of silver, even when solutions were acidified at pH 2 with nitric acid and kept in the dark, showed erratic concentration results in borosilicate glass and disappeared completely in a polypropylene vessel [45].

Significant studies in this area have been conducted using radiotracer techniques. Robertson used such methodology to conclude that seawater samples should be stored in polyethylene bottles after acidifying to pH 1.5 [46]. Published studies involved adsorptive losses of chromium [47], selenium [48], beryllium [49],and tin [50], among others.

Differences have been noted in the stability of trace mercury solutions in linear polyethylene bottles from different lots. The lots were found to differ significantly in composition [51]. Dilute hydrogen peroxide was found to be as good a preservative for trace levels of lead as dilute nitric acid. Moreover, hydrogen peroxide was found to provide an advantage in that it is a poor desorber for lead. Thus, unlike nitric acid, it does not leach previously adsorbed lead from borosilicate vessels [52].

6. MEASUREMENT OF THE BLANK

There comes a point when the analyst has done everything possible within practical constraints to minimize the analytical blank. At that stage his task is to measure it as accurately as he can. For trace and ultratrace analysis this generally means taking the average of several replicates. The nature of this procedure depends in large part upon the nature of the work that precedes measurement. For some methods one blank value is all that will be required. For others the total blank will be the sum of several distinct values—perhaps, a sample blank, a reagent blank, and an instrument blank, each, themselves, the average of several distinct readings.

It is very important that the blank accurately reflect all the important steps in the analysis. If a residue was filtered, ignited, fused in molten salt, and leached in acid, the blank should be designed to represent the effect of each of these steps as accurately as the analyst can manage it. Shortcuts may lead to serious errors.

If reliable certified reference materials are available they can be used as a valid guide to the establishment of the true blank value. Erratic blank values are the bane of the trace analyst. And often the blank represents a significant portion of the total analyte signal. In this case a suite of appropriate certified reference materials, which were analyzed at the same time and in the same manner as the blanks and the test portions, can be used to select among the blanks. Those blank values that yield correct results for the certified reference materials are most likely to be valid.

Needless to say, such an approach requires a separate set of validation standards that remain outside of the calibration "shell." These and these alone must be used to pass judgement on the validity of the analytical work.

The size of the blank in relation to the total measurement response provides a signal-above-background value representing the analyte response. Variability in both blank and total signal often brings the analyte response difference perilously close to the noise limit of the method. One must not accept these difference values uncritically. Statistical treatments should remind us that for much of this work we are at the edge of the possible.

REFERENCES

1. Gy, P.M. Sampling of Heterogeneous and Dynamic Material Systems, Elsevier, Amsterdam, 1992.
2. Gy, P.M. Sampling of Particulate Materials, Theory and Practice, Elsevier, Amsterdam, 1982.
3. Pitard, F.F. Pierre Gy's Sampling Theory and Sampling Practice, Vols. 1 and 2, CRC Press, Boca Raton, FL, 1989.
4. Gy, P.M. LC-GC, 12, 810 (1994).
5. Sampling and Analysis of Carbon and Alloy Steels, US Steel Corp., Rienhold, NY, 1938, p. 23.
6. Janicsek, L. Steelmaking Conference Proceedings, Vol. 73, Iron and Steel Society, Warrendale, PA, 1990, pp. 111–113.
7. Butin, G.; Hoffert, F.; Ravaine, D.; Gandar, J.L. Progress of Analytical Chemistry in the Iron and Steel Industry, Commission of the European Communities, EUR 14113, 1992, pp. 276–290.
8. Pella, P.A.; Marinenko, R.B.; Norris, J.A.; Marlow, A. Applied Spectroscopy, 45, 242 (1991).
9. "Standard Practice for Preparation of Metallic Samples by Electric Arc Remelting for the Determination of Chemical Composition," Annual Book of ASTM Standards, Vol. 03.06, American Society for Testing and Materials, West Conshohocken, PA, 1996, Designation E1306.
10. Sisco, F.T. Technical Analysis of Steel and Steel Works Materials, McGraw-Hill, NY, 1923, pp. 115–133.
11. Lundell, G.E.F.; Hoffman, J.I.; Bright, H. Chemical Analysis of Iron and Steel, John Wiley, New York, 1931, pp. 50–60.

12. "Standard Practice for Sampling Wrought Nonferrous Metals and Alloys for Determination of Chemical Composition." Annual Book of ASTM Standards, Vol. 03.05, American Society for Testing and Materials, West Conshohocken, PA, 1996, Designation E55.
13. "Standard Practice for Sampling Steel and Iron for Determination of Chemical Composition." Annual Book of ASTM Standards, Vol. 03.05, American Society for Testing and Materials, West Conshohocken, PA, 1996, Designation E59.
14. "Standard Practice for Sampling Nonferrous Metals and Alloys in Cast Form for Determination of Chemical Composition." Annual Book of ASTM Standards, Vol. 03.05, American Society for Testing and Materials, West Conshohocken, PA, 1996, Designation E88.
15. "Standard Practice for Sampling Copper and Copper Alloys for the Determination of Chemical Composition." Annual Book of ASTM Standards, Vol. 03.05, American Society for Testing and Materials, West Conshohocken ,PA, 1996, Designation E255.
16. "Standard Practice for Sampling Zinc and Zinc Alloys for Optical Emission Spectrochemical Analysis." Annual Book of ASTM Standards, Vol. 03.06, American Society for Testing and Materials, West Conshohocken, PA, 1996, Designation E634.
17. "Standard Practices for Sampling Aluminum and Aluminum Alloys for Spectrochemical Analysis." Annual Book of ASTM Standards, Vol. 03.06, American Society for Testing and Materials, 1996, Designation E716.
18. "Standard Practice for Sampling Molten Steel from A Ladle Using An Immersion Sampler to Produce A Sample for Emission Spectrochemical Analysis." Annual Book of ASTM Standards, Vol. 03.06, American Society for Testing and Materials, West Conshohocken, PA, 1996, Designation E1087.
19. Steel and Iron—Sampling and Preparation of Samples for the Determination of Chemical Composition, International Organization for Standardization, ISO/DIS 14284.
20. "Standard Practices for Sampling Ferroalloys and Steel Additives for Determination of Chemical Composition", Annual Book of ASTM Standards, Vol. 03.05, American Society for Testing and Materials, West Conshohocken, PA, 1996, Designation E32.
21. "Standard Method for Sampling and Sample Preparation of Iron Ores." Annual Book of ASTM Standards, Vol. 03.06, American Society for Testing and Materials, West Conshohocken, PA, 1996, Designation E877.
22. Ottley, D.J. World Mining, 19, 40 (1966).
23. Minkkinen, P. Analytica Chimica Acta, 196, 237 (1987).
24. Ingamells, C.O. Talanta, 21, 141 (1974).
25. Ingamells, C.O.; Switzer, P. Talanta, 20, 547 (1973).
26. Ingamells, C.O. Talanta, 23, 263 (1976).
27. Ingamells, C.O. Talanta, 25, 731 (1978).
28. Moody, J.R. Analytical Chemistry, 54, 1358A (1982).
29. Zief, M.; Mitchell, J.W. Contamination Control in Trace Element Analysis, John Wiley, New York, 1976.

30. "Clean Room and Work Station Requirements, Controlled Environment," Federal Standard 209a, General Services Administration Business Service Centers, Washington, DC, August 10, 1966.
31. Murphy, T.J. in Accuracy in Trace Analysis: Sampling, Sample Handling, and Analysis, Proceedings of the 7th IMR Symposium, NIST, Gaithersburg, MD, 1976, pp. 509–539.
32. Robertson, D.E. in Ultrapurity: Methods and Techniques, M. Zief and R.M. Speights, eds., Marcel Dekker, New York, 1972.
33. Freeman, D.J., ed. National Institute of Standards and Technology Technical Note 459, NIST, Gaithersburg, MD, 1968.
34. Moody, J.R.; Lindstrom, R.M. Analytical Chemistry, 49, 2264 (1977).
35. Tölg, G. Talanta, 19, 1489 (1972).
36. Reagent Chemicals, American Chemical Society, Washington, DC, 1995.
37. "Standard Specification for Reagent Water," Annual Book of ASTM Standards, Vol. 11.01, American Society for Testing and Materials, West Conshohocken, PA 1996, Designation D1193.
38. Knott, A.R. Atomic Absorption Newsletter, Sept.- Oct., 126 (1975).
39. Veillon, C.; Reamer, D.C. Analytical Chemistry, 53, 549 (1981).
40. Kuehner, E.C.; Alvarez, R.; Paulsen, P.J.; Murphy, T.J. Analytical Chemistry, 44, 2050 (1972).
41. Mattinson, J.M. Analytical Chemistry, 44, 1715 (1972).
42. Little, K.; Brooks, J.D. Analytical Chemistry, 46, 1345 (1974).
43. Mykytiuk, A.; Russell, D.S.; Boyko, V. Analytical Chemistry, 48, 1462 (1976).
44. West F.K.; West, P.W.; Iddings, F.A. Analytical Chemistry, 38, 1567 (1966).
45. Struempler, A.W. Analytical Chemistry, 45, 2251 (1973).
46. Robertson, D.E. Analytica Chimica Acta, 42, 533 (1968).
47. Shendrikar, A.D.; West, P.W. Analytica Chimica Acta, 72, 91 (1974).
48. Shendrikar, A.D.; West, P.W. Analytica Chimica Acta, 74, 189 (1975).
49. Gladney, E.S.; Goode, W.E. Analytical Letters, 10, 619 (1977).
50. Gladney, E.S.; Goode, W.E. Analytica Chimica Acta, 91, 411 (1977).
51. Heiden, R.W.; Aikens, D.A. Analytical Chemistry, 49, 668 (1977).
52. Issaq, H.J.; Zielinski, W.L., Jr. Analytical Chemistry, 46, 1328 (1974).

2

Chemistry for Trace Analysis: Dissolution

1. INTRODUCTION

Before we embark on a chapter-by-chapter survey of the major practical trace measurement techniques it is important to pause here and examine in some detail the chemical underpinnings of solution-based methods. Solution techniques are extremely important because they are either inherently immune from, or can be designed to circumvent, significant problems that beset a solids-based approach.

Some solids-based techniques, such as X-ray fluorescence spectrometry, are affected by the thermal and mechanical processing that an alloy has received. For example, a cast, a forged, and a compacted alloy powder test material of identical composition may exhibit three different instrumental responses. Solution-based methods are free from this effect. Most solids-based approaches have stringent requirements for the size, shape, and surface finish of the test sample. In particular small or irregularly shaped, cracked or porous pieces are usually impossible to analyze accurately. A solution-based approach handles such samples easily. If sample homogeneity is suspect the solids-based techniques are left with few, and usually inadequate, options. But the solution techniques can be linked to a rational sampling plan designed to accurately represent a moderately heterogeneous material.

Solution techniques are important for other reasons as well. Most solids-based approaches are linked to a profound dependence on matrix matched solid alloy reference materials. Solution techniques, on the other hand, can be calibrated by synthesizing any desired matrix from pure elemental solution standards. And while such a calibration approach is still one step removed from a definitive, standard-independent approach it can and does find use in the certification of metal reference materials. The "standardless" definitive methods, themselves—isotope dilution mass spectrometry, coulometry, gravimetry, and titration based on normality—are all solution-based techniques that play a critical role in reference material certification.

In trace analysis perhaps the most essential advantage of a solution-based approach derives from the options afforded by separation and preconcentration. By the judicious application of precipitation, solvent extraction, ion exchange, and other chemical laboratory techniques interferences can be removed, matrix background can be eliminated, and analyte sensitivity can be enhanced. By utilizing an appropriate separation and preconcentration scheme, conventional measurement techniques, such as UV/visible molecular absorption spectrophotometry and flame atomic absorption spectrophotometry can be extended into the trace realm. Moreover, there exists a great deal of largely untapped potential to extend highly sensitive techniques, such as graphite furnace atomic absorption spectrophotometry, to extremely low detection and quantitation limits.

There is, of course, an associated disadvantage to such schemes as well. This is the increased potential for contamination and analyte loss that derive from added laboratory manipulations. Indeed, it is always prudent to apply the minimal analytical sequence that will achieve the desired goal. In many industrial environments time is a constraint on the application of chemical separations. In addition the laboratory skills needed for the application of wet chemical procedures are becoming rare in some quarters. None of these concerns need prove to be an insurmountable obstacle when the quality of the work demands a chemical approach. Time can be found when there is sufficient economic incentive to do so, and laboratory techniques can be relearned.

In this chapter we will review the chemistry involved in preparing an aqueous solution from metals and alloys. In the next chapter we will investigate the uses of group separations and analyte-specific isolations as they are applied to trace analysis. We will also describe some specific trace methods in which separation and preconcentration play a key role. Finally we will conclude this two chapter summary with some examples of the chemical reactions involved in the trace measurement process itself.

2. DISSOLUTION

While there are important exceptions what is generally implicit in the dissolution of metals is the preparation of an acidic aqueous solution that faithfully represents the total analyte concentration in the metal or alloy. The most common means to this end is the reaction of a particulate test portion with acid. Millings, drillings, filings, and other chip forms are employed because their large surface area greatly enhances reaction rates. The evolution of reaction product gases and thermal convection currents from the heat of reaction (and externally applied heat) carry hydrated metal ion products away from the solid/liquid interface. This prevents the formation of a concentration layer that would slow or stop the reaction.

All metals will dissolve to some degree in some combination of acids, although an acid dissolution approach is not always analytically feasible. Reaction rate is one consideration, and the effect of acid anions on subsequent separation and measurement steps is another. Loss of analyte by precipitation, volatilization, or adsorption on vessel walls must be considered as well. Frequently an acid dissolution approach will leave undissolved inclusion compounds or metallic phases that can only be ignored if the analyst is certain that they contain an analytically insignificant amount of analyte. If that certainty is lacking the insolubles must be filtered, ignited in a muffle furnace, and then fused in an appropriate molten salt. The cooled melt is then leached in the filtrate.

Sometimes it is more practical to fuse the metal sample directly in molten salt, then leach the cooled melt in an appropriate aqueous medium. Another approach involves the electrochemical dissolution of a solid test sample using an applied potential. With appropriate instrumentation for the measurement of utilized current it is sometimes practical to use Faraday's Law to calculate the amount of sample dissolved. Such a scheme has been applied in the continuous flow injection analysis of metals, typically for simple alloys.

When it becomes important to distinguish analyte concentrations in inclusions and phases, highly specialized dissolution techniques are brought to bear. In these cases, typically, a hand-filed solid test sample is utilized to preserve the integrity of brittle or fragile structural entities. Occasionally a semi-empirical analyte value is required or accepted by convention—"acid soluble aluminum," for example. In these cases it is necessary that the analyst define and document the conditions under which the measured solution was prepared.

In trace analysis the effect of dissolution media on the blank should never be far from the analyst's thoughts since the use of the wrong reagent can have a dramatic debilitating effect on accuracy. The practical analyst will also realize that high purity reagents come at a significant price differential and reserve their use for occasions of real need.

2.1 Reaction Rate

The practicality of a metal alloy dissolution scheme depends to a large extent upon initiating and sustaining a vigorous but controlled reaction. Fine particles dissolve in acid more rapidly than coarse particles, while monolithic solid pieces may require a great deal of time. The application of heat nearly always accelerates reaction rates (for a notable exception see Section 3.3.1).

The problem is that heating does other things as well. It drives precipitation reactions and contributes to the loss of volatile species, including volatile acids.

Occasionally catalysts, such as platinum (IV) ion, will accelerate a dissolution such as the attack of hydrochloric acid on titanium or nickel. Platinum metal has been used to promote the dissolution of cadmium metal in sulfuric acid and hydrogen peroxide. And a drop of metallic mercury or a trace of mercury (II) chloride or copper (II) chloride will readily facilitate the dissolution of aluminum in hydrochloric or dilute sulfuric acid.

When it is necessary to dissolve the test portion in a controlled small volume of acid or to prevent the loss of a volatile compound of the analyte, the dissolution can be conducted in a reflux apparatus. Here volatiles are continuously condensed with a water-cooled condenser and returned to the reaction vessel by gravity. Such configurations allow metals and alloys to be completely dissolved in nearly stoichiometric volumes of acid and can prevent otherwise serious analyte losses, but they are only marginally faster than open vessel hotplate dissolutions.

Open vessel microwave oven acid dissolutions offer some advantage in speed over open vessel hotplate dissolutions. The microwaves heat the metal particles rapidly and intensely, producing localized regions of superheated acid. Similarly, ultrasonic agitation has been known to accelerate dissolution rates by producing localized superheating from the rapid collapse of microbubbles. Ultrasonics and microwaves also tend to prevent the coating of undissolved metal with reaction-stifling precipitates.

The greatest advantages in speed, however, come from the use of sealed high pressure vessels—either metallic armored acid dissolution bombs used in drying ovens or plastic closed vessels used in microwave ovens. Ever since the days of sealed glass Carius tubes it has been clear that the analytical chemist can utilize high pressure to great advantage. In a sealed vessel liquids superheat, reactant and product gases do not escape, and reaction rates soar. In some cases reactions occur under such conditions that are not observed at atmospheric pressure. Dissolutions which might be expected to take hours in an open vessel hotplate approach may be completed in minutes. Some dissolutions (high carbon ferrochromium in hydrochloric acid, for example) which are impossible in an open vessel are accomplished readily in a closed vessel microwave approach.

In both types of high pressure vessels safety considerations place restraints on the test portion weight and on the volume and type of acids and acid mixtures that can be employed. In particular perchloric acid and sulfuric acid/nitric acid mixtures should never be used due to an extreme explosion hazard. The maximum test portion size and acid solution volume are specified to keep the reaction pressure within the operating range of the device.

Armored acid digestion bombs consist of a Teflon TFE cup and lid which should never be more than 2/3 filled with the metal/acid charge. This assembly fits inside a stainless steel or nickel cylinder with a threaded lid. Some designs incorporate a safety rupture disk. Some are supplied with a special wrench assembly for tightening and opening the device.

The metal acid digestion bomb will be supplied with a nominal temperature limit which must be carefully observed. Designs vary in the maximum working pressure that can be safely tolerated. Organic or organic-bearing sample materials present special dangers and require small and carefully controlled test portion sizes, and acid volumes in a precise range. It is imperative that the user read and follow the supplied instructions.

When the sealed bomb has been heated in a drying oven for sufficient time to dissolve the test portion it must be removed and allowed to cool completely before being cautiously opened. In addition to the danger from hot escaping gases, premature venting is likely to result in the loss of volatile gases and liquid aerosols, and possibly contamination of the contents caused by contact of acid vapors with the metal parts of the bomb [1]. A major drawback of the technique is the heating and, particularly, the cooling time, because a sizeable thermal mass is involved.

Despite all these concerns the technique clearly demonstrates advantages for certain sample types. Dissolutions that might take days in an open vessel can sometimes be accomplished in hours. And certain inclusions, which with a conventional approach would require filtration-ignition-fusion-and recombination with the filtrate, can be directly dissolved in acid along with the metal matrix. Thus, for example, total aluminum values in alumina-bearing steels have been obtained by acid dissolution in a pressure bomb without further processing [2]. Not all inclusions yield to an acid pressure digestion, however. But for those cases where analyte-bearing inclusions dissolve the time savings are nearly always significant. And the fact that the resultant solution is free of molten salt flux anions and cations is sometimes an advantage in subsequent analytical steps.

Acid digestion bomb designs vary both within and between manufacturers, and some workers have designed and constructed their own devices, including a "cluster" apparatus capable of simultaneously dissolving six samples [3]. Manufacturers include Parr Instrument Company (211 Fifty-Third St.; Moline, IL 61265) and Uni-Seal Decomposition Vessels, Ltd. (P.O. Box 9463; Haifa, Israel).

Perhaps the most significant development related to sample decomposition in the last twenty years has been the introduction and acceptance of closed-vessel microwave technology. The first application of this approach was published by workers at the U.S. Bureau of Mines for dissolving slag samples [4]. The first published use for steel samples was by Fernando,

Heavner, and Gabrielli of Allegheny Ludlum Steel Corporation [5]. The early work with the technique utilized home appliance microwave ovens, which were subject to severe corrosion from open vessel work and from the venting of closed vessels (which were often merely polyethylene screw cap bottles). The first commercial laboratory microwave ovens were introduced in the late 1980s near the time of the publication of *Introduction to Microwave Sample Preparation*, edited by H.M. Kingston and L.B. Jassie (American Chemical Society, Washington, D.C., 1988) [6].

Today such devices have become commonplace and many useful innovative features have been incorporated. The sample and acid charge are sealed in a plastic (often Teflon PFA) vessel which is loaded into a carousel. A plastic rupture disk is mounted in-line with a vent tube directed to a central reservoir. One of the vessels loaded in the carousel may be instrumented with pressure and/or temperature sensors to monitor and provide closed-loop control to the heating program. The carousel rotates or oscillates at some fixed rate.

Microwaves are produced by a device known as a magnatron which is integrated into all commercial units. The magnatron consists of a cylindrical cathode surrounded by an anode which incorporates a series of resonant cavities. A magnetic field is superimposed on the magnatron and a potential of thousands of volts is imposed between the two electrodes. Electrons released from the cathode oscillate, imparting some of their energy to an antenna broadcasting microwaves down a waveguide and into the oven cavity, where a device known as a mode stirrer distributes them evenly. The moving carousel also helps ensure that heating is uniform. In laboratory microwave ovens a device called a terminal circulator is often incorporated to protect the magnatron from reflected (nonabsorbed) microwaves by directing them to a dummy load where they are dissipated as heat.

Laboratory microwave ovens, like those sold for kitchen use, operate at a frequency of 2450 MHz. Heating is caused by two mechanisms: ionic conduction and dipole rotation. The fact that the affected species change direction 2.5×10^9 times per second accounts for the efficiency of this heating process. Heating rate is controlled by cycling the magnatron. The duty cycle (i.e., the base time, a part of which the magnatron is off) is shorter for laboratory ovens than for home appliance ovens. Thus, for a "medium" setting the home oven may be off 5 sec on a duty cycle of 10 sec. For an equivalent setting a laboratory oven might be off 0.5 sec on a duty cycle of 1 sec. The laboratory oven thus shows less heat loss and can achieve closer temperature control.

In a typical microwave metal dissolution using a small volume of mixed mineral acids the contribution to the heating initially originates from dipole rotation of the water molecules and those acids with a net dipole moment. This absorption of microwave energy or net dielectric loss decreases with increasing temperature. Eventually the heating effect comes to be dominated by ionic conduction, which increases with increasing temperature.

Even a fully loaded carousel generally represents a very small mass of liquid for a (typical) 600 w oven. If the vessels are constructed of Teflon PFA they are virtually transparent to microwaves. This and other materials with similar "transparency" are said to exhibit a low dissipation factor, which is the ratio of the material's dielectric loss to its dielectric constant (in effect, the ratio of absorption to obstruction of microwave energy). Because of the low mass to absorb microwaves, reflected power is generally a concern in a laboratory oven and a good design will incorporate a terminal circulator or other device to deal with it.

Some solids show a high dissipation factor and are effectively resistance heated by microwaves. Examples include silicon carbide, carbon black, and certain ferrites and glasses. Microwaves also induce electron migration in metals. This points up a potential hazard in the use of this methodology since a significant quantity of hydrogen can be generated in the closed vessel microwave digestion of metals. Unless the charged vessel is purged with an inert gas enough atmospheric oxygen could be trapped in the sealed container to form an explosive mixture with the evolved hydrogen. An electric spark from a microwave induced current in undissolved metal chips could detonate the pressurized gas mixture [6].

The pressure limit for closed vessel microwave digestions is considerably lower than for acid digestion bomb dissolutions. One manufacturer whose design employs a Teflon PFA inner vessel and a polyetherimide outer vessel has established a 120 psig (830 kPa) pressure limit. Some designs use a mechanical capping station to apply reproducible torque in tightening the outer vessel seal. Depending upon the specific design metal-armored acid digestion bombs have maximum pressure ratings of ten times this level or higher. Microwave systems heat up much more quickly, of course, but both types of vessels require considerable time to cool before they can be opened. Teflon PFA is somewhat less porous than Teflon TFE (which is typically employed in metal digestion bombs); thus vessel contamination at high pressure may be less of a problem in certain trace analyses when microwave digestion is employed.

Both approaches suffer from another common problem—the complexity of method development. While the literature is replete with recipes for using these techniques to dissolve a host of different materials, there is no guarantee that any of them will be adequate for another type of material. Applying these techniques to a new type of alloy is largely a matter of systematic trial and error to find a combination of acids and a heating program that completely dissolves the needed test portion in a minimum amount of time. This problem is worse than the equivalent situation with open vessel hotplate digestion because: 1) reaction rate cannot be directly observed; 2) the vessels must be cooled completely before they can be opened and the contents observed; and 3) loading and cleaning the vessels is burdensome.

Closed vessel digestions in either bombs or microwave ovens are clearly not the answer to all trace analysis dissolution problems. When a suitable protocol has been established they can be extremely useful for a repetitive sequence of one alloy grade or one class of similar alloys. When it is necessary to limit the type or the volume of acid used to effect dissolution both approaches can prove invaluable. But for the laboratory that must dissolve many different types of alloys it may prove impractical to employ either approach routinely.

2.2 Loss of Precipitates

As we have suggested, merely efficiently achieving a complete dissolution of the test portion is hardly the complete answer to most analysis problems. The dissolution procedure must be chosen to prevent the addition of undesirable species, the loss of volatile analyte, and the loss of analyte as a precipitate. We will consider the last of these first.

Of the mineral acids commonly used to dissolve metals hydrochloric acid presents few problems in regard to the precipitation of analytes. If present in somewhat greater than trace amounts, silver ion will precipitate as $AgCl$; and even at the trace and ultratrace level it is wise to avoid the use of hydrochloric acid. Similarly, mercury (I) ion (Hg_2^{2+}) precipitates as Hg_2Cl_2, bismuth as $BiOCl$, gold as $AuCl_3$, and copper (I) as $CuCl$. Also, both lead and thallium(I) precipitate as the partially soluble $PbCl_2$ and $TlCl$, respectively. None of these analytes normally presents a problem since there are usually alternative dissolution routes to take. Hydrobromic acid and hydroiodic acid behave, for the most part, similarly to hydrochloric acid in regard to precipitating metal ions. Solders and other lead-base alloys dissolve to a clear solution in concentrated hydrobromic acid, but 1:1 $HBr:H_2O$ will precipitate $PbBr_2$.

A great many elements yield precipitates with nitric acid, taking the form of hydrous oxides: antimony, hafnium, molybdenum, niobium, tantalum, tin, titanium, tungsten,and zirconium. The presence of only a few drops of hydrofluoric acid is usually sufficient to prevent these elements from dropping out of solution. But hydrofluoric acid itself will precipitate an array of elements, notably the alkaline earths (magnesium, calcium, strontium, and barium), scandium, yttrium, lanthanum, and the lanthanides, as well as thorium and uranium. Hydrofluoric acid also has the potential to form complex metal fluorides and so-called "double-salts" (such as $MgAlF_5$ and K_2FeF_6), many of which have low solubilities. Even simple fluoride salts, such as FeF_3, NiF_2, and CrF_3 require patient warming and stirring to redissolve once they crystallize out of solution.

Sulfuric acid precipitates barium, strontium, lead, and calcium (as sparingly soluble $CaSO_4$). It also forms partially soluble double salts with potassium and another metallic element (such as chromium or aluminum). The hot, fuming acid effectively precipitates silicon as SiO_2. However, iron and nickel sulfates cause spattering, requiring special caution, when sulfuric acid solutions of iron- and nickel-base alloys are heated to fumes of SO_3. When water is added to the cooled salts and the solution is brought to a boil the sulfate salts dissolve slowly. Perchloric acid precipitates potassium, rubidium, and cesium as perchlorates (lithium and sodium do not precipitate). Fuming a solution with perchloric acid precipitates SiO_2 and brings down the niobium, tantalum, tungsten, titanium, tin, and antimony as hydrous oxides. Mixtures of perchloric and phosphoric acid will precipitate titanium as the phosphate [7].

Leaching a cooled molten salt fusion in neutral water sometimes results in the precipitation of insoluble compounds, but the analyte is rarely lost in this manner since the analyst has the option of acidifying the solution, which dissolves most of the precipitates. Sometimes such precipitation during leaching of the cooled melt is incorporated as a group separation in the analytical scheme. In such cases the neutral water leach is usually boiled and filtered. If it is the analyte that has initially precipitated the filter paper is washed and ignited and, perhaps, refused, then acid-leached. If, instead, one or more interferents have initially precipitated the filtrate bearing the analyte is retained and the insoluble residue is discarded.

Fusions which fall into this category are basic fusions—carbonates, hydroxides, and sodium peroxide. When a sodium, potassium, or lithium carbonate fusion is leached in hot neutral water some elements (like the alkali metals and uranium) are dissolved in the alkaline carbonate medium that forms and end up in the filtrate. Many other elements precipitate as either insoluble hydroxides or insoluble carbonates. Among the elements quantitatively precipitated are barium, beryllium, bismuth, calcium, hafnium, iron, magnesium, silver, strontium, titanium, zinc, zirconium, and the rare earths. These elements are collected on the filter for further processing. If the carbonate fusion is conducted in a platinum crucible without a closely fitting lid some slight oxidizing effect is likely to be exhibited and enough chromium will be oxidized to the base-soluble (+VI) oxidation state so that recovery of chromium in the precipitated residue will not be quantitative.

Sodium hydroxide, potassium hydroxide, or lithium hydroxide are sometimes applied to the molten salt fusion of metal related commodities. The elements precipitating in the water leach of the cooled melt are the same as those precipitating in a strong caustic separation: Cadmium, chromium (III), cobalt, copper, hafnium, indium, iron, magnesium, manganese, nickel, niobium, silver, tantalum, thallium (III), thorium, titanium, uranium, zirconium, and the rare earths. Precipitation of the majority of these elements is quantitative at a final concentration of about 4%(w/v) sodium hydroxide. However, niobium, tantalum, indium, and copper require a more dilute final concentration of about 0.25%(w/v) to ensure complete precipitation. The filtrate elements include aluminum, arsenic, beryllium, chromium (VI), gallium, lead, molybdenum, phosphorus, rhenium, tin, tungsten, vanadium, and zinc. Hydroxide fusions are often conducted in nickel crucibles, and in this case nickel hydroxide becomes a major component of the precipitate.

Sodium peroxide fusions are frequently incorporated into a precipitation separation scheme for chromium. The strongly oxidizing fusion ensures that all of the chromium is in the +VI state and the neutral water leach of the cooled melt produces a strongly basic solution in which the chromium (VI) is completely soluble. The other filtrate elements listed above for hydroxide fusions accompany chromium (VI) as well. The same hydroxide precipitates occur here also except for titanium which does not precipitate quantitatively due to the formation of the peroxytitanium complex. Nickel or zirconium, the two principal crucible materials, will always accompany the precipitate elements [8].

2.3 Loss of Volatiles

The trace analyst needs to be keenly aware of the potential for the loss of volatile forms of the analyte. While volatiles losses can occur at any point in an analysis the reactions involved in the dissolution of the test portion are usually responsible. Sometimes the formation of volatile compounds of the analyte can be avoided by merely substituting one dissolution medium for another. However, this course is not always open because of sample method requirements. Adjustment of the oxidation state of the analyte is often effective since in many cases a higher or lower oxidation state proves to be nonvolatile under the dissolution conditions. Complexing additives (like phosphoric acid or mannitol to prevent loss of boron halides) are sometimes useful. But the final recourse must sometimes be containment of some sort. This might involve thermal containment, as in dissolution under water-cooled reflux conditions, or pressure containment, as in the use of acid bomb digestion or closed-vessel microwave dissolution.

Hydrochloric acid, possibly the most generally useful dissolution agent for metals, also offers some of the greatest opportunities for analyte loss in open vessels. Antimony (III), arsenic (III), germanium (IV), phosphorus, selenium(IV), and sulfur are lost completely, even at low hotplate temperatures. Boron, indium, mercury (II), rhenium (VII), tellurium, tin (IV), and zinc losses have been reported as well (although the indium, tellurium and zinc losses are associated with extreme conditions). The application of oxidizing conditions, such as the use of a nitric acid/hydrochloric acid medium, will completely prevent the loss of antimony, arsenic, germanium, phosphorus, selenium, and sulfur. Acid digestion bombs and sealed microwave vessels greatly extend the utility of hydrochloric acid while ensuring that many volatiles are retained. Very low boiling species, such as BCl_3 (12°C), H_2S (−61°C), and PH_3 (−88°C), will be lost when the cooled vessel is vented, however.

Hydrofluoric acid, another extremely useful dissolution agent, contributes its own burden of volatile compounds. Arsenic (both +III and +V), boron, germanium, selenium, and silicon are all volatile at low temperatures. Extreme temperatures, such as those that occur when a solution is heated to dry salts, result in losses of antimony, tellurium, and even titanium. The addition of mannitol or phosphoric acid to a hydrofluoric acid solution of boron will retard boron loss at moderate temperatures. Evaporation to fumes of perchloric or sulfuric acid—commonly applied procedures following the use of hydrofluoric acid—will result in the loss of many elements. Fuming with either sulfuric or perchloric acid will evolve antimony, arsenic, boron, germanium, mercury, osmium, rhenium, selenium, and silicon. Fuming with perchloric acid will also evolve chromium and ruthenium.

Nitric acid, which is used alone for comparatively few metal alloy types, results in only two analytically important volatiles losses: osmium as OsO_4 and possibly ruthenium as RuO_4. Sulfuric acid, heated to fumes of its anhydride, SO_3, evolves mercury (II), osmium, phosphorus, rhenium (VII) (as Re_2O_7) and selenium (IV). Perchloric acid heated to fumes will always evolve some chromium as chromyl chloride (CrO_2Cl_2). Experimental studies suggest that 1–3% of the chromium present will be lost. Approximately 30% of the rhenium will be lost, and all of the selenium, osmium, and ruthenium will be lost. A significant portion of the antimony, arsenic, and mercury will also be evolved. Dissolving steel in perchloric acid will result in the evolution of sulfur as H_2S.

The use of acid mixtures and oxidizing additives is, of course, common and necessary in the dissolution of various alloys. However, the picture of volatiles losses that results is only partly predictable from behavior with individual acids. Mixtures of hydrobromic acid and bromine, commonly used to dissolve lead and lead alloys, evolve antimony, arsenic, selenium, and tin. If fuming with a "nonvolatile" acid follows, these losses will be complete, and some indium will be lost as well. Mixtures of hydrochloric and hydrofluoric acids are used for titanium and zirconium alloys as well as for many ores and silica-bearing commodities. However, antimony, arsenic, boron, germanium, phosphorus, rhenium, selenium, silicon, and sulfur are lost to some degree in open vessels, and minor losses of other elements are likely if heating is excessive. For this reason this mixture is often used in sealed vessels.

The 3:1 hydrochloric acid: nitric acid mixture known as *aqua regia* will result in the volatile loss of boron, germanium, osmium, and rhenium. Mixtures of hydrofluoric and nitric acid evolve boron, germanium, and silicon, while mixtures of hydrochloric, hydrofluoric, and nitric acids lose boron, germanium, osmium, rhenium, and silicon. If these mixtures are fumed with sulfuric acid or perchloric acid, of course, additional elements will be lost.

Losses of volatiles from molten salt fusions are not as well documented in the literature as are losses from acid dissolution. And as with acids there are some mutually contradictory reports. There is general agreement, however, that mercury and thallium are completely volatilized and that arsenic and selenium are partially volatilized from both carbonate and borate fusions. In pyrosulfate and bisulfate fusions arsenic, mercury, and selenium are lost, and if the sample contains any halogens antimony, germanium, tin, and possibly additional elements will be lost as well. In sodium peroxide fusions losses of arsenic, mercury, and nitrogen have been observed. Some work suggests, however, that arsenic will be retained if the reaction has been moderated by the addition of sodium carbonate. Losses from hydroxide fusions have not been extensively studied although it is known that refractory nitrides dissolve with the evolution of NH_3 [9–11].

2.4 Treatment of Insolubles

In the dissolution of metals and alloys for chemical analysis the analyst is frequently confronted with some resistant portion of the test material. This may amount to anything from a nearly invisible smudge on the bottom of a beaker to a sizeable fraction of the original test portion. Unless he has certain information to the contrary, analytical rigor requires that the analyst treat the undissolved residue as if it contains a significant amount of analyte. This enjoiner applies with special emphasis to certain trace work because in some alloys metallurgically important trace elements are associated with chemically resistant phases or inclusion compounds.

In the language of metallurgy a "phase" refers to a solid solution or an intermetallic compound that is intimately mixed with the base alloy but which remains as a distinct entity of fine particles. "Inclusions," in contrast, are crystals of inorganic compounds dispersed in the alloy matrix. They may consist of oxides, carbides, carbonitrides, sulfides, borides, or other types of molecules, occurring occasionally in pure form but more often in complex mixtures.

Inclusion compounds are sometimes categorized by their source of origin. Those created in the high temperature chemistry involved in the formation of the alloy are termed "endogenous inclusions." Those transferred to the alloy intact from charge materials, refractory furnace linings, and other external sources are termed "exogenous inclusions." We must also include here a third type—the intentionally added inclusion compound, such as are found in dispersion-strengthened alloys, in certain abrasion resistant alloys, and in the new class of materials known as metal matrix composites (MMCs).

Typically, an acid dissolution approach is applied to the test portion, and some amount of the material's metallic phases or inclusion compounds remain undissolved. In this case the solution is diluted (to prevent digestion of the filter medium) then filtered either through filter paper or through a microporous membrane filter with applied vacuum. Adherent particles must be carefully scrubbed out of the original vessel with a rubber policeman, and all rinsings must be carefully transferred to the filter. The residue is washed either with water or alternately with dilute acid and water until it is free of the original solution components. The filtrate must usually be retained and so it is essential that the collection vessel has been properly cleaned.

The filter is then transferred to a crucible (typically platinum) and ignited in a muffle furnace. A common procedure involves ignition at 1000°C for 1 hr, although certain analytes, such as molybdenum, require lower temperatures (625°C for Mo) and, perhaps, longer times. The crucible is cooled in a desiccator. At this point it is sometimes expeditious to remove silica by volatilization, especially if it is known to be a major component of the acid-insoluble residue. This may be accomplished in many cases by adding 1 mL of 1:1 sulfuric acid:water and 10–20 drops of hydrofluoric acid, then heating to dryness on a sandbath on a hotplate in a hood. For certain alloys the use of sulfuric acid is undesirable (due to the formation of insoluble sulfates, due to spattering, or because of the need to maintain oxidizing conditions) and another acid may be substituted. Thus, if molybdenum or tungsten is the analyte, 1 mL of nitric acid might be substituted for the dilute sulfuric acid.

After the volatilization of silica the crucible might be returned to the muffle furnace for a short time to ensure that acid species and other volatiles have been removed. The crucible is again cooled in a desiccator. Then an appropriate weight of a suitable flux is added.

Molten salt fluxes are often categorized as "acidic" or "basic" (in the Lewis sense), and as oxidizing or nonoxidizing. Selecting an appropriate flux depends upon having some knowledge about the nature of the acid insoluble residue. Table 2.1 offers some guidance, but trial-and error and the experienced knowledge that it engenders are often necessary. The amount of flux compound added will vary, of course, but 1 g is typical if the original test portion was 1 g. Fusions in platinum crucibles may be either conducted with a Meker burner (supporting the crucible with a tripod and porcelain triangle), or in a muffle furnace. A lid should be used in both cases. The Meker burner is the more versatile approach since it allows frequent cautious examination and swirling of the molten salt.

Upon cooling to room temperature the crucible and lid can either be leached in the filtrate from the filtration step or in a completely separate solution of an appropriate dilute acid. The former approach conveniently yields a solution that represents the total analyte concentration. The latter approach allows separate measurement of the acid soluble (filtrate) fraction and the acid-insoluble (leached fusion) fraction, thus providing valuable information about the metallurgical form of the analyte. Sometimes as well there are sound analytical reasons for keeping these fractions separate. For example, the total dissolved solids content of the combined fractions might exceed the tolerance of the measurement technique. Also, the molten salt flux and the acid matrix of the filtrate might be incompatible, as in the case of potassium pyrosulfate and perchloric acid, where a precipitate of potassium perchlorate would form.

Occasionally, a molten salt fusion fails to solubilize the last traces of acid-insoluble material. This may result from a faulty decision about the flux or the exact nature of the technique employed. In this case there is little recourse except to filter, wash, ignite, fuse, and leach again, perhaps modifying the protocol. Similar reiterations might follow failed procedures where the initial test portion was fused directly. It must be borne in mind, however, that each repeat fusion adds to the salt burden of the solution with its attendant viscosity effect and contribution to the blank. Frequently it proves prudent to restart the procedure instead, using a different dissolution approach—perhaps different acids, different fluxes, or even a sealed vessel dissolution [9].

Table 2.1 Some Suggested Fusions for Acid Insoluble Inclusions

Compound	Suggested Flux	Compound	Suggested Flux
Al_2O_3	A,B	Silicates	H,G
Aluminosilicates	A,B	TaC	E,J
AlN	C,D	Ta_2O_5	E,J
BN	C,D	TiC	I
Cr_3C_2	G,E,F	TiO_2	G,E,F
Cr_2O_3	G,E,F	TiN	I
CrN	I	VC	I
MoC	I	VN	I
NbC	E,J	V_2O_5	G,E,F
Nb_2O_5	E,J	WO_3	C,D
SiO_2	H,G		

KEY: A: 3:1 Na_2CO_3:H_3BO_3 (Pt crucible); B: 3:1 Na_2CO_3:$Na_2B_4O_7$ (Pt crucible); C: LiOH (Ni crucible); D: NaOH (Ni crucible); E: $K_2S_2O_7$ (Fused quartz or Pt crucible); F: $NaHSO_4$ + H_2SO_4 (Fused quartz or Pt crucible); G: Na_2O_2 (Zr crucible); H: Na_2CO_3 (Pt crucible); I: 2:1 Na_2O_2:Na_2CO_3 (Zr crucible); J: K_2CO_3 (Pt crucible);

2.5 The Acid Matrix

Before we begin a survey of industrial metal commodities and common ways each is dissolved we must consider one additional important aspect of the plan for obtaining a representative aqueous solution from the test portion. Once we have designed a chemical scheme that appears to dissolve the test portion completely in a reasonable amount of time we must verify that we have produced a solution that is compatible with the proposed measurement process. This seemingly obvious precaution is sometimes overlooked when a new procedure is being developed or a new alloy is being determined by an established method.

This admonition is not meant to imply that interfering species must never be added in the course of a dissolution, but merely that remedies to any created problem must be firmly in hand before the work begins. Added interferents may be chemically or physically separated, or masked in place, in a manner analogous to the treatment of interferences from the sample matrix itself.

A few examples should serve to illustrate these points:

A high temperature alloy resists complete dissolution in hydrofluoric/nitric acid media unless a small amount of hydrochloric acid is added. However, the analyte, lead, is to be measured at the trace level by graphite furnace atomic absorption spectrophotometry. It is known that if chloride is present some lead will be lost during the char cycle, resulting in low lead results when the measurement is taken during the atomize cycle. One relatively simple solution is to add ammonium nitrate or, better, ammonium EDTA to the sample as a matrix modifier and to lower the char temperature. This results in the evolution of volatile NH_4Cl during the char cycle, effectively removing the chloride interference.

Hydrofluoric acid is essential to dissolve certain alloys directly in acid, but its presence in solution interferes with many spectrophotometric procedures (usually bleaching the colored species). It also leaches many elements from borosilicate glass laboratory vessels. The best course is to use Teflon TFE labware and to take the samples to fumes of perchloric or sulfuric acid to remove the excess hydrofluoric acid after the sample has dissolved. This volatilization approach is less effective in the presence of significant amounts of stable fluoride formers, such as aluminum ion.

Sometimes the sample material requires a molten salt fusion to effect complete dissolution, but the measurement technique—ICP-MS, for example—is unusually sensitive to dissolved solids. Direct aspiration in ICP-MS is limited to about 0.1%(w/v) total dissolved solids. A 100 mg test portion fused in 1 g of flux, leached, and diluted to 1 l just exceeds this limit. But if a precipitation step is included in which the analyte is retained on a filter, most of the solids contribution from the salt flux can be washed out and discarded. The analyte is then dissolved off the filter paper with acid, diluted, and measured.

Nitric acid is very often essential to dissolve the test portion, but it, too, can be objectionable (as when it interferes with the TOPO-MIBK extraction of trace levels of lead, bismuth, thallium, and other elements). It can be conveniently removed by reducing the solution to a syrup and making dropwise additions of formic acid or a solution of urea to the hot sample.

Ion exchange schemes are sometimes necessary when a method requires removal of all of the dissolution acid anions. Such an approach has been applied to the volumetric determination of boron and should be readily adaptable to ICP-MS work, where isobaric interferences from chloride moieties is sometimes a problem.

The lesson here is that for trace analysis the analyst must be always cognizant of the effect of everything that he adds to the test portion, not only to control the blank but also to intercept separation and measurement problems. Frequently, there are many ways to achieve a clear, representative aqueous solution of the test portion, but, perhaps, not one of them offers a problem-free path to the measurement goal.

3. DISSOLUTION OF METAL COMMODITIES

In the remainder of this chapter we will offer descriptions of commercially traded metals and alloys and some commonly employed procedures for their dissolution. It is unlikely that even the most diversified and versatile laboratory will deal with all of these alloys or employ all of these procedures routinely. But reference sources like this one find their greatest utility *outside* of a laboratory's specialty areas. The steel laboratory which is asked to analyze a lead solder, or the aluminum laboratory which occasionally receives a request to analyze a die steel are the principal audience for such a compilation. In addition it is just possible that the expert might find some fresh insight here on his or her area of expertise.

3.1 Iron Base Alloys

Unlike some other alloy categories, an alloy system is often regarded as "iron-base" only down to a level of 50% iron. In most other categories (e.g., "cobalt-base") 40% or even 30% of the element in question still bestows the appellation which then derives from the principal component.

High iron materials begin with high purity iron at >99% iron (which is frequently surveyed for tramp elements) and includes white irons, cast irons, plain carbon and low alloy steels. So-called mild steels and medium alloy steels belong here as well. All of these materials contain more than 90% iron. Stainless steels range from about 60 to 80% iron. The category of tool steels overlaps this at about 60–95% iron. Iron-base high temperature alloys may be as low as 50% iron. Ferroalloys range from 50 to 70% iron, but with some important commodities significantly lower, providing a significant exception to our definition of "iron-base."

3.1.1 High Iron Alloys

High purity iron is a specialized commercial product produced either by vacuum refining techniques or by one of several direct reduction processes. It is sold in many forms, including wire and powder. The purity range of this product is rather wide with the purest material sold as exceeding 5 "9s" purity. However, contact with atmospheric oxygen is certain to produce surface oxides.

In marked contrast are the commodity irons. The high graphite material known as *pig iron* is produced in a blast furnace and used as a starting material for the production of both *wrought iron* and *cast iron.* Wrought iron contains a fine fibrous structure of entrained slag in a pure iron matrix and is resistant to both corrosion and mechanical shock. Wrought iron has been made for hundreds of years by a labor intensive hand puddling process, and in this century by the large scale Byers process. Cast iron is made by remelting, modifying with additives, and casting pig iron. Unlike the steel-making process, in producing cast iron no extensive attempt is made to refine out most elements. Carbon and silicon remain high and the iron content is typically about 93%. The highest carbon and silicon cast irons are known as *grey irons.* These relatively low strength materials contain a great deal of graphite. When both carbon and silicon have been reduced, graphite is replaced by cementite (Fe_3C) and the product is known as a *white iron.* This material is hard, brittle, and wear resistant, but also lacks great strength. Annealing white iron produces a strong, workable *malleable iron* by converting cementite to nodular graphite and ferrite. *Nodular cast iron* (or *ductile iron*) achieves similar properties in a high carbon/high silicon composition with the use of special additives (a nickel/magnesium/cerium innoculant, for example). Sometimes high levels of silicon or nickel are added to irons to bestow corrosion- or abrasion-resistant properties.

Steels, such as plain carbon and low alloy steels, are best delineated from irons by their manufacturing process. Steels are produced by a refining operation whose purpose is to *lower* the content of some starting material components (while, in some cases, adding other components). In terms of composition steels are usually higher in carbon than wrought iron and lower in carbon than grey cast iron. In general steels have much lower contents of sulfur, silicon, and phosphorus than irons. The elemental composition and the thermal and mechanical processes used to attain the finished steel are all directed toward a microstructural balance that imparts the desired properties. Ferrite, cementite, pearlite, austenite, and martensite are crystallographic phases whose intricate dance is choreographed by the metallurgist. And inclusion compounds and interstitial elements imbedded in the metal lattice play critical roles as well.

Plain carbon steels are defined by the American Iron and Steel Institute (AISI) as containing less than 1.65% manganese, less than 0.60% silicon, less than 0.60% copper, and smaller amounts of other elements. AISI and the Society of Automotive Engineers (SAE) have established a joint classification system for this broad category of alloys, as well as for many other types of steel. There are other classification systems as well.[*] In plain carbon steels, carbon is regarded as the primary alloying constituent.

In general terms, "alloy steel" refers to all iron-base systems with any degree of added constituents, up to a total of 50%. The term *low alloy steel* is reserved for those structural and automotive alloys containing a total of 1–4% alloying elements. Such steels are specified with a definite range (or sometimes a minimum limit) of a number of elements, which may include aluminum, chromium, cobalt, molybdenum, nickel, niobium, titanium, tungsten, vanadium, zirconium, and other metals. Nonmetallic elements, such as nitrogen, oxygen, hydrogen, boron, sulfur, and, of course, carbon may be narrowly specified as well. Also included under this heading are high strength quench and tempered alloys, and the high strength low alloy (HSLA) construction steels with complex low concentration compositions.

Maraging steel is a low carbon, high nickel steel with additions of molybdenum, cobalt, chromium, and other elements. These alloys have good working properties, and allow the finished part to be easily age-hardened. *Silicon steel* is used for its magnetic properties in motors, transformers, and generators. The silicon content ranges between 0.5 and 3.75% in a very pure iron matrix. Some compositions are sold in a directionally oriented morphology. *Tool steels* may be classed as low alloy tool steels, medium alloy tool steels, and high speed steels, reflecting an increase in total alloy content over the range 5–40%. These steels are classified by AISI with a leading letter designation that reflects the principal alloy component (M for molybdenum,

---

[*] A Unified Numbering System (UNS) developed jointly by SAE and the American Society for Testing and Materials (ASTM) has been published and periodically updated since 1975. It is an attempt to consolidate and simplify the classification of all types of metal alloys. For a discussion of classification systems see: J.E. Bringas *ASTM Standardization News*, September, 1994, 20–27.

T for tungsten), the intended processing mode (A for air hardening, W for water hardening, O for oil hardening), or the intended use (H for hot work, D for cold work, S for shock resistance) [12]. There are also *nickel-irons* which are sold for their electrical and magnetic properties, and some of these are predominantly iron.

Admittedly, some of these commodities are too highly alloyed to be correctly termed "high iron," but they all need to be considered together in this section because they all dissolve easily in acid. A more apt title for this section then might have been "Corrosion Susceptible Iron Alloys." All of these alloys rust easily in a moist atmosphere, except for certain HSLA's (like CORE-TEN™ from USX Corp.) which are designed to form an adherent protective oxide. If chips from any of these iron compositions are weighed into a moist beaker and allowed to stand overnight they will show noticeable oxide by morning.

Almost any strong mineral acid will attack these materials, but some acid media are more efficient than others. And in many cases there are acid insoluble inclusions to contend with. The plain carbon and low alloy steel industry is almost unique in sometimes using dilute perchloric acid to dissolve samples. This represents a time-savings since it is often necessary to bring the dissolved test portion to fumes of perchloric acid to aid in the breakdown of acid resistant chromium or vanadium carbides and to dehydrate silicon for its removal or its gravimetric determination. Since perchlorate ion has no complexing properties such an approach could represent a time advantage in certain separation and measurement schemes.

The carbon from high carbon alloys represents no known danger here, although safety considerations suggest that the addition of nitric acid between dissolution and fuming is most appropriate. If a clear solution does not result after the perchlorate salts have been dissolved in water the analyst must decide if it is necessary to take the time to filter, ignite, fuse, and recombine. And here it is essential that the ignited insoluble residue *not* be fused in potassium pyrosulfate because insoluble potassium perchlorate will form when the cooled melt is leached in the filtrate. A suitable substitute is "sodium bisulfate, fused," which is sold commercially. To minimize spattering, which is often a problem, ten drops of sulfuric acid should be added for each gram of flux.

Hydrochloric acid attacks all of these alloys, even the high tungsten high speed steels, although the addition of a few drops of nitric acid may hasten the process. With low alloys and carbon steels some procedures call for adding the dropwise nitric acid after all action with hydrochloric acid ceases. This causes a vigorous reaction due to the oxidation of ferrous iron. Such an approach is useful for steels which tend to passivate in nitric acid by forming an adherent, corrosion-resistant oxide layer, preventing further acid attack on the underlying metal. Among this group of alloys only the high tungsten high speed steels fall into this category.

Mixtures of hydrochloric acid with other reagents are also commonly applied for this class of steels. *Aqua regia* (3:1 HCl:HNO_3) is a valuable mixture that must be prepared immediately before use. In addition to its component acids it forms nitrosyl chloride (NOCl), chlorine gas (Cl_2), and NO_x gases. Among steels which do not passivate, there is likely to be less carbonaceous insoluble residue left after dissolution in *aqua regia* than if hydrochloric acid alone were used. Other ratios of hydrochloric acid to nitric acid are also employed. And sometimes the two acids are added to the test portion directly. To prevent volatile loss of phosphorus and sulfur the nitric acid should be added first in this case.

If silicon, itself, is not the analyte then silicon steels succumb easily to a mixture of *aqua regia* containing a few drops of hydrofluoric acid. This reaction can be conducted in a borosilicate beaker provided that boron, aluminum, or any of the other vessel components are not the analyte. Silicon steel may not dissolve completely in *aqua regia* alone since some of the metal matrix may be encapsulated by silica. If more than about four drops of hydrofluoric acid are needed, however, a Teflon vessel is a better choice.

For low level tramp determinations Teflon is always employed. Hydrofluoric acid in small amounts will also serve to prevent the precipitation of tungsten and molybdenum in high speed steel dissolutions. A similar benefit accrues for the minor amounts of niobium and titanium sometimes added to high-strength low alloys. Hydrofluoric acid also serves to dissolve silicate inclusions that may entrain the analyte element.

Sometimes a mixture of 1:1 hydrochloric acid:water and dropwise additions of 30% hydrogen peroxide is useful to dissolve this class of alloys. This approach has the combined advantage of preventing passivation and retaining minor amounts of silicon in solution. It is useful for the flame atomic absorption or plasma emission measurement of silicon content.

A few laboratories make extensive use of phosphoric acid/perchloric acid mixtures (occasionally known by the term "phosphodent"). These combinations dissolve most alloys and confer certain advantages. Under reflux conditions volatile boron loss is prevented, and tungsten, molybdenum, and niobium are prevented from precipitating. If the solution is heated to fumes, however, all of these benefits will be surrendered. Even without fuming, white titanium phosphate will usually precipitate if titanium is present as an alloying addition.

More useful, but surprisingly little used, are phosphoric acid/sulfuric acid combinations. The best approach here is to dissolve a 0.25 to 0.30 g test portion by warming with 25 mL of a 10:3:12 mixture of phosphoric acid: sulfuric acid:water. Then the temperature is increased until light fumes are evolved. After 1 min of fuming the sample is removed from the hotplate, allowed to cool for 1 min, then 25 mL of a 7:13 mixture of sulfuric acid:water is added cautiously and the vessel is swirled. The purpose of adding the dilute sulfuric acid while the solution is still hot is to dissolve a polymeric form of metaphosphoric acid which forms during fuming. The cooled solution can be diluted up to 100 mL with water without concern that hydrous oxide-forming elements will precipitate. This dissolution technique can be used to good effect when hydrofluoric acid cannot be tolerated in the analytical procedure. It is effective even with high tungsten tool steels. Titanium at the levels ordinarily encountered does not precipitate.

Among the remaining acid dissolution approaches for this class of steels dilute sulfuric acid is probably the most frequently employed. As with hydrochloric acid, dilute sulfuric acid leaves a significant amount of undissolved compounds when applied to certain plain carbon and low alloy steels. These must usually be filtered and ignited, but here they can be fused with potassium pyrosulfate, which spatters less than sodium bisulfate and is also a more effective flux for niobium and tantalum compounds. Boron inclusions are often ignited in the presence of calcium hydroxide and fused with sodium carbonate. In both cases the cooled melts are either dissolved in the filtrate or leached in dilute sulfuric acid and treated as a separate test solution, representing "acid-insoluble" analyte. Steels do not dissolve appreciably in concentrated sulfuric acid.

Mixtures of hydrobromic acid and bromine are very rarely applied to dissolve steels. They *do* allow the ready removal of potential interferences from antimony, arsenic, selenium, and tin by merely heating the solution to dryness, then redissolving the salts in hydrochloric acid. Concentrated nitric acid is effective for plain carbon steels, but creates passivation problems for higher alloys.

Direct molten salt fusions are seldom used or needed for this class of alloys. In exceptional cases potassium pyrosulfate fusion of finely divided alloy chips might be applied to save time if acid insoluble compounds need to be dealt with. However, it is in this alloy category that most frequent use is made of those specialized dissolution techniques that are designed to isolate inclusions and second phases for identification and analysis. This special subject will be treated in a later chapter [13–15].

3.1.2 Stainless Steels

The AISI defines a stainless steel as an iron-base alloy containing 4% or more of chromium. However, such a definition both includes alloys that are not particularly corrosion resistant and excludes others that are. Perhaps it is best to define this category as iron-base alloys that are sold and used for their corrosion resistance (among other properties), and which contain at least 4% chromium. At any rate most of what is normally recognized as belonging to this category falls into one of three types: *martensitic* alloys which are hardenable by heat treatment, and *ferritic* and *austentic* alloys which are not hardenable by heat treatment.

Martensitic and ferritic stainless steels are principally iron/chromium systems. Martensitic stainless grades range from 10.5–18% chromium, always with high carbon. Ferritic grades range from 10.5–30% chromium. Austentic stainless steels are iron/chromium/nickel systems, although for certain grades (the AISI 200 series) part of the required nickel is replaced with manganese and nitrogen. In general chromium may range between 16 and 26%, nickel may be as high as 35%, and manganese may reach 15%.

Other major additives for certain stainless steels include molybdenum, copper, titanium, niobium, and silicon. There are precipitation hardening martensitic and semi-austentic stainless steels, as well as duplex stainless steels that are composed of a mixture of ferrite and austenite.

AISI classifications include the 300 and 200 series of compositions for austentic grades and the 400 and 500 series for the ferritic and martensitic grades.

Stainless steels vary widely in their resistance to different corrosion mechanisms, in their yield strength, in their heat resisting properties, in their machinability, and in other properties. Some products find service in extreme conditions of heat or cold, or in aggressive chemical environments.

Some are used in food service or medical applications, where their "rustless" property facilitates hygienic cleaning.

In some cases the distinction between a stainless steel and an iron-base high temperature alloy is blurred. Defining these alloys by their intended use is not a good solution, however, since stainless steels are often utilized for their temperature resistance and iron-base high temperature alloys are often utilized for their corrosion resistance. AISI draws the line at about 28% chromium, although there are higher chromium levels in commercial alloys known as "stainless steels" and many iron-base high temperature alloys with much lower levels of chromium [16-18].

Even the most highly alloyed stainless steels are attacked to some degree by hydrochloric acid. In all cases the addition of dropwise portions of 30% hydrogen peroxide increases the rate of reaction dramatically. This approach may leave undissolved chromium (and other) carbides and nitrides, but it will not passivate the alloy surface. The advantage of this approach derives from the fact that hydrogen peroxide has an evanescent existence in the hot acid solution and allows the analyst a close control on the reaction rate. The free chlorine that forms by reaction with the hydrochloric acid and the oxygen that forms from the thermal decomposition of the hydrogen peroxide itself are both rapidly expelled from a boiling solution.

However, the most common dissolution media for stainless steels are mixtures of hydrochloric and nitric acids (generally, either 1:1 or 3:1 $HCl:HNO_3$). These are extremely effective for most grades, but when molybdenum levels reach 2% and higher there is a decrease in reaction rate. High chromium/high molybdenum grades of stainless steel are better dissolved by heating them in hydrochloric acid with periodic dropwise additions of nitric acid. Nitric acid alone is totally ineffective for stainless steels, usually passivating the sample surface to such an extent that even the subsequent addition of a large excess of hydrochloric acid will fail to restart the reaction.

If hydrochloric and nitric acids are added to the test portion separately it is important to always add the hydrochloric acid first to prevent possible passivation of the alloy. If trace levels of antimony, arsenic, phosphorus, or selenium are to be determined a freshly prepared mixture of hydrochloric and nitric acids must be added to the test portion to prevent the loss of volatile analyte during dissolution.

Dilute sulfuric acid is only moderately useful as a dissolution agent for stainless steels. Ferritic and martensitic grades dissolve readily, but certain austenitic grades, particularly those with high levels of nickel and molybdenum, such as type 316 (17Cr/12Ni/2.5Mo) are only slowly digested. Some more highly alloyed austenitic grades do not dissolve at all. The utility of dilute sulfuric acid here primarily relates to certain spectrophotometric methods for trace boron determination. For example, in the direct dianthrimide method for boron, if the alloy will dissolve in dilute sulfuric acid a result can be obtained in a very short amount of time as compared to more conventional procedures. Concentrated sulfuric acid does not appreciably attack stainless steels.

Occasionally the analyst has a reason to dissolve a stainless steel in a hydrofluoric acid/nitric acid medium (for example, as preparation for a fluoride/chloride system ion exchange separation). Reaction rates vary in this case, but slow reactions can be facilitated by the cautious dropwise addition of hydrochloric acid.

Fusions of finely divided stainless steels with potassium pyrosulfate are sometimes employed when chromium (or other) carbides and nitrides are known to be present and are known or suspected to be harboring some of the analyte. A direct fusion of the alloy saves the extra time involved in the filtration/ignition/fusion/leaching of acid dissolved samples. A Vycor or fused quartz crucible or flask is commonly used.

3.1.3 Iron Base High Temperature Alloys

This category is rather ill-defined and commonly includes many alloys that we have already considered as stainless steels. For our purposes here we can confine our discussion to alloys that are primarily used for their resistance to temperature and which fall into the 50–60% iron range. While there are several classification systems for these materials (including one by AISI) they are best known by trade names, some of which are legally protected as trademarks and some of which have fallen into the public domain.

Alloying additions are highly diverse in this class of materials. Nickel is always present in major amounts, and with certain exceptions chromium is nearly always present as well. Molybdenum is a common major component, and aluminum, cobalt, niobium, titanium, and tungsten occur frequently. High levels of copper, nitrogen, and tantalum are sometimes present. A near trace boron addition sometimes significantly enhances properties.

These alloys are designed with an austenitic face-centered cubic structure strengthened by solid solution hardening and precipitate forming mechanisms. Like the related nickel- and cobalt-base high temperature alloys, they are intended for applications requiring strength and corrosion and oxidation resistance at high temperatures. They must dependably resist failure under extreme conditions in service, yet be formable in manufacture.

Certain trace elements, such as antimony, bismuth, lead, selenium, silver, tellurium, and thallium, are considered undesirable tramp impurities since their presence can lead to catastrophic failures when these alloys are utilized. As a result a great deal of analytical effort is expended in ensuring that these and other elements remain absent or at acceptably low levels [18,19].

It should be noted parenthetically that there are many high temperature alloys with nickel or cobalt as the element of highest concentration whose specification lists iron as "balance." These will be regarded later under the section of this chapter devoted to the principal alloy component.

The most generally useful approach to dissolving this class of alloys is to begin by adding hydrochloric acid, then adding nitric acid slowly dropwise while heating until the last of the alloy particles have disappeared. This may sometimes be a lengthy process, but patience is usually rewarded with success. The analyst must resist the temptation to add nitric acid at a faster rate than that needed to sustain the reaction since many alloys in this category are easily passivated.

When hydrofluoric acid can be tolerated by the analytical procedure a very effective dissolution medium is a 2:1:1 mixture of hydrofluoric acid:nitric acid:hydrochloric acid. A Teflon vessel is, of course, necessary, and the reagents are added to the sample separately. A 1 g sample of iron-base high temperature alloy should dissolve in 20 mL of hydrofluoric, 10 mL of nitric, and 10 mL of hydrochloric acid.

Mixtures of perchloric and phosphoric acid are usually very effective as well, but titanium will precipitate if it is present in amounts much above 0.5%. Mixtures of sulfuric and phosphoric acid are generally applied initially with hydrochloric and nitric acids. When the alloy has dissolved the solution is taken to light fumes and fumed for 1 min. It is then cooled for 1 min. Then dilute sulfuric acid is added to solubilize polymeric metaphosphoric acid. In this approach titanium does not precipitate, but test portion size must be limited to about 0.3 g. Less effective but sometimes useful is an approach in which the sample is started in hydrochloric acid, followed by dropwise additions of 30% hydrogen peroxide. This technique requires gentle warming and sometimes extreme patience, but it avoids the use of nitric and hydrofluoric acids, which are often objectionable.

If it becomes necessary to resort to a molten salt fusion (to dissolve an acid resistant analyte-bearing phase in one step) 2:1 sodium peroxide:sodium carbonate is likely to be the most effective reagent. The alloy sample must be very finely divided and must be intimately mixed with the flux compounds in a zirconium crucible. A 15:1 mixed flux:test portion weight ratio is common, and the bottom of the crucible is often lined with a layer of sodium carbonate. A zirconium lid should be used. A 0.5 g or smaller test portion size is best.

The fusion is conducted on a tripod over a Meker burner behind a safety shield. The crucible should not be touched for the first two minutes. After that interval the lid is removed and the contents cautiously swirled. Undissolved particles should be easily observable. The lid is returned and heating is continued for two minutes more. This process of periodic swirling is continued until no undissolved material is visible. The crucible is then removed from the heat and allowed to cool to room temperature. The crucible is then gently rapped on a hard surface to loosen the salt cake. The bottom is wiped with a lint-free tissue, and the crucible and lid are dropped into 100 mL of water in a 600-mL beaker. When the reaction subsides hydrochloric or dilute sulfuric acid are added cautiously with stirring until a clear solution is obtained. The crucible and lid are removed with Teflon-coated forceps and rinsed into the beaker with water.

Naturally, if zirconium is the analyte, or if either phosphorus or arsenic is the analyte (these will precipitate), then a nickel or even an iron crucible and lid must be used instead of a zirconium crucible and lid. Fusions of these alloys with sodium peroxide and sodium carbonate should never exceed a 1 g test portion (using 10 g of Na_2O_2 and 5 g of Na_2CO_3, plus 2 g of Na_2CO_3 to line the crucible bottom), but, as mentioned above, smaller weights are much preferred.

The use of sodium peroxide as a molten salt flux is a potentially hazardous procedure since the presence of organic or easily oxidized inorganic substances in the test material could produce a violent reaction. It is not unknown that such fusions can erode through the bottom of the crucible, producing a dangerous situation with the potential for serious injury. It is prudent to weigh and then ignite in a muffle furnace totally unknown sample material or material known to be contaminated or reactive. The test material is then cooled and reweighed to obtain a loss (or gain) on ignition value. Then a portion of the ignited sample is weighed for sodium peroxide fusion. This procedure ensures that all reducing and carbonaceous substances are absent.

Potassium pyrosulfate fusion is another alternative for dissolving this class of alloys—much less hazardous, but also more likely to leave undissolved material. Such fusions are best conducted in a quartz erlenmeyer flask heated over a Meker burner. Sometimes a mixture of potassium pyrosulfate and concentrated sulfuric acid is used.

3.1.4 Ferroalloys

Here we mean all nonscrap high iron furnace charge materials that are sold for their alloy content. They are raw materials for steelmaking rather than end-use alloys. They are sold as bulk commodities, often as heterogeneous chunks accompanied by fines, and often in very large tonnage shipments. We have described some of the procedures for the sampling and sample preparation of these materials in the last chapter.

Ferroalloys are nearly as diverse as the elemental constituents of steels. The largest volumes sold are represented by ferrosilicon, low carbon ferrochromium, high carbon ferrochromium, ferrochromium silicon, ferromanganese, and ferromolybdenum. Other materials include ferroboron, ferrophosphorus, ferroniobium, ferrotungsten, ferrotitanium, and ferrovanadium. The industry which produces these materials typically also produces other types of charge materials (silicomanganese, chromium metal, and others).

The alloy content of ferroalloys varies widely. Sometimes the iron content drops as low as 15–20%, but it often ranges higher. In addition to the alloy constituent there are also usually significant amounts of minor elements. Silicon is nearly always present, and sometimes so is carbon. Trace levels of impurities are critical concerns with these materials since these elements will end up as tramps in the steels and irons which result from their use. Price and grade for a given ferroalloy are determined by alloy content and by the level of residual impurities present.

In addition to being hard and heterogeneous ferroalloys are also difficult to dissolve. The approach chosen is commodity-specific; however most dissolution schemes fall into a few major categories. For most ferroalloys there is an acid route and a molten salt fusion route to dissolution. This is fortunate since one or the other of these paths may be closed to the analyst due to the constraints of the analysis.

Finely divided test portions of all ferroalloys can be fused with sodium peroxide. The reaction is always moderated with the addition of sodium carbonate, and sometimes potassium carbonate as well. This moderation of the reaction is particularly important for ferrosilicon and ferromanganese which tend to be very exothermic.

In a typical fusion 2 g of sodium carbonate are weighed into a clean, dry zirconium crucible of about 50-mL capacity. This is followed by 15 g of sodium peroxide. The crucible is placed in a desiccator while a 0.5–1.0 g test portion of -100 mesh or finer ferroalloy is weighed on an analytical balance. The test portion is transferred to the crucible, 2 g of additional sodium carbonate are added, and the flux and sample are thoroughly mixed by stirring with a stiff platinum wire. A lid is added and the crucible is placed on a clay triangle on a tripod behind a protective plastic shield in a hood. A lit Meker burner is then placed under the crucible and the analyst steps back, allowing the reaction to proceed without intervention for 2 min. A runaway reaction is unlikely but possible at this point.

If a runaway reaction should occur, the bottom of the crucible may be breached, scattering sparks and molten material. In the event of a runaway reaction the analyst should shut off the gas supply and remain well clear until the reaction ceases. The likelihood of such a runaway reaction is greatly diminished after the first 2 min of heating.

After 2 min the analyst can cautiously remove the lid with tongs and swirl the crucible contents. The lid is returned and heating is then continued for another 2 min. This process of swirling every 2 min is continued until no undissolved material is observed. The crucible is then removed from the heat and allowed to cool to room temperature.

A 600-mL beaker containing 100 mL of water covered by a watchglass is placed in a hood. The cooled crucible is rapped sharply to loosen the cake and the bottom is wiped with a lint-free tissue. The crucible, lid, and contents are then transferred to the beaker and allowed to react to completion. Concentrated hydrochloric acid or 1:1 sulfuric acid:water are then added slowly with stirring until a clear solution results. The crucible and lid are then removed and rinsed into the beaker. Observation of the bottom of the beaker should show no undissolved material. The solution is then boiled for several minutes to expel gases and may be reduced somewhat in volume. The analyst must be aware in subsequent steps, however, that the salt burden of the solution is high.

As with all sodium peroxide fusions it is critical that the sample be free of easily oxidized foreign substances. The test portion should be at least -100 mesh and intimately mixed with the flux. The amount of sodium carbonate can be increased to moderate the reaction rate or decreased to help when the reaction is too slow and test material is otherwise left undissolved.

The use of other molten salt fluxes for ferroalloys is problematic. Potassium pyrosulfate has been successfully employed for ferroniobium in a flux:test portion ratio of 20:1 and is probably also suitable for certain other ferroalloys with low silicon contents. Because of its highly exothermic reaction with sodium peroxide some analysts dissolve ferrosilicon using a combined acid dissolution/molten salt fusion scheme. Here a 1 g test portion is reacted in a large platinum crucible with 1:1 sulfuric acid:water and hydrofluoric acid, heating to fumes of SO_3, and then to dryness to evolve out the silicon. The sample is briefly ignited in a muffle furnace, then cooled in a desiccator. Sodium tetraborate is then added and the crucible is returned to the muffle furnace. The cooled melt is then leached in dilute acid. Unfortunately, with this approach many important trace analytes are lost during the silicon evolution.

Direct acid dissolution of ferroalloys requires careful matching of the acid mix to the ferroalloy type, although groupings can be discerned. Low carbon ferrochromium, ferrotitanium, ferromolybdenum, and ferroboron will all dissolve in 1:1 hydrochloric acid:water, the latter two benefiting from the occasional addition of a few drops of nitric acid. Similarly, 1:1 sulfuric acid: water has been applied to all but the last of these materials, and is probably effective for it as well. High carbon ferrochromium, normally resistant to hydrochloric acid, dissolves readily under pressure in a microwave oven. Ferromanganese (both low and high carbon) will dissolve in either nitric acid or a mixture of nitric acid and 1:1 sulfuric acid:water. A 0.2 g sample of ferrovanadium can be dissolved in 30 mL of freshly prepared 1:1:1 nitric acid:sulfuric acid:water with dropwise additions of hydrochloric acid.

Most of the remaining effective acid dissolution schemes for ferroalloys require the use of hydrofluoric acid and thus require the use of Teflon labware. Ferroniobium dissolves best in a 2.5:3:2 mixture of hydrochloric acid:hydrofluoric acid:water with nitric acid added dropwise. Ferrophosphorus has been dissolved in 2:5:5 nitric acid:sulfuric acid:hydrofluoric acid. Ferrosilicon has been dissolved in hydrofluoric acid with dropwise additions of nitric acid [9]. Obviously, there are other acid routes as well. The analyst must be ever cognizant of the potential for volatiles loss in tramp level work, however. In some cases closed vessel dissolution, especially in a microwave oven, may prove to be a real boon for work with certain ferroalloys.

3.2 Nickel Base Alloys

Nickel alloys, by which we mean alloys with nickel as the element in greatest abundance, are largely high performance materials. Along with cobalt-base alloys, they are often manufactured by specialty steel producers since many of the melting and refining processes used to produce premium grade stainless steels and iron-base temperature resisting alloys are directly applicable to nickel alloys as well.

3.2.1 Nickel Base High Temperature Alloys

The largest category of nickel-base alloys both in the number of commercial compositions and in the volume of use are the superalloys. These find use in aircraft gas turbines, in power plant steam turbines, in reciprocating engines, in equipment for hot working and heat treating metals, in space vehicles, nuclear reactors, chemical and petrochemical processing equipment, and other applications where high temperature strength and corrosion resistance are important. Some nickel-base superalloys are routinely employed at 80% of their incipient melting temperature.

While there are various systematic numbering systems that have been applied to this category of alloys they are most often recognized by trade names, some of which are expired or legally genericized trademarks. In both cast and wrought compositions nickel-base superalloys commonly contain 10–12 intentionally added elemental components. Nearly as many other elements must meet maximum concentration levels to ensure freedom from their deleterious effects.

Among the alloy additions chromium typically ranges between 10 and 20%, although it is sometimes very low in materials designed to have low thermal expansion coefficients. It may also range as high as 30% in alloys designed for hot corrosion and oxidation resistance. Cobalt may reach 20%, but is commonly lower. Molybdenum usually falls between 1 and 10%; iron falls between 0.5 and 50%. Aluminum sometimes reaches 6%; titanium reaches 5%; tungsten may reach 14%. Most niobium additions are between 0.5 and 7%; tantalum, especially in casting compositions may, likewise, range from 0.5 to 7%. Boron and zirconium are frequently added, the former at .001–.2%, the latter at .01–.2%. Other elements that may be added at major and minor levels include lanthanum, yttrium, vanadium, copper, magnesium, and hafnium.

The exact composition of each of these alloys is determined by the need to achieve a delicate balance of metallurgical effects—phase formation and stabilization, the precipitation of carbides and carbonitrides, and other processes that impart desired properties.

Elements that are deleterious at low levels include nitrogen, oxygen, phosphorus, silicon, and sulfur. Elements that are deleterious at trace or ultratrace levels include antimony, arsenic, bismuth, cadmium, gallium, lead, selenium, silver, tellurium, tin, thallium, and zinc.

It should also be noted that current state-of-the-art developments for aircraft gas turbines—such as directionally solidified castings or single crystal castings—make use of quite different nickel-base compositions. These are also composed of 10–12 major elemental components, sometimes including hafnium, rhenium, and up to 12% tantalum. Dispersion-strengthened alloys represent another technology under development. These materials contain finely-dispersed rare earth oxides [19–20].

The dissolution of nickel-base high temperature alloys for analytical work cannot be accomplished by following a single straight-forward recipe.

But rather, this work requires a careful scrutiny of the base-alloy composition while keeping a watchful eye on the analytical goal.

If there is a "universal" acid solvent for this class of materials it is 2:1:1:1 hydrofluoric acid:nitric acid:hydrochloric acid:water. The author is unaware of a nickel-base high temperature alloy that will not eventually yield to this mixture, given enough time with gentle heating in Teflon. The limited solubility of nickel fluoride is rarely a problem, since the salts dissolve in added water with patient warming and stirring. This is far from a complete answer, however, since many volatiles are lost and the use of hydrofluoric acid is objectionable for some methods. In other methods hydrochloric acid is objectionable. Perhaps, more commonly, these materials are dissolved in 2:1 hydrofluoric acid:nitric acid with small amounts of hydrochloric acid sometimes added for alloys which contain more than about 15% chromium.

Hydrofluoric acid is valuable for its complexing power, which prevents the acid hydrolysis of many component elements that would otherwise precipitate as hydrous oxides. It *is* possible to dissolve certain of these alloys without its use, however. Hydrochloric acid with dropwise additions of nitric acid can be very effective provided that only enough nitric acid is added to sustain a reaction. Excess nitric acid will readily passivate many grades. And if the solution is subsequently reduced in volume, some or all of the "earth acids" (Hf, Mo, Nb, Ta, Ti, W, Zr) will likely precipitate. An even more limited number of these alloys will succumb to hydrochloric acid with dropwise additions of 30% hydrogen peroxide. There is also a lower, but nonzero chance that earth acid elements will precipitate when these solutions are reduced as well. These hydrofluoric acid-free approaches are most effective for the less heavily alloyed materials in this category (such as Inconel™ 600 and 601 or Nimonic™ 75, 80A, and 90).

Sometimes a hydrofluoric acid-free acid dissolution approach will leave an undissolved phase, which will usually have to be filtered, ignited, and fused (usually in either potassium pyrosulfate or fused sodium bisulfate), then leached in the filtrate or in a complexing solution. A totally different approach, which is often very effective, is to apply phosphoric acid as a complexing agent during dissolution. Here the sample size is generally limited to 0.3 g and the test portion should consist of small chips or fine powder. First, add 25 mL of hydrochloric acid, then 5 mL of nitric acid, followed by 25 mL of water, then 25 mL of a 10:3:12 mixture of phosphoric acid:sulfuric acid:water (which has been previously prepared by cautiously adding the acids to the water with stirring and cooling). At first gently warm the test portion with the acids, then heat more strongly until dissolved. Heat to fumes of phosphoric acid and continue fuming for 1 min. Remove from the heat and allow the solution to cool for 1 min. Then cautiously, with swirling, add 25 mL of 7:13 sulfuric acid: water (also previously prepared, as above). When cooled to room temperature the sample can be diluted up to 100 mL with water.

Other acid mixtures have been applied to this class of alloys with limited success. Perchloric acid/phosphoric acid mixtures are among the more effective, but they are often slow and suffer from the precipitation of titanium phosphate (which does not trouble the sulfuric acid/phosphoric acid approach described above). Moreover, fuming such a mixture will likely precipitate some niobium, tantalum, tungsten, or other earth acids despite the presence of phosphoric acid.

The use of a closed vessel microwave oven approach with hydrofluoric and nitric acid can greatly speed up that protocol. But such methods must be tailored to the alloy. A laboratory that analyzes a wide range of nickel-base high temperature alloys must rely on a universal oven program that will handle all or most of its varied workload. That may require 90 min or more and is wasteful of time for alloys that dissolve more readily. A more suitable approach in that situation is to utilize 2.5:10:7.5 hydrofluoric acid:nitric acid:water with screw-cap Teflon PFA vessels that can sustain a light pressure. These can be warmed on sandbaths on a hotplate for varying amounts of time, depending on the alloy.

Molten salt fusions are not common for dissolving these alloys, but they can be used and are sometimes necessary. Potassium pyrosulfate, with or without the addition of sulfuric acid, or fused sodium bisulfate (always with the addition of sulfuric acid—to limit spattering) are best weighed into a small quartz flask. The test portion should be finely divided and preferably 0.5 g or smaller. Flux:sample ratios should range between 5:1 and 10:1. The flask is held by the neck in tongs and swirled over a Meker burner in a hood. Copious fumes are evolved as the sample dissolves. When the salt has been converted to sulfate SO_3 evolution will diminish and the melt will thicken and congeal. At this point the flask should be removed from the heat and allowed to cool to room temperature. The salt cake is then dissolved by adding 1:1 hydrochloric acid:water or 4%(w/v) sodium oxalate (which will help prevent the hydrolysis of the earth acids) and warming on a hotplate. Not all alloys in this category will succumb to this treatment, but many will. The potassium pyrosulfate salt is probably slightly more effective than the fused sodium bisulfate salt, especially for niobium- and tantalum-bearing alloys.

Sodium peroxide fusions are sometimes a last resort to dissolve nickel-base high temperature alloys that fail to yield to acid treatment. For example, the determination of trace or low levels of boron in heavily alloyed nickel-base superalloys might normally utilize a dissolution with perchloric acid/phosphoric acid under reflux conditions. But certain alloy compositions are impervious to this treatment. The use of hydrochloric or hydrofluoric acid are precluded because of the volatility of BCl_3 (b.p. 12°C) and BF_3 (b.p. −101°C). A useful alternative is to weigh 2 g of sodium carbonate into a new zirconium crucible, followed by 10 g of -20 mesh (<0.850 mm) or finer sodium peroxide. The crucible is stored in a desiccator. A 0.5 g test portion of -100 mesh

(<0.150 mm) or finer test material is weighed and transferred to the crucible. The charge is stirred with a stiff platinum wire, covered with a zirconium lid, and placed on a tripod in a hood behind a suitable shield.

A lit Meker burner is placed under the crucible and the setup is left undisturbed for 2 min. The lid is then cautiously removed with forceps and the contents cautiously swirled to dislodge undissolved particles. The lid is then replaced and the heating is continued for an additional 2 min. This process is continued until no undissolved particles are observed. The crucible is then removed from the heat and allowed to cool to room temperature.

The cake is dislodged by rapping on a hard surface. The crucible bottom is wiped with a lint-free tissue, and the crucible, salt cake, and lid are transferred to a 600-mL covered beaker containing 100 mL of water in a hood. When the reaction ceases 1:1 sulfuric acid:water is added with stirring until a clear solution is obtained. The crucible and lid are rinsed into the beaker, and then the solution is boiled for 1 min.

A nearly identical approach is applicable to phosphorus determination except that a new nickel crucible must be used to circumvent the precipitation of zirconium phosphate.

3.2.2 Other Nickel Alloys

Other commercial alloys that contain more nickel than any other component represent a wide range of applications. One of the largest categories is the Monels™: generally 60% nickel/30% copper with additions of iron, manganese, silicon, carbon, and other elements. They are used for their corrosion resistance, especially to seawater in marine applications. With a 3% aluminum addition, "Monel K" can be precipitation hardened.

Certain nickel alloys are used in thermocouples and as resistance elements in heating devices. These are primarily nickel/chromium compositions, sometimes with a significant amount of iron. Cerium, zirconium, and other elements may be added to impart thermal cycle fatigue resistance.

Magnetic alloys are represented by the nickel/irons which are close to being binary alloys of nickel and iron, but with minor additions. These materials are utilized for their closely controlled magnetic permeability which is intimately related to chemical composition. Controlled expansion (or glass-sealing) alloys are another important category. These are also nickel/irons with significant alloy additions.

Certain nickel/titanium compositions are known as shape-memory alloys. These return to their original configuration after deformation when they are heated to a specific temperature. There are also nickel-base charge materials—nickel/niobium and others which are sometimes used to introduce alloying additions into nickel-base high temperature alloys [22].

Monels dissolve in *aqua regia* and hydrochloric acid/hydrogen peroxide. They have also been dissolved by potassium pyrosulfate fusion. Nichrome™, Chromel™, and other heat-resisting and thermocouple nickel/chromium alloys dissolve in the same media. The nickel/irons dissolve in most acids, including hydrochloric acid. The shape-memory alloy known as Nitinol™ will dissolve in hydrochloric acid or dilute sulfuric acid. Nickel/niobium requires hydrofluoric acid/nitric acid and may benefit from a closed-vessel microwave approach.

There are other nickel-base compositions: a 60% nickel/platinum-palladium-vanadium alloy which should eventually dissolve in *aqua regia*; and high-purity nickel, itself, which is widely used in a variety of forms, and which dissolves easily in nitric acid. There are also the so-called "low alloy nickels," as well as nickel-base welding alloys and filler metals. Most of these will yield to either nitric acid or *aqua regia*.

3.3 Cobalt Base Alloys

Like nickel alloys, although representing a much smaller market, cobalt alloys find use as high temperature materials. Smaller amounts are sold as magnet alloys and "hard metal" carbides.

3.3.1 Cobalt Base High Temperature Alloys

This is a much more limited category than the equivalent nickel-base high temperature alloy group. In general, cobalt high temperature alloys exhibit superior thermal fatigue resistance, better hot corrosion resistance at high temperatures (possibly due to their higher chromium contents), and better weldability than their nickel-base equivalents. Chemical composition is at least as complex as among nickel-base materials. Chromium ranges from 20–30%; nickel is typically in the 10–35% range. Iron is often 2–9%, molybdenum 0–10%, tungsten 0–15%. Sometimes niobium, tantalum, and aluminum are added in major amounts, and zirconium, boron, and titanium are added in minor amounts. Most of the same elements that cause trouble in trace and low amounts in nickel-base high temperature alloys have similar effects here. Sometimes yttrium or beryllium is added to certain compositions. Alloys designated "MP" are multiphase compositions (e.g. MP35N, which has corrosion resistance applications) [19–21].

Cobalt high temperature alloys present special challenges to the analytical chemist. They resist some of the common dissolution operations that are effective with nickel- and iron-base high temperature alloys. Particularly notable is the effect of heating with hydrochloric/nitric acid mixtures. Even with limited additions of nitric acid these alloys often appear inert on the hotplate. The problem can be solved by not heating the solution. To a 1 g test portion add 50 mL of hydrochloric acid and 5 mL of nitric acid and allow the beaker to stand at room temperature overnight. The next morning any remaining undissolved material will usually yield to gentle warming. Evidently, initially heating a hydrochloric acid/nitric acid mixture produces some species that passivates these materials.

Cobalt high temperature alloys can also be dissolved in mixtures of hydrofluoric and nitric acids and in hydrofluoric/hydrochloric/nitric acid mixtures. Here, again, excessive temperatures must be avoided. The phosphoric/sulfuric acid approach described for nickel-base high temperature alloys is also effective here. Cobalt forms a particularly gummy polymer after the initial fuming, and so it is especially important to add the dilute sulfuric acid while the fumed solution is still warm, and to swirl vigorously.

Sodium peroxide fusions are occasionally used for cobalt-base high temperature alloys. Here moderation with at least 2 g of sodium carbonate is important. After the melt is cooled, cobalt peroxide salts react with water with particular vigor, and so caution is advised. It is prudent to work deep in a hood for the leaching step.

3.3.2 Other Cobalt Alloys

Cobalt-rare earth compositions, in particular cobalt-samarium alloys, are widely used permanent magnet materials. These dissolve readily in hydrochloric/nitric acid mixtures. Similarly, there are orthopedic implant and dental prosthesis alloys based on cobalt (e.g., Vitallium™, a 60Co/30Cr/5Mo composition) which will dissolve in *aqua regia*. The same is true of cobalt alloys used for their magnetic properties (e.g., Vicalloy™, 52Co/10V), and those with controlled expansion or high temperature dimensional stability.

The so-called "hard metals" and cobalt alloys sold for their abrasive resistance properties are another matter. The Stellite alloys contain complex metal carbides dispersed in a cobalt-base superalloy matrix. The principal matrix alloying elements are chromium and tungsten. These materials find use where hardness at elevated temperatures is critical. They sometimes replace tungsten carbide due to their superior ductility [23,24].

Stellite™ alloys can be dissolved by adding 5 mL of hydrofluoric acid to the 1 g test portion in a Teflon beaker, then adding nitric acid dropwise with gentle heating on a sandbath. A 0.5 g test portion of finely divided sample can be fused in 5 g of potassium pyrosulfate. The melt is cooled, 10 drops of concentrated sulfuric acid is added and the salt is fused a second time. Stellites can also be fused in sodium peroxide and sodium carbonate in a zirconium crucible. Here, a 0.5 g finely divided test portion is intimately mixed with 10 g of sodium peroxide and 5 g of sodium carbonate and fused with a lid over a Meker burner, as described for nickel alloys in section 3.2.1. Since cobalt alloys fused in this way are particularly reactive with water the melt should be completely cooled to room temperature before leaching it.

If, instead of acidifying the water-leached fusion, the crucible and lid are removed and policed and rinsed into the beaker, it is possible to obtain a clean separation of the tungsten and molybdenum by boiling and filtering the basic solution. The cobalt hydroxide precipitate is retained on a hardened filter paper (such as Whatman No. 54 or equivalent), while tungsten and molybdenum (and chromium and vanadium) pass through into the filtrate.

3.4 Copper Base Alloys

Pure copper, itself, which is soft and ductile, is an important commodity sold in a wide variety of forms and grades for electrical, electronic, and mechanical fabrication applications. Copper alloys are referred to by an argot of terms that include the well-defined *brasses* and ill-defined *bronzes*, as well as many others. Copper alloys are generally lower in strength and higher in cost than steels, but they are valued for their unique properties, including thermal and electrical conductivity, deep drawability, and corrosion resistance.

3.4.1 Pure Copper and Dilute Copper Alloys.

Commercially pure copper should always contain less than 0.7% total impurities and some "coppers" are very pure. Deoxidizer elements are often found in low or trace amounts (these might be aluminum, beryllium, boron, calcium, carbon, magnesium, manganese, phosphorus, silicon, or zinc). Certain high purity grades require certification testing for a long list of tramp elements. The *dilute copper alloys* are very high copper alloys with minor additions of low solubility elements, such as cadmium, beryllium, iron, or chromium.

The Copper Development Association (CDA) has developed a classification numbering system for these and all other commercial copper compositions. Both wrought and cast versions of most compositions are commercially practical, which allows the end-product designer to select an alloy before selecting a manufacturing process.

The *beryllium-copper alloys* have unique properties that prove invaluable in certain specialized applications. For example, they can exhibit high strength and hardness, but are nonmagnetic. In this class of materials beryllium can range from 0.2 to nearly 3.0%. Typically a small amount of cobalt or nickel is included to control grain growth during annealing. *Cadmium-copper alloys* may range from 0.1–1.0% cadmium. There are also some *copper-tellurium alloys* with tellurium levels of up to 1% or higher. Some dilute copper alloys are dispersed with alumina at from 0.2–1.1%. And some are specified with four or more alloying additives (e.g., C19500: 97.0Cu/1.5Fe/0.6Sn/0.10P/0.80Co) [25–29].

All copper alloys dissolve in nitric acid. Usually it is necessary to use 1:1 nitric acid:water instead of the concentrated acid to slow the reaction, especially when the test portion is finely divided. Another approach which is sometimes used for copper-beryllium alloys is to apply freshly prepared mixtures of dilute sulfuric and nitric acids. Crude copper is sometimes dissolved in 1:4 sulfuric acid:water with the dropwise addition of 30% hydrogen peroxide.

Hydrochloric acid dissolves copper alloys slowly, and even the dropwise addition of 30% hydrogen peroxide does not generally produce high reaction rates. Interestingly, hydrobromic acid dissolves copper alloys rapidly due to the highly stable bromide complexes which form, and, perhaps also to the traces of bromine which are always present in hydrobromic acid. Occasionally mixtures of hydrochloric and nitric acids are employed [9–11,29].

3.4.2 Brasses and Bronzes.

All "brasses" are alloys of copper and zinc. Copper ranges between 60 and 85%, zinc generally between 15 and 40%. Sometimes lead is added to improve machinability (selenium and tellurium can perform a similar function). Tin and lead are major components of naval brass, leaded yellow brass, and leaded red brass. There are also silicon brasses and aluminum brasses with the named element as a major component. In addition there are other copper-zinc alloys known by a variety of names. Admiralty metal, used in marine applications, is 70% copper/29% zinc/1% tin. Muntz metal is 60% copper/40% zinc.

"Bronzes" are commonly thought of as copper/tin alloys, but the name is also applied to many other highly alloyed copper-base materials. Copper content is, in general, higher than in the brasses, typically ranging from 85–95%; although it can be as low as 58% in some alloys referred to as a "bronze." The phosphor bronzes all contain tin as their principal alloying element along with some level of phosphorus between 0.01–0.50%. There are also aluminum bronzes with 2–14% aluminum as the principal additive and sometimes as much as 5% iron, among other elements. The silicon bronzes may contain from 0.4–4% silicon.

The nomenclature clouds further when we consider "manganese bronze," which is a copper-zinc alloy with other elements and only a minor addition of manganese. Similarly, "commercial bronze" and "jewelry bronze" are copper-zinc alloys [25–27].

All brasses and bronzes dissolve in 1:1 nitric acid:water; however, with high tin alloys SnO_2 will precipitate. If tin will prove troublesome in the analytical procedure the test portion can be dissolved in a 9:1 hydrobromic acid:bromine mixture. When dissolution is complete perchloric acid is added and the solution is heated to strong fumes of $HClO_4$. This will expel volatile tin (and antimony) bromides. Arsenic and selenium will be lost as well. Also, lead precipitated as a bromide will be converted to soluble lead perchlorate. Some copper-base industry laboratories are not equipped with perchloric acid hoods, however, and must, therefore forego this option. Under no circumstances should perchloric acid be employed without a properly utilized perchloric acid hood.

Up to 1 g of tin- and lead-bearing copper alloys can be dissolved in 5 mL of fluoboric acid, 5 mL of nitric acid, and 5 mL of water. Neither tin nor lead will precipitate under these conditions. A similar effect will be achieved with hydrofluoric acid in place of fluoboric acid, but the presence of HF is objectionable for many of the analytical sequences that are likely to follow [9–11,29].

3.4.3 Copper-Nickels and Nickel Silvers

These two distinct alloy categories are both alloys of copper and nickel. Copper-nickels may range from 70 to nearly 90% copper and from 10-30% nickel. There may also be a percent or so of iron in the alloy, and, perhaps, another additive or two in low amounts. These alloys are frequently used as tubing or piping where corrosion resistance and good thermal exchange properties are needed. Such tubes and pipes are used in condenser, distillation, evaporator, and heat exchanger systems. Also, nickel coinage metal in the United States technically belongs to this category; it is 75% copper/25% nickel.

Nickel silvers (sometimes known as "German silvers") generally contain more zinc than nickel. Copper ranges from 55–88%, zinc from 10–30% (if it is present at all) and nickel from 8–18%. In a few grades lead or iron, chromium, and manganese may be added. Among other applications, these alloys are used as a base metal for silver plating [25–27].

The best approach for many analytical procedures is to dissolve these alloys in *aqua regia*. Some analysts prefer a freshly prepared "solvent–acid mixture" prepared by adding 150 mL of sulfuric acid to 700 mL of water with stirring and cooling. When cool, 150 mL of nitric acid is added with additional stirring and cooling. Such a medium is ideal both for dissolving these alloys and also electrodepositing the copper (which may function as a separation step for nickel determination by dimethylglyoxime precipitation) [29]. Sometimes 1:1 nitric acid:water is utilized as the dissolution medium, especially for trace work by GFAA or ICP-MS.

3.5 Light Alloys

The term "light alloys" is used here to designate aluminum and magnesium and their alloys. They are linked by their relatively high strength to weight ratio. Titanium will be considered in the next section (Refractory and Reactive Alloys) primarily because of its special corrosion resistant and high temperature properties.

3.5.1 Aluminum Alloys

Aluminum and its alloys combine low density with physical and mechanical properties that make both wrought and cast forms major items of commerce. One third as dense as steel, some aluminum alloys can exceed some steels in strength. Less electrically conductive than copper, aluminum shows nearly twice the electrical conductivity of copper on an equivalent weight basis. In addition to their structural and electrical properties aluminum alloys are utilized for their corrosion resistance, heat transfer properties, ease of fabrication, and their many attainable forms of surface finish [30].

Wrought aluminum alloys are designated by a four digit classification code. The first digit specifies the alloy type: 1 is unalloyed aluminum, 2 means copper is the principal alloying element. Similarly, 3 means manganese, 4 represents silicon, 5 indicates magnesium, 6 represents magnesium and silicon, 7 is zinc. Other elements are all designated with an 8. The second digit indicates a modification of the basic alloy. If it is zero no modification is indicated. The last two digits indicate the exact alloy composition, or the purity if the aluminum is unalloyed. Thus, "1095" indicates unalloyed aluminum that is at least 99.95% aluminum.

Cast aluminum alloys are also identified by a four digit system. Again the first digit specifies an alloy type: 1 is unalloyed aluminum, 2 means copper is the major additive, 3 means silicon with some copper and/or magnesium, 4 means silicon, 5 is magnesium, 7 is zinc, 8 is tin, 9 is all other elements. Note that 6 is unused. The next two digits indicate the exact alloy (or the purity of the unalloyed aluminum). The last digit is separated from the others by a decimal point. It indicates an ingot if it is a 1 and castings if it is a zero. More than three hundred compositions are commonly classified by this system which was developed by the Aluminum Association.

Following the compositional designation of wrought and cast alloys is a hyphen followed by a capital letter, or a letter and numbers. These identify the thermal and mechanical treatment that the material has received [31].

Nearly all aluminum alloys contain between 75%–99% aluminum. Copper may range from 2–10%, manganese may reach nearly 2%. Silicon greatly improves the casting properties of aluminum alloys; it may reach 25% in some compositions. Magnesium imparts strength and other desirable properties; it may reach 5.5% in some wrought alloys and 10% in castings. Magnesium also imparts beneficial properties as a minor additive in many aluminum alloys where it is not the principal alloying element. Other important additives, beneficial in certain alloys, include antimony, beryllium, bismuth, boron, cadmium, chromium, iron, lead, lithium, silver, tin, titanium, zinc, and zirconium [32,33]. The aluminum–lithium alloys are very low density structural materials that can reduce the weight of aeronautical and aerospace designs. They contain between 1–3% lithium, between 2–5.5% copper, a small amount of zirconium, and a few other additives, such as silver and magnesium [34]. Metal-matrix composites (MMCs) produced by powder metallurgy techniques consist of an aluminum alloy matrix with finely dispersed particles of ceramic materials. Silicon carbide (SiC), alumina (Al_2O_3), and titanium boride (TiB_2), among others have been used [35].

Aluminum and aluminum alloys are readily dissolved in mixtures of hydrochloric, nitric, and sulfuric acid. Alone, hydrochloric acid is a reasonably good solvent, but it benefits from the addition of a small drop of mercury, which accelerates the reaction by forming a galvanic cell which inhibits the formation of a passive oxide layer. Nitric acid alone is not recommended. It is slow to react and may create an impervious passivation layer unless catalyzed with a drop of mercury. Sulfuric acid diluted 2:1 with water attacks aluminum vigorously at first, then quickly slows down. Again, a drop of mercury restores and sustains the reaction.

The mercury drop can in each case be replaced by a small amount of mercury or copper salt (a few drops of a 1g/100 mL solution of mercuric chloride or copper (II) chloride are usually sufficient). However, the mercury drop is advantageous in that it can removed by filtration or decanting after the test portion has dissolved, resulting in little or no contamination of the solution.

Mixtures of sulfuric acid, nitric acid, and water; hydrochloric acid and 30% hydrogen peroxide; and 3:1:6 sulfuric acid: phosphoric acid: water have all been utilized for aluminum alloys.

Aluminum ion in aqueous solution behaves amphiprotically (amphoterically) by forming soluble aluminates in strong base. Aluminum alloys dissolve in strong base with the evolution of hydrogen gas. On certain occasions this approach is analytically useful. It may serve as the basis for a precipitation separation of certain alloy components, such as magnesium, that form insoluble hydroxides. As much as 3 g of aluminum can be dissolved in 20 mL of 20% (w/v) sodium hydroxide solution.

To avoid contamination from borosilicate glass vessels, Teflon TFE beakers should be employed . Alternative reagents for this approach are potassium hydroxide and sodium carbonate solutions. Weak bases such as ammonium hydroxide, do not dissolve aluminum to any useful degree [9,10,36].

3.5.2 Magnesium Alloys

Magnesium alloys average about 2/3 of the density of aluminum alloys, show good machinability, and have adequate strength for many applications.

Structural parts for automobiles, industrial machinery and materials handling equipment are common uses. Small parts that must operate at high speeds are a frequent application. Consumer goods manufactured from magnesium alloys include luggage, ladders, and hand tools. In particulate form magnesium is sold for use in pyrotechnics, and as a reagent for industrial scale application of the Grignard reaction in organic synthesis. Many other metals industries employ magnesium as an alloying element, an oxygen scavenger, a desulfurizer, or a reducing agent. Nodular cast iron, for example, owes its properties to magnesium additions.

The classification of magnesium alloys is based on two alphabetic characters that designate the two principal alloying elements, followed by two digits which indicate the respective rounded percentages. Then a letter designates which alloy in a series is represented. Finally a temper designation is included. Thus, AZ63B-T6 indicates the second alloy of a 6Al/3Zn composition with a "T6" thermal and mechanical history.

Aluminum is a common addition and may range from 1–10%; zinc ranges from 1–7%. Thorium, when present, occurs in concentrations of 1–3%. Manganese and zirconium are both typically added at less than a percent. Some alloys may contain a percent or two of silver, copper, silicon, or the neodymium/praseodymium rare earth mixture known as "didymium." Other rare earth mixtures, as well as pure yttrium, are also added to some alloys. Bismuth, cadmium, iron, lithium, nickel, lead, chromium, tin, and antimony are other possible additives [37].

Magnesium and its alloys dissolve readily in dilute mineral acids, 1:1 hydrochloric acid:water being frequently used. Other dissolution mixtures that are sometimes employed include 1:1 sulfuric acid:water, 1:1 nitric acid:water, 1:1:5 sulfuric acid:nitric acid: water, and 1:1 hydrochloric acid:water with dropwise additions of 30% hydrogen peroxide. Dilute sulfuric acid and bromine water and also dilute hydrobromic acid with a trace of added bromine have also both been used. One must be on guard to avoid concentrated reagents since magnesium alloys may react too vigorously.

Unlike the other alkaline earth elements (calcium, strontium, and barium) magnesium does not react with water. Instead it forms a passivated oxide surface film. Magnesium metal will dissolve in methanol, however.

3.6 Refractory and Reactive Metals

This is a rather broad category loosely bifurcated into a group of materials utilized for their excellent heat resisting properties and a group utilized primarily for other unique features. Currently, the refractory metals for which there is significant commerce include molybdenum, niobium, tantalum, tungsten, and rhenium, and their alloys. As the pure metals these materials have the highest melting points and lowest vapor pressures of any elements outside of the platinum group [38]. The so-called "reactive metals" include titanium, zirconium, and hafnium. The appellation "reactive" here is meant to contrast their high temperature corrosion and oxidation resistance to the "refractories," but it is unfortunate since they all exhibit serviceable qualities in that regard.

Titanium and its alloys have the largest volume of commerce of all the materials in this section. Some titanium compositions possess the tensile strength of steel with 56% of its density, making them important materials for high performance aircraft. Titanium alloys are also used for essential parts in jet engines, and titanium metal is utilized by the chemical industry for its excellent corrosion resistance. Zirconium and hafnium are almost exclusively nuclear industry materials, although both are used as alloying additions to superalloys and other alloy compositions.

3.6.1 Molybdenum Alloys

More than 95% of molybdenum is used as an alloying additive to steels, cast irons, and iron-, nickel-, and cobalt-base superalloys. In pure form or as a molybdenum-base alloy this material is used for tooling that must perform at high temperatures (e.g., boring bars, and equipment for truing grinding wheels). Molybdenum is used for resistance heating elements for high temperature electric furnaces, rocket nozzles, and for the tips of resistance welding electrodes. Its resistance to hydrochloric acid finds application in the chemical industry.

Only two alloys find widespread use: Mo-0.5Ti, which contains 0.5% titanium and 0.02% tungsten, and TZM which is identical except for the addition of 0.1% zirconium [39].

Molybdenum and molybdenum-base alloys are soluble in concentrated sulfuric acid with heating and the cautious dropwise addition of nitric acid. They will also succumb to hydrochloric acid, again with the dropwise addition of nitric acid. Excess nitric acid will lead to the precipitation of hydrous MoO_3, however. Molybdenum foil will dissolve slowly in 1:1 hydrochloric acid:water with the addition of bromine water.

A surprisingly effective dissolution medium is 30% hydrogen peroxide. About 50 mL of this reagent will dissolve 1 g of finely divided molybdenum powder with gentle warming overnight. This reaction is occasionally useful where acid concentration must be strictly controlled [9,40].

3.6.2 Tungsten Alloys

Most commercially produced tungsten is used as tungsten carbide (often with a cobalt binder to form "cemented carbides"). These products are used for cutting and abrasive resistance applications. The second largest use is as an alloying addition to steels, especially high speed tool steels. Pure tungsten and tungsten-base alloys account for most of the remainder. These products are used for counterweights, governors, X-ray targets, light bulb filaments, thermocouples, and radiation shields, among other applications.

Tungsten–molybdenum thermocouples can be used to measure temperatures in excess of 2,000°C, provided that they are not exposed to oxygen. The "tungsten heavy metals," produced by powder metallurgy, are 95–98% tungsten with a binder phase of nickel and copper, or, more usefully, nickel and iron. The majority of tungsten-based alloys contain more than 90% tungsten with additions of molybdenum, tantalum, rhenium, silver, or dispersed thoria (ThO_2). A few alloys contain 50–90% tungsten; many of these are alloyed with copper for electrical contacts, or with rhenium for ultra-high temperature uses [41].

Tungsten and tungsten-base alloys will dissolve in hydrofluoric acid with the cautious dropwise addition of nitric acid. Interestingly, they can also be dissolved by the procedure described in Section 3.1.1 and 3.2 of this chapter. A 0.25–0.30 g test portion of powder or fine chips is treated first with 25 mL of hydrochloric acid, then 5 mL of nitric acid, then 25 mL of water, then 25 mL of a 10:3:12 mixture of phosphoric acid:sulfuric acid: water. The solution is warmed until the metal dissolves, then heated to light fumes. After 1 min of fuming the solution is removed from the hotplate and allowed to cool for 1 min. Then 25 mL of a 7:13 sulfuric acid:water solution is added with swirling. When cooled to room temperature this solution can be diluted up to 100 mL with water and will remain stable to hydrolysis for weeks. Finely divided tungsten powder will dissolve overnight in 30% hydrogen peroxide with very gentle warming. Some labs prefer to oxidize tungsten powders to WO_3 (by heating them at 700°C in air or oxygen) then fuse the oxide with sodium carbonate or mixtures of sodium and potassium carbonate.

Tungsten carbides and cobalt-bearing cemented carbides are attacked by hydrofluoric/nitric acid mixtures but may leave a residue which must be filtered using a plastic funnel and beaker, then washed, ignited, fused with sodium carbonate, and leached into the filtrate. A better approach is a direct fusion of a finely divided 1 g test portion with 10 g of sodium peroxide and 5 g of sodium carbonate using a zirconium crucible and lid [9].

3.6.3 Niobium Alloys

After molybdenum and tungsten, more niobium is sold in pure and alloy form than any other refractory metal. Its uses as an alloy additive in the form of ferroniobium and nickel-niobium are significant, but it has a niche market as a high performance material on its own.

Alloy C-103 is 87.2%Nb/9.8%Hf/0.91%Ti/0.18%Zr; it is used for rocket components. Alloy Nb-1Zr is used in nuclear reactors. Cb-752 (10W/2.5Zr/Bal.Nb), C-129Y (10W/10Hf/0.15Y/Bal.Nb), FS85 (27.5Ta/11W/1Zr/Bal.Nb), and others are used for leading edges on hypersonic flight vehicles, rocket nozzles, and guidance structures on reentry vehicles, among other applications. Both the older niobium-tin (Nb_3Sn) and newer, more fabricable Nb-46.5Ti superconducting alloys are sold commercially. There is also C-3009 alloy which contains 30% Hf, 9-15% W, and 5% Ti or less. Some older alloy compositions include up to 7.5% vanadium and 5% molybdenum [42].

Niobium and niobium-base alloys require hydrofluoric/nitric acid mixtures for acid dissolution. The best approach is to add hydrofluoric acid first, then nitric acid dropwise, while warming. The best fusion procedure is to use sodium peroxide/sodium carbonate in a zirconium crucible with a lid. For most alloys a 1 g test portion should yield to 10 g of sodium peroxide and 5 g of sodium carbonate. Potassium pyrosulfate is less effective as a flux, but may serve for small test portions. The addition of sulfuric acid to the potassium pyrosulfate is often advantageous [9,43].

3.6.4 Tantalum Alloys

Pure tantalum is used in electrolytic capacitors and in the chemical process industry where its excellent resistance to hydrochloric, nitric, and sulfuric acids is employed to great advantage. Tantalum is also used as heating elements and other parts for high temperature vacuum furnaces, for surgical implants, and as a component of many superalloys. Certain cemented carbides contain tantalum carbide. Ta-10W (10% tungsten alloy) has found use in rocket nozzles and leading edge structures for hypersonic vehicles. There is a Ta-40Nb (40% niobium alloy) and several alloys with varied additions of tungsten and hafnium [44].

Tantalum and its alloys require the same dissolution procedures as niobium, but they may take somewhat longer to reach completion. Some laboratories ignite tantalum carbide in air or oxygen, then fuse the resultant oxide with potassium pyrosulfate. In general potassium salt fluxes prove more effective than sodium salt fluxes.

3.6.5 Rhenium Alloys

Rhenium is rare and expensive but does find application where its unique properties are needed. About 85% of the approximately 8 tons consumed by the U.S. in 1988 was employed in platinum-rhenium reforming catalysts for the petrochemical industry. Rhenium finds use in heating elements, X-ray tube targets, filaments for mass spectrographs and ion gages, and in rhenium/

tungsten thermocouples. Rhenium is an alloying element in certain nickel-, molybdenum-, and tungsten-base alloys [45].

Rhenium dissolves vigorously in dilute nitric acid. Both the pure metal and its alloys have been dissolved by making them the anode in an electrolytic cell containing a concentrated base and passing 10 mA/cm^2 through the circuit [9].

3.6.6 Titanium Alloys

This category is by far the largest and most important commercially of the "Refractory and Reactive Metals." Titanium alloys are sold as wrought, cast, and powder metallurgy products. The aerospace uses of titanium alloys—both for airframe components, and as disks, blades, and vanes in jet turbines—account for the greatest volume of applications. Other important uses include parts and machinery that must resist corrosion for the chemical, pulp and paper, and other industries; surgical implants, and certain consumer goods (such as sports equipment).

Titanium and its alloys are very reactive in the molten state, a property adding to the expense and expertise required to produce them. Titanium alloys exhibit two distinct allotropic forms: the alpha phase: a hexagonal close-packed (hcp) structure and the beta phase: a body-centered cubic (bcc) structure. Commercially pure titanium (98.635 to 99.5% Ti) is alpha at room temperature and transforms to beta at 883° C. The two phases have different properties and can be manipulated by alloying additions and by thermal and mechanical processing. Thus, there are alpha alloys, alpha + beta alloys, and beta alloys.

Many grades of commercially pure titanium are recognized. Specifications set maximum levels for carbon, iron, nitrogen, oxygen, and hydrogen. In a few grades a small amount of molybdenum and nickel, or palladium are intentionally added. Most of these grades are utilized for their corrosion resistance.

Alpha alloys contain alpha-stabilizing elements such as aluminum. They are higher in strength but lower in corrosion resistance than commercially pure titanium. They cannot be further strengthened by heat treatment. Beta alloys contain elements like vanadium, chromium, iron, molybdenum, and niobium which lower the temperature of the alpha to beta transformation. They show better fracture toughness than alpha alloys at the same strength level. They also show good forgeability and hardenability. Alpha+beta alloys contain both alpha and beta stabilizers. The properties of these alloys can be adjusted over a wide range by employing heat treatments and hot working.

Nomenclature for titanium alloys is largely self-evident. Thus, Ti-5Al-2.5Sn contains 5% aluminum and 2.5% tin. Alpha (and so-called "near alpha") alloys may contain 2–8% aluminum, 2–12% tin, 1–6% zirconium, 0.4–2% molybdenum. Vanadium, niobium, tantalum, and silicon may be present in small amounts. Some have added copper or palladium. The alpha + beta alloys also contain 2–8% aluminum, but lower levels of tin and zirconium. Some may have up to 6% molybdenum, and most will have between 2–6% vanadium. Beta alloys contain 2–6% aluminum and 9–15% vanadium. Some will have as much as 12% chromium or 13% molybdenum. Tin, zirconium, and iron are also added to certain grades [46–48].

Titanium and its alloys dissolve in dilute sulfuric acid forming the dark blue titanous sulfate. A drop or two of nitric acid will discharge the color so that any undissolved material can be readily observed. Excess nitric acid will cause hydrous TiO_2 to precipitate, and thus should be avoided. Other useful acid dissolution media include hydrochloric acid, 1:1 hydrochloric acid:water with dropwise additions of hydrofluoric acid, and 1:6:10 nitric acid:hydrofluoric acid: water [9]. Hydrofluoric acid tends to react vigorously with titanium and its alloys and is best moderated by adding water to the test portion before the addition of HF.

The use of at least a small amount of hydrofluoric acid will tend to prevent hydrolysis, but it necessitates plastic labware and may interfere in some procedures. High molybdenum, chromium, or vanadium alloys may require some hydrochloric acid to complete the dissolution in a reasonable amount of time. It can also be noted here that some laboratories prepare the surface of solid test pieces for oxygen analysis by briefly pickling them in hydrofluoric acid to which some 30% hydrogen peroxide has been added. The piece is then rinsed in distilled water, and then in a low residue solvent.

3.6.7 Zirconium and Hafnium Alloys

These two metals owe much of their commercial development to the nuclear industry. Zirconium is transparent to thermal neutrons, and thus finds use as nuclear fuel cladding. Conversely hafnium has a high-capture cross section for thermal neutrons and at one time was heavily used in reactor control rods. Both beryllium and magnesium have an even lower thermal neutron cross section than zirconium but neither material can approach its corrosion resistance and strength. Thus, with a density similar to steel, and other properties similar to titanium, zirconium and its alloys remain the material of choice for some nuclear reactor designs. It also finds use in the chemical industry, where its corrosion resistance and strength are needed. Both zirconium and hafnium are utilized as alloying additions in diverse metals—nickel-base superalloys, titanium alloys, copper alloys, aluminum alloys, and magnesium alloys.

Like titanium, pure zirconium exhibits an hcp alpha phase at room temperature that converts to a bcc beta phase at 870° C. Unalloyed reactor grade zirconium is carefully checked for oxygen and other impurities that have a large effect on the phase transition. Some alloys, like Zircaloy-2, contain small additions of tin, iron, chromium, and nickel. Others may contain niobium. Because hafnium occurs in nature with zirconium, and is chemically nearly identical, a maximum hafnium level is often specified for zirconium alloys. Hafnium is sold primarily as the pure metal [49].

Zirconium, zirconium alloys, and hafnium will dissolve in the same acids and acid combinations as titanium. As with titanium alloys, the reaction with hydrofluoric/nitric acid mixtures is very vigorous and is best moderated by adding some water first, then the hydrofluoric acid, then nitric acid cautiously dropwise. Dilute hydrofluoric acid alone (15 mL water and 2 mL hydrofluoric acid) will dissolve a 1 g test portion. If hydrofluoric acid is to be avoided, 1:1 sulfuric acid:water or concentrated sulfuric acid with cautious dropwise additions of 30% hydrogen peroxide will do a slower but adequate job. Both solutions should be eventually heated to strong fumes of SO_3 to ensure complete dissolution.

If an acid fusion is employed, fused sodium bisulfate (with a small amount of sulfuric acid) is a better choice than potassium pyrosulfate since the sodium double salts that form are more soluble than the potassium double salts [50].

3.7 Precious Metals

These consist of silver, gold, and the six platinum-group metals (platinum, palladium, osmium, iridium, ruthenium, and rhodium). Silver and gold have important industrial uses, but a significant portion of their demand is for jewelry, coinage, and decorative arts. The platinum group and their alloys, on the other hand, are primarily industrial materials with uses as catalysts, thermocouples, glow plugs, resistance wire, crucibles, spinnerettes for the textile industry, and other applications [51].

There are many different alloys of the precious metals, many of which are highly resistant to common dissolution procedures. The entire subject of preparing these materials for analysis is somewhat of a specialized art, but a few general suggestions can be made here.

3.7.1 Silver Alloys

Commercially pure silver (including various grades of "fineness" at 99.9% Ag and higher) is used for photographic emulsions, electrical contacts, batteries, mirrors, and catalysts. Sterling silver must be at least 92.5% silver; the balance is generally copper. Coinage silver is 90% silver/10% copper. Silver brazing filler metals may contain 24–73% silver, 14–47% copper, and up to

30% zinc. Some contain major amounts of cadmium or minor amounts of tin, nickel, manganese, or lithium. An alloy of silver with 0.25% magnesium (max.) and 0.25% nickel, which is hardened by heating in an oxygen atmosphere, finds applications in relays and sliding electrical contacts. Silver dental amalgam is formed by grinding silver–tin alloy powder (65Ag min. and 29Sn/6Cu/3Hg/2Zn—all max.) with liquid mercury. The product typically contains 40–50% mercury, 20–35% silver, 12–15% tin, 2–15% copper, and <1% zinc [52].

Silver and its alloys all dissolve readily in nitric acid; often 1:1 nitric acid:water is used. The "parting" of fire assay dore beads is commonly performed with 1:4 nitric acid:water, although the use of hot concentrated sulfuric acid prevents the dissolution of platinum that occurs when nitric acid is used to dissolve out the silver. Palladium dissolves in either case [53].

3.7.2 Gold Alloys

Commercial fine gold typically ranges above 99.95% gold (it is called "proof gold" above 99.99% Au). Lower purity gold (down to 89.9% Au), including coin gold containing only copper as a hardener, is also sometimes traded as "commercial gold." High purity gold is used for surface decoration on glass and china, as a deposit on glass to produce light filters, as electroplated surfaces in high voltage, microwave, and semiconductor components and circuitry, as a freezing point standard, and as both high melting and low melting solders, as well as in dental applications. Gold/silver/copper alloys are jewelry alloys whose properties are determined largely by the silver to copper ratio. Thus, low-karat alloys can be converted to high-karat alloys of similar properties by adding pure gold. Gold/nickel/copper/zinc alloys are the jewelry "white golds." A modern 14 karat version is 58.33% gold/12.21% nickel/23.47% copper/5.99% zinc. A gold-platinum alloy (70Au/30Pt) is used for the spinnerettes from which rayon fibers are extruded. Gold-base brazing filler metals may contain from 30–80% gold alloyed with copper, palladium, or nickel and sometimes with both palladium and nickel. They are used for critical applications, such as in brazing jet engine components [51,52].

Aqua regia was so named for its ability to dissolve gold, at least in hot, concentrated form. Gold also dissolves by a number of other strategies involving free chlorine gas, especially with molten sodium chloride in an open tube furnace or a sealed glass ampoule. However, a simpler but slower approach is to add 30% hydrogen peroxide to a finely divided gold sample in concentrated hydrochloric acid. Potassium hydroxide fusions (in graphite crucibles, for example) will dissolve gold alloys, and the leached and acidified melt produces a clear solution. Sodium hydroxide, on the other hand, yields a melt which is difficult to dissolve. Unlike the six platinum group metals gold shows no volatility in an oxygen atmosphere at high temperatures. However, precipitated gold may gain as much as 1% in weight by heating in oxygen due to adsorbed oxygen [53].

3.7.3 Platinum Alloys

Most commercially important platinum group alloys are either platinum or palladium based. The platinum-base alloys represent the larger market at present. Commercially pure platinum is at least 99.95% Pt. It is used for thermocouples, for the production of certain chemotherapy drugs used in the treatment of cancer, as a catalyst for the production of sulfuric and nitric acids, vitamins, and high-octane gasoline.

Various other platinum group metals are added as alloying additions, primarily to harden pure platinum, but also to alter its electrical properties, or to add to its already formidable corrosion resistance. Such alloys are used in cardiac pacemakers, laboratory ware, in the production of high purity optical glass and synthetic fibers, and for jewelry. Platinum-palladium alloys are used for jewelry in Europe (but not in the U.S.), for electrical contacts, and for anodes used in marine applications. In most media the corrosion resistance of platinum is not diminished up to about 25% palladium. Platinum-iridium alloys show great room temperature dimensional stability and permanence. Thus, a 10% iridium/90% platinum alloy is used for the international standards of length and mass. At elevated temperatures, however, volatile iridium oxide is lost. In contrast rhodium is not volatilized at high temperature from platinum-rhodium alloys. Thus, a 10% rhodium/90% platinum alloy (versus pure platinum) thermocouple has been used to define the International Practical Temperature Scale. Other alloys for high temperature applications contain up to 40% rhodium.

Platinum–ruthenium alloys are used for electrical contacts, hypodermic needles, jewelry, laboratory electrode stems, and aircraft spark plugs. Most compositions range between 5–14% ruthenium. A special composition (known as Alloy No. 851) is 79% platinum/15% rhodium/6% ruthenium. It shows high hardness, strength, corrosion resistance, and weldability. It is used for wire-wound potentiometers, bridgewires, catalytic glow plugs, and other specialized applications. Platinum–tungsten alloys (2–8% W) are currently used for potentiometer wire, bridgewire, and heating elements, as well as in biomedical applications. Platinum–nickel alloys (1–20% Ni) are used in electrical and electronic applications where use is made of their extremely low hysteresis. Platinum–cobalt permanent magnet alloys (23.3% Co) show an extremely high coercive force. Their cost is justified where the length to thickness ratio of other magnet alloys is a problem: hearing aids, magnetic phonograph cartridges, rotors in miniature motors, etc. [51,52].

The platinum alloys present many dissolution problems. The effectiveness of an acid attack is greatly influenced by the presence of alloying elements and impurities, as well as by the thermomechanical processing that the alloy has experienced. Most platinum alloys are dissolvable in *aqua regia*, although large solid pieces are attacked quite slowly. Some workers evaporate the *aqua regia* to dryness, then add 1:1 hydrochloric acid:water and evaporate to dryness repeatedly until a clear solution is obtained. Finely divided platinum alloy combined with ten times its weight of sodium chloride in a porcelain boat can be heated in a tube furnace in a flowing chlorine gas atmosphere. At 675°C, for example, a platinum-iridium alloy treated in this manner will form a product which is soluble in *aqua regia*, although some platinum will have been volatilized.

Fusions of very corrosion resistant alloys with zinc metal in a quartz crucible is one approach sometimes used. The melt is stirred cautiously from time to time with a graphite rod. Upon cooling, the zinc melt is dissolved in hydrochloric acid leaving the platinum group metals as a finely divided insoluble residue. These can be filtered off, but they must be ignited cautiously at low temperature since they are often dangerously pyrophoric (and platinum group oxides are volatile). They can sometimes then be fused with sodium bisulfate. Another method involves very cautiously heating a 10 mg test portion of platinum alloy with 5 g of potassium pyrosulfate and 0.5 g of potassium chloride in a sealed tube at 520°C for 2 hours. Platinum alloys can also be dissolved anodically, using alternating current superimposed on the direct current [9,53].

3.7.4 Palladium Alloys

Commercially pure palladium is 99.85% Pd. It is similar to platinum in ductility, but somewhat lower in melting point and corrosion resistance. It is also less expensive per unit weight. It is used in light-duty electrical relays and as a catalyst to remove oxygen from gas streams and for the hydrogenation of organics. Sulfur-free hydrogen will diffuse through palladium and emerge highly purified. With ruthenium it forms a white jewelry alloy. Palladium-base alloys are also employed in dentistry. Palladium-silver alloys are used for contacts and as a contact finish for connectors in the electronics industry, and for brazing heat-resisting alloys. 60Pd/40Cu alloy is used for electrical contacts; 40Pd/30Ag/30Cu is used in dentistry, as is 40Ag/30Pd/30Cu with the addition of indium or tin. 95.5% palladium/4.5% ruthenium is used in jewelry; higher ruthenium levels are used for electrical contacts [51,52].

Most palladium alloys dissolve more readily in *aqua regia* than do most platinum alloys. Anodic dissolution using alternating current or direct current with a repeated reversal of polarity has been employed. Palladium is slowly attacked by most mineral acids if heat is applied and oxidizing conditions prevail. For example, a mixture of nitric and sulfuric acid is effective. However, *aqua regia* remains the reagent of choice. Palladium can also be fused with sodium peroxide to yield a soluble salt [9,53].

3.8 Zinc Base Alloys

Its use as a coating for steel, both before and after the fabrication of products, is the largest use for zinc. Hot dip galvanizing, in which the coating is applied from a molten bath, traditionally has employed a 0.20% aluminum alloy. Recently, new coatings at 5% aluminum (Galfan™) and 55% aluminum (Galvalume™) have been introduced for improved corrosion protection. After-fabrication galvanizing utilizes a composition that contains up to 1% lead. Electrogalvanizing processes employ pure zinc, zinc–nickel, and zinc–iron systems. In the metallizing processes used for heavy coatings either pure zinc or a 15% aluminum alloy are vaporized by a high temperature flame or electric arc and propelled onto the steel surface by a high speed jet of gas.

Zinc base casting alloys fall into the hypoeutectic (<5% aluminum) or hypereutectic (>5%—up to 28% aluminum) categories. The latter are higher strength materials developed for gravity casting, while the former are traditional die casting alloys. Both alloy systems include a small amount of magnesium, and copper ranges from 0–3%. In both systems iron, lead, cadmium, and tin are harmful, and so maximum limits are set for these elements.

Zinc castings are used in myriad applications as parts and housings in the automotive, domestic appliance, and computer and business machine industries. Wrought products are largely produced from pure zinc or zinc with minor alloying additions; these include rolled sheet and drawn wire. Superplastic zinc (21–23% aluminum/0.4–0.6% copper) shows an extreme formability, such as that of molten glass or plastics. Two forging alloys are in general use: 14.5Al/0.75Cu/0.02Mg/Bal.Zn and 1.0Cu/0.1Ti/Bal.Zn [54].

All zinc alloys dissolve easily, but care must be taken to select a mineral acid that does not result in the loss of volatile or precipitated analytes.

3.9 Tin Base Alloys

As with zinc, the largest use of tin is in coating steel; 90% of this tinplate is used for tin cans for holding food, beverages, and other products. Most tinplate is produced electrolytically using either a basic or acidic electrolyte. Binary coatings (tin–zinc, tin–cadmium, tin–copper, tin–nickel, tin–lead, and tin–cobalt) and ternary coatings (copper–tin–lead and copper–tin–zinc) are also electroplated for specialized applications. Hot dip tin coatings are used on food handling equipment and on wire. "Terne plate" is a lead–tin steel coating also produced by a hot dip process.

Commercially pure tin is 99.8% Sn or more. It is used for soldering the side seam of tin cans. Maximum residual levels of antimony, arsenic, bismuth, cadmium, cobalt, copper, iron, nickel, lead, sulfur, and zinc are generally specified. A 92% tin/8% zinc alloy is used as foil (e.g., for wine bottle capsules), while a 99.6% tin/0.4% copper alloy is used for collapsible tubes for medicinal and cosmetic salves and artist's paints. "White metal" is a 92% tin/8% antimony alloy used for costume jewelry. Tin is an important component of solders since it wets many metals far below their melting points. High tin solders (60–95% tin) are heavily used in electronics applications due to their superior electrical conductivity. Antimony and silver are alloyed to 95% tin solders, while lead makes up the balance of the 60–70% tin solders. General purpose solders are 50Sn/50Pb and 40Sn/60Pb. Tin is added to many other solder compositions as well.

Modern pewter is a tin-base alloy containing 1–8% antimony and 0.25–3.0% copper; it is used for decorative housewares and costume jewelry. Tin-base babbit alloys are used for bearing applications; they contain 84–91% tin, 4.5–8% antimony, and 3–8% copper. Maximum levels of aluminum, arsenic, bismuth, iron, lead, and zinc are specified. Tin is an important component of type metals (where it is alloyed with lead and antimony) and of the tin-lead alloys traditionally used for organ pipes (where the tin content is related to the "brightness" of the tone). Many fusible alloys used in fire sprinkler systems and other automatic fire control devices are tin-base compositions with lead, cadmium, zinc, silver, bismuth, antimony, indium, or other metals [55].

Tin-based alloys dissolve in concentrated sulfuric acid, but not in dilute sulfuric acid. A very common approach is to weigh the test portion into a 125-mL quartz flask, add sulfuric acid and potassium pyrosulfate, and heat slowly over a Meker burner. It is important that the tin alloy not melt before it is completely dissolved since it may partially "alloy" with the quartz. 1.5 g of tin can be dissolved in 20 mL of sulfuric acid and 5 g of potassium pyrosulfate. Some approaches specify dissolution in a 1:9 bromine:hydrobromic acid mixture. Low heat is used to prevent the loss of bromine. Once dissolved and cooled, perchloric acid can be added and the solution heated to fumes of $HClO_4$ to expel antimony, arsenic, selenium, and tin as the volatile bromides, and to decompose any $PbBr_2$ precipitate which may have formed. A 1 g sample of tin alloy can be dissolved at room temperature in 50 mL of 3:2:5 nitric acid:fluoboric acid:water using a Teflon beaker in an ultrasonic bath. The solution must be diluted with the same mixture to prevent the precipitation of metastannic acid [9,56,57].

3.10 Lead Base Alloys

Lead metal itself is sold under some unique names derived from present and past uses: "corroding lead" (or pure lead) is at least 99.94% Pb.

"Common lead" is also at least 99.94% Pb but contains higher levels of silver and bismuth. "Chemical lead" is at least 99.90% Pb and contains a residual level of copper and silver retained from the ore. "Acid copper lead" is, likewise, 99.90% Pb, but it is produced from highly refined lead by adding copper.

Lead-antimony alloys range from 0.5–15% antimony with some containing minor to major additions of tin, arsenic, cadmium, copper, and other elements. These materials are used for battery grids, type metal, pipe, sheet, bearings, and ammunition. Lead-calcium alloys range from 0.03–6% calcium and may contain tin, copper, silver, aluminum, or other additives. They are used for battery grids and bearings, among other applications. Lead-tin alloys include an important group of low and intermediate temperature solders; they range from 2 to 50% tin and may contain additions of antimony, arsenic, cadmium, or copper.

"Terne-coated" steel sheet derives its name from the French word for dull or tarnished (*terne*), which contrasts with the bright appearance of tinplate. The hot dip alloy is a lead-base with 3–15% tin. Terne-coated steel can be readily soldered and offers excellent corrosion protection unless the coating is breached. It is used to manufacture automobile gasoline tanks among other applications.

Lead-base babbitt alloys, used for bearing applications, contain 1–10% tin, and most contain major amounts of antimony and minor amounts of arsenic and copper. There are also lead-base babbitts that contain controlled amounts of calcium or other alkaline earth elements [58].

Lead-base alloys dissolve in nitric acid. Sometimes perchloric acid is added, the sample is heated to fumes of $HClO_4$, and then cooled and diluted. A small amount of hydrochloric acid is then added to precipitate $PbCl_2$ for its removal from solution. Sometimes a lead alloy is dissolved in dilute nitric acid in the presence of tartaric acid to complex analytes such as antimony, bismuth, and tin, preventing their precipitation. Lead also dissolves in concentrated sulfuric acid, but not in dilute sulfuric acid. Mixtures of hydrobromic acid and bromine or hydrochloric acid and bromine at 9:1 or 10:1 ratios are sometimes used. If such solutions are then fumed in perchloric acid all of the antimony, arsenic, selenium, and tin will be evolved out of the solution. If lead alloys are dissolved in 1:1 hydrobromic acid:water, however, $PbBr_2$ will precipitate.

Lead-base alloys have also been dissolved in mixtures of fluoboric acid and hydrogen peroxide, and in mixtures of glacial acetic acid and hydrogen peroxide. Other combinations that have been used are: nitric acid and fluoboric acid, nitric acid with sodium fluoborate, and perchloric acid and phosphoric acid [9].

3.11 Other Metals and Alloys

Obviously, there are many other types of metal commodities. We can only briefly touch upon a few more categories. There are the *permanent magnet alloys* that seem to belong in a category by themselves. These include the Alnicos™—alloys of iron, aluminum, nickel, cobalt, and sometimes copper or titanium. Many of these are by strict definition binary-base materials with nearly equal amounts of iron and cobalt. This category also includes samarium–cobalt, neodymium–iron-boron, platinum–cobalt, and steel magnet materials, as well as ceramic materials such as the barium and strontium ferrites [59]. All of these materials except the platinum-cobalt composition dissolve easily in acids. The platinum–cobalt alloy will yield with time to *aqua regia*.

The *rare earth metals* are most often defined as consisting of scandium, yttrium, and the lanthanide series [lanthanum, cerium, praseodymium, neodymium, promethium (which has no stable isotopes), samarium, europium, gadolinium, terbium, dysprosium, holmium, erbium, thulium, ytterbium, and lutetium]—17 in all. Misch metal is a mixture of 50% cerium, 30% lanthanum, 15% neodymium, and 5% praseodymium which is obtained from the reduction of monazite or bastnasite ore without separation of the rare earth components from each other. It is used widely as a metallurgical additive—to steels, ductile irons, magnesium, aluminum, copper, and zinc alloys, and as a 50–75% alloy with iron in lighter flints.

Purified individual rare earths are also important alloy additives. Cerium, lanthanum, or yttrium are added to cast irons, and to nickel-, cobalt-, and iron-base high temperature alloys. Oxide dispersion strengthened nickel-base superalloys utilize either Y_2O_3 or La_2O_3. Erbium and yttrium as either metals or oxides have been added to titanium alloys in major amounts. In addition to the permanent magnet applications mentioned above intermetallic rare earth compounds with iron, nickel, and cobalt have been used for many exotic applications, such as magnetostrictive materials, as magnetic refrigerants, and as magnetooptic disks [60].

The rare earths are rarely encountered by the analyst as metals. They react readily with air to form oxides that dissolve easily in dilute hydrochloric acid. Hydrofluoric acid is to be avoided since they all form insoluble fluorides.

Uranium as it occurs in nature contains 0.7% fissionable ^{235}U. Once this has been largely removed for nuclear applications the depleted metal can be employed for industrial purposes (although it typically still retains 0.2% of the fissionable isotope). The nonnuclear uses of uranium make use of its great density (19.1g/cm^3) as counterweights, radiation shielding, and armor-piercing projectiles. Depleted uranium is employed unalloyed and also alloyed with titanium, niobium, or molybdenum. More general use is limited by the hazards posed by radioactivity, toxicity, and pyrophoricity [61].

Mixtures of hydrochloric, hydrofluoric, and nitric acids have been used to dissolve some uranium alloys. Anodic dissolution at a potential of 6V in 6M sulfuric acid has also been employed. Dissolution in sodium hydroxide solution with the addition of hydrogen peroxide has also been reported [9].

Beryllium is an extremely light, high stiffness metal, transparent to high energy electromagnetic radiation. Most beryllium is produced by a powder metallurgy process. Commercial grades of beryllium range from 94–99% Be with controlled amounts of BeO, which controls grain size during powder consolidation. Beryllium is used in high precision instruments, such as inertial guidance systems, in optical components where precise dimensional stability is required, and also as X-ray windows. Beryllium is highly toxic. Acute and chronic lung disease caused by respired particles of beryllium is an extreme hazard [62].

Beryllium dissolves easily in dilute hydrochloric, sulfuric, or nitric acids. It is an amphiprotic element, also dissolving in concentrated sodium hydroxide solutions. The BeO content can be determined by isolating it after dissolving the metal in bromine and methanol [9].

Bismuth and indium are both used in low melting solders and in fusible alloys. In addition bismuth is sometimes used as an additive to aluminum alloys and to steel to improve machinability. Indium is used as gaskets and seals and as compounds in electrical and electronic applications [63].

Both dissolve easily in mineral acid. Bismuth is usually dissolved in nitric acid or sulfuric acid, and then often fumed in sulfuric acid. Sometimes, however, tartaric acid is added to dilute nitric acid and the fuming step is dispensed with. Hydrochloric acid is not used due to the precipitation of BiOCl. Indium, on the other hand, can be dissolved in hydrochloric acid, hydrobromic acid, or dilute nitric acid [9].

Germanium is not a metal but a semimetal, like silicon. It is included here because it resists many common acid dissolutions. It ushered in the age of the transistor, but today its principal use is in infrared optics. Other applications include fiber optics, gamma ray detectors, and catalysis. The metal reacts with hydrofluoric acid/nitric acid mixtures, but the base element is partially volatilized. The oxide and raw materials may be fused in sodium carbonate to which a little sodium peroxide has been added [64,9].

Gallium in metallic form has limited use, principally in high temperature thermometers. As semiconducting compounds, however, it is used for integrated circuits and optoelectronic devices. The metal dissolves in dilute *aqua regia* or dilute hydrochloric acid. The oxide dissolves in concentrated hydrochloric acid [65,9].

Chromium metal or "electrolytic chromium" is hard and brittle. In addition to its use as an alloy additive chromium is electrodeposited as a hard, corrosion-resistant, shiny layer on steel and other metals. "Chromizing" is the high temperature surface diffusion of chromium into an alloy to produce a resistant surface layer. Chromium metal is soluble in 1:1 sulfuric acid:water and in hydrochloric acid. It is passivated by nitric acid.

Mercury forms amalgams with many metals, but not with platinum or iron. It represents a health hazard due to its volatility and toxicity. It dissolves readily in dilute nitric acid and in hot concentrated sulfuric acid.

Silicon "metal" is commonly seen as a gray crystalline solid. It reacts at high temperatures with oxygen to form SiO_2, with nitrogen to form Si_3N_4, and with chlorine to form volatile $SiCl_4$ (which is a liquid at room temperature). Silicon metal can be dissolved by adding 1:1 sulfuric acid:water, hydrofluoric acid, and nitric acid. A 1 g test portion should be started in a Teflon beaker with 1 mL 1:1 sulfuric acid:water and 20 mL HF; then nitric acid is added dropwise until 3 mL have been added. The sample is cautiously heated to fumes of SO_3, cooled to room temperature, then 5 mL of 1:1 sulfuric acid:water, 1 mL of hydrofluoric acid, and 5 mL of nitric acid are added, and the sample is again heated to strong fumes of SO_3.

Silicomanganese, an important steel-making charge material, is dangerously reactive with sodium peroxide, and so the best course to its dissolution is to use an acid approach. A 1 g test portion will dissolve in 30 mL of nitric acid and 2 mL of hydrofluoric acid in a Teflon beaker.

Cadmium metal is silverish and ductile but becomes brittle when heated to 80°C. It is a major component of several fusible alloys, including Wood's metal (12.5Cd/50Bi/25Pb/12.5Sn). Cadmium dissolves in 1:1 nitric acid:water, hydrobromic acid/bromine mixtures, and also in dilute sulfuric acid/hydrogen peroxide (with the aid of a small piece of platinum as a catalyst).

Selenium occurs in many different allotropic forms, a red and a gray form being most commonly observed. It is commonly dissolved in concentrated nitric acid. *Tellurium* has two allotropes—a gray and a black form. It is commonly dissolved in *aqua regia* [9,66].

REFERENCES

NOTE: In this chapter extensive information was also obtained from entries on the various alloy systems in *Van Nostrand's Scientific Encyclopedia*, 4th ed., D. Van Nostrand, Princeton, NJ, 1968. Also consulted were: *Woldman's Engineering Alloys*, 7th ed., (J.P. Frick, ed.), ASM International, Materials Park, OH, 1990; Parr Instrument Co. literature (Moline, IL); *Metal Selector* from *Steel*, Oct. 7, 1968; and *Annual Book of ASTM Standards*, Vols. 03.05 and 03.06; American Society for Testing and Materials, West Conshohocken, PA, 1996.

1. Dolezal. J.: Lenz. J.: Suleck. Z. Analytica Chimica Acta. 47. 517 (1969).
2. Fernando. L.A. Analytical Chemistry. 56. 1970 (1984).
3. Zief. M.: Mitchell. J.W. Contamination Control in Trace Element Analysis. John Wiley. New York. 1976. pp. 166–167.
4. Mathes. S.A.: Farrell. R.F.: Mackie. A.J. Technical Progress Report TPR120. U.S. Bureau of Mines. 1983.
5. Fernando. L.A.: Heavner. W.D.: Gabrielli. C.C. Analytical Chemistry. 58. 511 (1986).
6. Kingston. H.M.: Jassie. L.B. (eds.) Introduction to Microwave Sample Preparation. American Chemical Society. Washington. DC. 1988.
7. Hillebrand. W.F.: Lundell. G.E.F.: Hoffman. J.I.: Bright. H.A. Applied Inorganic Analysis. 2nd ed.. John Wiley. New York. 1953.
8. Lundell. G.E.F.: Hoffman. J.I. Outlines of Methods of Chemical Analysis. John Wiley. New York. 1938.
9. Bock. R. A Handbook of Decomposition Methods in Analytical Chemistry. John Wiley. New York. 1979.
10. Dolezal. J.: Povondra. P.: Sulcek. Z. Decomposition Techniques in Inorganic Analysis. Elsevier. New York. 1966.
11. Sulcek. Z.: Povondra. P. Methods of Decomposition in Inorganic Analysis. CRC Press. Boca Raton. FL. 1989.
12. The Making. Shaping. and Treating of Steel. 9th ed.. USX Corp.. Pittsburgh. 1971.
13. Lundell. G.E.F.: Hoffman. J.I.: Bright. H.A. Chemical Analysis of Iron and Steel. John Wiley. New York. 1931.
14. Sampling and Analysis of Carbon and Alloy Steels. USX Corp.. Pittsburgh. 1938.
15. Standard Methods of Analysis of Iron Steel. and Ferroalloys. 4th ed. (United Steel Companies. Ltd.: Sheffield. UK): Lund. Humphries & Co.. London. 1951.
16. "Wrought Stainless Steels" (revised by S.O. Washko and G. Aggen) Metals Handbook. 10th ed.. Vol I. ASM International. Materials Park. OH. 1990. pp. 841–908.
17. "Cast Stainless Steels" (revised by M. Blair) Metals Handbook. 10th ed.. Vol I. ASM International. Materials Park. OH. 1990. pp.908–930.
18. "Elevated Temperature Properties of Stainless Steels" Metals Handbook. 1oth ed.. Vol I. ASM International. Materials Park. OH. 1990. pp. 930–949.
19. Stoloff. N.S. "Wrought and P/M Superalloys" Metals Handbook. 10th ed.. Vol. I. ASM International. Materials Park. OH. 1990. pp. 950–976.
20. Erickson. G.L. "Polycrystalline Cast Superalloys" Metals Handbook. 10th ed.. Vol.I. ASM International. Materials Park. OH. 1990. pp. 981–994.
21. Harris. K.. Erickson. G.L.. Schwer. R.E. "Directionally Solidified and Single Crystal Superalloys" Metals Handbook. 10th ed.. Vol.I. ASM International. Materials Park. OH. 1990. pp. 995–1006.
22. Mankins. W.L.: Lamb. S. "Nickel and Nickel Alloys" Metals Handbook. 10th ed.. Vol.II. ASM International. Materials Park. OH. 1990. pp. 428–445.
23. Dawson. R.J.: Foley. E.M. "P/M Cobalt-Base Wear Resistant Materials" Metals Handbook. 10th ed.. Vol.I. ASM International. Materials Park. OH. 1990. pp. 977–980.

24. Crook, P. "Cobalt and Cobalt Alloys" Metals Handbook, 10th ed., Vol.II, ASM International, Materials Park, OH, 1990, pp. 446–454.
25. Tyler, D.E. "Introduction to Copper and Copper Alloys" Metals Handbook, 10th ed., Vol.II, ASM International, Materials Park, OH, 1990, pp. 216–240.
26. "Wrought Copper and Copper Alloy Products" (revised by D.E. Tyler) Metals Handbook, 10th ed., Vol.II, ASM International, Materials Park, OH, 1990, pp. 241-264.
27. "Selection and Application of Copper Alloy Castings" (revised by R.F. Schmidt and D.G. Schmidt) Metals Handbook, 10th ed., Vol.II, ASM International, Materials Park, OH, 1990, pp. 346–355.
28. Harkness, J.C.; Spiegelberg, W.D.; Cribb, W.R. "Beryllium-Copper and Other Beryllium Containing Alloys" Metals Handbook, 10th ed., Vol.II, ASM International, Materials Park, OH, 1990, pp.403–427.
29. Elwell, W.T.; Scholes, I.R. Analysis of Copper and Its Alloys, Pergamon, New York, 1967.
30. Rooy, E.L. "Introduction to Aluminum and Aluminum Alloys" Metals Handbook, 10th ed., Vol.II, ASM International, Materials Park, OH, 1990, pp. 3–14.
31. Cayless, R.B.C. "Alloy and Temper Designation Systems for Aluminum and Aluminum Alloys" Metals Handbook, 10th ed., Vol.II, ASM International, Materials Park, OH, 1990, pp.15–28.
32. Bray, J.W. "Aluminum Mill and Engineered Wrought Products" Metals Handbook, 10th ed., Vol.II, ASM International, Materials Park, OH, 1990, pp. 29–61.
33. "Aluminum Foundry Products" (revised by A. Kearney and E.L. Rooy) Metals Handbook, 10th ed., Vol.II, ASM International, Materials Park, OH, 1990, pp. 123–151.
34. James, R.S. "Aluminum-Lithium Alloys" Metals Handbook, 10th ed., Vol.II, ASM International, Materials Park, OH, 1990, pp. 178–199.
35. Pickens, J.R. "High-Strength Aluminum P/M Alloys" Metals Handbook, 10th ed., Vol.II, ASM International, Materials Park, OH, 1990, pp. 220–215.
36. Churchill, H.V.; Bridges, R.W. Chemical Analysis of Aluminum, Aluminum Company of America, New Kensington, PA, 1935.
37. "Selection and Application of Magnesium and Magnesium Alloys" (revised by S. Housh, B. Mikucki, and A. Stevenson) Metals Handbook, 10th ed., Vol.II, ASM International, Materials Park, OH, 1990, pp. 455–479.
38. Lambert, J.B. "Refractory Metals and Alloys" Metals Handbook, 10th ed., Vol.II, ASM International, Materials Park, OH, 1990, pp. 557-585.
39. Johnson, W.A. "Molybdenum" Metals Handbook, 10th ed., Vol.II, ASM International, Materials Park, OH, 1990, pp. 574–577.
40. Busev, A.I. Analytical Chemistry of Molybdenum (translated by J. Schmorak), Ann Arbor-Humphrey, Ann Arbor, MI, 1969.
41. Johnson, W.A. "Tungsten" Metals Handbook, 10th ed., Vol.II, ASM International, Materials Park, OH, 1990, pp. 577-581.
42. Gerardi, S. "Niobium" Metals Handbook, 10th ed., Vol.II, ASM International, Materials Park, OH, 1990, pp. 565–571.
43. Elwell, W.T.; Wood, D.F. Analysis of the New Metals, Pergamon, New York, 1966.

44. Pokross, C. "Tantalum" Metals Handbook, 10th ed., Vol.II, ASM International, Materials Park, OH, 1990, pp. 571-574.
45. Grobstein, T.; Titran, R.; Stephens, J.R. "Rhenium" Metals Handbook, 10th ed., Vol.II, ASM International, Materials Park, OH, 1990, pp. 581-582.
46. Destefani, J.D. "Introduction to Titanium and Titanium Alloys" Metals Handbook, 10th ed., Vol.II, ASM International, Materials Park, OH, 1990, pp. 586–591.
47. Lampman, S. "Wrought Titanium and Titanium Alloys" Metals Handbook, 10th ed., Vol.II, ASM International, Materials Park, OH, 1990, pp. 592–633.
48. Eylon, D.; Newman, J.R.; Thorne, J.K. "Titanium and Titanium Alloy Castings" Metals Handbook, 10th ed., Vol.II, ASM International, Materials Park, OH, 1990, pp. 634–646.
49. Webster, R.T. "Zirconium and Hafnium" Metals Handbook, 10th ed., Vol.II, ASM International, Materials Park, OH, 1990, pp. 661–682.
50. Elinson, S.V.; Petrov, K.I. Analytical Chemistry of Zirconium and Hafnium (translated by N. Kaner), Ann Arbor-Humphrey, Ann Arbor, MI, 1969.
51. Robertson, A.R. "Precious Metals" Metals Handbook, 10th ed., Vol.II, ASM International, Materials Park, OH, 1990, pp. 699–719.
52. "Properties of Precious Metals" Metals Handbook, 10th ed., Vol.II, ASM International, Materials Park, OH, 1990, pp. 699–719.
53. Beamish, F.E. The Analytical Chemistry of the Noble Metals, Pergamon, New York, 1966.
54. Barnhurst, R.J. "Zinc and Zinc Alloys" Metals Handbook, 10th ed., Vol.II, ASM International, Materials Park, OH, pp. 527-542.
55. "Tin and Tin Alloys" (revised by W.B. Hampshire) Metals Handbook, 10th ed., Vol.II, ASM International, Materials Park, OH, 1990, pp. 517-526.
56. Annual Book of ASTM Standards, Vol. 03.06, American Society for Testing and Materials, West Conshohocken, PA, 1996.
57. Hwang, J.Y.; Sandonato, L.M. Analytical Chemistry, 42, 744 (1970).
58. "Lead and Lead Alloys" (revised by A.W. Worcester and J.T. O'Reilly) Metals Handbook, 10th ed., Vol.II, ASM International, Materials Park, OH, 1990, pp. 543–556.
59. "Permanent Magnet Materials" (revised by J.W. Fiepke) Metals Handbook, 10th ed., Vol.II, ASM International, Materials Park, OH, 1990, pp. 782–803.
60. Gschneidner, K.A.; Beaudry, B.J.; Capellin, J. "Rare Earth Metals" Metals Handbook, 10th ed. Vol.II, ASM International, Materials Park, OH, 1990, pp. 720–732.
61. Eckelmeyer, K.H. "Uranium and Uranium Alloys" Metals Handbook, 10th ed., Vol.II, ASM International, Materials Park, OH, 1990, pp. 670–682.
62. Stonehouse, A.J.; Marder, J.M. "Beryllium" Metals Handbook, 10th ed., Vol.II, ASM International, Materials Park, OH, 1990, pp. 683–687.
63. Stevens, L. G.; White, C.E.T. "Indium and Bismuth" Metals Handbook, 10th ed., Vol.II, ASM International, Materials Park, OH, 1990, pp. 750–757.
64. Adams, J.H. "Germanium and Germanium Compounds" Metals Handbook, 10th ed., Vol.II, ASM International, Materials Park, OH, 1990, pp. 733–738.

65. Kramer, D.A. "Gallium and Gallium Compounds" Metals Handbook, 10th ed., Vol.II, ASM International, Materials Park, OH, 1990, pp. 739–749.
66. Annual Book of ASTM Standards, Vols. 03.05 and 03.06, American Society for Testing and Materials, West Conshohocken, PA, 1996.

3

Chemistry for Trace Analysis: Preconcentration, Separation, Masking, and Measurement

1. INTRODUCTION

Once the analyst has prepared an aqueous solution representative of the metal or alloy sample, several options present themselves, nearly all of them involving some form of chemistry. The simplest course is, of course, to proceed directly to the analyte measurement. And even here we confront chemistry. In UV/visible spectrophotometry a molecular complex is formed and its light absorbance is measured. In atomic absorption and emission, valence shell transitions are the signature of a physicochemical process. Diffusion-controlled redox reactions categorize voltammetric techniques. And even mass spectrometry is a merger of high temperature chemistry with physics.

But traditional aqueous chemistry may have a critical role to play between dissolution and measurement. It can amplify analyte sensitivities and extend detection limits, moderate or eliminate interferences from the sample matrix, and undo some deleterious effects caused by the dissolution process. There are two cautionary notes that the analyst must consider, however, before deciding how and when to play this "middle game." First, there is wedded to every sequence of manipulative steps the possibility of losing analyte or adding contamination. Second, the modern industrial analyst must balance the time these extra steps will cost against the improvements they will buy. In some cases the advantage will be marginal, the time excessive. In others the required result will be greatly improved with only a minor degree of effort. In still others the required result will be obtainable in no other way.

What are these intermediate steps between dissolution and measurement, and how are they employed? *Preconcentration* is a catchall concept that includes all techniques that increase the mass of analyte per unit volume of solution. It encompasses a wide range of processes from simple evaporation to the most sophisticated separation strategies. The term *separation* connotes a different, only partially overlapping, concept. Many separations concentrate the analyte but some may actually dilute it. The emphasis here is on either of two analytical goals: 1) the isolation of the analyte, either alone or as part of a select group of elements, from the remainder of the sample matrix; or 2) the removal of one or more interferences from the analyte. The analytical processes employed are highly diverse; they include precipitation, distillation, ion exchange, and solvent extraction, among others. They may be applied as a single intervention prior to the measurement step or as a long chain of interlinked processes comprising a complex separation scheme.

Finally, we use the term masking in this chapter's title in a general sense to denote all other chemical means of treating interferences. In this sense it includes the addition of complexing agents in UV/visible spectrophotometry, ionization suppressors in flame atomic absorption, and matrix modifiers in graphite furnace atomic absorption, among other techniques. All three paradigms—preconcentration, separation, and masking—can be effectively combined in an analytical scheme finely tuned to a specific need.

The chemistry involved in preconcentration, separation, masking, and measurement is usually complex and often imperfectly understood. As a result the working analyst sometimes ends up supplementing theory with lore and rigor with windage. In trace analysis, as in other areas of metals analysis, success here must be judged with the best performance-based criteria we can apply. Validation standards, such as certified reference materials, are one important mainstay of this continuing process. And the comparison of results from independent analytical methodologies is another.

2. PRECONCENTRATION

"Preconcentration," although a widely used term with a well-defined meaning, is actually a semantic redundancy. Certainly, one would not concentrate the analyte *after* its measurement. However, we appear to be stuck with a pointless prefix if we do not wish to be accused of creating neologisms.

Any process, physical or chemical, is covered by this category if it results in a solution more concentrated in the analyte than the starting solution. Sometimes all or part of the sample matrix will be concentrated along with the analyte. Sometimes components of the dissolution medium will be concentrated as well. In other cases preconcentration processes lead to very "clean" solutions that are rich in analyte.

2.1 Evaporation

This simplest approach to preconcentration cannot be applied indiscriminantly. Some analytes in certain oxidation states and/or in certain solution media will prove volatile, even below boiling water temperatures. For example, certain spectrophotometric methods for trace levels of boron employ chloride or fluoride media, and many versions of the silicomolybdate spectrophotometric method for trace silicon employ a dilute fluoride medium. Such solutions cannot be concentrated by evaporation because even the mildest heating in an open vessel will lead to serious analyte loss as a volatile species. There are numerous other examples. The analyst needs to remain aware of these stable but "fragile" analytical conditions when they occur.

Another problem concerns the total dissolved solids loading limits associated with certain trace measurement techniques. In concentrating the analyte by solvent evaporation one is usually also concentrating sample matrix components and dissolution medium components, and sometimes separation medium components as well. Molten salt fusion leachates are a particular problem in this regard. Certain precipitation separations also contribute significantly to the dissolved salt burden of the solution. These approaches severely limit the usefulness of simple evaporation to preconcentrate the analyte.

Sometimes there are severe hazards associated with an evaporative scheme that might be contemplated. For example, the filtrate from an ammonium hydroxide separation of a perchloric acid fumed sample must never be heated because explosive ammonium perchlorate salts will form. Sometimes there are unanticipated mechanisms for analyte loss. For example, if the analyte has been extracted into a volatile organic solvent it will likely be lost as a volatile compound unless some acidified water is added to the solvent before it is heated. This entire approach is hazardous, however, since most light solvents are flammable. Gentle warming on a hot water bath is best. The analyst must be cautious with aqueous solution evaporations as well. Salts may crystallize, causing superheating of the solution and "bumping," with the potential for analyte loss.

Instrumental approaches that are particularly limited in the total dissolved solids content that they can tolerate notably include all those that employ a nebulizer to produce a fine spray of solution droplets. In flame atomic absorption spectrophotometry there is some danger of a flashback above 15 g/L, although instrument designs vary considerably in this regard. Many laboratories operate routinely and safely at a total dissolved solids content of 20 g/L. With direct current plasma (DCP) and inductively-coupled plasma (ICP) optical emission spectrometers 10 g/L is the normal recommended level, although molten salt fusion solutions have been aspirated at levels as high as 30 g/L without extinguishing the plasma. DCP-OES is somewhat more tolerant of high total solids content than ICP-OES. It is generally recognized that ICP-mass spectrometers are limited to 1 g/L of total dissolved solids. Here, it is not the torch tip or the nebulizer that is the limiting component but rather the cone orifices on the spectrometer interface. These tend to clog with deposited salts, although the process seems to be somewhat element-specific. Higher total dissolved solids content samples can be analyzed by ICP-MS using a flow injection analysis sample introduction approach.

Sometimes evaporative preconcentration is intrinsic to an analytical procedure. This true of all electrothermal atomic absorption techniques and whenever a graphite furnace is used in optical emission or mass spectrometry. These devices are used to create a discrete analytical pulse, as opposed to an continuous signal, by evaporating, charring, then atomizing a precisely metered volume of sample solution. A similar scheme is applied in some older optical emission procedures in which a prepared sample is evaporated drop-by-drop on an electrode, which is then arced or sparked in the spectrometer. For example, at one time rare earths were routinely analyzed by optical emission after evaporating the prepared solution on a copper electrode.

2.2 Preconcentration by Group Separation

It is difficult to discuss preconcentration without jumping ahead to the separation topics of Section 3 in this chapter. Here we must confine the discussion to the general role these techniques play in a preconcentration scheme and postpone details for Section 3. Group separations are extremely useful to remove a major amount of the sample matrix from the analyte. This, in turn, allows further evaporative concentration of the analyte.

A good example is the use of the mercury cathode (discussed in Section 3.2.5) which very effectively removes a host of elements from solution but leaves in solution aluminum, boron, hafnium, niobium, tantalum, titanium, tungsten, vanadium, and zirconium, as well as the alkali and alkaline earth metals, the rare earths, and the actinides. Using only this one group separation it is often possible to usefully extend the lower quantification limit for these analytes. For example, the lower limit for aluminum in steel by flame atomic absorption using a 1 g test portion and diluting to 100 mL (10g/L) is about 0.005%. The mercury cathode offers a simple alternative, however. A 1.5 g test portion is dissolved, fumed in perchloric acid, then cooled, diluted, and electrolyzed on the mercury cathode. The sample is then filtered and the filtrate is fumed to near dryness, then cooled and treated with 5 mL of hydrochloric acid and 2 mL of nitric acid. The solution is then diluted to the mark in a 50-mL volumetric flask and measured by flame atomic absorption. The lower quantification limit in this case has been reduced to 0.001%.

As with the use of the mercury cathode, there are other group separations that allow for the effective preconcentration of analyte without an excessive solids burden from the alloy matrix or the reagents. Ion exchange techniques can be employed in this way, as can solvent extraction schemes.

For example, a strongly basic anion exchange resin column can be used to simultaneously elute chromium, cobalt, copper, iron, nickel, and manganese, while retaining most other major components of high temperature alloys. Trioctylphosphine oxide (TOPO) in methylisobutyl ketone (MIBK) will simultaneously extract trace amounts of antimony, bismuth, lead, silver, thallium, and tin from an aqueous iodide solution. But, perhaps the most generally useful class of group separations are *precipitations*, and especially in the case of trace analysis, *coprecipitations*.

Precipitations are frequently used to remove major matrix elements, leaving trace analytes in solution. This proves feasible in many cases, although intuitive judgement might suggest that significant amounts of tramp analytes would be lost by occlusion. It is true, however, that certain precipitations, such as those which result in a hydrophillic colloidal precipitate are notoriously contaminated. But precipitations that produce hydrophobic colloids, such as silver chloride, show little tendency to occlude tramp elements. The purity of crystalline precipitates can be optimized by adjusting initial conditions to produce slow crystal growth. Precipitation from homogeneous solution (in which the reactant is formed in the solution) is a nearly ideal way to produce a pure crystalline precipitate.

The converse of this process is coprecipitation. In this case a carrier element is added (or occasionally it constitutes part of the sample matrix) which brings down a group of tramp elements with it as it precipitates. The mechanism of coprecipitation is only dimly understood but is generally agreed to involve surface adsorption, occlusion (entrainment of impurity ions during rapid crystal formation), and the formation of polyatomic-substituted crystals. A classic example of the use of coprecipitation to preconcentrate a group of trace elements is the collection of antimony, bismuth, thallium, and tin in a manganese dioxide precipitate [1].

In each of these examples it becomes possible to achieve an enrichment of a select group of analytes while reducing the burden of sample matrix and sometimes also reagent matrix ions. Often the group separation is followed by an evaporation step that might have been otherwise impractical due to limitations imposed by the presence of excessive total solids.

2.3 Preconcentration by Specific-Element Separation

Sometimes it is feasible to isolate the analyte completely from the alloy matrix components. Boron as the analyte presents three examples. Traces of boron can be distilled and collected as the methyl borate ester. In a different approach a mixed bed ion exchange resin column can be used to simultaneously remove matrix cations and acid digestion anions from boron as boric acid. In a third method boron is retained in solution on the mercury cathode while many typical alloy matrix components are deposited in the mercury. In each of these techniques a very clean separation of boron results in a dilute solution that can be readily concentrated by evaporation (provided that suitable precautions against the loss of volatile boron compounds have been taken).

Specific organic precipitants, such as dimethylglyoxime for nickel or palladium, p-bromomandelic acid (PBMA) for zirconium or hafnium, or α-benzoinoxime for molybdenum are also useful for preconcentration. In some cases a carrier element can be added. For example, hafnium can serve this role for zirconium (and vice versa) with PBMA. And the addition of a large excess of molybdenum makes α-benzoinoxime a useful collector for tungsten. The washed precipitate can sometimes be wet ashed with nitric and perchloric acids, or sometimes with nitric, perchloric, and sulfuric acids, then cooled and diluted to some appropriate volume. Alternatively, it is sometimes practical to ignite the precipitate to an oxide, then fuse it with molten salt.

Sometimes reduction to elemental form is employed. Selenium, tellurium, gold, and mercury are chemically reduced; copper and cobalt are electroplated. Each of these reduced elements can then be dissolved in a small volume of acid and appropriately diluted. In the case of selenium and tellurium, arsenic can be an effective carrier for trace amounts.

Some ion exchange systems are designed to be highly selective. The work of L.L. Lewis and W.A. Straub [2], for example, described a chloride system anion exchange procedure in which nickel, cobalt, and iron are separated by three separate dilutions of hydrochloric acid. The fluoride/chloride anion exchange procedure refined by workers at the National Institute of Standards and Technology (formerly the National Bureau of Standards) [3–6], and elaborated by S. Kallmann et al. [7,8] includes clean separations for molybdenum, niobium, tantalum, and rhenium.

Antimony, arsenic, germanium, osmium, selenium, and tin can be distilled as the halides, leading to the isolated element. Similarly, nitrogen can be steam distilled as ammonia from strong caustic solution and fluorine can be steam distilled as hydrofluosilicic acid ($H_2SiF_6 \cdot xH_2O$) from dilute perchloric acid solution.

There are comparatively few solvent extraction procedures that lead directly to an isolated analyte which can be conveniently further concentrated. Often the best approach is to return the analyte to an aqueous medium before evaporating. If volume reduction must occur in the organic phase, acidified water should be added before proceeding. Caution should always be observed when evaporating an organic liquid. Vapors are always noxious and often flammable. A hot water bath in a hood is essential here.

Some examples of specific-element extractions are the extraction of molybdenum (with iron) into diethyl ether from 6M hydrochloric acid solution; the extraction of zirconium (and hafnium) from 7M nitric acid with TOPO/

cyclohexane; and the extraction of nickel dimethylglyoximate from an ammoniacal citrate medium with chloroform.

2.4 Merits and Drawbacks of Preconcentration

We have already touched upon some of these but it seems useful to briefly review and expand this discussion. In trace analysis it is usually important to create the most concentrated solution of analyte feasible to achieve that generates no problems for the measurement process. We say "usually" because even in the trace realm, modern instrumentation sometimes has capabilities that exceed our needs. In ICP-MS work, for example, the signal generated by 1 ppm of analyte may exceed 1 million counts per second. Thus, it may sometimes be necessary to *dilute* trace samples rather than concentrate them.

More typically, however, we are seeking to extend the lower limit of quantification as far down the concentration scale as possible, and to maximize the instrument response to all low levels of analyte. A simple evaporation is a relatively "painless" way to achieve this but it may sometimes be more of a problem than a cure. Total solids loading may be high enough to clog nebulizers, extinguish plasmas, block sampling cones, or produce a flashback explosion. All of this results in costly downtime to correct what the sample has done to the equipment.

Group or element-specific separations will always eliminate some (sometimes all) of the alloy sample matrix. In certain instances they will eliminate some or all of the dissolution matrix as well. However, many of these separations add their own array of cations and anions to the sample solution which may adversely affect the measurement outcome.

The analyst must remain on guard for deleterious effects to the analyte during evaporative concentration. The potential for the volatilization of analyte is not always obvious. Sometimes the evaporation of an organic solvent bearing a metal-organic complex can lead to loss of the metal analyte (although the presence of water, and especially acidified water, often prevents such losses). Certain inorganic compounds, not normally considered volatile from aqueous solution, may become volatile as the highly concentrated, salt-laden solution superheats. Salting-out of the analyte as a partially soluble compound is sometimes a problem. Soluble fluorides and sulfates, for example, may salt-out during an evaporative preconcentration. And, of course, analyte contamination from room air and vessel walls are very real concerns during evaporation.

If the preconcentration scheme involves a series of separations the analyst must weigh the expected increase in sensitivity against any anticipated problems with an increased blank. In general the analyst should reflect on the analytical "balance sheet": "Is the background really cleaner, or have I substituted reagent and impurity ions for sample matrix ions?"

3. SEPARATION

This topic deserves the lion's share of space in this chapter, not only because it is a broad and involved subject, but also because it can be made to well serve the needs of trace analysis for metals. Unfortunately, but perhaps, understandably, separation techniques are no longer widely used in industry for trace work. The reasons are twofold: time and expertise. There is no denying that trace analyte separations require significantly more effort and analytical skill than loading an autosampler and typing some entries into a dedicated computer. Our argument here is that they are often *worth* the expensive time and training they require.

3.1 General Considerations

Clearly, if a goal *can* be achieved simply, it *should* be achieved simply. There are two aspects to this paradigm. Sometimes it is possible to *buy* a simple solution to an analytical problem. Suppose that laboratories A and B need to determine 0.5 ppm lutetium in a cobalt-base superalloy. Laboratory A can afford to buy an ICP-MS instrument which can do the job directly on a 0.1 g acid-dissolved test portion diluted to 100 mL. Laboratory B cannot justify such a purchase and must start with a 10 g test portion, separate the rare earth fraction by ion exchange and evaporate the eluent onto an electrode for conventional DC arc optical emission determination. Laboratory A invested a large amount of money in an instrument which may find wide use for other work (or may sit idle much of the time). Laboratory B invested in chemical analysis expertise which may be flexible enough to solve other tough problems for them.

The second aspect to this paradigm is the situation where it is *not* possible to buy a solution to an analytical problem. Laboratories A and B need to determine 50 ppb selenium in high purity nickel. Laboratory A is thwarted because its quadrupole ICP-MS cannot resolve the isobaric overlap of $^{62}Ni^{16}O^+$ on $^{78}Se^+$. Laboratory B builds on its separation experience by coprecipitating selenium with an added arsenic carrier and measuring the analyte by graphite furnace atomic absorption.

The lesson is not that expensive, sophisticated instrumentation lacks versatility but that separation science allows for even more. In reality both are needed—the instruments for speed and throughput, and the separations for solving really tough analytical problems.

In the sections that follow we will limit our discussions to practical problem solving, avoiding examples of separations for which a direct measurement approach is best. For example, the alkali metals are seldom separated since there are good, sensitive ways to measure them in their alloy matrix.

We will also forego any attempt at an exhaustive survey of all separation techniques that could be useful for the trace analysis of metal alloy systems. Such a project would undoubtedly prove futile and its attempt would occupy many more pages than we can devote here. What must serve are a series of illustrative examples of how the separation categories have been used in selected metal industries. It is hoped that these will encourage the trace analyst to begin thinking in terms of wet analytical chemistry, if not in the first, or even the second, attempt at solving a problem, but perhaps in the third. When the instrumental and mathematical answers fall short a chemical separation scheme may be the best answer.

3.2 Precipitation and Coprecipitation

In *Textbook of Quantitative Inorganic Analysis* [9], Kolthoff and Sandell refer to coprecipitation as the precipitation of elements that would remain in solution were it not for the presence of the carrier element. This is not the precise sense in which the term "coprecipitation" is used in trace metal analysis. Here we mean the (nearly) *complete* precipitation of trace elements that would not precipitate *completely* were it not for the presence of the carrier element [1]. This is an important distinction that helps us define the application of both precipitation and coprecipitation in trace analysis.

If we wish to remove a major concentration interferent, such as the nickel from an austenitic stainless steel or the tin from a bronze by *precipitation* then coprecipitation is an objectionable phenomenon (often referred to erroneously as the narrower term "occlusion"). If on the other hand, our goal is the collection of trace elements, we are most likely to select a precipitant that reacts with the trace analytes and with the carrier in the same manner. We may also select a carrier for its approximate match to the ionic radius of the analyte. By this definition of coprecipitation we are using every device at our disposal to favor the incorporation of all trace analyte in the massive carrier precipitate—by the formation of mixed-element crystals, by physical occlusion as crystals rapidly form in a supersaturated solution, and by adsorption of trace analyte on the surface of the carrier precipitate. We will first treat the subject of precipitations, then return to a discussion of coprecipitations.

3.2.1 Group Precipitations

This is an area of limited application in which we remove several major or minor components of the matrix at once while retaining the tramp analytes in solution. Many common group precipitants (such as ammonium hydroxide) are unsuitable for matrix removal in trace work because the precipitates they form occlude, adsorb, or form mixed crystals with the tramps.

Barium, lead, and strontium can be precipitated as sulfates with good reason to believe that most tramp elements will remain in solution. Precipitation is best conducted in a very weak acid [<1%(v/v) HCl] solution in which any iron present has been reduced by boiling with hydroxylamine hydrochloride. A solution of ammonium sulfate is slowly stirred in to the warm solution. The solution is then allowed to stand at low hotplate temperature for at least one hr, then at room temperature overnight. Filtration requires a double layer of fine porosity filter paper (such as Whatman No. 42) and filter pulp. The residue is washed with water.

Silver, bismuth, lead, mercury (I), and thallium (I) can be precipitated as chlorides (and oxychlorides) with many tramp elements (e.g., cobalt, iron, nickel, and zinc) being retained in the filtrate. Molybdenum and tungsten can be separated with α-benzoinoxime, and zirconium and hafnium can be separated with para-bromomandelic acid. In both cases trace elements such as aluminum, antimony, barium, bismuth, cadmium, calcium, chromium, cobalt, copper, iron, magnesium, manganese, nickel, silver, titanium, vanadium, and zinc are retained in the respective filtrates.

For the chloride separation it is best to precipitate with hydrochloric acid from dilute nitric acid solution. The silver chloride precipitate must be protected from photoreduction by the use of actinic glass (or Teflon TFE) vessels, or by allowing the precipitate to settle in the dark. In a related reaction bismuth and thallium iodide can be precipitated as BiI_3 and TlI while retaining certain elements in solution—cadmium, iron, and nickel, among others [11].

Tungsten accompanies molybdenum quantitatively in the α-benzoinoxime precipitate but will not precipitate quantitatively in the absence of molybdenum. Niobium, palladium, silicon, and tantalum may contaminate the precipitate but it should be free of many other low level and tramp elements. A 10% (v/v) sulfuric acid solution containing 50–150 mg of molybdenum is cooled to ice bath temperature. Then 5 mL of 2% (w/v) α-benzoinoxime in methanol is added with stirring for every 10 mg of molybdenum believed to be present. Then 5 mL of bromine water is added with stirring. Then after 1 minute 10 mL additional α-benzoinoxime solution is added and the solution is stirred until the precipitate's white color returns. The solution is allowed to stand in the ice bath for 10 min then filtered with suction using a fine porosity filter paper and pulp. The residue is washed with chilled 2% (v/v) sulfuric acid to which 2% (v/v) of the prepared α-benzoinoxime solution has been added. The filtrate will contain a large excess of reagent which will crystallize out of solution as it warms.

Zirconium and hafnium can be precipitated from strong hydrochloric acid solution by adding 0.5 g of para-bromomandelic acid to a solution that contains 20–100 mg of zirconium plus hafnium and heating in a boiling water bath for 20 min. The solution is then cooled, diluted to about 50% (v/v) hydrochloric acid with water and allowed to stand overnight. It is then filtered through a hardened low porosity filter (such as Whatman No. 542) with filter pulp and washed with water. Here, again, the filtrate contains excess reagent.

3.2.2 Specific Element Precipitations

In this category we are removing one major or minor constituent while retaining the trace elements in solution. Examples of reasonably clean separations include the reduction of mercury to the metal using 5% (v/v) formic acid (where traces of bismuth, cadmium, cobalt, copper, iron, magnesium, manganese, nickel, lead, thallium, and zinc are retained in solution) and the precipitation of basic bismuth nitrate $[Bi(OH)_x(NO_3)_y]$ at pH 4 (where aluminum, cadmium, copper, magnesium, manganese, nickel, lead, and silver are retained in solution). Other examples are the precipitation of palladium as the dimethylglyoximate, thallium (I) as the thiocyanate, and even aluminum as $AlCl_3 \cdot 6H_2O$ by saturation of the solution with hydrogen chloride gas. Less satisfactory is the precipitation of platinum as the metal with formic acid from a pH 2 solution (where aluminum, chromium, iron, manganese, nickel, and lead are cleanly separated, but copper and zinc contaminate the platinum). The reduction to the element of selenium and tellurium (with hydroxylamine hydrochloride, sulfurous acid, or stannous chloride in the case of selenium, and with sulfurous acid and hydrazine hydrochloride in the case of tellurium) may both yield a clean separation from certain tramps, but conclusive work has not yet appeared [1].

The possibility of unwanted coprecipitation is always present in work of this type (i.e., in both group and specific element precipitation), especially if the reagent is added too quickly, in too concentrated a form, or with insufficient stirring. The analyst should endeavor to prevent the sudden localized creation of a supersaturated concentration of reaction product. This strategy in turn prevents the formation of fine contaminated crystals and favors the formation of coarse, slowly grown, high purity crystals.

Colloidal precipitates form as molecular clusters agglomerate to larger and larger sizes, finally reaching filterable dimensions. When filtering agglomerated colloids one must guard against "peptization" of the precipitate as the stabilizing coagulant ions are washed out. This is reversion to a colloid which then passes through the filter. To prevent this effect, electrolytes are added to the wash solution. All the colloidal precipitates that yield clean separations are hydrophobic. Hydrophillic colloidal precipitates (such as iron or aluminum hydroxide) are nearly always traps for tramp elements [1].

3.2.3 Coprecipitation

As stated above there is much about the phenomenon of coprecipitation that is incompletely understood. The three mechanisms that have been described: mixed crystal formation, physical occlusion, and chemical adsorption, all appear to be favored by rapid reaction from a supersaturated solution. Mixed crystals are favored when the carrier and trace analyte have similar atomic radii. Although they *can* form as anomalous crystals with dissimilar-sized cations, in such a case the extent of coprecipitation is limited by morphological constraints. Moreover, when large differences in atomic radii create dislocations in crystalline precipitates the effect of so-called "aging" or "ripening" phenomena (recrystallization equilibria with the supernatent solution upon prolonged warming) is that some of the out-sized coprecipitated ions are likely to be expelled from the precipitate, producing low recoveries.

It is difficult to completely isolate the individual effect of occlusion and adsorption phenomena since it is generally recognized that the occlusion of a trace analyte species in a carrier precipitate is directly proportional to its adsorbability on the surface of the carrier precipitate. The Paneth-Fajans-Hahn Rule may be summarized as follows: ions are adsorbed (and occluded) by a precipitate to an extent that is directly proportional to the insolubility of the compound they form with a precipitate lattice ion. Thus, if our carrier is barium ion, Ba^{2+}, and we precipitate with an excess of sulfate, SO_4^{2-}, we will adsorb (and occlude) more lead, Pb^{2+}, than calcium, Ca^{2+}, because $PbSO_4$ (0.004mole/100g H_2O @ 25°C) is less soluble than $CaSO_4$ (0.20mole /100g H_2O @ 25°C). The Freundlich isotherm describes the adsorption of trace analyte on a carrier precipitate with the expression:

$$C_p = \alpha C_O^{\,n} \tag{1}$$

where, C_p is the concentration of trace analyte in the carrier precipitate; C_O is the original concentration of analyte in solution, and α and n are constants [1].

Another effect is often at work—postprecipitation. This is the precipitation of a trace impurity *after* the carrier has completely precipitated. And there are what appear to be catalytic effects as well. The classic example here, which is frequently cited, is zinc sulfide, which remains in supersaturated solution under conditions which should favor its precipitation, but which precipitates quantitatively when mercuric sulfide precipitates under the same conditions [9,10].

Clearly, there is much more at work here than the theorist understands or that the practical analyst needs to know. For here, as in so many places in trace analysis, the working chemist has little option but to be performance driven. Some process either succeeds or fails as decided by the best means to assess the results—the standards yield the certified values or an independent method yields a similar result. There is usually little time to determine exactly why something worked. And so the practical analyst dutifully adds to the knowledge bank and reserves judgement about the theoretical mechanism behind success.

3.2.3.1 Manganese Dioxide. One of the oldest and best-known carriers for the coprecipitation of trace elements is manganese dioxide (MnO_2). First used as a scavenger for iron [11], it was developed for the determination of antimony in copper by Blumenthal [12]. Other workers extended the technique to bismuth [13,14], tin and molybdenum [15], and lead [16] in copper alloys. It has been used for arsenic, antimony, bismuth, and tin in lead alloys [17–19], and for antimony [20], and tin and cerium [21] in cast irons. By varying the acid concentration it was shown that traces of antimony and tin could be separated from bismuth [22]. The technique has been applied to ores (for antimony, bismuth, and tin) [23], to the separation of gallium from aluminum and zinc [24], and to the preconcentration of radiotraces [25,26], as well as for the collection of traces of thallium [27,28].

In 1970 Burke published a method for traces of antimony, bismuth, lead, and tin in nickel utilizing a manganese dioxide preconcentration and a flame atomic absorption finish [29]. The paper also includes an extensive bibliography on the use of manganese dioxide coprecipitation. In 1973 Blakeley, Manson, and Zatka published a modified version of the procedure that improved lead recoveries [30]. In 1974 Vassilaros published an optical emission method for traces of bismuth and lead in iron-base and high temperature alloys based on preconcentration using a manganese dioxide carrier [31].

The reaction proceeds with the generation of hydrogen ions:

$$\underset{(+II)}{3Mn^{2+}} + \underset{(+VII)}{2MnO_4^-} + 2H_2O \rightarrow \underset{(+IV)}{5MnO_2} + 4H^+$$

In a typical application for the collection of tramps a 1-5 g alloy sample is dissolved in dilute nitric acid or dilute nitric and sulfuric acids, if that is feasible. Otherwise, it is dissolved in hydrochloric and nitric acids and then fumed with 20 mL of perchloric acid to low volume. If nickel is the matrix the solution is diluted to 250 mL and adjusted to pH 4–4.5 with ammonium hydroxide, otherwise it is diluted to 250 mL with 1.2M nitric acid. Then 5 mL of 5% (w/v) manganese (II) sulfate solution and 10 mL of 0.25M potassium permanganate solution are added with stirring. The solution is then slowly heated to boiling, the heat is reduced to approximately 90°C, and the solution is maintained just below the boiling point for 1 hr, then allowed to stand at room temperature overnight.

The manganese dioxide precipitate is collected using a low porosity filter paper (e.g., Whatman No.42) and filter pulp. The residue is washed three times with a nitric acid solution that matches the solution medium used

(0.008M nitric acid in the case of nickel, 1.2M nitric acid in the case of iron-base alloys). The original beaker is placed under the funnel and 10 mL of a 2:1:1 hydrochloric acid: hydrogen peroxide (30%): water mixture is poured over the residue to dissolve it. The filter is then washed alternately with small volumes of 6M hydrochloric acid and water. Care should be taken to dissolve all particles of precipitate on the paper and on the beaker walls. The solution is then reduced to an appropriate volume, transferred to a volumetric flask, and diluted to the mark. As an alternative approach it is also possible to utilize the dried manganese dioxide residue directly for optical emission analysis using a crater electrode and measuring manganese/tramp element intensity ratios.

Blakeley et al. [30] discovered that in high nickel solutions the collection of trace levels of lead is favored by adjusting the pH of the solution to 4.5, and by adding the permanganate ion to a room temperature solution of the sample containing manganous ion. As the sample is slowly heated to boiling, then maintained just below the boiling point, a flocculent form of manganese dioxide forms which scavenges lead effectively. Recoveries of 94-99% of the lead in up to 10 g test portions of nickel metal were obtained with a single precipitation.

In the case of iron-base alloys, however, it is evident that a significant amount of iron is coprecipitated with manganese dioxide. The work by Vassilaros [31], which utilized precipitation from a 1.2M nitric acid medium, demonstrated that the collection of lead was not a problem with iron-base materials. Good recoveries were obtained from both low alloy and stainless steel certified reference materials, and comparative data from an alternative technique showed similar lead levels over the range 0.0006—0.23%. This suggests that iron serves as a secondary carrier for tramp elements.

Dulski and Bixler [32] utilized a slightly modified version of the Vassilaros procedure as a preconcentration approach for graphite furnace atomic absorption (GFAA) analysis of iron-, and nickel-base alloys. Lead, bismuth, and thallium were successfully collected and measured by this means. Comparable data were obtained for all three elements at ppm levels from a 70Ni/17Cr/4Fe/2Nb high temperature alloy, using manganese dioxide coprecipitation, TOPO-MIBK solvent extraction, and direct GFAA measurement.

3.2.3.2 Iron Hydroxide. The precipitation of flocculent, brown ferric hydroxide by the addition of ammonium hydroxide to a dilute hydrochloric acid solution containing iron (III) will carry down a large number of trace elements quantitatively. First we must consider those that would ordinarily precipitate, even in the absence of iron, if they were present in significant concentration and could be brought down in a filterable form. This group consists of aluminum, beryllium, chromium (III), gallium, hafnium, indium, niobium, the rare earths, thallium (III), thorium, tin, titanium, uranium, and zirconium. Next, there are the true coprecipitating elements—these are brought down completely by the iron (III) hydroxide but would not precipitate completely if present by themselves, at any concentration. This group includes antimony (III), arsenic (V), germanium, phosphorus (V), selenium, tellurium, and vanadium (V).

There are also other elements that will accompany the ferric hydroxide which are much less likely to be quantitatively removed from solution. These include: bismuth, boron, gold, mercury, lead, the platinum group metals, and silicon. However, under specific conditions and in specific concentration ranges certain of these elements may also prove to be completely recovered.

Donaldson has utilized an iron (III) hydroxide coprecipitation for traces of arsenic in copper metal and copper alloys. A test portion containing a maximum of 2 mg of arsenic is dissolved in nitric acid/hydrochloric acid/water and then fumed in sulfuric acid. The solution is diluted to 100 mL, then 5 mL of hydrochloric acid and 80 mg of iron in the form of an iron (III) sulfate solution are added. The solution is heated to dissolve the salts, then ammonium hydroxide is added to blue litmus plus 5 mL excess. The solution is brought to a boil, then allowed to settle. While still warm the solution is filtered through a medium porosity filter paper (e.g., Whatman No. 40) and washed with 5% (v/v) ammonium hydroxide. The original beaker is placed under the funnel and the residue is dissolved with hot 15% (v/v) hydrochloric acid. The iron is then precipitated with ammonium hydroxide a second time and the rest of the procedure is repeated [33].

The same author has published methods for trace levels of tellurium in brasses [34], and bismuth in copper and copper-base alloys [35] that utilize the same preliminary coprecipitation with iron (III) hydroxide.

Zörner et al. describe a rapid method for the coprecipitation of traces of aluminum, antimony, chromium, vanadium, and zirconium in steel using equimolar quantities of iron (II) hydroxide and iron (III) hydroxide. Such a mixture is attracted by an electromagnet placed under the beaker, facilitating settling and decanting off the supernatant solution [1,36]. In other work a flotation approach has been utilized to facilitate the removal of the iron (III) hydroxide precipitate. A hot solution of paraffin dissolved in 99.5% ethanol is added to the precipitated solution and a stream of small nitrogen bubbles is used to create an easily removed foam. This approach has been applied to the determination of traces of tin in high-purity zinc metal [37].

3.2.3.3 Lanthanum Hydroxide. Lanthanum hydroxide is a useful carrier precipitate for a number of ions. Marczenko applied it to the collection of traces of aluminum, bismuth, gold, indium, iron, lead, thallium, and titanium in silver metal [38]. Reichel and Bleakley reported quantitative recoveries for antimony, arsenic, bismuth, iron, lead, selenium, tellurium, and tin in anode copper. Their approach utilized a 20 g test portion which was dissolved in an 800-mL vessel with 25 mL of water and 90 mL of nitric acid added in small increments. The solution is boiled to expel NO_x fumes, cooled, and diluted to 300 mL. Then 20 mL of a 5% (w/v) lanthanum nitrate solution is added. With constant stirring, 190 mL of ammonium hydroxide is added. The solution is allowed to stand for 1 min, and then filtered washed, and redissolved in a manner analogous to that described for iron (III) hydroxide above. At the 20 g test portion size about 20 mg of copper were found to be carried down by the lanthanum hydroxide even with essentially no standing time before filtration. Recoveries for the eight tramp elements tested were complete, however. It was also found that the complete recoveries of the coprecipitated analytes require a final basicity between pH 9 and pH 10 [39].

Lanthanum offers an advantage over iron and manganese when the coprecipitated analytes are to be measured by flame atomic absorption since lanthanum acts as an ionization suppressor, significantly increasing the sensitivity for many elements. Of course, this approach often also effectively removes enough of the matrix to prevent "matrix effect" interferences. Lanthanum, itself, unlike iron or manganese, shows very low sensitivity by atomic absorption and forms few colored compounds in UV/visible molecular absorption spectrophotometry.

However, Donaldson utilized 50 mg *each* of iron and lanthanum to completely coprecipitate antimony from a copper metal sample. A 0.1—0.5 g test portion (containing about 2 mg of antimony) was used. The method can tolerate up to about 25 mg of aluminum, tin, or zirconium, as well as milligram amounts of chromium and vanadium [40].

3.2.3.4 Other Carriers. There are a host of other "scavenger" compounds that have been successfully employed to carry down trace analytes. Continuing for the moment with inorganic carriers, there are several effective approaches involving the coprecipitation of elemental forms with an elemental carrier. Luke used arsenic as an added carrier for traces of selenium or tellurium in lead or copper metal. The (5 g) copper test portions were dissolved in a freshly prepared 5:1 perchloric acid: nitric acid mixture and brought to copious fumes of $HClO_4$. Dilute hydrochloric acid, 2 mg of arsenic, and 15 mL of hypophosphorous acid are used to bring down the carrier and analyte. The solution is boiled for 5 min, cooled to room temperature, and filtered through a low porosity filter paper. The (10 g) lead test portions are dissolved and fumed in the same way, but lead is then precipitated as $PbCl_2$ with dilute hydrochloric acid, filtered and discarded. Then 2 mg of arsenic and 10 mL of hypophosphorous acid are added to the filtrate, which is then boiled and filtered [41].

Vassilaros applied a similar approach for traces of arsenic and selenium in low alloy and stainless steels. In this work 0.1 to 1 g test portions were dissolved in a 7:1 hydrochloric acid:nitric acid mixture, reduced to 25 mL; 300 μg arsenic carrier was added if selenium was to be determined, or 300 μg selenium was added if arsenic was to be determined. Solutions were then diluted to 40 mL and reduction was accomplished with 10 mL of a 1g/mL solution of stannous chloride in concentrated hydrochloric acid. After 10 min the solutions were filtered on a micropore membrane filter (5 μm) and washed with 3M hydrochloric acid and water [42].

Coprecipitation phenomena are common for organic reagent precipitations but they are infrequently utilized to advantage in the metals industries. We have previously mentioned the fact that a-benzoinoxime will completely precipitate tungsten *only* in the presence of an excess of molybdenum. It is proper to regard this widely used effect as a true coprecipitation. Nickel and palladium both precipitate with dimethylglyoxime, the former under ammoniacal basic conditions, the latter under slightly acidic conditions. However, it has been shown that traces of palladium can be carried down using nickel dimethylglyoxime as a carrier [1,43].

Perhaps the most widely used organic precipitant for metals is cupferron (the ammonium salt of N-nitrosophenylhydroxylamine). It is commonly used as a 6% (w/v) aqueous solution which is filtered, then kept refrigerated. It is prepared fresh weekly. It is a known carcinogen and must be handled with proper protective equipment in an efficient fume hood. Cupferron precipitations are usually conducted in 10% sulfuric or hydrochloric acid sample solutions which have been chilled in an ice bath. The precipitate is filtered with suction, then the beaker and the filter paper are washed with ice cold 5% (v/v) hydrochloric acid. The filter and residue may be wet ashed in the original beaker with 100 mL of nitric acid, 30 mL of 1:1 sulfuric acid:water, and 10 mL of perchloric acid, heating immediately to boiling, then bringing the solution to strong fumes. Alternatively, the filter and residue may be transferred to a large platinum crucible. The crucible is heated under an infrared lamp in a hood for 2 hr to dry the precipitate then placed immediately inside the open door of a muffle furnace in a hood set at 600°C. When the paper and residue are well-charred the crucible is placed deeper into the furnace, the door is closed, and the temperature is raised to 1000°C. Ignition is usually complete in 1 hr. The ignited residue may then be fused with a suitable molten salt flux.

Cupferron precipitates gallium, hafnium, iron (III), niobium, palladium, tantalum, tin, titanium, uranium (IV), vanadium (V), and zirconium. It completely coprecipitates chromium, cobalt, manganese, mercury, nickel, and silver. And it partially brings down antimony (III), copper, iron (II), molybdenum, thorium, thallium (III), and tungsten. As in most such cases, specific conditions may exist under which the "partial" elements become quantitative.

The reagent known variously as "oxine," "8-quinolinol," or "8-hydroxyquinoline" can be used to precipitate two distinct groups of elements at two distinct sets of analytical conditions. At pH 4–6 in an acetic acid/acetate medium aluminum, bismuth, cadmium, copper, vanadium(V), and zinc are quantitative. At a pH >9 in an ammoniacal medium magnesium is quantitative. Many additional elements are typically incomplete in both cases. At pH 4–6 the list includes antimony, cobalt, hafnium, iron, lead, manganese, mercury(II), molybdenum, nickel, niobium, silver, tantalum, titanium, uranium, vanadium(IV), and zirconium. At a pH > 9 barium, beryllium, calcium, strontium, and tin, as well as all the previously mentioned elements, excepting silver, may be partially carried down.

This large array of reactant elements is the reason that 8-hydroxyquinoline is primarily utilized as a precipitant after a series of preliminary separations. However, several workers have used its eclectic reactivity to advantage as a carrier for the coprecipitation of a large number of trace elements by combining it with other carriers, such as thionalide and tannic acid, using aluminum, iron (III), and indium as the "collector" reactants [1]. Such broad-range coprecipitation is useful for the spectrochemical survey of tramp elements in a "clean" matrix but is not generally useful for most commercial alloys.

3.2.4 Adsorbants and Filtration Aids

There are a number of analytical processes connected with the collection of tramps in a solid residue that strictly belong in a separate miscellaneous category. Sometimes the reactant is a solid with limited solubility in the sample solution but whose large surface area acts as an aid to the filtration of the analyte. Tertipis and Beamish separated rhodium from iridium by reacting the mixture in 1.0M hydrochloric acid with freshly reduced copper powder. Rhodium was reduced to the metal and filtered off along with the excess copper. Iridium was reduced to a lower oxidation state but remained in solution [44].

The reagent known as "tannin", "tannic acid," or "gallotannic acid" is a natural product commonly used in combination with other precipitants to ensure complete quantization. Its mechanism of action is poorly understood and it is best regarded as a filtration aid. Cinchonine, used in the hydrolysis of tungsten oxide, is another natural product whose exact function has not been characterized. Physical or chemical adsorption appears to play some role in the application of both of these reagents, since they both are large molecules with limited solubility in acidic sample solutions.

There is also a group of related analytical techniques that utilize the addition of two reagents—one to form either soluble complexes or insoluble or sparingly soluble compounds with trace analytes, and the other to adsorb these products and effectively remove them from solution in a filterable form [11].

Potassium ethyl xanthate has been reacted with trace analytes which were then adsorbed on activated carbon on a filter. The analytes are readily stripped from the activated carbon with nitric acid. The technique has been applied to traces of bismuth, cadmium, cobalt, copper, indium, iron, lead, mercury, nickel, silver, and thallium in high purity zinc [45], and to a similar array of traces in pure manganese metal [46]. Traces of silver and bismuth in cobalt and nickel metal were separated by reacting them at pH 3.5–4.5 with ammonium pyrrolidinedithiocarbamate (APDC) and similarly collecting them on activated carbon [47]. The book by Minczewski, Chwastowska, and Dybezynski [11] discusses numerous examples of the use of two organic reagents—one an analyte reactant, and the other an adsorbent, coprecipitation carrier, or filter aid [52–56].

3.2.5 Electrolytic Deposition

The mercury cathode apparatus is a device whose greatest utility lies in the area of matrix and interference removal. Undesirable elements are reduced to the metallic state and amalgamated with the mercury pool, leaving the trace analyte elements in the supernatant solution. Although the general concept is over a century old the essential form of the modern instrument dates from 1951 [48]. It utilizes tall-form beakers with a stopcock designed to allow aqueous solution withdrawal at a height just above the mercury pool. There are 35 mL and 70 mL mercury pool designs. The mercury pool becomes the anode of a dc circuit by contact with a platinum wire fused into a borosilicate tube filled with mercury. A similar connection is made with the spiral-shaped cathode which is suspended in the sample solution level with, and about 7 mm above, the mercury surface.

A magnet under the beaker serves to rotate the mercury in one direction and the sample solution in the other, promoting the speed of the electrolysis. The magnet also draws ferromagnetic amalgam (such as that produced from an iron-base alloy) down to the bottom of the mercury pool, further increasing extraction efficiency. A glass cooling water loop prevents boiling from the heat of electrolysis, which might interfere with the process of electrolysis by agitating the mercury and possibly shorting the gap between the mercury and the anode.

The mercury cathode apparatus plates out the following elements quantitatively: bismuth, cadmium, chromium, cobalt, copper, gallium, germanium, gold, indium, iridium, iron, mercury, nickel, palladium, platinum, rhenium, rhodium, silver, thallium, tin, and zinc. Arsenic and osmium are quantitatively removed from solution, but partially volatilized. Selenium and tellurium are quantitatively removed from solution, but form elemental particles in the aqueous electrolyte. Lead is, likewise, quantitatively removed, but may partially plate as PbO_2 on the anode. The amount of manganese and molybdenum removed is subject to solution conditions. In some alloy/acid combinations they may be completely eliminated, in others one or both elements may be largely retained. Antimony and ruthenium are partially retained and partially volatilized.

While many methods have been published describing analytical applications for the recovery of the elements amalgamated in the mercury pool the practical trace metal analyst must confine his attention to the elements remaining in the electrolyte. Quantitatively retained elements thus include: the actinides, the alkali metals, the alkaline earths, aluminum, boron, hafnium, niobium, the rare earths, tantalum, titanium, tungsten, vanadium, and zirconium. Mercury cathode electrolysis represents an excellent means for freeing these analytes from potentially interfering elements, or from the bulk of a matrix which may be suppressing measurement sensitivity.

Electrolysis is usually conducted at about 13 amp from a 5% (v/v) solution of perchloric or sulfuric acid. Sometimes phosphoric acid is also present. A typical mercury cathode electrolysis requires 1–4 hr. It is often advantageous to "fume out" chromium as volatile chromyl chloride (see Section 3.5.1) prior to the electrolysis, since the removal of large amounts of chromium on the mercury cathode is slow. Large amounts of molybdenum, titanium, tungsten, or vanadium may act as a redox buffer during electrolysis, preventing the complete removal of certain difficult elements (such as chromium). Sometimes adding a few drops of 30% hydrogen peroxide from time to time during the electrolysis remedies the problem. If manganese is very high manganese dioxide may precipitate during the electrolysis, but it can be redissolved with a few drops of sulfurous acid.

The electrolye must always be filtered upon completion of the electrolysis. This is best accomplished with a double layer of a high porosity filter paper (such as Whatman No. 41). This will remove the mercury compounds and metallic mercury that are likely to be present. Mercury vapor is extremely hazardous so it is essential that all operations with mercury be conducted in an efficient fume hood, or that mercury be kept covered at all times with a layer of water. Waste mercury from the electrolysis cell should be stored in a sealed approved container until it can be returned to a refining company for redistillation.

Table 3.1 shows some typical examples of the use of mercury cathode separation.

It is also sometimes possible to utilize the spent electrolyte from an applied potential electrogravimetric determination of a major element, such

Table 3.1 Selected Examples of Mercury Cathode Use in Trace Metal Analysis

Trace Analyte	Matrix	Interferences Removed	Measurement Technique
Aluminum	Co-base alloy	Matrix Elements	8-Quinolinol Colorimetric
Boron	Steel	Iron, cobalt	ICP-OES: 2496Å
Calcium	Stainless Steel	Chromium	ICP-OES: 3179Å
Hafnium	High temperature alloy	Iron, chromium	ICP-OES: 2738Å
Magnesium	Zinc alloy	Matrix zinc	FAA—improved sensitivity
Rare earths	High purity copper	Matrix copper	XFS—fluorides on filter membrane
Tantalum	High temperature alloy	Iron, cobalt	ICP-OES: 2400Å
Titanium	Ni-base alloy	Matrix Elements	Diantipyrlmethane colorimetric
Tungsten	Copper alloy	Copper, zinc, chromium	ICP-OES: 2079Å
Vanadium	High-strength low alloy steel	Matrix iron	FAA—improved sensitivity

as copper, cobalt, or nickel, which are deposited as the metal on a platinum cathode, or even lead, which is deposited as an oxide on a platinum anode. In these cases the electrolyte can be reduced in volume, or sometimes taken through additional separation steps, prior to the measurement of tramp elements.

3.3 Ion Exchange

It is probably safe to say that, despite many decades of development and an abundant literature, the full potential of ion exchange chromatography for use in the trace analysis of metals has not yet been realized. The reason may lie in the difficulties associated with "fine tuning" a procedure for a specific matrix, which in the most rigorous studies is accomplished using radiotracer techniques. It may also lie in an unfamiliarity with the mechanisms at work in ion exchange. The first objection has some validity in the sense that there are of necessity some development costs associated with adapting a procedure to a specific application. However, the task is sometimes no greater than preparing a few synthetic mixtures to test elution rates and volumes for a procedure published for a related type of sample matrix.

The second objection may stem from the often daunting academic literature on the subject of ion exchange. There *are*, in fact, many different physicochemical mechanisms that can be brought to bear to achieve separations, and the associated theoretical treatments are not light reading, but the practical inorganic trace analyst need only be concerned with the *manipulative* details of one or two of these processes.

The archetypical ion exchange process is the metal trace analyst's primary concern. Synthetic ion exchange resins prepared and sold for analytical applications are the media that require our attention (although ion exchange processes are associated with a variety of other types of materials as well). These polymers consist of a "backbone" material, such as polystyrene, and a cross-linkage structure, such as divinylbenzene. This molecular framework is, itself, inert. However, attached to it at regular intervals are chemically reactive functional groups with an associated charge. An exchangeable counterion of opposite sign balances this charge so that electroneutrality prevails.

Analytical ion exchange materials may be strong acid or weak acid resins. These are cation exchangers, sold commercially in hydrogen ion, sodium ion, or ammonium ion form. They may also be strongly basic or weakly basic resins. These materials are anion exchangers, sold commercially in chloride ion, hydroxide ion, acetate ion, or formate ion forms. There are also chelating resins, ion retardation resins, and others.

The degree of cross-linkage is a critical parameter. It typically ranges from 2–12%. Small amounts of cross-linkage are used in organic and biochemical applications because the resultant open structure permits large molecules to permeate the resin bed. Such resin formulations swell in water and shrink in the presence of strong electrolytes. Inorganic ion exchange applications require a much higher degree of cross-linkage with its correspondingly greater exchange capacity. Resins with 8% cross-linkage are commonly used. Here ion size is generally not an issue, and volume change with changing electrolyte composition is minimal.

Ion exchange resins may be employed in a "batch reaction" mode in which the resin is stirred with the sample solution and then filtered off. They may also be coated onto a filtration medium through which a dilute sample solution is poured. But these are processes with specific and limited utility. The primary mode of ion exchange separation is column chromatography in which the sample solution is passed through a glass or plastic tube packed with ion exchange resin, and then followed by one or a series of eluent solutions.

Ion exchange resins for analytical applications are sold as spherical beads with a narrow range of diameters. Typically useful ranges are 50-100 mesh (0.300–0.150 mm), 100-200 mesh (0.150 - 0.075 mm), and 200-400 mesh

(0.075–0.0375 mm). As might be expected for a given column length, the finer "cuts" of mesh sizes correspond to improvement in resolution and reduction in flow rate. Even prescreened analytical grade resins may contain some fines. These can be easily removed by preparing a batch slurry with water or an eluent solution, allowing the beads to settle, and decanting and discarding the supernatant. After several such treatments the resin should be ready for packing.

Glass columns with stopcocks can be purchased, but plastic columns for use with hydrofluoric acid are best assembled in-house from polymethylmethacrylate tubing. A one-hole, paraffin-coated rubber stopper with a polyethylene tube and stopcock is inserted in the bottom and sealed in place first with paraffin, then with an acid-resistant sealant, then firmly secured in place with wire and clamps. Some glass columns are sold with a porous glass frit disk at the bottom, but this design may hamper flow. An open bottom with an inserted Pyrex wool plug is a better choice to support the resin column. For plastic hydrofluoric acid columns Teflon wool (available from gas chromatography supply houses) or SEF modacrylic fiber (Monsanto Chemical Co.) must be used instead.

The resin slurry is poured into the column and allowed to settle in a series of increments until the desired column height is obtained. A second Pyrex wool plug, or SEF modacrylic fiber plug in the case of the plastic column[*] is inserted firmly on top of the settled resin bed. About 2.5 cm of solution should be kept in the headspace above the plug at all times.

Most ion exchange procedures require some conditioning of the resin column. This consists of one or more eluents that are passed through the column at a selected rate and then discarded. When the column has been conditioned, the sample solution is applied to the column. The procedure will specify the analyte and matrix concentration that can be tolerated; it will also specify the exact reagent composition of the solution. Once the test portion has been transferred to the column a series of eluents are passed through the resin bed.

In the most frequently encountered ion exchange process, counterions on the conditioned column are displaced by some counterions from the test solution that have a greater affinity for the functional groups on the resin. Displaced counterions and low affinity ions from the test solution are washed out with the flowing eluent. By changing eluents it is possible to selectively remove individual analytes or groups of analytes from sites on the column, collecting them in prescribed volumes of eluent. In a given electrolyte environment higher affinity counterions displace lower affinity counterions.

---

[*] Teflon wool floats and is not suitable.

Admittedly, this explanation ignores a host of other processes that may be simultaneously at work. Ionic radius, ionic charge, and the electrostatic effects between fixed and diffusible charge all play a role in what ultimately happens. But ion exchange as an *empirical* separation process need not fixate on exactly modeling at the molecular level how a working system works. For the reader who wishes to pursue a theoretically derived approach to method development, there is a valuable discussion of the application of plate theory to practical separation problems in *Ion Exchange in Analytical Chemistry* by W. Rieman and H.F. Walton (Pergamon, New York, 1970, pp. 120–121).

A few general points can be touched upon here, but their applicability must not be unquestioned. In the use of ion exchange for real samples simple answers are usually simplistic answers.

1. In dilute solution ions with a higher absolute value of charge are bound more strongly to the resin than ions of a lower charge. In high concentration solution lower charge ions are bound more strongly.
2. Uncomplexed anions (e.g., the halides) show greater differences among themselves in their affinities for anion exchangers than uncomplexed cations (e.g., the alkali metal and alkaline earth cations) show among themselves for cation exchangers.
3. Large ions of low charge (such as ReO_4^-, $AuCl_4^-$, or ClO_4^-) distort the hydrogen-bonded structure of water to locate themselves at exchange sites on the resin, and thus they show high affinity.
4. A series of ions that show a range of the degree of lyophillic association with water molecules often separate well by ion exchange.
5. If two elution curves overlap with the ion of lower charge emerging first and the ion of higher charge emerging second, repeat the test using an eluent of lower concentration to increase resolution.

There are hundreds of synthetic ion exchange resins available from sources worldwide. The majority of these are production grade commodities designed for use in water treatment or by the chemical process industry. There remain, however, a large number of reagent grade resins that are potentially useful in the trace analysis of metals. Table 3.2 lists some examples of the common types [1,49].

3.3.1 Anion Exchange Systems

For metals analysis applications anion exchange systems (often based on chloride or fluoride metal complexes) have proven to be the most useful. The chloride system was first studied extensively by Kurt A. Kraus [50] at Oak Ridge National Laboratory and by D. Jentzsch [51] in Germany. Kraus and coworkers conducted a series of elution studies for a large number of elements on strongly basic polystyrene resins with quaternary amine functional groups using a range of hydrochloric acid concentrations from 0–12 molar.

Table 3.2 Some Ion Exchange Resins Useful in the Trace Analysis of Metals

Trade Name	Backbone	Cross-linkage	Functional Group	Manufacturer
Strong Acid (Cation Exchange) Resins				
Amberlite IR-120	Polystyrene	4% DVB	$-SO_3^-$	Rohm & Haas
Amberlite IR-122	Polystyrene	12% DVB	$-SO_3^-$	Rohm & Haas
Dowex 50W-X8	Polystyrene	8% DVB	$-SO_3^-$	Dow
Dowex 50W-X10	Polystyrene	10% DVB	$-SO_3^-$	Dow
Weak Acid (Cation Exchange) Resin				
Amberlite IRC-50	Poly(acrylic) acid	(Macroporous)	$-COO^-$	Rohm & Haas
Strong Base (Anion Exchange) Resins				
Amberlite IRA-400	Polystyrene	8% DVB	$-N(CH_3)_3^+$	Rohm & Haas
Amberlite IRA-410	Polystyrene	8% DVB	$-N-(CH_3)_2^+$ with C_2H_4OH	Rohm & Haas
Dowex 1-X8	Polystyrene	8% DVB	$-N(CH_3)_3^+$	Dow
Dowex 2-X8	Polystyrene	8% DVB	$-N-(CH_3)_2^+$ with C_2H_4OH	Dow
Weak Base (Anion Exchange) Resin				
Amberlite IRA-45	Polystyrene	DVB copolymer	$-NH_2$, $-NH(R)$, and $-N(R)_2$	Rohm & Haas
Chelating Resin				
Dowex A-1	Polystyrene	DVB copolymer	$-N(CH_2COO^-)_2$	Dow

Note: Commercial resins are sieved and purified for analytical chemistry use by secondary suppliers. For example, Bio-Rad Laboratories markets Dowex resins as "AG" (Analytical Grade) versions. Thus, Dowex 50W-X8 is sold as AG 50W-X8. Dowex A-1 is sold as Chelex 100.

For each eluent they calculated the distribution coefficient, D_v, which is a measure of the equilibrium ratio of adsorbed to eluted analyte.[*] The exact relationship is given by:

$$D_v = v/Ad - i \qquad (2)$$

where v is the volume of eluent (in mL) that moves a band of adsorbed[†] analyte a distance, d (in cm) in a column of cross-sectional area, A (in cm^2). The letter i is a correction term for the fractional interstitial volume between the resin beads (usually with a value of about 0.4).

For each element tested they plotted the log of D_v versus the molarity of hydrochloric acid. A typical plot first shows a rising value of D_v with increasing acid concentration. The shape of this low molarity part of the curve is influenced by the charge on the ionic complex which is adsorbed. The curve typically reaches a maximum and then drops because the adsorbed complex is displaced by chloride ions at high molarity hydrochloric acid concentrations. Another reason for the drop in D_v at high acid concentrations is the Donnan invasion of the column by chloride ion—that is, the diffusion of chloride ion into the internal solution of the resin bead [49,52].

The data that Kraus and coworkers assembled has been invaluable in the design of ion exchange schemes in hydrochloric acid media. Nickel, aluminum, the rare earths, thorium, the alkali metals, and the alkaline earths are not adsorbed at any concentration of hydrochloric acid. These elements pass through the column immediately, thus affording an excellent means of separation from other matrix elements. The most strongly adsorbed species are antimony (V), bismuth (III), gallium (III), gold (III), iron (III), mercury (II), and thallium (III), which all have D_v values greater than 10,000. Numerous other elements show D_v values above 10, including antimony (III), cadmium (II), cobalt (II), copper (II), germanium (IV), hafnium (IV), indium (III), iridium (IV), lead (II), molybdenum (VI), osmium (III), palladium (II), rhenium (VII), rhodium (III), ruthenium (IV), silver (I), tin (II and IV), titanium (IV), uranium (IV and VI), vanadium (V), zinc (II), and zirconium (IV) [49,53].

By studying the shape of these curves—in particular at the low acid concentrations where D_v is near zero—it is possible to plan an eluent sequence to achieve many desired separations. Kraus and Moore [50] used the separation of nickel, manganese (II), cobalt (II), copper (II), iron (III), and zinc, as

---

[*] In common practice values of D_v are obtained by a batch method in which the resin is shaken or stirred with a standard solution of the analyte and acid, then filtered. The amount of analyte is then determined in the resin and in the solution.

[†] Rieman and Walton (Ref.49) object to the term "adsorption" since the ion exchange resin bead has a three dimensional structure. They prefer "absorption," but find themselves in a minority. We have adopted "adsorption" in keeping with the common parlance.

an example. This mixture is added to the strongly basic anion exchange column in 12 molar hydrochloric acid. More 12M HCl is added until the nickel is completely removed, then manganese (II) is removed with 6M HCl, cobalt (II) with 4M HCl [which enters the column as blue $CoCl_4^{2-}$ and exits as pink $Co(OH_2)_6^{2+}$], copper (II) with 2.5 M HCl, iron (III) with 0.5 M HCl, and zinc with 0.005M HCl.

Hague et al. [54] applied this separation as part of an involved quantitative scheme to determine nickel, manganese, cobalt, and iron in high temperature alloys. Lewis and Straub [55] used a modified version of this approach to rapidly separate and determine nickel and cobalt in high temperature alloys and stainless steels. A 0.2 g test portion was dissolved in 15 mL of 2:1 hydrochloric acid:nitric acid. The solution was evaporated to dry salts which are heated until NO_x fumes are absent. The residue is cooled then dissolved in 10 mL of hydrochloric acid. The sample is then transferred to a 14-cm high bed of Dowex 1-X8 (100–200 mesh) anion exchange resin in a 2-cm O.D. glass column that has been preconditioned by passing 50 mL of hydrochloric acid through it. The sample beaker is washed with 9M hydrochloric acid and the rinsings added to the column. Flow through the column is adjusted to 2 mL per min, collecting the eluent in a 400-mL beaker.

Additional 9 M hydrochloric acid is passed through the column until the green band containing the nickel has been eluted (about 75 mL). If copper is absent cobalt is eluted next with about 75 mL of 4 M hydrochloric acid. In the presence of copper 7 M hydrochloric acid is added until the copper and cobalt bands overlap, then cobalt is eluted with 4 M hydrochloric acid.

Nickel, copper, zinc, and cadmium have been separated from a silver solder sample using the same system. The test portion is dissolved in nitric and hydrochloric acids, evaporated to dryness and taken up in hydrochloric acid. The silver chloride is filtered off and discarded and the filtrate is transferred to the anion exchange column in 6 M hydrochloric acid (which elutes nickel). Copper is then eluted with 1 M hydrochloric acid, followed by zinc with 0.01 M hydrochloric acid. Finally cadmium is eluted with neutral water [56].

Gallium in the concentration range of 0.001–0.05% can be separated from aluminum alloys. H.J. Seim published a procedure in which a 1 g test portion is reacted with 30 mL of 1:1 hydrochloric acid:water. When the reaction ceases 2 mL of 30% hydrogen peroxide is added and the solution is boiled for 5 min. If a silica residue is observed it is filtered off, ignited, and volatilized with hydrofluoric and nitric acids. The residue is heated to dryness, dissolved in 1 mL of 1:1 hydrochloric acid:water and combined with the previously reserved filtrate. The solution is then evaporated to about 10 mL. Then 40 mL of 6M hydrochloric acid is added and the solution is warmed to dissolve the salts, then cooled to room temperature. The solution is then transferred to a 150-mm column of Dowex 1-X8 (50–100 mesh) in a 2-cm O.D. glass tube which has been preconditioned with 50 mL of 6 molar hydrochloric acid. The beaker is rinsed with 6M HCl and then 3 10-mL portions of additional 6M HCl are passed through the resin bed, maintaining a 3–5 mL/min flow rate. Then 200 mL of 4 molar hydrochloric acid is passed through the column.

All the eluate collected up to this point (containing all the aluminum and most of the copper) is discarded. Then 150 mL of 0.5 molar hydrochloric acid is passed through the column to elute gallium. Iron accompanies gallium but does not interfere in its flame atomic absorption determination. Zinc is then released from the anion exchange resin with 200-400 mL of boiling water at a flow rate of 5–7 mL/min [57,58]. If zinc is the analyte a modified version of the method is recommended [58]. Lead is separated by dissolving a 1 g test portion of aluminum alloy in 6 M hydrochloric acid and diluting to 50 mL with water. The solution is transferred to the column with 1.75M HCl. Lead is then eluted by passing first 50 mL of 0.5M HCl, then 100 mL of hot water through the resin bed [57].

The maximum D_v for lead (i.e., its maximum affinity for column sites) is around 1.5M HCl, but its range is short, dropping to zero below 0.5M HCl and above 6M HCl. While the aluminum alloy method described above releases lead with low acid conditions, a similar method for lead in low alloy steel places the sample solution on the column with 1:11 hydrochloric acid: water (1M HCl) and releases it with 12M HCl [59].

Kraus et al. [60] observed an interesting parallel between the adsorption of anionic chloride metal complexes on strongly basic resin and the extractability of such complexes with ether. For example, the iron (III), gallium (III), and thallium (III) chloro-complexes are strongly adsorbed on the resin and readily extracted into the solvent. Aluminum (III) is not adsorbed or extracted; and indium (III) is both poorly adsorbed and poorly extracted. It is believed that a stable complex of the form, MCl_4^-, is required for ether extraction. The ions, $FeCl_4^-$, $GaCl_4^-$, and $TlCl_4^-$, are stable, but $InCl_4^-$ is in equilibrium with other forms, and aluminum forms no negatively charged chlorocomplex. However, while the MCl_4^- ion will adsorb on strongly basic ion exchange resin, other species (e.g., MCl_4^{2-}, MCl_6^{2-}) which do not extract, will also adsorb. Palladium (II) and platinum (IV) complexes are examples of this effect. Trivalent species, such as MCl_6^{3-} ,do not extract or adsorb, as illustrated by an iridium complex (III).

Even after the half century that has passed since the early work with chloride system anion exchange, much of its rich potential remains to be exploited by the practical analyst. The line interference from molybdenum in the ICP-OES determination of aluminum at 3961A, for example, is a natural application. Matrix removal for the trace analysis of the rare earths, the alkali metals, and the alkaline earths by ICP-OES is another. Table 3.3 lists the

Table 3.3 Molarity of Hydrochloric Acid at Maximum and Zero Adsorption on Strongly Basic Anion Exchange Resin

Element	Maximum Adsorption	No Adsorption
Antimony (V)	12	2
Cadmium (II)	2	0
Cobalt (II)	10	6
Copper (II)	7	4
Gallium (III)	6	2
Germanium (IV)	12	6
Hafnium (IV)	12	8
Iron (III)	10	0.5
Lead (II)	1.5	0 and 6
Molybdenum (IV)	4	1
Silver (I)	0	10
Uranium (IV)	12	1
Uranium (VI)	12	1
Vanadium (V)	12	8
Zinc (II)	2	0
Zirconium (IV)	12	7

No zero adsorption: antimony (III), bismuth (III), gold (III), iridium (IV), mercury (II), osmium (III), palladium (II), platinum (IV), rhenium (VII), ruthenium (IV), thallium (III), Tin (II and IV).
No adsorption: alkali metals, alkaline earths, aluminum, nickel, thorium, rare earths.
Source: Ref. 52.

hydrochloric acid molarity at which maximum and zero D_i values occur for a selected group of potential analytes.

Weakly basic anion exchange resins have not been extensively studied for inorganic trace analysis, but they have found utility in the separation of platinum and palladium from base metals in a dilute thiocyanate solution [61]. One paper surveyed the distribution coefficients for 40 elements for a weakly basic chloride system [62].

There are, of course, many published variations on the pure chloride system. In particular, some of those elements irreversibly adsorbed in hydrochloric acid media can be stripped from the column by the use of other reagents. For example, in the separation of nickel, cobalt, and iron from complex high iron alloys (such as Alnico magnet alloys) iron may not be quantitatively removed except with 1M nitric acid. This approach is effective but has a deteriorating effect on the resin even with careful regeneration to the chloride form. McCracken et al. [63] using short columns removed matrix copper from copper alloys using 2M HCl, then stripped tin (along with cadmium and zinc) using 1M nitric acid. The difficult separation of zinc and cadmium, referred to above, is greatly improved by eluting with 0.01M hydrochloric acid which is 10% in methanol. In this case zinc, then cadmium, are eluted with the same eluent [49,64].

The use of mixtures of hydrofluoric acid and hydrochloric acid greatly expands the useful range of anion exchange on strongly basic resins. Faris [65] published distribution coefficients of 50 elements using straight hydrofluoric acid media over the range 1M to 24M HF. Combined fluoride and chloride systems have proven much more generally useful, however. As early as 1954, Hague et al. published work which illustrated the utility of hydrofluoric/hydrochloric acid mixtures for the separation of titanium, tungsten, molybdenum, and niobium [66]. In 1959 Hague and Machlan published a comprehensive fluoride/chloride anion exchange separation directly applicable to the separation of titanium, zirconium, niobium, and tantalum from steels[67]. Kallmann et al. used a modified version to separate niobium and tantalum from minerals, ores, and concentrates [68]. Kallmann and Oberthin later discovered that rhenium was retained on the column throughout the standard sequence of fluoride/chloride eluents but could be stripped off at the end with 10% (v/v) perchloric acid [69].

The "standard approach" for the separation of iron-, nickel-, and cobalt-base high temperature alloys follows this sequence:

1. Dissolve a 0.5–1.0 g test portion in a Teflon beaker in either HF/HNO_3 or HF/HCl/HNO_3 (the latter may be necessary if chromium exceeds 15%). Heat to dryness, add 50 mL of 4%HF/1%HCl* and again heat to dryness. Add 50 mL of 4%HF/1%HCl and warm and stir with a Teflon rod to dissolve the salts. Cool to room temperature.

2. If the solution is clear, transfer it directly to a 17.5-cm column of Dowex 1-X8 (200–400 mesh) or equivalent in a 1-in. (2.54-cm) I.D. plastic tube (pretreated by eluting 250 mL of 4%HF/1%HCl). If a suspect residue is evident filter the solution directly onto the column through a close-textured, hardened filter paper (such as Whatman No. 542) using a small plastic funnel. Wash the paper with several small volumes of 4%HF/1%HCl. Ignite the paper and residue at 525°C in a platinum crucible, cool in a desiccator, then fuse over a burner with 1g of sodium carbonate, using a platinum lid. Cool, then leach in 50 mL of 4%HF/1%HCl and transfer to the column. Rinse vessels with 4%HF/1%HCl and transfer to the column.

3. Elute 350 mL of 4%HF/1%HCl into an 800-mL Teflon beaker. This solution contains most of the aluminum and all of the chromium, cobalt, copper, iron, manganese, and nickel. These can be measured collectively by a spectrometric approach or further separated by chloride system ion exchange. If the later course is chosen the solution is reduced to 50 mL, then 25 mL of

---

* All percentages in this procedure are (v/v) unless otherwise stated.

HNO_3 and 15 mL of $HClO_4$ are added and the sample is heated to strong fumes, then to dryness. Heat the dry salts for at least 15 min, then cool and dissolve in HCl, warming if necessary. Then transfer to a 14-cm long column of Dowex 1-X8 (100–200 mesh) in a 1-cm O.D. glass column which has been pretreated with 50 mL of HCl. Proceed as described above for chloride system anion exchange.

4. Elute 350 mL of 10%HF/50%HCl into an 800-mL Teflon beaker. This solution contains all the hafnium, titanium, tungsten, vanadium, and zirconium. These can be measured by a multi-element approach or aliquotted for separate analyte measurement. If the latter course is chosen it is often best to first eliminate the fluoride. Heat the solution to reduce it to 50mL. Add 25 mL of HNO_3 and 30 mL of 1:1 H_2SO_4:H_2O to the cooled solution and heat to fumes of SO_3. Cool, dilute to volume, and remove aliquots.

5. Elute 350 mL of 20%HF/25%HCl into an 800-mL Teflon beaker. This solution contains all the molybdenum.

6. Elute 300 mL of 4%HF/14%(w/v)NH_4Cl into an 800-mL Teflon beaker. This solution contains all the niobium.

7. Elute 350 mL of 4%(w/v)NH_4F/14%(w/v)NH_4Cl into an 800-mL Teflon beaker. This solution contains all the tantalum.

8. Elute 400 mL of 10% $HClO_4$ into an 800-mL Teflon beaker. This solution contains all the rhenium.

The resin is then prepared for the next sample by regeneration with 3M HCl.

Of course, other anion exchange schemes for separating these same elements with fluoride/chloride eluant systems have been proposed and used as well [49,70,71]. High purity niobium metal (0.5 g) was dissolved in 5 mL of hydrofluoric acid and 1 mL of nitric acid, then evaporated to dryness. The residue was dissolved in 10 mL of 7MHF/5MHCl and passed through a 20-cm (by 0.8 cm O.D.) column of AG 1-X8 (100-200 mesh) resin followed with 90 mL of 7MHF/5MHCl. The eluate contains the trace impurities, chromium, magnesium, manganese, molybdenum, titanium, and tungsten. The base metal niobium was eluted next with 60 mL of 0.2MHF/7MHCl. Finally, the trace impurities, tantalum and iron were eluted next with 50 mL of 3M H_2SO_4 containing 1% (v/v) H_2O_2.

Other systems have also been used for anion exchange. For example, some investigators have separated rare earths from cast steels using a thorium coprecipitation with fluoride, then with oxalate. The thorium is then separated from the rare earth analytes by anion exchange. The oxalates are wet ashed with nitric and perchloric acids, and the cooled solution is transferred to a 4-cm bed of Dowex 1-X10 (in a 10-mm O.D. glass column) with 8M nitric acid. The rare earths are eluted with 8M nitric acid, and thorium is retained [73]. Faris and Buchanan [74] published distribution coefficients for many elements in a nitric acid anion exchange system. Thorium, uranium, protactinium, and, in particular, neptunium and plutonium, are strongly adsorbed.

Strelow and Bothma [75] published distribution coefficients for about 50 elements in a sulfuric acid anion exchange system. Chromium(VI), hafnium(IV), molybdenum(VI), and zirconium(IV) are among the most strongly adsorbed species. Earlier, Hague and Machlan had proposed a sulfuric acid anion separation scheme for the chemically similar hafnium and zirconium[76,77], but Strelow and Bothma caution that solutions should be fumed in sulfuric acid, then cooled, diluted to 1.25M H_2SO_4 and immediately passed through the column. Prolonged standing or excessive dilution evidently results in a polynuclear polymerization between zirconium and hafnium complexes and poor ion exchange resolution [75].

Bandi et al. attempted to circumvent the use of hydrofluoric acid in the separation of molybdenum, niobium, tantalum, titanium, tungsten, and zirconium from iron- and cobalt-base alloys by employing oxalate-, citrate-, and hydrogen peroxide-containing eluents [78]. A mixture of nitric acid, methanol, and acetic acid was used to adsorb cerium from cast iron, steel, and ferrosilicon-magnesium sample solutions onto Dowex 1-X10 (50–100 mesh) [79]. A similar approach was used to adsorb cerium, lanthanum, and praseodymium from carbon steels [80]. A combined HF/HCl/H_2SO_4 anion exchange and HF/HNO_3 cation exchange procedure has been attempted for traces of bismuth, cadmium, lead, silver, and zinc in high nickel alloys [81].

Alumina of chromatographic quality is sometimes utilized to effect anion separations. The most strongly adsorbed species are the hydroxyl (OH^-), fluoride (F^-), bisulfate (HSO_4^-), and dihydrogen phosphate ($H_2PO_4^-$) ions. Alumina columns have proven useful in the separation of low and trace amounts of sulfur and phosphorus. The column of chromatographic grade alumina is conditioned with ammonium hydroxide or 1.5M sodium hydroxide solution then rinsed with water. Suction with a vacuum flask to collect the eluate is the preferred approach.

Dissolved steel samples may be fumed in perchloric acid and chromyl chloride volatilized. The solutions are cooled, diluted, and any residual chromium (VI) is reduced with iron (II) perchlorate. Dilute hydrochloric acid is passed through the column just before the sample solution. Then additional dilute hydrochloric acid is passed through the column, then water. The vacuum flask is replaced with a clean one and the sulfur or phosphorus is stripped from the column with ammonium hydroxide or 1.5M sodium hydroxide solution, followed by water. Variations on these procedures have been employed for the gravimetric determination of sulfur in iron and steel [82,83], the colorimetric determination [84], and the ICP-OES determination of phosphorus in steels, and the ICP-OES determination of phosphorus in copper-base alloys

[86]. The ICP-OES methods are thus free of the serious interferences from copper and iron which plague phosphorus emission lines at 2136, 2149, and 2536Å.

3.3.2 Cation Exchange Systems

While some cation exchange separations are performed without the use of complexing media (notably separations of the alkali metals and the alkaline earths) most work requires complex formation. Landmark investigations conducted between 1943 and 1945 as part of the Manhattan Project and published in 1947 involved the cation exchange separation of the rare earth elements using citrate complexing buffers [87–92]. Today, the separation of the rare earths continues as one of the most important uses of cation exchange. In most elution schemes the lanthanide elements elute "backwards" in order of *decreasing* atomic number. In fact, they are eluting in order of *increasing* atomic radius, due to the so-called "lanthanide contraction" [i.e., lanthanum ($Z = 57$) has the largest atomic radius, while lutetium ($Z = 71$) has the smallest atomic radius]. This phenomenon is due to a steady decrease in nuclear shielding by 4f shell electrons with each incremental rise in atomic number.

Carboxylic acid-type complexing media are typically required to separate the rare earths. In addition to citric acid [$COOH–CH_2–C(OH)(COOH)–CH_2–COOH$], lactic acid [$CH_3–CHOH–COOH$], and, in particular, 2-hydroxyisobutyric acid [$(CH_3)_2–COH–COOH$] have been used. Story and Fritz used the latter for the flow injection analysis of 13 lanthanides using gradient elution with increasing pH on Dowex 50W-X8 (250–325 mesh) and 4-(2-pyridylazo)resorcinol (PAR) as the colorimetric reagent [93]. Hwang, et al. used a liquid chromatograph and Li Chrosorb KAT cation exchanger (10 μm) with a concentration gradient of 2-hydroxyisobutyric acid to obtain similar results for 14 rare earths [94].

Similar results had been published earlier by Elchuck and Cassidy using different liquid chromatographic packings. They concluded that in some instances conventional strong acid resins offered some advantages over the bonded phase resins recommended for LC use [95]. Lactic acid media results using LC columns have also been reported [96]. A concentration gradient elution using 2-hydroxyisobutyric acid, ammonium n-octylsulfonate, and methanol produced a separation of the lanthanides plus peaks for yttrium, thorium, and uranium [97]. The resolution of yttrium and dysprosium was not complete, however. Strelow and Victor used hydroxyethylenediaminetriacetate to separate yttrium and neodymium from the heavier lanthanides using column chromatography [98].

Since spectrometric methods generally allow good quantification of the separate rare earths in mixtures, some efforts have been expended on cation exchange separation of the rare earths as a group from their sample matrix. Using AG 50W-X12 (200–400 mesh) Strelow isolated thorium on the column, eluting the rare earths and zirconium with 4M hydrochloric acid. Thorium was determined by ashing the resin [99]. Crock et al. isolated the rare earths (including yttrium) from potentially interferring matrix elements in geological materials using AG 50W-X8 (100–200 mesh). First, 2M hydrochloric acid, then 2M nitric acid eluted matrix elements, then the rare earths were eluted by volumes of 6M nitric acid and 8M nitric acid (which were combined for measurement) [100].

The rare earths aside, the utility of cation exchange systems for trace metal analysis has not yet been convincingly demonstrated. An interesting approach with potential application to the trace analysis of high-purity materials was developed by Tera et al.in the 1960s. The matrix element was precipitated as a chloride compound of sparing solubility on strongly acid ion exchange resin and trace impurities were eluted with concentrated hydrochloric acid before "breakthrough" of the matrix element occurred. The initial work utilized Dowex 50-X8 (100–200 mesh) and was successfully applied to barium, strontium, sodium, potassium, and silver matrices [101].

The process was the refined by using hydrochloric acid/solvent mixtures as eluents on Dowex 50W-X2 (100-200 mesh) and the list of matrices was extended to include lithium, rubidium, magnesium, calcium, scandium, yttrium, lanthanum, nickel, chromium, manganese, lead, and aluminum [102]. Some cation exchange work for specific analytes has also been published Examples include a method for tungsten in steels [103], and a procedure for the separation of gold from base metals [104].

3.3.3 Chelating Resin Systems

Chelating resins are weakly acid cation exchangers. Common types, such as Dowex A-1 and Chelex 100 are a styrene/divinylbenzene copolymer with pairs of iminodiacetate ions as the chelating functional group. These materials show an extreme degree of selectivity for divalent cations. Thus, the alkaline earths and divalent transition metals can be effectively concentrated from seawater and commercial chemicals like caustic soda. Their utility in the trace analysis of metals is limited, although some applications have been found for the precious metals. Platinum, iridium, osmium, palladium, rhodium, and gold have been selectively retained on a specially synthesized chelating resin [105].

3.3.4 The Effect of Solvents

The addition of polar organic solvents to eluent media often has a profound effect on the results in both cation and anion exchange systems. Methanol and other alcohols, acetone, and dioxane have been utilized to improve selectivity. In the separation of noncomplexed cations (such as the alkali metals) the addition of solvent appears to diminish the force of hydration by water and enhance the attractive force of the functional groups on the resin. In the case of complexed ions the addition of solvent increases the attractive strength holding the complex together. This stabilization of the complexed species has the effect of allowing their adsorption on anion exchange resins at lower concentrations of the complexing agent than occurs in purely aqueous systems. It also alters the relative selectivities of the complexed species, sometimes drastically altering the order of elution from that in the equivalent water-based system. In cation exchange systems metals that form complexes are eluted more readily in complexing media dosed with solvent, than with the same eluent in a purely aqueous medium [49].

3.4 Solvent Extraction

Also known as "liquid-liquid extraction," "solvent extraction" applies to all those practices that involve the preferred transfer of analyte between immiscible or nearly immiscible liquids. The distribution of analyte between the two liquid phases under equilibrium conditions was derived by Nerst in 1891 [106]:

$$k = c_1/c_2 \qquad (3)$$

where k is the distribution constant, c_1 is the concentration of analyte in one liquid phase, and c_2 is the concentration of analyte in the other liquid phase. Nearly always one phase (c_2 in the expression) is water. A more accurate treatment was derived by Morrison and Freiser [107] based on the partial molar Gibbs free energy of the analyte in both liquid phases. However, both theoretical formalisms assume that the analyte remains in the same form in both liquid phases. This, in fact, rarely occurs, and so a more practical function is the distribution coefficient, D.

$$D = \Sigma(C_A)_O/\Sigma(C_A)_w \qquad (4)$$

where $\Sigma(C_A)_O$ represents the sum of the concentrations of all forms of analyte A in the organic phase, and $\Sigma(C_A)_w$ represents the sum of the concentrations of all forms of analyte A in the aqueous phase, under equilibrium conditions [1].

Solvent extraction is analytically feasible for three classes of analyte: 1) those that form effectively neutral ion-association complexes; 2) those that form stable chelates; and 3) those that form extractable neutral compounds. Each of these categories of molecular species exhibits no net charge, and each must be relatively free of coordinated water molecules. The relative ease with which solvating water molecules are displaced from an analyte ion depends inversely on the ion's charge density (the ratio of ionic charge to ionic radius). A more practical concept, specific charge (the ratio of ionic charge to the total number of atoms in a complex ion) relates to hydration energy in the same manner. Anionic halide complexes with certain metals are readily extractable because the metal ion is enclosed by species that are weakly hydrated [1].

The effect of acidity is always important. If one considers the generalized chelate extraction as a chemical reaction:

$$M^{n+} + nHA_{(O)} \rightarrow MA_{n(O)} + nH^{+}$$

where the subscript (O) refers to a species in the organic phase, then it is possible to write:

$$K_{ex} = \frac{[MA_n]_{(O)}\,[H^+]^n}{[M^{n+}]\,[HA]_{(O)}^n} \quad (5)$$

Here, M^{n+} is the analyte metal ion, HA is the organic phase-soluble chelating reagent, MA_n is the extractable chelate, and K_{ex} is the equilibrium constant for the extraction.

From the definition of the distribution coefficient, assuming no intermediate complexes occur:

$$D = \frac{[MA_n]_{(O)}}{[M^{n+}]} \text{ and } K_{ex} = \frac{[H^+]^n}{D\,[HA]_{(O)}^n} \quad (6)$$

Rearranging this expression and converting to a logarithmic relationship:

$$\log D = \log K_{ex} + n\text{pH} + n\log[HA]_{(O)} \quad (7)$$

and $$\text{pH} = (\log D/n) - (\log K_{ex}/n) - (\log[HA]_{(O)}) \quad (8)$$

Thus, it can be seen that the distribution coefficient depends in a sensitive way upon both the pH and the charge on the analyte ion. In the absence of complicating side reactions, D increases ten-fold for a unit increase in pH for a monovalent analyte ion, 100-fold for a divalent ion, and 1000-fold for a trivalent ion. Some books refer to a parameter termed $pH_{1/2}$. This is the pH at which D = 1. This value is specific for a given extraction system. The greater the difference in $pH_{1/2}$ values for two ions, the more completely they are separated by the solvent extraction scheme [1].

It should be obvious that the stability of the extracted complex is critical to the efficiency of the extraction. Moreover, complexes with large stability constants can be extracted from solutions of lower pH, resulting in a larger distribution coefficient. The effect of the solvent is much more difficult to define and may range from completely inert to highly active in the extraction process. For example, partially hydrated complexes extract much more effectively with polar solvents that can displace coordinated water molecules. The effect of the solvent's dielectric constant on the extraction process is real but difficult to assess usefully because of many concurrent effects. In some cases reagents, such as the organophosphorus compounds, dimerize in nonpolar solvents, but do not dimerize in polar solvents. Tri-octylphosphine oxide (TOPO) shows different extractive properties in nonpolar cyclohexane than it does in polar MIBK, for example [1].

Before an extraction is attempted it is usually necessary to make a series of additions to the sample solution. The oxidation state of the species to be extracted may require adjustment. And undesired species may require oxidation state adjustment to ensure that they do *not* extract. Masking agents may be added to prevent the formation of undesired extractables. And pH adjustment or the addition of pH buffer solutions are often necessary.

Unlike coprecipitation, coextraction is a comparatively rare phenomenon. When it *has* been observed it is usually attributed to the formation of some type of ion association complex (although other mechanisms are possible). Synergistic effects, in which an added reagent increases the speed of an extraction, are much more common. The mechanism at work here appears to involve the formation of a mixed complex of analyte/chelate/additive. The additive may displace coordinated water, enhancing the extractability of the analyte/chelate. Organophosphorus compounds are noted synergistic additives. TOPO is particularly useful in some extraction systems. There are also antagonistic effects that slow the process of attaining an extraction equilibrium. Masking agents, such as EDTA, added to eliminate the extraction of undesired species, tend to have this effect [1].

Solvent extraction is primarily a diffusion-dependent mechanism. When two immiscible solvents are agitated together by shaking in a separatory funnel, either manually or mechanically, innumerable thin-film interfaces are created between the two liquids. The extracted species must diffuse toward, then transfer across these interfaces. The time required for quantitative extraction thus depends on some combination of diffusion and extraction kinetics. One consequence of this physical arrangement is that excessively vigorous shaking causes the two phases to move together rather than with opposed relative velocities. This does little to enhance the extraction and should be avoided [1,107].

Overenthusiastic shaking can also contribute sometimes to the formation of emulsions. However, this problem may occur at other times, especially when one or both phases are high in viscosity. Emulsions can sometimes be avoided by adding inert particles (which accumulate at the interface) or by adding neutral salts to the aqueous phase. If they *do* form they can usually be effectively broken by pouring them onto a silicone-treated phase separating filter paper.

When the initial extraction is completed it may be necessary to remove undesired coextracted species from the analyte-bearing organic phase by back-extracting with an aqueous solution designed to remove them. Alternatively, if the analyte has been retained in the aqueous phase in the initial extraction it may be necessary to wash the organic phase with a prepared aqueous solution to collect traces of analyte that may have been co-extracted. In both cases the aqueous phase may be washed with organic solvent as well. The protocol chosen will depend upon the exact analytical requirements.

Sometimes when the initial extraction results in an analyte-bearing organic phase it is necessary or desirable to transfer it to an aqueous solution. The best approach is to back extract it with an appropriate aqueous medium. An alternative approach, suitable if the solvent is low boiling (b.p. < 100°C), is that the organic phase can be cautiously evaporated using a boiling water bath. Water, or in the case of a chelate complex, dilute acid, must first be added to prevent loss of the analyte. When the solvent is gone, residual organic material may remain and may require a wet ashing treatment. Both the solvent evaporation and the wet ashing present potential fire and explosion hazards that are avoided if a suitable back extraction approach can be found [107].

3.4.1 Chelate Complexes

A few metals are extractable into organic solvents as inorganic compounds. Germanium, for example, forms $GeCl_4$, which can be extracted into carbon tetrachloride. Most metal cations, however, must form ion pairs or chelate complexes before they can be extracted. We will consider chelation systems first.

There is a gray area between electrovalent and covalent bonds sometimes referred to as "semicovalent." It is here that coordination ions and compounds of all types are categorized. In these species, ligand groups donate electron pairs to a central atom, creating a chemically stable aggregate structure. When the charge on the central ion is neutralized, free coordination sites may still exist. If all sites are occupied the complex is said to be coordinatively saturated. In some cases, such as with the heteropoly acids, the role of the central atom is played by a complex ion. Chelates are coordination complexes in which at least two coordination sites are simultaneously occupied by a polydentate ligand. Most chelating agents simultaneously occupy two sites and, thus, are termed "bidentate." A common chelating agent which occupies six sites is EDTA.

A chelate complex may be positively or negatively charged, but only those with a neutral net charge are ordinarily extractable. Even in this case some coordination sites may be unoccupied by ligands. Such coordinately unsaturated chelates will likely have water molecules residing in the free coordination sites, impeding extraction. Thus only neutral, coordinatively saturated chelates are readily extractable. Sometimes charged chelate complexes are purposely formed with interferent elements to prevent their extraction.

There is, however, one mechanism by which a charged chelate complex can be extracted. Certain charged chelates can form an ion pair association to yield an effectively neutral aggregate which is extractable. For example, the iron (III)-o-phenanthroline complex has a positive charge, but it can be extracted as an ion pair. We have already mentioned mixed ligand complexes—that is, chelates in which more than one type of chelating agent binds to the central atom. These sometimes show markedly improved selectivity and extraction rates (the synergistic effect) over uniform complexes of either ligand alone [1].

Extraction equilibrium with a purely chelate system frequently requires more time than the typical ion association system. When the extracted chelate must first form in the aqueous layer before it can be extracted into the organic layer a solvent with a slightly lower value of the distribution coefficient may actually favor a higher rate of extraction [1].

It is obviously impossible to discuss every chelate system that has ever been applied to trace metal analysis since this separation approach lends itself to a great many useful analytical schemes. Many of the chelates used in solvent extraction are familiar as organic precipitation reagents when applied under different conditions. Some systems are strongly linked to UV/visible molecular absorption measurement since the extract is highly colored and obeys Beer's Law. Others have found use in the direct atomic absorption or ICP-OES measurement of the organic phase. All are potentially useful as a procedural step in a larger separation scheme.

3.4.1.1 Beta-Diketones. These important chelate-formers can all exist as tautomeric mixtures of keto and enol forms:

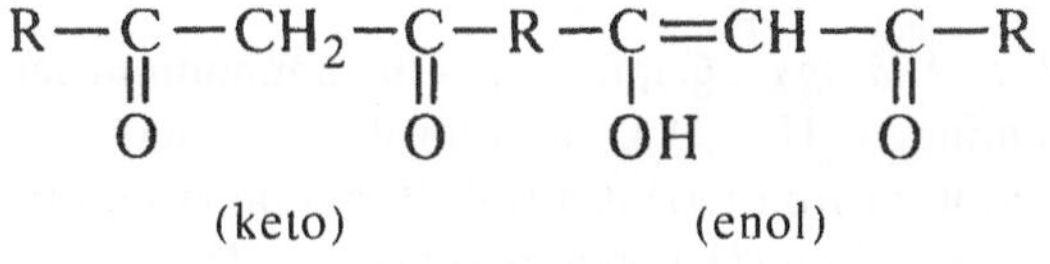

$$\underset{\text{(keto)}}{R-\underset{\underset{O}{\|}}{C}-CH_2-\underset{\underset{O}{\|}}{C}-R} \qquad \underset{\text{(enol)}}{R-\underset{\underset{OH}{|}}{C}=CH-\underset{\underset{O}{\|}}{C}-R}$$

Removal of hydrogen from the enol form generates the bidentate ligand that forms a six-membered chelate ring structure:

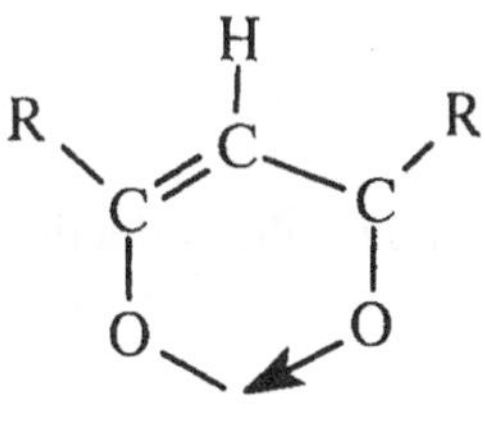

```
           H
           |
    R      C       R
     \\  //  \\   /
       C       C
       |       |
       O       O
        \\  <-/
```

The best known reagent of this class is acetylacetone (2,4-pentanedione, $C_5H_8O_2$). It is a liquid, slightly soluble in water (0.17g/mL @25°C), in which medium it behaves as a weak acid. It is often used directly as an extractant, but mixtures with polar solvents are advantageous since they lower the water solubility of the chelate complexes. Acetylacetone will extract about 50 metals, most notably aluminum, beryllium, gallium, indium, iron (III), manganese (III), molybdenum (VI), palladium (II), thallium(III), thorium (IV), uranium (IV), and vanadium (III and V) for which extraction has been reported at 100%. The extraction efficiency is often pH sensitive. For example, molybdenum(IV) is best extracted from 2M hydrochloric acid, while iron (III) extracts best at pH 1. However, some metals are extractable over a broad range of acid concentrations. Acetylacetone forms colored complexes with many metals which can be used for spectrophotometric measurement, although sensitivity is not high enough for most trace and ultra-trace work [1,107,108].

E.M. Donaldson has employed an acetylacetone extraction at pH 6.5 (following a mercury cathode separation) to isolate aluminum from iron, steel, nickel–chromium, nickel–copper, copper-, tin-, zinc-, and cobalt-base alloys [109,110]. J.P. McKaveney utilized the bleaching effect of free hydrofluoric acid on the iron (III) acetylacetone complex to measure the free HF in hydrofluoric/nitric acid stainless steel pickling baths [111].

Thenoyltrifluoroacetone ($C_8H_5O_2SF_3$) forms chelate complexes that are very stable in highly acid solution. However, it is no more selective than acetylacetone. As with two other widely studied β-diketones, benzoylacetone ($C_{10}H_{10}O_2$) and dibenzoylmethane ($C_{15}H_{12}O_2$), extraction equilibrium is reached after a much longer time than with acetylacetone [1,107,108].

The considerable thermal stability of this category of chelates has led to numerous studies involving gas chromatographic separation of β-diketone chelates. Aluminum and iron have been successfully determined in a nickel-copper alloy utilizing a GC separation of the trifluoroacetylacetone complexes [112,113].

3.4.1.2 Cupferron and Its Analogs. Cupferron is the ammonium salt of N-nitrosophenylhydroxylamine ($C_6H_9N_3O_2$). In chloroform it acts as a bidentate ligand for the extraction of iron (III), tin (IV), titanium (IV), uranium (IV), vanadium (V), and zirconium (IV), among other metals:

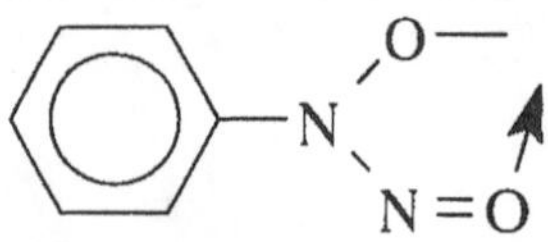

It is useful both for the removal of matrix interferences and for the collection of tramps in the organic phase. It has limited potential as a spectrophotometric reagent since most cupferrate extracts are colorless. Cupferron solutions deteriorate when exposed to light and air and should be prepared fresh a short time prior to use and kept cold. The compound and its analogs are known or suspected carcinogens and should be used with appropriate safety precautions. One common use is for the removal of residual iron in the spectrophotometric (8-hydroxyquinoline) method for trace aluminum [59].

N-benzoyl-N-phenylhydroxylamine (BPHA, N-phenylbenzohydroxamic acid, $C_{13}H_{11}NO_2$) is structurally similar to cupferron but more stable at high concentrations and largely impervious to photodegradation, air oxidation, or low heat. It forms 5-membered ring chelates:

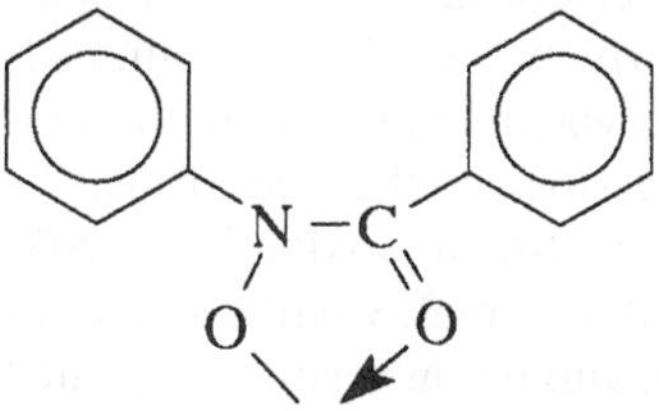

BPHA reacts with most elements and over 40 metals have been extracted at near 100% recovery. Optimum extraction conditions vary from 8 molar acid to pH 12. BPHA also forms extractable colored complexes with many metals. Donaldson applied the reagent to the extraction/spectrophotometric determination of vanadium in iron, steel, aluminum-, and nickel-base alloys [114], as well as to the determination of vanadium in high purity niobium, tantalum, titanium, and zirconium [109]. Other workers have applied it to the extraction of aluminum from uranium fuel [115]; molybdenum [116] and tin [117] from geological samples; and titanium from alloys [108,118]. A more selective, related compound, N-benzoyl-N-o-tolylhydroxylamine (BTHA, $C_{14}H_{13}NO_2$) is almost a specific reagent for vanadium (titanium interference can be masked with fluoride) [108,119].

3.4.1.3 8-Hydroxyquinoline and Its Derivatives. This reagent is known by three names: 8-hydroxyquinoline, 8-quinolinol, and oxine. It has the structural formula:

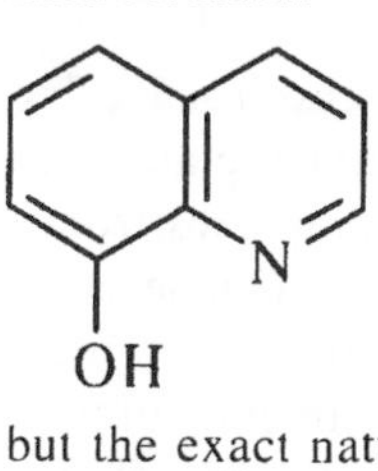

but the exact nature of its bidentate ligand reactions are imperfectly understood. It *is* known to exhibit different complexing behavior with cations of different valency. A great number of metals can be extracted with this compound (usually into chloroform), but selectivity can be enhanced by masking agents, such as cyanide, which eliminates interference from traces of iron, copper, molybdenum, and nickel. It forms colored complexes with many metals. Aluminum in iron and steel is determined by a procedure that involves preliminary separations (mercury cathode and cupferron extraction), then 8-hydroxyquinoline extraction (at pH 9.0 in the presence of cyanide ion) into chloroform, followed by spectrophotometric measurement of the complex [59]. Magnesium in electronic nickel can be determined without preliminary separations by chloroform extraction of the 8-hydroxyquinoline complex from dilute ammonium hydroxide/cyanide/ethylene glycol monobutylether solution [120]. Other elements that extract quantitatively include: bismuth, cadmium, cobalt, hafnium, indium, manganese, molybdenum, thallium, tin, uranium, vanadium, zinc, zirconium, the alkaline earths and the rare earths.

A related compound, 8-hydroxyquinaldine (2-methyloxine, $C_{10}H_9NO$) was originally reported to not form an extractable complex with the aluminum cation due to steric hindrance. Even though aluminum is not precipitated with this compound it has been found that aluminum is partially extracted in the presence of acetate [108,121]. Other investigated derivatives include methyl-8-hydroxyquinoline and 8-mercaptoquinoline.

3.4.1.4 Sodium Diethyldithiocarbamate and Analogs. This reagent is also known as Cupral or NaDDTC ($C_5H_{10}NS_2Na \cdot 3H_2O$). It is a bidentate univalent ligand:

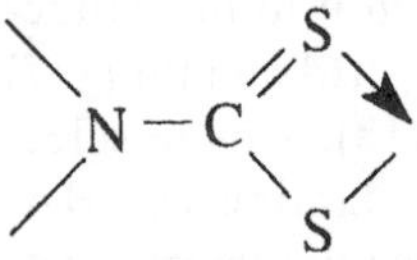

S
N — C
S

that reacts with many elements to form extractable complexes. Bismuth, copper, and nickel (among others) form colored extracts suitable for spectrophotometric measurement. The careful choice of pH range and masking agent will greatly improve sensitivity with this reagent. Strong acid conditions must be avoided due to reagent degradation.

Rooney determined trace levels of lead and bismuth in cast iron by first extracting iron from hydrochloric acid solution with isobutyl acetate, then extracting lead and bismuth as the diethyldithiocarbamate complexes from a cyanide/ammoniacal tartrate solution with chloroform [122]. Goode et al. separated trace impurity levels of cadmium, cobalt, copper, iron, lead, manganese, nickel, and zinc in high-purity beryllium by solvent extraction of their diethyldithiocarbamates [123]. Dean and Cain extracted copper, nickel, and manganese from aluminum alloy sample solutions [124]. Donaldson used the reagent to isolate bismuth from copper metal and from copper-, tin-, and lead-base alloys [125,126]. Antimony, cadmium, copper, iron, lead, manganese, nickel, tin, and zinc are all removed from a copper alloy matrix (in preparation for an aluminum determination) using a NaDDTC extraction at pH 5.5 [127].

Diethylammonium diethyldithiocarbamate ($C_9H_{22}N_2S_2$) is an advantageous alternative reagent because it is soluble in the solvent phase, limiting acid attack on the chelate-forming agent. Ammonium pyrrolidinedithiocarbamate (APDC, $C_5H_{12}N_2S_2$) is water soluble but much more stable than NaDDTC. Methylisobutylketone is usually employed as the solvent. APDC is widely used as a preconcentration extractant for dilute samples, such as water, although it has been applied to dissolved metal samples [132,133]. The silver analog, silver diethyldithiocarbamate is used as a specific color reagent for trace arsenic by the evolution method.

The xanthates, especially potassium ethylxanthate ($C_3H_5OS_2K$), have been used to extract numerous metals [130]. Donaldson has applied this separation to antimony [40], arsenic [33], bismuth [131], selenium [132], and tellurium [133] in various nonferrous alloys, and to molybdenum in iron and steel [134].

3.4.1.5 Dithizone. Dithizone (1,5-diphenylthiocarbazone, $C_{13}H_{12}N_4S$) is a purple-black solid that dissolves readily in organic solvents to form a mixture of two tautomers:

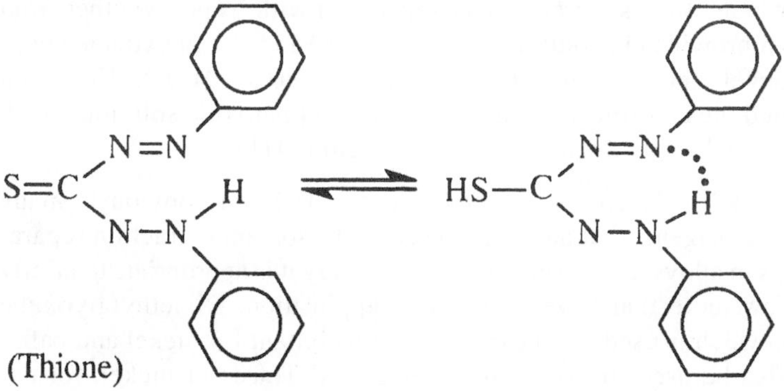

(Thione)

It forms colored complexes useful for sensitive spectrophotometric measurement with over 20 elements, in some cases forming different complexes with different oxidation states of the same metal. The commercial reagent often requires purification before use. This can be accomplished by dissolving it in 0.8 M ammonium hydroxide, extracting repeatedly with carbon tetrachloride, then reprecipitating the reagent from the aqueous phase with dilute sulfuric acid. The precipitate is filtered, then dried in a vacuum desiccator.

Complexes of copper, gold, mercury, platinum, palladium, and silver are extracted from strong acid solution. Complexes of bismuth, indium, tin, and zinc are extracted from weakly acidic solution. And a neutral or basic solution is required for cadmium, cobalt, iron, lead, manganese, nickel, and thallium. Preliminary separations, pH control, and the use of masking agents, such as salts of citrate, EDTA, or cyanide, can all be used to "fine tune" the selectivity. In the presence of thiosulfate and cyanide, for example, only zinc and tin(II) are extractable (at pH 4.5).

Both chloroform and carbon tetrachloride are used as solvents. However, the reagent itself and most metal complexes it forms are more soluble in chloroform. Kinetic effects can sometimes be manipulated by solvent choice. For example, mercury and copper dithizone complexes are both extracted at pH 1.4 with carbon tetrachloride, but due to slower kinetics in the case of copper, only mercury is extracted by chloroform [107].

Lead can be extracted from many different types of alloy samples using dithizone. Methods have been published for slab zinc [135], cadmium [136], magnesium alloys [137], various ferroalloys [138], and steel, among others [139]. These generally utilize cyanide ion as a masking agent during the extraction step. Bismuth in excess of 0.005% will interfere, but can usually be removed by backwashing the chloroform layer with a 10g/L solution of hydroxylamine hydrochloride which has been adjusted to pH 3.0 with dilute hydrochloric acid.

Collins and Kanzelmeyer developed a procedure for indium in zinc and zinc oxide which utilizes a preliminary separation with isopropyl ether from 6 molar hydrobromic acid solution [140]. Hubbard and Green extracted copper, lead, nickel, and zinc from high-purity tungsten at pH 9.2. They then back-extracted the chloroform layer with a benzoyl peroxide solution which had been adjusted to pH 2 with hydrochloric acid [141].

3.4.1.6 Miscellaneous Chelating Agents. There are obviously many more chelating reagents that have been used in the solvent extraction separation of tramps in alloys. Before closing this summary it is appropriate to briefly mention a few more that have found wide application. Dimethylglyoxime ($C_4H_8N_2O_2$), which is used as a near-specific precipitant for nickel and palladium, can also be used to extract these elements. Traces of nickel, for example, have been extracted with chloroform from magnesium and magnesium alloys from a weakly ammoniacal citrate solution (after bromine or iodine oxidation) [142]. A reagent sometimes used to precipitate cobalt, 1-nitroso-2-naphthol ($C_{10}H_7NO_2$), can also be used to extract palladium. Palladium can be separated from titanium and titanium alloys by extraction with this reagent and toluene from an aqueous solution containing ammonium and citrate ions [143]. Toluene-3,4-dithiol ($C_7H_8S_2$) is often used to separate traces of molybdenum or tungsten. Machlan and Hague developed a procedure for traces of tungsten in steel and in titanium alloys [144]. Donaldson extracted molybdenum from iron and steel with an α-benzoinoxime solution in chloroform [134].

3.4.2 Ion Association Complexes

The other major means by which a metal can transfer from an aqueous solution to an organic solvent is as an ion association which, considered as a unit, is effectively neutral. The extracted species may be complexes with anions or with coordination species. As in the case of neutral chelates, extraction efficiency requires that coordinated water be displaced. In the case of anion complexes, such as those between a metal cation and a halide (F^-, Cl^-, Br^-, I^-), or pseudohalide (CN^-, SCN^-), the central metal cation is encircled by species that do not interact strongly with water molecules. Such anionic complexes, associated with protons, extract best into oxygenated solvents which can act like Lewis bases. Here, the solvent plays a very active role in the removal of coordinated water. Morrison and Freiser have speculated that when iron (III) is extracted by diethyl ether from hydrochloric acid solution the species actually extracted is $\{(C_2H_5)_2O{:}H^+, FeCl_4[(C_2H_5)_2O]_2^-\}$.

Another type of ion association forms when an anionic complex associates, not with a proton, but with one or more large, positively charged molecules. A third type occurs when the metal cation reacts to form a large, charged metallorganic complex ion, which associates with an anion or cation to form an extractable species.

Some methods call for the addition of a large amount of nonextractable salt to the aqueous solution prior to extraction. This has the effect of increasing the activity of the analyte species and decreasing its solvation by water molecules. It is best to utilize a salt whose cation has a large radius (and charge). Ammonium ion is a frequent choice, although sodium, lithium, magnesium, aluminum, and other cations have been used. This technique has been successfully utilized in halide, nitrate, and thiocyanate systems.

Theoretical treatments of ion association extractions remain incomplete because the ionic concentrations generally are in a high range where ionic activities are influenced by many factors, and because many complex equilibria are always involved [1,107,108].

3.4.2.1 Chloride Systems. The solvent extraction of chloride complexes may be the most widely used liquid–liquid extraction technique in metals analysis. Between 1892 and 1901 the extraction of iron (III) from hydrochloric acid solution using diethyl ether developed from fundamental studies into a useful analytical tool [145–148]. If one plots the logarithm of the distribution coefficient for iron (III) over a range of hydrochloric acid concentrations the function describes a roughly parabolic curve with a maximum around 6M HCl. The same plot for diisopropyl ether peaks at around 8M HCl.

For both curves, the steep rise from low aqueous phase hydrochloric acid concentrations represents a sharp increase in the presence of the extractable $FeCl_4^-$ complex and a simultaneous decline in the activity of water molecules. Above the acid concentration corresponding to the maximum distribution coefficient value the ether begins to dissolve in the aqueous phase at a significant rate. This has the effect of lowering the effective hydrochloric acid concentration in the aqueous phase by dilution. Since diisopropyl ether is less soluble in water than diethyl ether the dilution effect is less pronounced and thus the extraction maximum occurs at a higher acidity.

In addition to iron (III), gallium (III), gold (III), molybdenum (IV), and thallium (III) are extracted by diethyl ether from 6M hydrochloric acid. Other metals extract best under somewhat different conditions: antimony (V) from 7.5M HCl with diisopropyl ether; arsenic (III) from 11M HCl with benzene; germanium (IV) from 10.5M HCl with carbon tetrachloride; mercury (II) from 0.1M HCl with a 3M solution of trichloroacetic acid in n-butyl acetate; niobium (V) from 11M HCl with pre-equilibrated diisopropyl ketone; platinum from 3M HCl with added stannous chloride. Impurities in gallium and gallium arsenide will be retained in the aqueous phase when 6-8M HCl sample solutions are extracted with diisopropyl ether. Adventitious application of this separation was utilized in early work with high-purity semiconductor materials [149–151]. The extraction of molybdenum from aluminum (from 6M HCl with diethyl ether) is particularly useful in eliminating the molybdenum line overlap interference when aluminum is determined using the 3961Å line in ICP-OES.

3.4.2.2. Other Halide Systems. The bromide system shows good diethyl ether extractability for thallium (III) over a broad range of hydrobromic acid molarities (1M–6M). Other elements are at optimal extraction efficiency over much narrower acidity ranges. Gold (III) is best at 3M HBr, gallium (III) at 5M HBr, indium (III) at 4M, antimony (V) at 5M, tin (II and IV) at 4M, and iron (III) at 4.5M.

Iodide systems are more widely utilized, often as potassium iodide in dilute sulfuric acid since hydroiodic acid is even more subject to photodecomposition than hydrobromic acid. Antimony (III), cadmium (II), gold (III), mercury (II), and tin (II) extract well from 6.9M HI with diethyl ether, while indium (III), cadmium (II), and tin (II) extract best from 1.5M KI/1.5N H_2SO_4. Under these latter conditions iron (II) and aluminum (III) are not extracted, and copper only to a minor extent.

Luke {152] extracted traces of antimony, bismuth, cadmium, copper, indium, and lead from iron and steel, zinc, magnesium, nickel, cobalt, titanium metal, and titanium alloys, using the iodide system. The sample materials were dissolved, then reacted with sodium iodide in the presence of a small amount of hydrochloric acid. Ascorbic acid was added and the solutions were extracted with methylisobutyl ketone (MIBK) [152]. Headridge and Richardson separated bismuth from cast irons and stainless steels by extraction with MIBK from a solution which was 2.3M HCl/0.09M ascorbic acid/0.2M potassium iodide. Antimony, copper, and tellurium were found to be quantitatively extracted as well under these conditions. Arsenic, lead, and molybdenum accompanied these elements but were incompletely extracted [153]. Donaldson separated tin from iron, steel [154], and copper-, zirconium-, titanium-, aluminum-, and zinc-base alloys by a toluene extraction from a 3M H_2SO_4/1.5M potassium iodide solution containing both tartaric and ascorbic acids. The tin is then back-extracted into an aqueous phase with a solution which is 8% H_2SO_4/4% HNO_3/16% HCl (all v/v) [109]. An alternative approach, suitable for trace levels of tin, using 2M H_2SO_4/1.5M potassium iodide and a back extraction with dilute sodium hydroxide was applied to the same array of alloys (except titanium-base) [109,155].

Fewer elements are extracted from fluoride solutions than from all other halide or pseudohalide media. MIBK which has been pre-equilibrated with the solution medium will extract niobium (V) and tantalum (V) from 10M HF/6M H_2SO_4/2.2M NH_4F. Diethyl ether will extract tin (II) from 4.6M HF and tin (IV) over the range 1.2-4.6M HF. Some significant amount of arsenic (III) will extract at 4.6M HF, and a significant amount of rhenium (VII) will extract at 20M HF, but neither is quantitative. Schlewitz and Shields used an MIBK extraction from 0.6M HCl/1.4M HF to separate low levels of tantalum from samples of zirconium and zirconium alloys [156].

3.4.2.3 Nitrate Systems. This solvent extraction medium is used intensively in the nuclear industry since, in addition to uranium (VI), other actinides extract efficiently [notably plutonium (VI), americium (VI), and neptunium (VI)]. Ether will extract cerium (IV) and gold (III) from 8M HNO_3. The addition of salts to the aqueous phase is usually very effective in increasing extraction efficiency in this system. These effects have been noted for scandium (III), and thorium(IV). Salts, such as ammonium nitrate, are commonly added in the ether extraction of uranium from nitrate solutions [107].

3.4.2.4. Thiocyanate Systems. In 0.5 M hydrochloric acid beryllium extracts best into diethyl ether at a CNS concentration of 7M, gallium (III) and scandium (III) also at 7 molar, titanium (III) and zinc (II) at 3M, and iron (III) at 1M. Tin (IV) and molybdenum (V) extract with high efficiency across the range 1–7M. Cobalt (II), indium (III), and uranium (VI) also extract, but incompletely. Many thiocyanates are strongly colored and suitable for spectrophotometric measurement of the extracted complex. Examples are iron (red), molybdenum(red), tungsten (yellow), and rhenium (yellow); the latter three in the presence of stannous chloride.

3.4.2.5. Organophosphorus Complex Systems. The usefulness of organophosphorus compounds for the solvent extraction of metals has been known since the early days of the nuclear industry. As with ion exchange, it was developments in the nuclear field that drove research with this class of reagents. The organophosphorus acids are one important subcategory. These compounds show various degrees of polymerization in different solvents. Mono-2-ethylhexylphosphoric acid [$(C_8H_{17}O)PO(OH)_2$] has been applied to the extraction of the rare earths and the actinides. Dibutylphosphoric acid [$(C_4H_9O)_2POOH$] and di(2-ethylhexyl)phosphoric acid [$(C_8H_{17}O)_2POOH$] have, likewise, found application in the nuclear field, but elements such as beryllium, hafnium, indium, titanium, and zirconium show such good recoveries with both reagents that wide applications are feasible. The use of these two compounds to extract the rare earths is also of considerable importance.

The trialkylphosphates are best exemplified by tri-n-butyl phosphate [TBP, $(C_4H_9O)_3PO$]. The alkaline earths, aluminum, antimony(V), bismuth, cadmium, chromium (III), cobalt (II), copper (II), gallium (III), germanium (IV), gold (III), hafnium, indium, iron (III), manganese (II), molybdenum (VI), niobium, lead, palladium, platinum, the rare earths, selenium (IV), tantalum, tin (IV), titanium (IV), tungsten (VI), uranium (VI), vanadium (V), and zirconium have all been extracted from some aqueous solution medium at greater than 90% efficiency. TBP is a thin liquid, nearly insoluble in water, and thus is often used without a diluting solvent. For some applications it is dissolved in a non-polar solvent, or occasionally in a C_4 or C_8 alcohol [1]. Triisooctyl thiophosphate [TOTP, $(C_8H_{17}O)_3PO$] has been used to extract silver traces [157] and to remove a silver matrix for the trace determination of copper and iron [158].

The most widely useful class of organophosphorus extractants in trace metals analysis is the phosphine oxides. By far the most important compound has proven to be tri-n-octylphosphine oxide [TOPO, $(C_8H_{17})_3PO$], a stable white powder, soluble in many solvents. Like the other organophosphorus extractants TOPO is a nonselective group extractant, but it shows efficiencies and group selectivities that are extremely useful. In general, emulsions are less of a problem than with TBP. TOPO exhibits different extraction properties with polar and nonpolar solvents.

The most popular nonpolar solvent is cyclohexane. A 0.1M solution of TOPO in cyclohexane will completely extract the following from 1M HCl: chromium (VI), gold (I), hafnium (IV), iron (III), molybdenum (VI), tin (IV), uranium (VI), and zirconium (IV). At 7M HCl the following are added to the list: antimony (III), gallium (III), titanium (IV), and vanadium (IV) [107]. Nitric acid media is also used, sometimes with the addition of nitrate salt to the aqueous phase. Young and White [159] developed a procedure for the extraction of zirconium, which was applied by Wood and Jones [160] to reactor grade niobium, and by Ratcliffe and Byford to steels [161].

Burke first applied a 5% (w/v) solution of TOPO in MIBK to the extraction of trace levels of antimony, bismuth, lead, and tin from a 10% (v/v) HCl/ 6% (w/v) KI/ 2% (w/v) ascorbic acid solution [162]. The initial work was applied to a broad range of nickel- and iron-base alloys, as well as to aluminum alloys. Later publications confirmed the utility of this separation approach [163] and expanded the scope to include the extraction of traces of thallium [164] and silver [165] from 10% HCl/9% KI/2% ascorbic acid. Bedrossian later applied the separation to trace tellurium in steels [166].

If nitric acid is used to dissolve the sample it must be removed by reducing to low volume and cautiously adding formic acid or a urea solution to the hot sample until no NO_x gases are evolved. High copper samples present difficulty because copper(I), which forms from the reduction by ascorbic acid, reacts to form insoluble CuI. A preliminary ammonium hydroxide precipitation separation (with added iron for very low iron alloys) removes copper into the filtrate. However, some workers have used minute amounts of nitric acid with micro-sized test portions of copper-base metals and avoided any difficulties for the subsequent measurement [167].

Some work has been done with tris(2-ethylhexyl)phosphine oxide [TEHPO, $(C_8H_{17})_3PO$], an oily liquid, soluble in many solvents. It is an isomer of TOPO which extracts fewer elements due to steric hindrance. Gold (I), tin (IV), and uranium (VI) extract from 1M HCl, and gallium (III), hafnium (IV), iron (III), molybdenum (VI), tin (IV), uranium (VI), and zirconium (IV) extract from 7M HCl [107]. Ross and White applied the reagent to the determination of tin in lead-, zinc-, copper-, iron-, and zirconium-base alloys. Tin is extracted with a 0.01M solution of TEHPO in cyclohexane from an aqueous phase that is 1M HCl/2.5M H_2SO_4 [168].

3.4.2.6. Other Ion Association Systems. The heteropoly acids formed by molybdenum, phosphorus, vanadium, arsenic, silicon, and tungsten are all extractable by oxygen-bearing solvents. Frequently employed solvents are 1-butanol, and isobutanol. Sometimes the extracted species is reduced and then back-extracted into the aqueous phase. These extractions are all used very extensively in metals analysis. Other large molecule association systems are also frequently applied to metals. Copper, iron, and nickel have been extracted from uranium with bathophenanthroline/MIBK [169]. Copper is extracted from steel [59], nickel [170], magnesium alloys [142], and cadmium [136], among other matrices, with neocuproine/chloroform.

There are many other important ion association systems—high molecular weight amines, carboxylic acids, and many large molecular complexes. For additional information the reader is referred to [107] and [108].

3.4.3 Safety Considerations

While all laboratory operations carry potential hazards, those associated with solvent extraction are too often underestimated. Among the most important hazards in this work are the flammability of many of the solvents and the chronic and acute toxicity hazards associated with many of the reagents, including solvents.

Laboratories in the U.S. are required to maintain a complete file of Material Safety Data Sheets (MSDSs) which contain detailed hazard information on reagents. The analyst needs to become familiar with Threshold Limit Values (TLVs) and Permissible Exposure Limits (PELs), as well as flashpoint temperatures for the relevant compounds. An efficient hood, gloves, and appropriate safety equipment should be used when working with any substance with a published TLV of less than 50 ppm, or with reagents whose reaction could generate such a substance. It is prudent to regard all chemicals as hazardous, however, and extreme care should be taken to avoid all possibility of inhalation, ingestion, or skin contact.

When working with flammable solvents the room should be posted with appropriate signs and all flames and spark sources (including motors and switches) should be secured. Ethers are a particular concern due to their high vapor pressure and extremely low flashpoints. An ignition source even a long distance away is an imminent danger. Ethers also form explosive peroxides in storage. Stocks should be kept low and old reagent properly disposed of.

Three excellent sources of information on reagent safety are: *Prudent Practices for Handling Hazardous Chemicals in Laboratories* (National Research Council, National Academy Press, Washington, DC, 1981); *Prudent Practices for Disposal of Chemicals from Laboratories* (National Research Council, National Academy Press, Washington, DC, 1983); and *Safe Storage of Laboratory Chemicals* (D.A. Pipitone, ed., John Wiley and Sons, NY, 1991).

3.5 Volatilization

In this section we will briefly discuss those reactions which are intended to produce a volatile compound, either to remove an interferent, or to evolve, collect, and thus preconcentrate the analyte.

3.5.1 Removal of Interferences

Silicon is frequently a problem for the tramp analysis of metal alloys since it tends to remain as an insoluble oxide with many acid dissolution schemes. Its presence tends to clog filtering media, adhere to glassware, and sometimes encapsulate undissolved particles of the alloy test portion. Adding a few drops of hydrofluoric acid to the sample solution, then fuming with sulfuric or perchloric acid is a most effective means of volatilizing it as hydrofluosilicic acid ($H_2SiF_6 \cdot xH_2O$). When sulfuric acid and hydrofluoric acid are heated with a dry, silica-bearing residue the volatilized species is silicon tetrafluoride (SiF_4). Sometimes sulfuric acid must be avoided because the sample matrix contains significant amounts of elements that form insoluble sulfates, which could spatter, resulting in analyte loss. Sometimes both perchloric and sulfuric acids must be avoided because of a spattering problem. In these cases substituting nitric acid is sometimes effective.

Chromium may be undesirable because of its many oxidation states, or because of the strong background color it contributes to the sample solution. Or it may represent a line interference in ICP-OES (as it is, for example, on Ca 3179Å). It can be removed to a large extent by heating the sample solution to strong fumes of perchloric acid, then, while fuming, introducing chloride ion to generate chromyl chloride (CrO_2Cl_2), a dark brown vapor at such temperatures.

The most efficient means of introducing chloride is as dry hydrogen chloride gas from a cylinder using neoprene tubing connected to a bent glass tube which is cautiously inserted just above the fuming liquid. A constant boiling azeotrope generator is another approach. This is just a flask containing hydrochloric acid and some glass boiling beads, fitted with a neoprene stopper and glass tube for inserting above the fuming sample. Dropwise additions of concentrated hydrochloric acid directly into the fuming solution will work, but spattering is a danger, and each addition cools the sample to such an extent that it is necessary to wait to attain strong $HClO_4$ fumes again before the addition can be repeated. Adding small portions of sodium chloride crystals is very effective without excessively cooling the sample, but it adds sodium to the solution, which may be a problem in some methods.

In the analysis of high tin alloys it is sometimes advantageous to remove the tin since it tends to hydrolyze as a white, insoluble hydrous oxide. This can be accomplished by first dissolving the alloy in hydrobromic acid, or in a mixture of hydrobromic acid and bromine, then adding perchloric acid to the cooled sample and heating to strong $HClO_4$ fumes. Tin is removed as stannic bromide ($SnBr_4$).

While tin can be removed as a bromide in either the +II or +IV oxidation states, arsenic, which is an interferent for phosphomolybdate spectrophotometric procedures must be in the +III state. This can be accomplished by cooling a previously fumed perchloric acid solution to room temperature, then adding 1:4 hydrobromic acid:water and heating very gradually to strong fumes again. Germanium, which also interferes in some heteropoly blue procedures for phosphorus, can be removed by the same technique [171].

3.5.2 Preconcentration by Distillation

Table 3.4 summarizes the distillation procedures that have found the broadest use in metals analysis. All are applicable to trace levels of the analyte with appropriate analytical parameters. Most require some type of specialized distillation apparatus.

Omitted from Table 3.4 are the hydride techniques in which a solution of sodium borohydride is employed to generate gaseous hydrides of antimony, arsenic, bismuth, germanium, lead, selenium, and tin. Since these are either measured immediately or frozen out in a cold trap and then vaporized and measured, they are best discussed when atomic absorption is treated. Also omitted are two elements which form volatile compounds, but which are seldom separated in this manner—namely ruthenium and rhenium. Ruthenium has a volatile tetroxide, RuO_4, which decomposes below its boiling point. The oxyfluoride, $RuOF_4$, however, boils at 184°C and can be evolved away from fuming perchloric acid solution. Rhenium forms a volatile oxide, Re_2O_7 (b.p.362°C) 30% of which will be lost from a fuming perchloric acid solution. Typically, little analytical use is made of these phenomena, however.

3.6 Other Separations

There are a few other techniques applicable to the separation of analytes from their matrix, or matrix interferents from the analyte. Perhaps the most impor-

Table 3.4 Important Distillations in the Trace Analysis of Metals

Element	Oxidation State	Distilling Medium	Volatile Species	Application Example	Reference
Antimony	+III	$HCl/H_2SO_4/H_3PO_4$	$SbCl_3$	Minerals	174
Arsenic	+III	HCl/H_2SO_4	$AsCl_3$	Pb-,Sn-, Cu-Alloys	172,173
Boron	+III	H_2SO_4	$B(OCH_3)_3$	Steel	59
Fluorine	−I	$HClO_4$	$H_2SiF_6 \cdot xH_2O$	Slags, Minerals	174
Germanium	+IV	HCl/Cl_2	$GeCl_4$	Minerals, ZnO	174
Nitrogen	−III	NaOH	NH_3	Steel, Nitrides	174
Osmium	+VIII	$HClO_4$	OsO_4	Platinum Group	175
Selenium	+IV	HCl(gas)/ dilute H_2SO_4	$SeCl_4$	Minerals	174
Sulfur	−II	HCl/ HCOOH	H_2S	Steel (ultra-trace)	176
Tin	+II,+IV	$HCl/HBr/H_2SO_4$	$SnBr_2$, $SnBr_4$	Steels, Minerals	174

tant of these is the comparatively new approach, extraction chromatography. In this methodology a porous support is coated with an extractant (the stationary phase). The porous support may be an organic polymer, diatomaceous earth, silica gel, or other inert hydrophobic material. The stationary phase may be any extractant—tributyl phosphate (TBP), tri-n-octylphosphine oxide (TOPO), α-benzoinoxime, dithizone, and many more. The combined effect of chromatographic separation and chemical selectivity can produce a high level of resolution. Unfortunately, with the exception of nuclear applications, this methodology seems to have arrived too late for industry, where the needed development work, in many cases, can no longer be accomplished due to lack of trained personnel. Perhaps in some future renaissance of wet analytical chemistry this technology will return to become a dominant theme.

Even more recent is the use of liquid ion exchangers as liquid–liquid extractants. High molecular weight amines have been described as anion exchangers (e.g., tri-n-octylamine) and high molecular weight acids [such as bis-(2-ethylhexyl)phosphoric acid] have been described as cation exchangers. Work in this area is expected to expand.

By contrast, fire assay is a technique that is 3,000 years old, antedating the modern science of analytical chemistry by 28 centuries. "Modern" versions of the basic technique still are considered the last word on the assay of gold and silver. And trace impurities are primarily what are removed by the entire artful process.

4. DEALING WITH INTERFERENCES

There are, of course, alternatives to separation. Many of these are instrumental or mathematical, as we shall observe in the ensuing chapters, but some are *chemical*.

4.1 Masking Agents

A masking agent is a reagent added to a solution to prevent an undesired reaction from occurring. The term, thus, has meaning only in a specific chemical context. Generally, the term is restricted to refer to an additive that undergoes an aqueous-based chemical reaction to perform its intended function. The masking reaction may be a precipitation (such as the addition of silver ion to remove chloride), but more typically it is a complexing reaction.

Many chromophoric agents used in molecular absorption spectrophotometry are nonspecific. The solution conditions, such as pH and ionic strength are, of course, optimized for the chromogenic reaction with the desired analyte, but generally there remain some other ions from the sample or the reagents that will also react to form a color. If the absorption spectrum of the interferent is similar to that of the analyte there is a problem. And if the background at the analytical wavelength is large it may be impossible to obtain accurate results, even with the careful measurement of reagent and sample blanks. The answer is often the addition of a complexing agent that reacts with the interferent, preventing its reaction with the chromophore. An alternative situation occurs when some species from the sample or the reagents prevents the completion of the desired analyte reaction. And again a masking agent to block this effect can often be added.

Masking agents are also used in precipitation separations and gravimetric analysis to prevent the precipitation of undesired species and in solvent extraction to ensure that the extract is not contaminated with an unmanageable interferent. Masking agents are also utilized in titrimetry, voltammetry, and other techniques where a chemical reaction is linked to the analyte measurement.

There are hundreds of reagents which in one context or another can serve as masking agents, but strong complex formers are the most frequently used. Examples of these are: EDTA, cyanide, thiocyanate, citrate, tartrate, oxalate, thiosulfate, and pyrophosphate.

4.2 Matrix Modifiers

This term arose with the development of graphite furnace atomic absorption techniques, although a good case could be made that optical emission workers had been doing something very similar for decades. Matrix modifiers are solution additions that react with interferents, usually to form a volatile species that is removed during the char cycle of a GFAA analysis. They may also be used to condition the graphite furnace walls or platform, or to form a compound with the analyte that has good atomization characteristics. Here, we can apply the term to any substance added to the sample matrix to improve the characteristics of a physical measurement step.

In graphite furnace work a good example is the use of ammonium nitrate or ammonium-EDTA to evolve out ammonium chloride during the char. This eliminates the interference due to chloride since, otherwise, volatile chloride compounds of the analyte might be lost. Nickel nitrate is sometimes added if nickel is low in the sample matrix. Nickel contributes by forming intermetallic compounds with the analyte (such as nickel selenide, for example) that show clean, reproducible peaks during atomization. Magnesium nitrate, palladium, and certain phosphates are frequently recommended as additives. Furnace conditioning agents are generally reagents that form refractory carbides or are reduced to a refractory metal by the carbon in the walls or platform.

One could argue that optical emission spectrographers have been applying similar strategies for a long time. So-called spectrographic "buffers" and "carriers" are blended with powder samples before they are packed into crater electrodes and arced. These are reagents designed to produce favorable vaporization characteristics so that a reproducible plume of analyte-bearing plasma is produced.

4.3 Ionization Suppressors

These are compounds that are added to a sample solution intended for flame atomic absorption (AA) analysis in order to prevent, or at least moderate, ionization interference during the measurement. In flame AA, certain analytes tend to be excited beyond the desired ground state vapor to an ionized form which emits, rather than absorbs light. Ionization suppressors are compounds that contribute readily ionized elements to the sample matrix. These elements tend to prevent the ionization of analyte by being more readily ionized themselves. In effect they absorb sufficient flame energy to allow all or most of the analyte to remain in the ground state.

Typical ionization suppressors are lanthanum, strontium, and sodium compounds. All show bright visible emission in the atomic absorption flame. Obviously, ionization suppressors must be added in the same form and concentration to samples, calibration standards, and blanks.[*]

4.4 Internal Standards

This term is used in many analytical methodologies. In inorganic trace analysis it applies primarily to ICP-OES, DCP-OES, and ICP-MS techniques, where it is used to indicate a correction scheme for the effect of differences in the mass transfer rate of aspirated solution samples. In these techniques calibration and measurement are based on a rate signal which relates to the mechanical transfer of a solution to the spray chamber and the corresponding transfer of the nebulized droplets to the plasma.

In many trace methods a series of sample solutions are fumed in sulfuric or perchloric acids, or combinations of these with phosphoric acid. In such situations after dilution it is nearly impossible to arrive at a series of solutions with precisely the same viscosity. Differences in viscosity lead to different solution uptake rates, which introduces an uncontrolled variability in the rate signal. Also, sometimes it is convenient or necessary to utilize calibration

[*] There are other "suppressors" in trace analysis, notably the oxygen-wave suppressors, like gelatin, used in DC polarography.

standards that are prepared in a somewhat different solution medium than the test samples. This frequently leads to the same sort of uncontrolled variation.

The answer to these difficulties is to add to samples, blanks, and standards the same precise amount of a substance that exhibits a clean spectral line that can be measured for correction of the analyte response. The internal standard must be an added substance that is not expected to occur in the sample, or something present in all samples and standards in an exactly known concentration.

In work with many materials scandium or indium are frequently used as ICP-OES additives, although, depending on the analyte matrix, other elements may serve. In ICP-MS several different internal standards are commonly added to provide reference peaks at regular mass to charge intervals throughout the entire scanned spectrum. Variations in the internal standard response are used to correct the analyte response, effectively canceling the error due to viscosity effects. In steel analysis sometimes the response of the iron as the base metal is used as the internal standard. A similar approach has been applied in other types of alloys where one element constitutes 90% or more of the matrix. This is a weaker use of the technique since minor variations in the base metal content, although they represent a small correction error, produce an uncontrolled variation in the result. Also, the base metal content of the calibration materials cannot be known as accurately as an added aliquot of indium, for example, and so the accuracy of the result suffers.

4.5 Matrix Matching

An important way around interference problems in solution-based OES techniques is matrix matching. This is the preparation of blanks and calibration solutions that closely match the concentrations of the sample and reagent components. This sounds obvious and simple in principle, but it presents a dilemma for the analyst. How closely must the matrix be matched to produce the desired accuracy in the analytical result? Usually it is impossible to say with complete confidence until results are obtained on certified reference materials. But if closely matching certified reference materials were available there would be no need to resort to the preparation of a synthetic matrix.

The best approach is one which is guided by informed knowledge about interference effects. While much useful information can be gleaned from published texts that list emission lines and illustrate spectra, actual measurements with pure solutions at appropriate concentration levels are essential. This means measuring the effect of each matrix component at the analyte wavelength. With multi-element polychromators this task should be a part of the original instrument setup. A large data array is thus formed from known concentrations of every element for which a channel is available. "Apparent concentrations" are recorded from all the nonanalyte channels. With a scanning monochromator instrument, the same sort of array using frequently used wavelengths can be likewise generated, albeit with a somewhat greater effort.

Guided by these apparent concentrations one can make good judgements about which matrix elements need to be in the blanks and calibrants, and how closely they must match their concentrations in the sample solution. For large apparent concentrations, matrix matching is a more accurate approach than a mathematical correction utilizing so-called interelement correction factors (IECs).

5. THE CHEMISTRY OF MEASUREMENT

It remains to say a few general words about the chemistry involved in the measurement process itself. As we address each measurement category in the subsequent chapters much more detail will be added.

In UV/visible spectrophotometry the optical density of a colored solution is related to analyte concentration by a Beer's Law calibration curve. At the physicochemical level the source radiation, generally in the 200–800 nm range, promotes changes in the electron energy levels of the colored molecular species and is thus partially absorbed. Careful blanking procedures are usually necessary to compensate for absorbance by the reagents and the sample matrix. In certain spectrophotometric procedures a chemical reaction is used to destroy the colored species after it is measured so that an accurate blank can then be measured.

The measurement step in voltammetric techniques is, of course, electrochemical. The analyte undergoes a redox reaction driven by an applied potential, and the current flow is measured. Diffusion-based reactions are necessary, and so many approaches take precautions to avoid even inadvertent stirring of the solution. Dissolved oxygen from the atmosphere is an important interferent in some methodologies, producing an oxygen wave that obscures the analyte response. Besides using additives to suppress this effect, a nitrogen sheath over the surface of the solution is often employed.

Atomic absorption, like molecular absorption, involves the promotion of electrons to higher energy levels. Here, the absorbing medium is a plume of ground state atoms. A great deal of high temperature chemistry is involved in producing that plume, whether it is in a premixed gas flame, or confined in a resistance heated graphite tube. Additives, sample solution, and flame or furnace parameters must be "fine tuned" to produce just the right conditions to maximize the analyte's ground state atom population without producing harmful amounts of emission.

The opposite situation prevails in optical emission measurement. Conditions must be chosen to produce the most stable population of highly excited or ionized analyte ions. As these species drop back to lower or ground state levels they emit the characteristic measured wavelengths. The time required to achieve a stable emission is the preburn, which precedes the initiation of the measurement. Sometimes additives in powder/crater electrode techniques can contribute here. The carrier distillation approach and other related techniques are designed to create conditions that produce a stable, sensitive analytical measurement.

In ICP-MS the reagent matrix is a serious potential source of interference since anions such as chloride form polyatomic ions which often intrusively obscure analyte peaks. The solution matrix in ICP-MS must be carefully designed to avoid potential problems. The high temperature chemistry in the plasma will directly effect the quality of the resolved spectrum. ICP-MS is also limited in its ability to handle high dissolved solids loading. Orifice clogging problems limit most systems to 0.1g/100mL. To obtain a representative test portion from a moderately inhomogeneous sample an aliquotting scheme is thus required. Internal standards are necessary, sometimes four or more, to adequately correct the measurements for an array of analytes.

REFERENCES

1. Minczewski, J.; Chwastowska, J.; Dybczynski, R. Separation and Preconcentration Methods in Inorganic Trace Analysis (trans. ed.: M.R. Mason), Ellis Horwood/John Wiley, NY, 1982.
2. Lewis, L.L.; Straub, W.A. Analytical Chemistry, 32, 96 (1960).
3. Hague, J.L.; Brown, E.D.; Bright, H.A. Journal of Research of the National Bureau of Standards, 53, 261 (1954).
4. Hague, J.L.; Machlan, L.A. Journal of Research of the National Bureau of Standards, 62, 11 (1959).
5. Hague, J.L.; Machlan, L.A. Journal of Research of the National Bureau of Standards, 65A, 75 (1961).
6. Machlan, L.A.; Hague, J.L. Journal of Research of the National Bureau of Standards, 66A, 517 (1962).
7. Kallmann, S.; Oberthin, H.; Liu, R. Analytical Chemistry, 34, 609 (1962).
8. Kallmann, S.; Oberthin, H. Analytical Chemistry, 37, 280 (1965).
9. Kolthoff, I.M.; Sandell, E.B. Textbook of Quantitative Inorganic Analysis, 3rd ed., Macmillan, NY, 1952.
10. Day, Jr., R.A.; Underwood, A.L. Quantitative Analysis, 2nd ed., Prentice-Hall, Englewood Cliffs, NJ, 1967.
11. Stokes, H.N.; Cain, J.R. Journal of the American Chemical Society, 29, 409 (1905).

12. Blumenthal, H. Zeitschrift für das analytische Chemie, 74, 33 (1928).
13. Kameyama, N.; Makishima, S. Journal of the Chemical Society (Japan), 36 346 (1933).
14. Yao, Y. Industrial and Engineering Chemistry, Analytical Edition, 17, 114 (1945).
15. Park, K.B. Industrial and Engineering Chemistry, Analytical Edition, 6, 189 (1934).
16. Pyburn, C.M.; Reynolds, G.F. Analyst, 93, 375 (1968).
17. Luke, C.L. Industrial and Engineering Chemistry, Analytical Edition, 15, 626 (1943).
18. Tsuiki, M. Journal of the Electrochemical Society (Japan), 29, 42 (1961).
19. Reynolds, G.F.; Tyler, F.S. Analyst, 89, 579 (1964).
20. Rooney, R.C. Analyst, 82, 619 (1957).
21. Amsheev, A.A. Zavodskaya Laboratorya, 34, 789 (1968).
22. Ogden, D.; Reynolds, G.F. Analyst, 89, 538 (1964).
23. Kallmann, S.; Pristera, F. Industrial and Engineering Chemistry, Analytical Edition, 13, 8 (1941).
24. Biskupsky, V.S. Analytica Chimica Acta, 46, 1496 (1969).
25. Katzin, L.I.; Stoughton, R.W. Journal of Inorganic Nuclear Chemistry, 3, 229 (1956).
26. Yamagata, Y.; Iwashima, T. Nature, 200, 52 (1963).
27. Geilmann, W.; Neeb, K. Zeitschrift für das analytische Chemie, 165, 251 (1959).
28. Luke, C.L. Analytical Chemistry, 31, 1680 (1959).
29. Burke, K.E. Analytical Chemistry, 42, 1536 (1970).
30. Blakeley, St.J. H.; Manson, A.; Zatka, V.J. Analytical Chemistry, 45, 1941 (1973).
31. Vassilaros, G.L. Talanta, 21, 803 (1974).
32. Dulski, T.R.; Bixler, R.R. Analytica Chimica Acta, 91, 199 (1977).
33. Donaldson, E.M. Talanta, 24, 105 (1977).
34. Donaldson, E.M. Talanta, 23, 823 (1976).
35. Donaldson, E.M. Talanta, 26, 1119 (1979).
36. Zörner, A.; Krath, E.; Feucht, H. Techn. Mitt. Rheinhausen, 4, 237 (1955).
37. Mizuike, A. Hiraide, M. Analytica Chimica Acta, 69, 231 (1974).
38. Marczenko, Z. Chemical Analysis (Warsaw), 11, 347 (1966).
39. Reichel, W.; Bleakley, B.G. Analytical Chemistry, 46, 59 (1974).
40. Donaldson, E.M. Talanta, 26, 999 (1979).
41. Luke, C.L. Analytical Chemistry, 31, 572 (1959).
42. Vassilaros, G.L. Talanta, 18, 1057 (1971).
43. Marczenko, Z.; Krasiejko, M. Chemical Analysis (Warsaw), 9, 291 (1964).
44. Tertipis, G.G.; Beamish, F.E. Analytical Chemistry, 32, 486 (1960).
45. Kimura, M. Talanta, 24, 194 (1977).
46. Berndt, H.; Jackwerth, E.; Kimura, M. Analytica Chimica Acta, 93, 45 (1977).
47. Kimura, M.; Kawanami, K. Talanta, 26, 901 (1979).
48. Center, E.J.; Overbeck, R.C.; Chase, D.L. Industrial and Engineering Chemistry, Analytical Edition, 23, 1134 (1951).
49. Rieman III, W.; Walton, H.F. Ion Exchange in Analytical Chemistry, Pergamon, Oxford, 1970.

50. Kraus, K.A.; Moore, G.E.: Journal of the American Chemical Society, 75, 1460 (1953).
51. Jentzsch, D. et al. Zeitschrift für das analytische Chemie, 144, 8 (1955).
52. Kraus, K.A.; Nelson, F. "Metal Separations in Ion Exchange" Proceedings of the Symposium on Ion Exchange and Chromatography in Analytical Chemistry, ASTM STP195, American Society for Testing and Materials, West Conshohocken, PA, 1958.
53. Kraus, K.A.; Nelson, F. Proceedings of the First International Conference on the Peaceful Uses of Atomic Energy, Vol.7, United Nations, NY, 1956, pp.113,131.
54. Hague, J.L.; Maczkowski, E.E.; Bright, H.A. Journal of Research of the National Bureau of Standards, 53, 353 (1954).
55. Lewis, L.L.; Straub, W.A. Analytical Chemistry, 32, 96 (1960).
56. Jones, S.L. Analytica Chimica Acta, 21, 532 (1959).
57. Seim, H.J. "The Application of Ion Exchange to the Determination of Impurities in Aluminum and Aluminum Alloys" Chemical Analysis of Metals, ASTM STP944 (F.T. Coyle, ed.), American Society for Testing and Materials, West Conshohocken, Pa, 1987, pp. 17-30.
58. Annual Book of ASTM Standards, Vol.03.05, American Society for Testing and Materials, West Conshohocken, PA, 1996, Designation E34.
59. Annual Book of ASTM Standards, Vol.03.05 , American Society for Testing and Materials, West Conshohocken, PA, 1996, Designation E350.
60. Kraus, K.A.; Nelson, F.; Smith, G.W. Journal of Physical Chemistry, 58, 11 (1954).
61. Ishida, K.; Kiriyama, T.; Kuroda, R. Analytica Chimica Acta, 41, 537 (1968).
62. Kuroda, R.; Ishida, K.; Kiriyama, T. Analytical Chemistry, 40, 1502 (1968).
63. McCrackan, J.D.; Vecchione, M.C.; Longo, S.L. Atomic Absorption Newsletter, 8, 102 (1969).
64. Berg, E.W.; Truempner, J.T. Analytical Chemistry, 30, 1827 (1958).
65. Faris, J.P. Analytical Chemistry, 32, 520 (1960).
66. Hague, J.L.; Brown, E.D.; Bright, H.A. Journal of Research of the National Bureau of Standards, 53, 261 (1954).
67. Hague, J.L.; Machlan, L.A. Journal of Research of the National Bureau of Standards, 62, 11 (1959).
68. Kallmann, S.; Oberthin, H.; Liu, R. Analytical Chemistry, 34, 609 (1962).
69. Kallmann, S.; Oberthin, H.K. Analytical Chemistry, 37, 280 (1965).
70. Wilkins, D.H. Talanta, 2, 355 (1959).
71. Dixon, E.J.; Headridge , J.B. Analyst, 89, 185 (1964).
72. Imakita, T.; Fudagawa, N.; Kubota, M. Analyst, 115, 1185 (1990).
73. Bornong, B.J.; Moriarity, J.L. Analytical Chemistry, 34, 871 (1962).
74. Faris, J.P.; Buchanan, R.F. Analytical Chemistry, 36, 1158 (1964).
75. Strelow, F.W.E.; Bothma, C.J.C. Analytical Chemistry, 39, 595 (1967).
76. Hague, J.L.; Machlan, L.A. Journal of Research of the National Bureau of Standards, 65A, 75 (1961).
77. Machlan, L.A.; Hague, J.L. Journal of Research of the National Bureau of Standards, 66A, 517 (1962).
78. Bandi, W.R.; Buyok, E.G.; Lewis, L.L.; Melnick, L.M. Analytical Chemistry 33, 1275 (1961).

79. Green. H. Metallurgia. 76. 223 (1967).
80. Kashuba. A.T.: Hines. C.R. Analytical Chemistry. 43. 1758 (1971).
81. Kirk. M.: Perry. E.G.: Arritt. J.M. Analytica Chimica Acta. 80. 163 1975).
82. Nydahl. F. Analytical Chemistry. 26. 580 (1954).
83. ISO International Standard 4934: 1980. International Organization for Standardization (ISO).
84. Davies. J.E.: Grzeskowiak. R.: Mendham. J. Journal of Chromatography. 201. 305 (1980).
85. McLoed. C.W.: Cook. I.G.: Worsfold. P.J.: Davies. J.E.: Queay. J. Spectrochimica Acta. 40B. 57 (1985).
86. Epstein. M.S.: Koch. W.F.: Epler. K.S.: O'Haver. T.C. Analytical Chemistry. 59. 2872 (1987).
87. Tompkins. E.R.: Khym. J.X.: Cohn. W.E. Journal of the American Chemical Society. 69. 2769 (1947).
88. Spedding. F.H.: Voigt. A.F.: Gladrow. E.M.: Sleight. N.R. Journal of the American Chemical Society. 69. 2777 (1947).
89. Marinsky. J.A.: Glendenin. L.E.: Coryell. C.D. Journal of the American Chemical Society. 69. 2781 (1947).
90. Spedding. F.H. et al. Journal of the American Chemical Society. 69. 2786 (1947).
91. Harris. D.H.: Tompkins. E.R. Journal of the American Chemical Society. 69. 2792 (1947).
92. Ketelle. B.H.: Boyd. G.E. Journal of the American Chemical Society. 69. 2800 (1947).
93. Story. J.N.: Fritz. J.S. Talanta. 21. 892 (1974).
94. Hwang. J-M: Shih. J-S: Yeh. Y-C: Wu. S-C Analyst. 106. 869 (1981).
95. Elchuk. S.: Cassidy. R.M. Analytical Chemistry. 51. 1434 (1979).
96. Kuroda. R. Journal of Chromatographic Science. 28. 606 (1990).
97. Barkley. D.J.: Blanchette. M.: Cassidy. R.M.: Elchuk. S. Analytical Chemistry. 58. 2222 (1986).
98. Strelow. F.W.E.: Victor. A.H. Talanta. 37. 1155 (1990).
99. Strelow. F.W.E. Analytical Chemistry. 31. 1201 (1959).
100. Crock. J.G.: Lichte. F.E.: Riddle. G.O.: Beech. C.L. Talanta. 33. 601 (1986).
101. Tera. F.: Ruch. R.R.: Morrison. G.H. Analytical Chemistry. 37. 358 (1965).
102. Ruch. R.R.: Tera. F.: Morrison. G.H. Analytical Chemistry. 37. 1565 (1965).
103. Bottei. R.S.: Trusk. Bro. Ambrose Analytica Chimica Acta. 37. 409 (1967).
104. Pitts. A.E.: Beamish. F.E. Analytical Chemistry. 41. 1107 (1969).
105. Koster. G.: Schmuckler. G. Analytica Chimica Acta. 38. 179 (1967).
106. Nerst. W. Zeitschrift für das Phys. Chemie. 8. 110 (1891).
107. Morrison. G.H.: Freiser. H. Solvent Extraction in Analytical Chemistry. John Wiley & Sons. NY. 1957.
108. Cheng. K.L.: Ueno. K. Handbook of Organic Analytical Reagents. CRC Press. Boca Raton. FL. 1982.
109. Donaldson. E.M. Some Instrumental Methods for the Determination of Minor and Trace Elements in Iron. Steel. and Non-ferrous Metals and Alloys. Monograph 884. Energy. Mines. and Resources Canada. ottawa. Canada. 1982.
110. Donaldson. E.M. Talanta. 28. 461 (1981).
111. McKaveney. J.P. Analytical Chemistry. 40. 1276 (1968).

112. Scribner, W.G.; Treat, W.J.; Weis, J.D.; Moshier, R.W. Analytical Chemistry, 37, 1136 (1965).
113. Morie, G.P.; Sweet, T.R. Analytica Chimica Acta, 34, 314 (1966).
114. Donaldson, E.M. Talanta, 17, 583 (1970).
115. Villarreal, R.; Krsul, J.R.; Baker, S.A. Analytical Chemistry, 41, 1420 (1969).
116. Roy, N.K.; Das, A.K.; Ganguli, C.K. Atomic Spectroscopy, 7, 177 (1986).
117. Roy, N.K.; Das, A.K.; Ganguli, C.K. Atomic Spectroscopy, 6, 166 (1985).
118. Tanaka, K.; Takagi, N. Bunseki Kagaku, 12, 1175 (1963).
119. Jeffery, P.G.; Kerr, G.O. Analyst, 92, 763 (1967).
120. Annual Book of ASTM Standards, Vol.03.05, American Society for Testing and Materials, West Conshohocken, PA, 1996, Designation E107.
121. Zolotov, Yu. A.; Demina, L.A.; Petrukhim, O.M. Journal of Analytical Chemistry (USSR), 25, 1283 (1970).
122. Rooney, R.C. Analyst, 83, 83 (1958).
123. Goode, G.C.; Herrington, J.; Bundy, J.K.; Analyst, 91, 719 (1966).
124. Dean, J.A.; Cain Jr., C. Analytical Chemistry, 29, 530 (1957).
125. Donaldson, E.M. Talanta, 26, 1119 (1979).
126. Donaldson, E.M. Talanta, 25, 131 (1978).
127. Annual Book of ASTM Standards, Vol.03.06, American Society for Testing and Materials, West Conshohocken, PA, 1996, Designation E478.
128. de la Guardia, M.; Vidal, M.T. Atomic Spectroscopy, 4, 39 (1983).
129. Skidmore, P.R.; Taylor, K. Analytica Chimica Acta, 92, 405 (1977).
130. Donaldson, E.M. Talanta, 23, 411 (1976).
131. Donaldson, E.M. Talanta, 25, 131 (1978).
132. Donaldson, E.M. Talanta, 24, 441 (1977).
133. Donaldson, E.M. Talanta, 23, 823 (1976).
134. Donaldson, E.M. Talanta, 27, 79 (1970).
135. Annual Book of ASTM Standards, Vol. 03.05, American Society for Testing and Materials, West Conshohocken, PA, 1996, Designation E40.
136. Annual Book of ASTM Standards, Vol.03.06, American Society for Testing and Materials, West Conshohocken, PA, 1996, Designation E396.
137. Annual Book of ASTM Standards, Vol.03.05, American Society for Testing and Materials, West Conshohocken, PA, 1996, Designation E137.
138. Annual Book of ASTM Standards, Vol.03.06, American Society for Testing and Materials, West Conshohocken, PA, 1996, Designations E361, E362, E363, E368.
139. Sandell, E.B. Colorimetric Determination of Traces of Metals, 3rd. ed., Interscience, NY, 1959.
140. Collins Jr., T.A.; Kanzelmeyer, J.H. Analytical Chemistry, 33, 245 (1961).
141. Hubbard, G.L.; Green, T.E. Analytical Chemistry, 38, 428 (12966).
142. Annual Book of ASTM Standards, Vol.03.05, American Society for Testing and Materials, West Conshohocken, PA, 1996, Designation E35.
143. Annual Book of ASTM Standards, Vol.03.05, American Society for Testing and Materials, West Conshohocken, PA, 1996, Designation E120.
144. Machlan, L.A.; Hague, J.L. Journal of Research of the National Bureau of Standards, 59, 415 (1957).

145. Rothe, J.W. Chem. News, 66, 182 (1892).
146. Langmuir, A.C. Journal of the American Chemical Society, 22, 102 (1900).
147. Speller, F.N. Chem. News, 83, 124 (1901).
148. Kern, E.F. Journal of the American Chemical Society, 23, 685 (1901).
149. Oldfield, J.H.; Mack, D.L. Analyst, 87, 778 (1962).
150. Oldfield, J.H.; Bridge, E.P. Analyst, 86, 267 (1961).
151. Owens, E.B. Applied Spectroscopy, 13, 105 (1959).
152. Luke, C.L. Analytica Chimica Acta, 39, 447 (1967).
153. Headridge, J.B.; Richardson, J. Analyst, 95, 930 (1970).
154. Donaldson, E.M. Talanta, 27, 499 (1980).
155. Donaldson, E.M. Mineral Sciences Division Bulletin TB195, Mines Branch, Energy, Mines, and Resources Canada, 1974.
156. Schlewitz, J.H.; Shields, M.G. Atomic Absorption Newsletter, 10, 43 (1971).
157. Chao, T.T.; Ball, J.W. Nakagawa, H.M. Analytica Chimica Acta, 54, 77 (1971).
158. Mathieu, G.; Guiot, S. Analytica Chimica Acta, 52, 335 (1970).
159. Young , J.P.; White, J.C. Talanta, 1, 263 (1958).
160. Wood, D.F.; Jones, J.T. Analyst, 90, 125 (1965).
161. Ratcliffe, D.B.; Byford, C.S.; Analytica Chimica Acta, 58, 223 (1972).
162. Burke, K.E. Analyst, 97, 19 (1972).
163. Thornton, K.; Burke, K.E. Analyst, 99, 469 (1974).
164. Burke, K.E. Applied Spectroscopy, 28 234 (1974).
165. Burke, K.E. Talanta, 21, 417 (1974).
166. Bedrossian, M. Analytical Chemistry, 50, 1898 (1978).
167. Kharbade, B.V.; Agarwal, K.C. Atmic Spectroscopy, 14, 13 (1993).
168. Ross, W.J.; White, J.C. Analytical Chemistry, 33, 424 (1961).
169. Sparks, R.W.; Vita, O.A.; Walker, C.R. Analytica Chimica Acta, 60, 222 (1972).
170. Annual Book of ASTM Standards, Vol 03.05, American Society for Testing and Materials, West Conshohocken, PA, 1996, Designation E39.
171. Boltz, D.F.; Lueck, C.H. Colorimetric Determination of Non-Metals (D.F. Boltz, ed.), Interscience, NY, 1958, p. 31.
172. Annual Book of ASTM Standards, Vol.03.05, American Society for Testing and Materials, West Conshohocken, PA, 1996, Designation E46.
173. Annual Book of ASTM Standards, Vol.03.05, American Society for Testing and Materials, West Conshohocken, PA, 1996, Designation E54.
174. Hillebrande, W.F.; Lundell, G.E.F.; Bright, H.A.; Hoffman, J.L. Applied Inorganic Analysis, 2nd ed., John Wiley & Sons, NY, 1953.
175. Beamish, F.E. The Analytical Chemistry of the Noble Metals, Pergamon, Oxford, 1966.
176. Annual Book of ASTM Standards, Vol.03.06, American Society for Testing and Materials, West Conshohocken, PA, 1996, Designation E1473.

4

UV/Visible Molecular Absorption Spectrophotometry

1. INTRODUCTION

Colorimetry may be the oldest of the "modern" techniques. There once was a period in which colorimetric techniques swept through the metals industry. They promised simplicity and high levels of sensitivity. And they satisfied most of the chemists' expectations. The term "colorimetric"—today, a descriptor for all forms of UV/visible absorption methods—is particularly appropriate in discussing this field's origins since "colorimetry" was once confined to those early methods that made use of the analyst's own color perception in matching a standard with an unknown sample.

It was not too long before devices were developed to improve and standardize the chemist's judgement call: cylindrical, flat-bottomed Nessler tubes with fixed-volume fill-lines, and color comparators, such as the Dubosq, which illuminated standard and sample solutions in the same visual field. Colorimetry at this level is limited by the color acuity of the human visual system. For example, John Dalton (1766–1844), father of our modern atomic theory, could not have practiced it since he was color blind. One can imagine the chief chemist of a metals lab in the early part of this century testing potential applicants for this affliction.

Furthermore, even normal average human color vision is limited to the wavelength range 400 to 700 nm.[1] And within this range apperceptions vary. An object (such as a clear solution) exposed to transmitted white light appears colored if it absorbs one or more regions of the visible spectrum. The perceived color is the complement of the color absorbed. Thus a yellow solution has absorbed blue light and a blue solution has absorbed yellow light.

---

[1] Color acuity may range as low as 380 nm or as high as 780 nm in certain individuals.

But books that list wavelength ranges for colors and their complements seldom agree completely because individuals and their word for a given percept vary. What is "orange" to one person may be "red" to another.

Fortunately, it wasn't long before instruments were developed which were capable of directly measuring the radiant power of transmitted (and by difference, absorbed) light. Filter photometers are reasonably simple devices that generally utilize a bandpass filter to select a narrow range of wavelengths from a white light source and a photovoltaic cell to measure the radiant power absorbed by the sample. Except for the most elementary applications these have been largely supplanted by spectrophotometers.

Spectrophotometers utilize a prism or a diffraction grating and a set of slits to select an extremely narrow bandpass—in effect a single wavelength—from a continuum source. The source is generally a tungsten lamp for the visible region and a high pressure hydrogen or deuterium lamp for the ultraviolet. Detection is by means of a phototube, photomultiplier, or photodiode array. There are dual channel, automatic scanning versions, as well. And, of course, software innovations and cheap computer power have been liberally employed.

These instruments have removed the subjectivity from colorimetry. Seven decades of refinements have allowed the full potential of UV/visible molecular absorption spectrophotometry to be realized. In this time interval in the metals analysis laboratory many other measurement approaches have been introduced. Some of these are even more sensitive and require less chemical manipulations, but none can match the investment return from a basic spectrophotometer. Spectrophotometric applications abound in the minor and trace concentration regions. Higher concentrations are amenable to a differential calibration approach. And the ultratrace realm can often be reached using long path-length cells or a preconcentration technique.

In this chapter we will explore this basic measurement technique both as a subdiscipline of analytical chemistry and as a powerful practical tool for the trace analysis of metals and alloys. First, we will examine our current understanding of the electronic transitions that lie behind molecular absorption of UV/visible radiation, briefly return to color as a human percept/concept, outline current nomenclature, and describe some quantitative laws and their exceptions. Next, we will cover instruments, both "ancient" and modern, as well as instrument components such as sources, dispersion elements, filters, detectors, and sample cells. We will also briefly describe a few instrumental techniques with largely undeveloped potential. Next, we will examine calibration modes and sources of error, then describe the important use of UV/visible spectrophotometry in qualitative analysis. Then we will examine some of the practical details of UV/visible spectrophotometric quantitative measurement. Finally, we will survey, element-by-element, some of the spectrophotometric trace methods which have been routinely applied to metals and alloys.

2. PRINCIPLES

2.1 Molecular Absorption Spectra

Both atoms and molecules absorb electromagnetic radiation, and, as might be expected, the molecular behavior is the more complex to describe. Most analytical texts begin here, however, because molecular absorbance measurements historically preceded atomic absorption measurements as a routine analytical tool. Once molecular behavior is grasped, atomic absorption follows readily, as we will see in the subsequent chapter.

2.1.1 Energy Levels [1–5]

Aside from its translational energy (which is a separate paradigm, as it were), a molecule exhibits three types of internal energy: rotational, vibrational, and electronic. Rotational energy stems from the fact that a molecule may rotate about one or more axes. These energy levels constitute a very closely spaced fine structure. Quantized changes between rotational states are induced by the absorption of minute amounts of energy. Thus, a purely rotational spectrum is produced by the absorption of microwave and far infrared radiation (wavelengths of 0.1–100 mm).

Vibrational energy results from the periodic motion of atoms and groups of atoms in a molecule. This stretching and bending of various parts of the molecule represent higher energy transitions, with rotational fine structure superimposed at each vibrational level. Thus, when energy in the infrared region (wavelengths of 2–100 μm, or, more commonly, 5000–100 cm^{-1}) is absorbed a complex band structure results. The embedded rotational fine structure is seldom evident in routine work except in the infrared measurement of gases, such as ammonia, where a series of narrow rotational lines form broad vibrational envelopes. However, the many vibrational modes in organic molecules and their interaction with each other produce spectra rich in structural information. This forms the basis of infrared absorption spectrophotometry, one of the organic analyst's most powerful tools.

When electromagnetic radiation of even higher energy, representing the UV/visible region, is absorbed transitions in the electron energy levels begin to occur. This wavelength range is commonly regarded as 200–750 nm, often separated into the near (or "quartz") ultraviolet (200–380 nm) and the visible (380–750 nm). Some work requires inclusion of the near infrared (750–2500 nm). Highly specialized studies may also be conducted in the vacuum ultraviolet (10–200 nm).

When quantized absorption occurs in the UV/visible energy range it is generally bonding electrons that are promoted to higher states in the molecule's characteristic electronic structure. And superimposed on this bond-related "signature" is a fine structure of both vibrational and rotational states. With solutions the resultant spectrum is nearly always observed as a small number of broad, smoothly curved peaks. Specialized high-resolution spectrophotometers may be capable of resolving some of the fine structure of these bands from liquid samples. And in the vapor phase where molecules can rotate freely and vibrational states are not affected by the force field of solvent molecules a good laboratory spectrophotometer may be capable of resolving fine structure detail. But such studies are of little practical significance to the metals analyst.

It should be noted that the absorption being recorded is really a rate phenomenon since the individual excited electrons spontaneously return to lower states within 10^{-8}–10^{-9} seconds, emitting quanta of energy as they do. These ultimately take the form of an immeasurably small amount of heat or, more rarely, longer wavelength radiation (fluorescence*). Another possibility is that the absorbing molecule may dissociate.

2.1.2 Electronic Selection Rules

Quantum mechanics imposes constraints on energy level transitions. That is, just because an incoming photon has an energy that exactly matches the energy spacing between two electronic levels it does not mean that an electron transition will occur. Recalling that electron pairs occupying the same energy level must have opposite spin, quantum mechanics states that energy level transitions that result in a change in electron spin are forbidden. There are also symmetry and orbital overlap requirements in order for electron transitions to occur with high probability. On the other hand, quantum mechanically forbidden transitions do occur to some degree in many molecules because force field perturbations between molecules and between different parts of the same molecule mitigate the "forbidden" edicts [2].

The Franck-Condon principle states that the movement of atomic nuclei in a polyatomic molecule is negligible in the time (about 10^{-15} sec.) required by an electron to move between energy levels. It further states that electron transitions are favored when the distances between nuclei after a transition are largely the same as they were before a transition, and that electron transitions are favored when the nuclei have minimal velocity [2].

---

* If the re-emission of light is delayed the term is phosphorescence.

2.1.3 Electronic Transitions [2–4]

There are six generally recognized forms of electronic transition that occur due to the absorption of ultraviolet and visible radiation by stable molecules. Some of these, such as those involving double-bonded molecules, are low energy transitions that occur in the visible region. A few are high-energy phenomena, such as the bond rupture of ethane by far UV light.

It will be recalled that covalent bonding occurs when two atomic electron orbitals overlap to form a low energy bonding molecular orbital. In a high energy state, however, a corresponding antibonding molecular orbital forms. Single bonds in organic molecules consist of sigma (σ) orbitals (two electrons), while double bonds consist of a sigma orbital and a pi (π) orbital (four electrons). The excited state antibonding orbitals for single bonds are designated σ^*, and for double bonds, π^*. Nonbonding electrons in an organic molecule are designated n.

$\sigma \rightarrow \sigma^*$ transitions require the input of large amounts of energy such as that corresponding to the deep ultraviolet wavelengths. Ethane, for example, shows an absorption maximum at 135 nm (a region that requires specialized instrumentation to study). $n \rightarrow \sigma^*$ transitions occur in molecules containing oxygen, sulfur, nitrogen, or halogens. These "hetero" atoms contain nonbonding electron pairs that can be promoted to form σ^* orbitals by electromagnetic radiation in the 150–250 nm wavelength region. Absorption maxima for these types of transitions are highly dependent on the type of "hetero" atom involved and less dependent on the overall molecular structure. The maximum absorbance wavelengths for oxygen–carbon and chlorine–carbon bonds occur below 200 nm, while maxima for most other $n \rightarrow \sigma^*$ transitions occur above 200 nm.

The presence of any multiple bond in a molecule means that π bond electrons are present and the $\pi \rightarrow \pi^*$ transition can occur. The additional presence of atoms with nonbonding electron pairs (oxygen, nitrogen, sulfur, and the halogens) means that the transition $n \rightarrow \pi^*$ can occur. These are both often comparatively low energy transitions, occurring in a wavelength region (200–700 nm) where the absorbance is easily measured, especially when the molecule contains a sequence of alternating single and multiple bonds. While the presence of several multiple bonds in a molecule ordinarily produces a simple additive effect on absorbance, their arrangement in an alternating sequence with single bonds delocalizes the π^* orbitals, lowering their energy and their antibonding character. Such conjugated π electron systems are common in strongly colored organic compounds (which, of course, absorb intensely in the visible wavelength region).

The polarity of the solvent affects the absorption maximum for both the $n \rightarrow \pi^*$ and the $\pi \rightarrow \pi^*$ transitions. Increasing solvent polarity blue-shifts the n → π* maximum to a shorter wavelength and usually red-shifts the π → π* maximum to a longer wavelength. Aromatic molecules show three absorbance maxima attributed to π → π* transitions. Benzene shows maxima at 184, 204, and 254 nm. Ring substituents with nonbonding electron pairs (e.g., –OH or $-NH_2$) tend to stabilize the π* state, lowering its energy and red-shifting the absorbance maxima. The n → π* transition occurs for a number of inorganic anions, as well. Thus, nitrate shows an absorbance maximum at 313 nm, nitrite shows maxima at 360 and 280 nm, and carbonate at 217 nm.

So far we have looked at four electronic transitions that can be promoted by the absorbance of electromagnetic radiation in the UV/visible region (σ → σ*, n → σ*, π → π*, and n → π*). There are two more. *Charge transfer* absorbance results in an electron being transferred from a "donor" region to an "acceptor" region of the same molecule. In the case of the iron (III) thiocyanate soluble complex, for example, a visible light photon is absorbed resulting in the transfer of an electron from a thiocyanate molecular orbital to an iron (III) atomic orbital. The short-lived excited state consists mostly of an iron (III) associated with a neutral thiocyanate radical. Sometimes, as in the case of the iron (III)/1,10-phenathroline complex, the metal is the electron donor and the organic ligand is the electron acceptor. Other examples of charge transfer absorbance are evidenced by the yellow peroxytitanium complex and the ferrous ferricyanide molecule known as "Prussian Blue." Charge transfer is also responsible for the ultraviolet absorbance maxima of certain hydrated ions, such as $Cl(H_2O)_n$ and $Fe^{2-}(H_2O)_n$.

The last electronic transition we need to mention is that produced by *ligand-field* absorbance. For the most part, the characteristic colors exhibited by the transition metals, the lanthanides, and the actinides are due to ligand-field absorbance. In the case of the transition metals transitions of the 3d and 4d atomic orbitals are involved. When a transition metal is complexed with ligands (water molecules, for example) its d orbitals are split into a number of different orbitals of diverse energies, affording many opportunities for energy-promoted transitions. The *pattern* of this splitting depends upon the symmetry of the complex formed. Ligands may form octahedral, tetrahedral, tetragonal, or square planar structures with the central atom. The *extent or degree* of the orbital splitting depends on the nature of the ligand, the charge on the central atom, and the size of the d orbitals. The so-called *spectrochemical series* of ligands is a rough measure of *increasing* ligand field strength and *increasing* orbital splitting: I^-, Br^-, Cl^-, F^-, OH^-, $C_2O_4^-$, C_2H_5OH, H_2O, CNS^-, NH_3, ethylenediamine, NO_2^-, 1,10-phenathroline, CN^-, CO. *Higher* charge on the central metal ion and *larger* d orbitals both *increase* splitting.

The transition metals show broad absorbance curves whose shape and maxima are highly influenced by the reagent solution environment. The lanthanides and actinides, on the other hand, show narrow, fairly well resolved peaks which are not strongly affected by the nature of the ligand present. The electron transitions for the lanthanides are believed to involve 4*f* orbitals, and those for the actinides to involve 5*f* orbitals. This explains both aspects of the spectra—the sharp peaks and the immunity to the chemical environment—because these *inner shell* transitions are shielded from extra-atomic influences.

The theoretical summary that we have just completed must not mislead the reader into a belief that the physicochemical laws governing molecular absorption in the UV/visible are perfectly understood, or that absorbance behavior is exactly predictable from first principles for all (or any) compounds. There are, in fact, sets of rules that can be applied to certain distinct classes of compounds, which do a fair job of predicting UV/visible absorbance maxima from structural features. There are also many cases that presently elude such treatments. The reverse exercise—extracting structural information from UV/visible absorbance spectra—is extremely difficult or impossible because the output is lean in informational detail. Organic qualitative analysts generally utilize a UV/visible absorption spectrum as a supplement or support for conclusions drawn from other, information-rich technologies, such as infrared absorption spectrophotometry or nuclear magnetic resonance spectrometry.

2.2 Concepts and Percepts of Color [6]

The phenomenon of color as a human perception is problematic since it involves physical, physiological, and psychological factors. One must not think of the *percept* of a color as a wavelength but rather as a segment of a range of sensation that includes white, black, and grays. Its attributes are brightness, hue, and saturation. Chromatic colors (those excluding white, black, and grays) exhibit hue; and the purity or brilliance of hue is the color's saturation. If the color has low hue and high brightness it is a tint. If it has low hue and low brightness it is a shade.

We have already pointed out that the wavelength ranges perceived as violet, blue, green, yellow, orange, or red are ill-defined due to variation in an individual's visual response. We also mentioned that the perception of transmitted color is the complement of the absorbed wavelength. Thus, the absence of green from white light appears purple, the absence of red appears blue-green, and the absence of blue appears yellow. These transmitted light perceptions are indistinguishable from "monochromatic" purple, blue-green, and yellow light, respectively.

Further, it is possible to produce the perception of a color by mixing two other colors. Thus, a mixture of green light and red light appears yellow with no yellow wavelengths present. And, in fact, any color sensation can be reproduced by mixtures of red, green, and blue light. White light can be produced not only by equal intensities of these three "primary" colors, but also by adding suitable proportional intensities of complementary color pairs. It should be remembered that color as a psychophysiological percept has more to do with the "hardwiring" of our neural/visual system than it has to do with a narrow band of wavelengths in the electromagnetic spectrum.

2.3 Nomenclature [2,4,8]

The quantitative laws of molecular absorption spectrophotometry, which will be described in the next section, require a consistent nomenclature. In the past there existed a plethora of jargon with many synonyms for the same concept. In 1952 the journal *Analytical Chemistry* published the report of the Joint Committee on Nomenclature in Applied Spectroscopy, comprised of members from the Society for Applied Spectroscopy (SAS) and the American Society for Testing and Materials (ASTM) [7]. These recommendations are now widely accepted. In addition, the use of SI units, recommended by the International Union of Pure and Applied Chemistry (IUPAC) and by the National Institute of Standards and Technology (NIST) has gained considerable consensus. Unfortunately, there remain several concepts for which ASTM and IUPAC recommended different terms. Table 4.1 summarizes the current state of the nomenclature.

2.4 Quantitative Laws [2,4]

The quantitative relationship between analyte concentration and absorbed monochromatic light has a long history that begins almost three centuries

Table 4.1 UV/Visible Spectrophotometry Nomenclature

ASTM Name [8]	Symbol	IUPAC Name	Definition	Discouraged
Absorbance	A	Absorbance	$-\log T$	Absorbancy, optical density
Absorptivity	a	Specific absorption coefficient	A/bc[a]	Extinction coefficient
Molar Absorptivity	ε	Molar absorption coefficient	A/bc[b]	Molar extinction coefficient
Transmittance	T	Transmittance	P/P_0	Transmittancy
Nanometer	nm	Nanometer	10^{-9} m	Millimicron (mμ)

Note: b = sample path length, in cm; P = transmitted radiant power; P_0 = incident radiant power

[a] c = concentration of analyte in kg/m^3

[b] c = concentration of analyte in moles/L

ago. Bouguer* in 1729 first described the relationship between the length of the monochromatic light path through an absorbing medium and the diminution of radiant power in the emerging beam. If we consider P as the emergent beam's radiant power and b as an infinitesimally short sample path length:

$$- dP/db = k_1 P \qquad (1)$$

Rearranging, and integrating between the limits P_0 (incident radiant power) and P, and between a path length of zero and b:

$$- \int_{P_0}^{P} dP/P = \int_{o}^{b} k_1 db \qquad (1)$$

$$dP/P = k_1 db$$

$$-(\ln P - \ln P_0) = k_1 b$$

$$\ln P_0 - \ln P = k_1 b$$

$$\ln (P_0/P) = k_1 b$$

Or, since base 10 logarithms are usually employed:

$$\log (P_0/P) = k_2 b \qquad (2)$$

Over a hundred years later in 1859 Beer described an analogous relationship between concentration (c) and absorbance for a fixed path length:

$$-dP/dc = k_3 P \qquad (3)$$

which, by the same mathematical treatment reduces to:

$$\log (P_0/P) = k_4 P \qquad (4)$$

Noting that Bouguer's equation (2) depends on a fixed value for concentration and that Beer's equation (4) depends on a fixed value for path length, we can write these as functions:

$$\text{Bouguer: } \log (P_0/P) = f(c)b \text{ ;and Beer: } \log (P_0/P) = f(b)c \qquad (5)$$

Since the two laws apply simultaneously:

$$f(c)b = f(b)c$$

Separating the variables:

$$f(c)/c = f(b)/b$$

Two functions of independent variables are equal if they both equal a constant:

$$f(c)/c = f(b)/b = K$$

Thus, $f(c) = Kc$ and $f(b) = Kb$ (6)

* Lambert formulated the same relationship in 1768 and frequently has been mistakenly given sole credit for the discovery.

Substituting in the two equations in (5) yields the same result for both:

$$\log (P_o/P) = Kbc = A \tag{7}$$

This is the combined Bouguer/Beer Law, an expression which is set equal to A, the absorbance.

The units ordinarily used for b are centimeters. The units for c may be either grams per liter or moles per liter, requiring different constants. Thus:

$$A = abc \tag{8}$$

where: A = absorbance; a = absorptivity constant; b = path length in cm; and c = analyte concentration in g/L

and $$A = \varepsilon bc \tag{9}$$

where: A = absorbance; ε = molar absorptivity constant; b = path length in cm; and c = analyte concentration in moles/L.

It follows that ε is simply a multiplied by the molecular weight of the absorbing analyte. The value of ε, the molar absorptivity, is an independent characteristic of the absorbing species for given solvent and wavelength. It is *not* dependent upon path length or analyte concentration.

Most spectrophotometers and filter photometers generate a signal that is linearly proportional to the transmittance, T, which is the fraction of the incident radiant power transmitted:

$$T = P/P_o \text{ and } \%T = P(100)/P_o \tag{10}$$

Since: $A = \log (P_o/P)$

then: $$A = \log (1/T) \tag{11}$$

Thus, according to these laws, absorbance (A) is directly and linearly proportional to concentration, and (log T) is inversely and linearly proportional to concentration (for a constant path length).

Failure to respond linearly in this manner is considered a deviation from (or "failure of") Beer's Law. In fact, many of these deviations from linearity result from failure to control either the solution chemistry or the wavelength of the incident radiation. For example, an analyte species, J, which absorbs at the analytical wavelength with one value of the molar absorptivity constant, ε_j, may be in equilibrium with another chemical form (a dissociation product, or a different oxidation state, or an isomer) say, Q, which has a different molar absorptivity constant, ε_q, at the measured wavelength. There are two basic solutions to this dilemma: 1) solution conditions (pH, ionic strength, temperature, redox conditions, etc.) can be adjusted to suppress the formation of Q (e.g., the dissociation of a weak acid analyte can be suppressed by lowering the pH), or 2) measurements can be made at the *isobestic wavelength*, which is defined as a wavelength where two (or more) species in equilibrium exhibit the same ε value. Unfortunately, the isobestic wavelength seldom provides optimum sensitivity for the analyte species.

Failure to provide monochromatic radiation of the correct wavelength will also lead to "deviations" from Beer's Law. Filter photometers, which provide a range of wavelengths, lead to non-linear response in some situations. Older design instruments—both filter photometers and spectrophotometers—are subject to nonlinear calibration effects due to instability in illumination sources, fatigue in detectors, and deterioration effects in amplifier and readout circuitry. Modern instruments are much better in regard to both long-term and short-term stability. Stray light may still be a problem, however, especially near the UV limit of the instrument. This produces a negative deviation from a linear Beer's Law curve.

Also, if the absorbance peak is narrow, or if an analytical wavelength is selected that is on the steep side of an absorbance curve, deviations from Beer's Law are expected since wavelength stability and the reproducibility of dial settings may be insufficient. It is generally prudent to select an analytical wavelength at a location where absorbance does not change dramatically with wavelength. It is also wise, if possible, to select a wavelength where the absorbance due to excess colorimetric reagent is minimal, while that due to the analyte chromophore is maximal.

There are two *real* deviations from Beer's Law behavior, but they are seldom encountered as problems in practical analytical situations. If the refractive index of a substance increases the value of ε will decrease. The refractive index of a solution increases with solute concentration. Thus, we expect a negative deviation of the calibration curve at very high concentrations of analyte. This effect is ordinarily very minute and generally below the level of detectability. Secondly, *saturation* occurs when only a small number of analyte molecules are present and the incident radiation is very strong. Here, all the molecules of the chromophore are promoted to an excited state by a small fraction of the available photons, and any further increase in photon flux has no further effect on absorbance. This effect is, likewise, rarely encountered.

There are two additional assumptions of the Bouguer-Beer Laws that have not been mentioned. The first is that the absorption process takes place within a volume of uniform cross-section. That is, the beam through the absorbing medium must not be diverging or converging. Optically flat surfaces for the sample cuvettes help to prevent any "lensing" effects. The second assumption is that two or more absorbing substances simultaneously present will behave independently and additively. This allows the simultaneous determination of two or more analytes in certain situations, a subject that will be treated in Section 4.5. It also allows for proper correction of sample and reagent blank effects.

3. INSTRUMENTATION

3.1 Color Comparators [6,9]

Color comparators were the first devices applied systematically to the measurement of analyte concentration based on the absorbance of light. A rack of flat-bottomed Nessler tubes, filled to the same volume mark, and viewed from above against a white, evenly illuminated base was one of the first attempts at standardizing color comparisons between "unknowns" and standards. It was usually necessary to prepare many standard solutions to ensure that the unknown would match one of them, or would fall between two closely spaced standard concentrations. In routine work sometimes artificial standards—either liquids in sealed viewing tubes or even colored glasses were employed. Attempted compensation for the effect of a sample blank was accomplished by preparing a sample solution minus the color reagent and placing it in line with the comparison standard.

It was recognized, however, that the most accurate work required many carefully prepared, closely matching comparison standards. The Duboscq colorimeter—a major advance in its day—was one means of simplifying this process. Two solution cups, one for the sample and one for a single matching standard were viewed from above through an eyepiece against the same evenly illuminated visual field. Individually controlled mechanisms were used to raise and lower the solution cups in relation to fixed optical elements immersed in the solutions to alter the viewing depth until the color intensities appeared to match. The relative cup heights could then be read from a graduated scale and related to analyte concentration in the unknown solution.

3.2 Filter Photometers [1,3]

These were the first instruments that removed the subjective element from colorimetry. While color comparison methods are estimated to have a minimum error of ± 5%, filter photometers can usually do much better. Because they can limit the wavelength range to a region where absorption is at a maximum for the analyte and at a low level for interferences, we can say with assurance that their sensitivity and selectivity are superior to the color comparators.

Nomenclature for many physical measurement instruments is overlapping and inconsistently applied. For example, "photometer" is both a general term for any device for measuring radiant flux density or radiant power, and, simultaneously, a specific synonym for these simple devices, which use filters to select a narrow band of wavelengths.

For descriptions and illustrations of color comparators of various designs, as well as early versions of filter photometers and spectrophotometers, see: F.D. Snell and C.T. Snell *Colorimetric Methods of Analysis*, D. Van Nostrand, NY, 1948.

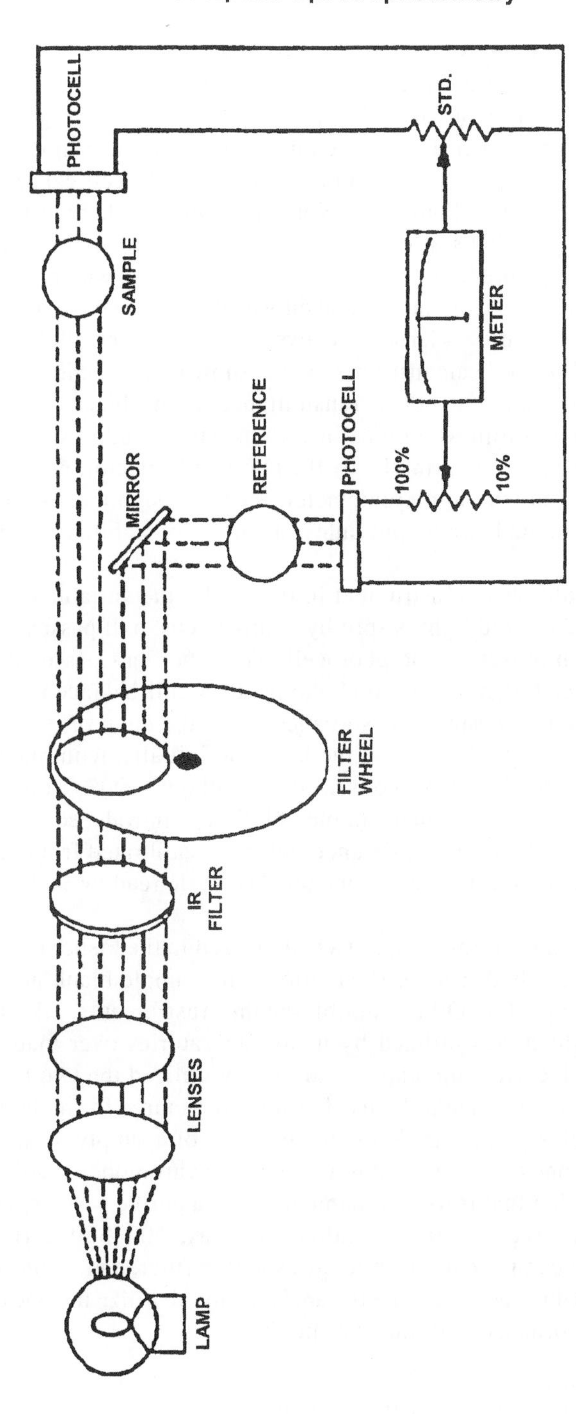

Figure 4.1 The Fisher Electropholome4er II. (Courtesy Fisher Scientific: used with permission.)

By some of today's standards a filter photometer may be considered a relatively unsophisticated instrument, but there are many cases (where the chromophore exhibits a broad peak in the visible wavelength region) in which these devices can deliver accuracy equivalent to the most modern spectrophotometer design. Figure 4.1 illustrates an optical and electrical configuration for a double-beam filter photometer. Note that a tungsten lamp is the radiation source for the design. This is reflective of the fact that most filter photometers are designed to operate in the visible wavelength region where tungsten filament lamps emit reasonably intense, albeit wavelength sensitive, radiation. The pictured device also utilizes lenses to convert divergent to parallel light waves.

In related single-beam instruments (not shown), a variable iris diaphragm is used to set the readout to 100% transmittance when a blank sample is in the light beam.* This is followed by an interchangeable bandpass filter, then the sample compartment, and, finally, by the photocell detector, whose electrical output is measured with a microammeter. Such a design makes no provision for fluctuation in the lamp output unless a battery or stabilized line source is employed.

In the double-beam instrument in Figure 4.1 the variable diaphragm is not needed. The filtered light is split by a mirror. One half passes through the sample cell then passes to one photocell. The other half passes through the blank solution cell, then to a second photocell. A null-balance potentiometer is used to measure the sample absorbance. With the lamp off the galvanometer is adjusted to zero. Then with the lamp on, initially, with blank solution in both cells a variable resistance is used to "null out" 100% T on the readout variable resistance scale. Then sample solution is introduced in the sample (top) light path and variable resistance (which is calibrated from 10 to 100% *T*) is used to again null the galvanometer. The scale reading is the %*T* of the sample solution.

The advantage of this design (which is used in the Fisher Electrophotometer II™) is that both source fluctuations and sample blank are automatically compensated for. Other "double-beam" instruments, like the Klett-Summerson, which was utilized by many laboratories over many decades, lacked the blank correction feature and merely utilized the two beams to allow unfiltered line voltage to be used as a power source for its lamp.

If the analyst is ever in doubt about which of a supply of bandpass filters to employ, one need merely observe the pure chromophore solution color and select the filter that is its complement. Thus, a purple permanganate solution requires a green filter (a bandpass of, say, 500–560 nm), a yellow peroxytitanium complex solution requires a blue filter, and so on. If a selection of similar filters are available the analyst should utilize the one that yields the greatest absorbance with the chromophore.

---

* In other designs lamp voltage is adjusted to obtain 100% T.

There are also UV sensitive filter photometers, but they currently find application only as detectors for liquid chromatography. These generally utilize a mercury vapor lamp and UV filters to pass the 254 nm line. Fiberoptic probe visual range instruments are also available. These are used for field portable colorimetric work, for colorimetric endpoint titrations, and for the continuous monitoring of process streams. The Brinkmann instruments use either selectable insert filters or a color wheel. The light is conducted by fiber optic cable to a glass or stainless steel probe with a fixed gap between the cable end and a concave mirror. When inserted in a solution the filtered light beam traverses the gap to the mirror and back again for an effective path length of double the gap size. The center of the fiber optic cable returns the light to a photocell for readout. In Section 3.4 some of the components and accessories that are common to both filter photometers and spectrophotometers will be described.

3.3 Spectrophotometers [1–4]

While the term "spectrometer" applies to any device for measuring the electromagnetic spectrum, as well as the energies and masses of particles, the term "spectrophotometer" is always limited to the UV-visible-IR range of wavelengths. A spectrophotometer is defined as an instrument which utilizes a monochromator to select the light that is presented to and absorbed by the sample. Instead of a band of wavelengths, which is typical of a filter photometer, even the lowest resolution spectrophotometer utilizes light with a bandpass of less than 10 nanometers. The monochromator in these instruments consists of an entrance and an exit slit, a dispersion element (prism or grating), and an array of mirrors to direct the light.

The relationship between resolution, bandpass, and slit width needs some discussion. The resolution of a monochromator is ordinarily described as its ability to distinguish two closely spaced spectral lines. However, when used as a source of monochromatic illumination, as in UV/visible absorption, resolution is best equated with the *monochromaticity* of the emergent light beam. Generally, the width of the entrance and exit slits is set to the same value (in millimeters). Multiplying this value by the dispersion of the grating (in nm/mm) gives a somewhat optimistic estimate of the resolution, which is always degraded because of light losses in the optics. The intensity of the light source and the sensitivity of the detector affect resolution as well—the more of each, the better the resolution.

The optical bandpass of an instrument is a measure of how much of the spectrum can be extracted as pure wavelengths. The *nominal wavelength* in the center of the slit is the wavelength set on and recorded from the instrument. The *effective bandpass* is taken as the width in nanometers at one-half height of the peak transmittance. This gap defines the region of pure wavelengths. Twice this value is called the *spectral slit* (often listed in manufacturer's literature). The spectral slit is related to the *mechanical slit* (the actual slit gap in millimeters) by the grating dispersion and the focal length of the monochromator. It is generally true that two distinct absorbance (or emission) peaks can be resolved if their nominal wavelengths differ by an amount that is at least equal to the spectral slit.

In the case of a prism instrument, the dispersion is highly wavelength dependent. While a grating shows the same dispersion across the entire useable wavelength range a prism instrument shows markedly greater dispersion at the low wavelength side of the spectral range, with the red end of the visible being quite crowded. Thus, even a simple prism-based instrument uses a variable slit control.[*] The slit can be opened very wide at the UV end and still achieve good wavelength resolution. At the red and near IR end the slit must be narrowed to achieve equivalent wavelength resolution. However, there are two limits to this process: the theoretical Rayleigh diffraction limit (diffraction effects from the slit edges) and the more important practical light throughput requirements for the detector. In high quality instruments it may be possible to narrow the slit to 0.01 mm or less.

Despite the drawbacks of nonlinear dispersion, prisms offer a freedom from spectral overlap. Diffraction gratings, on the other hand, show some overlap of spectral orders. A combination grating and prism arrangement can be employed to help sort spectral orders.

Besides the presence of a monochromator, another distinguishing feature of a spectrophotometer is a sensitive detector. While filter photometers often make do with a simple photovoltaic cell, spectrophotometers employ phototubes, photomultipliers, or even photodiode arrays; and future instruments will likely utilize charge transfer solid state detectors.

A design problem that must be addressed by all instrument manufacturers is stray light reaching the detector. Stray light is radiation of any wavelength, either from the source or from light leaks, which is scattered and reflected from surfaces in the monochromator and makes its way to the exit slit. Instrument designers employ light baffles and black matte paint on interior surfaces to trap reflected stray light. Stray light is a special problem in measuring absorbance when low amounts of spectral energy are reaching the detector and may increase with time as instruments age.[†] Manufacturers quote stray light as a "less than" percent transmittance at one or more wavelengths (often 220 nm and 340 nm).

---

[*] The entrance and exit slits are adjusted to the same setting with a ganged control or by an "over-and-under" optical design that allows the entering and emerging beams to use the same slit.

[†] See ASTM Standard Test Method E387 (*Annual Book of ASTM Standards*, Vol.03.06, American Society for Testing and Materials, West Conshohocken, PA, 1996) for a procedure for estimating the stray light in transmittance-recording double-beam spectrophotometers.

3.3.1 Single-beam Instruments

These are instruments that pass the selected wavelength through one solution at a time. Single-beam spectrophotometers require three tasks of the analyst: 1) with the light shutter closed, the dark current (due to thermal emission of electrons by the cathode of the phototube) must be adjusted to read out infinite absorbance (or zero percent transmittance); 2) with the reference solution in the open light path, the amplifier gain, the slit width, or the source intensity must be adjusted to read out zero absorbance (or 100% transmittance); 3) with the sample solution in the open light path, the experimental absorbance (or % transmittance) must be read out.

With microprocessor controlled instruments (e.g., the Spectronic 501™ and 601™ manufactured by Milton Roy) much of this occurs at the touch of a button. Steps 1) and 2) must be repeated each time the wavelength is changed (also now commonly accomplished from a keyboard). Since these three steps must occur sequentially, both the intensity of the source and the sensitivity of the detector must remain very stable. This places design and performance constraints on these instruments which, for example, generally require some form of line stabilization.

At one time there were two basic configurations of single-beam spectrophotometers: direct reading and null-balance. Today, digital readout designs constitute a third category. All of these instruments are well suited to typical trace metal analysis situations where only a few wavelengths are accessed for absorbance measurements. Scanning, qualitative work *can* be performed as well, but it is tedious and lengthy because the monochromator must be manually stepped through a series of wavelength increments and the reference solution must be reset to zero absorbance at each before the experimental solution absorbance is measured. More recently, it has become possible to achieve rapid scanning qualitative and multicomponent quantitative results with a computer controlled single-beam instrument utilizing a photodiode detector array. This will be discussed in the next section since it is, in effect, a virtual double-beam system.

Single-beam instruments have traditionally served as the workhorse of spectrophotometric trace analysis as it relates to metals and alloys. Direct reading single-beam instruments are relatively inexpensive and have been for decades. Until relatively recently they represented a compromise in accuracy related to panel meter readout inadequacies. The logarithmic absorbance scale on such meters was sometimes difficult to read and mechanical lag in the pointer mechanism could also affect results.

For a better quality single-beam instrument most laboratories purchased a null-balance design, which could be read out much more precisely. Today, the use of digital circuitry and digital displays has allowed instrument manufacturers to produce "low end" direct reading instruments with none of the

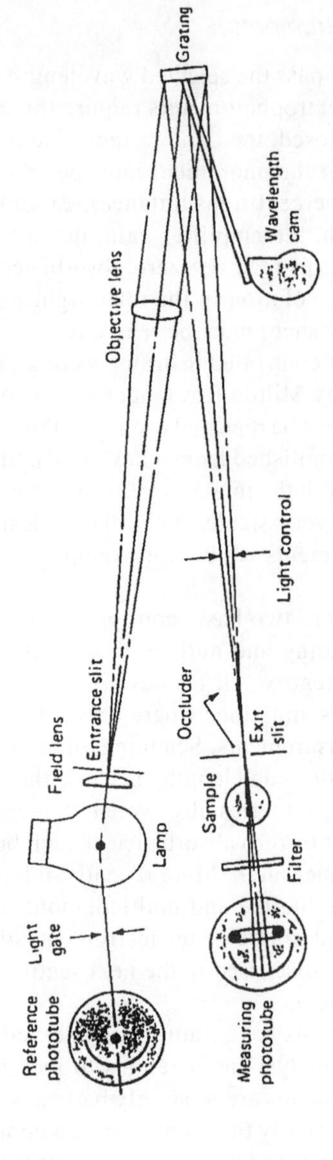

Figure 4.2 The Spectronic 20 (Courtesy of Bausch & Lamb. Inc., used with permission.)

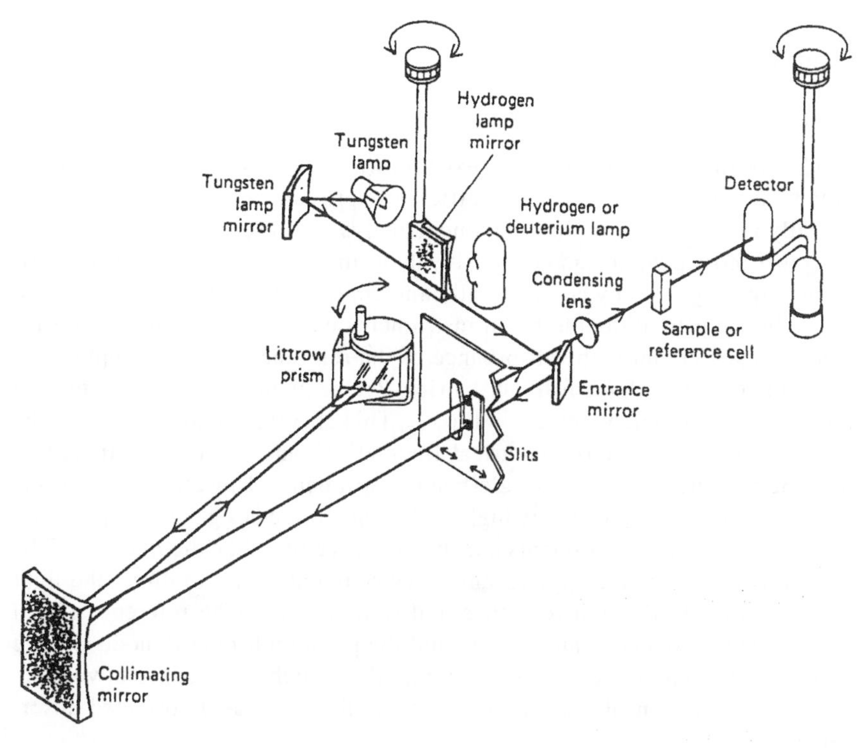

Figure 4.3 The DU-2. (Courtesy of Beckman Coulter, Inc., used with permission.)

read out difficulties that plagued earlier designs. We will return to the subject of readability and precision in Section 4.

Figures 4.2 and 4.3 illustrate the optical plan of two single-beam spectrophotometers that have been used in laboratories for more than half a century. The Spectronic 20™ (Figure 4.2) was originally designed and sold by Bausch & Lomb and is now manufactured as a digital readout instrument by Milton Roy. It uses a diffraction grating to cover the visible range. The DU-2™ (Figure 4.3) was originally designed and sold by Beckman Instruments and is now manufactured by Gilford Instruments as a digital readout instrument. It uses all quartz optics, including a mirror-coated Littrow prism, and both tungsten and deuterium lamps to cover the UV and visible ranges. It is a credit to these optical designs that they are still sold and used after a long period that has seen enormous changes in analytical instrumentation.

There are many other single-beam instruments currently being sold. These range from hand-held battery operated devices for field testing to research grade designs with a full complement of quantitative software programs, including multiwavelength/multicomponent routines. Among the manufacturers of single-beam spectrophotometers are: ATI Unicam, Cecil Instruments,

Cole-Parmer, Fisher Scientific, Gilford, Hach, Milton Roy, Perkin Elmer, Shimadzu, and Turner Instruments.

3.3.2 Double-beam Instruments

Double-beam spectrophotometers excel in wavelength scanning as applied to organic qualitative work, but also efficiently provide quantitative results, especially in the case of multi-component analysis. These instruments are considerably more complex and expensive than single-beam instruments of comparable quality. As they automatically and continuously vary the wavelength of the incident light double-beam instruments either continuously, or nearly continuously, compare the absorbance of the reference and the sample solutions, in effect simultaneously measuring P_o (the incident radiant power) and P (the emergent radiant power) in Eq. (7). This effortlessly solves most of the concerns about short term electrical fluctuations and source variations that must be designed around in single-beam instruments. Thus, the need for line stabilization and the use of very high quality electronic designs is less pressing.

Double-beam instruments may be direct reading, like the Beckman DB-G™, which utilized a mirror assembly vibrating at 35 Hz to direct the light beam alternately through reference and sample cells. The reference signal was compared to a constant voltage and the photomultiplier dynode voltage was trimmed to produce a reference signal that matched the constant voltage. The sample portion of the output was then displayed as a corrected absorbance.

Double-beam instruments may also be optical-null devices, like the Perkin Elmer 202™. Here, the light beam is split by a rotating chopper mirror which alternately directs it through reference and sample. The photomultiplier receives both absorbances alternately, generating an ac signal which is used to operate a servomotor that drives a logarithmic optical wedge into the reference beam. This proceeds until the light intensity of the two beams are identical. At this point the ac signal is zero. Movement of the optical wedge is linked to the recorder output, representing corrected absorbance. To prevent inaccuracies due to low light levels at the detector an automatic gain control circuit keeps the photomultiplier output voltage constant.

Many double-beam spectrophotometers are potentiometric null instruments. Some designs, like the Perkin Elmer 450™, utilize two matched detectors to simultaneously measure reference and sample beams. Others, like the Cary 14™, use a single detector which receives the chopped reference and sample beams alternately, generating an ac signal. This time the ac signal is used to drive a potentiometric slide wire until the difference in the two halves of the signal matches an accurate reference voltage. The excursion of the slide wire drives the recorder output.

Modern double-beam instruments, like the Perkin Elmer Lambda 900™ are controlled by both "on-board" microcomputer electronics and an "outboard" PC. This instrument can scan over the wavelength range 185–3300 nm with an incrementally variable bandpass (0.05–5.00 nm in the UV/visible and 0.2–20 nm in the near IR). Spectra can be stored, displayed, and manipulated in a number of ways.

The Hewlett-Packard 8453A™ is a version of that manufacturer's rapid scanning single-beam instruments. This unit behaves like a double-beam instrument since it can very rapidly scan its wavelength range of 190–1100 nm for both reference and sample. To accomplish this feat photodiode array detector technology has been employed. Software is used to subtract the reference spectrum from the sample spectrum and to perform complex multicomponent quantitative analysis. Rapid scanning is a feature of spectrophotometers that is of particular value in the study of short-lived chemical phenomena, such as in enzyme kinetics and for certain electrochemical and photochemical studies.

3.4 Components and Accessories [1–5]

3.4.1 Sources

From 325–3000 nm the commonly used light source is a tungsten filament incandescent lamp. This is called a *thermal source* because the photon flux is caused by the high temperature of a resistance heated tungsten filament in an evacuated envelope. The emission is a continuum because electronic and vibrational energy levels are severely overlapped by the interaction of closely packed tungsten atoms. In many lamp designs a low pressure vapor of bromine or iodine and a quartz envelope are employed. The halogens extend the life of the tungsten filament and the quartz glass allows more short wavelength light to pass.

The light emitted by a tungsten lamp operated at its normal temperature of 3000 K is a smooth black body continuum with nearly 85% of its total radiant power emitted in the infrared region. This leaves only about 15% for use in the visible part of the spectrum. Increasing the current to the lamp increases its temperature and shifts more radiant power to the visible,[*] but it also ruins the lamp rather quickly, since tungsten melts at 3600K.

The voltage supplied to the lamp needs to be very stable, especially for conventional single-beam instruments. A constant voltage power supply is usually incorporated in the instrument electronics. An external stabilized power supply to supply line voltage to the instrument is usually also called for. Older instruments sometimes employed a 9V lead-acid storage battery to supply steady dc current.

[*] According to the Wien displacement law $\lambda_{max} \cong 3.0 \times 10^6/T$, where T is the source temperature in Kelvins and λ_{max} is the wavelength of maximum emission, in nanometers.

Lamp heat is sometimes a problem for nearby optics and some spectrophotometer designs therefore incorporate heat sink cooling fins, a cooling fan, or even connections for cooling water. On rare occasions a carbon arc (at 4000°C) may be substituted for a tungsten source when more light intensity in the visible is required. But this is a specialized adaption, requiring special provisions for protecting the optical and sample compartments from the heat.

Below 325 nm emission from thermal sources is too weak and some type of *electric-discharge source* is generally used. As the name implies, these sources depend upon an electrical discharge through a high pressure gas to produce a flux of photons with high amounts of radiant power in the ultraviolet region. At very low gas pressures line spectra would be emitted, but at higher gas pressures band spectra or a continuum results from the interaction of highly excited molecules.

Hydrogen-discharge lamps are filled to 0.2–5 mm (of mercury) pressure and typically operate at around 40 V of dc current, although much higher voltage ac lamps have been designed. The envelope is glass but a quartz or fused silica window is provided. The former allows transmission down to 200 nm, the latter down to 185 nm. Above about 360 nm the hydrogen continuum is weak and distinct hydrogen emission lines begin to emerge. Deuterium-filled discharge lamps produce a similar continuum but at 2–5 times the intensity of an equivalent hydrogen lamp. They also show 2–5 times the lifetime of a hydrogen lamp.

Most modern spectrophotometers designed to cover the UV and visible range are fitted with both a tungsten–halogen and a deuterium source which are selected by some switch-controlled mirror or shutter mechanism. Alternative UV sources, such as the xenon-discharge lamp and the mercury arc lamp, have drawbacks for routine work. The xenon lamp shows an intense structured background in the visible that could lead to stray light problems in the UV, while the mercury arc shows a series of sharp bands superimposed on the continuum, making it unsuitable for broad-range scanning work.

3.4.2 Optical Components

The parts of the optical bench that refract, reflect, or transmit light must not absorb a significant amount of the wavelengths of interest. This figure is taken to be 0.200 absorbance units. Optical losses greater than this amount are generally deemed unacceptable. Silicate glasses are useable between 350 and 2,300 nm. Special borosilicate glass can be used down to 300 nm. Quartz is useable down to 190 nm, and some fused silica material can be used slightly lower. Prisms and lenses in visible range instruments are made of glass, while in UV/visible range instruments they are made of quartz.

The sample cuvettes (of whatever size and shape) are part of the optical system and must be made of suitable material. Cell materials include borosilicate glass, quartz, and polymethylmethacrylate.[*] Cell path lengths vary from 1 mm in thin microcells to 5 cm or more in cylindrical-shaped cells with optically flat end windows. The most commonly used cell is a square cross-section design with a 1-cm optical path.

In addition to their use in filter photometers absorbance filters are used in spectrophotometers to limit stray light and to reduce some spectral orders produced by diffraction gratings. They may be either cutoff filters, which pass radiation above or below a certain wavelength, or bandpass filters which pass a small, distinct segment of the spectrum. Two cutoff filters in combination can be selected to produce the effect of a single bandpass filter. Absorbance filters are generally made by tinting glass with a pigment. Another type, interference filters, make use of destructive interference between transmitted and reflected light to produce a very narrow bandpass.

Spectrophotometers should be installed in an area reasonably free of vibration, temperature variation, and acid and solvent fumes. The optics should be cleaned periodically by a trained professional. One common problem is lens and mirror fogging due to the formation of a polymerized film caused by UV light from the source acting on solvent vapors in the room air.

3.4.3 Detectors

Photovoltaic cells (sometimes called barrier-layer cells) are simple semiconductor devices that usually consist of a thin layer of selenium deposited on an iron or aluminum substrate. The selenium is then coated with a very thin layer of silver which is somewhat transparent to light. A glass cover protects the optically sensitive surface. Incident photons excite electrons at the interface of the selenium and silver causing a net transfer of electrons from the selenium (base) electrode to the silver (collector) electrode. A low resistance circuit shows a current flow that is closely related to the radiant power of the impinging light.

Photovoltaic cells suffer from a number of shortcomings that limit their usefulness. The response time of these devices is slow, precluding any modulation/noise reduction optical or electronic designs. Their sensitivity to incident photons is also not impressive when compared to other detectors. Photovoltaic cells also show a diminishing response to a given light flux, reaching a steady state after some minutes. These devices also age with time and use, gradually losing sensitivity. However, none of these limitations is serious enough to preclude their use in inexpensive filter photometers, where high light thoughput is typical. And the fact that these devices need no external power supply helps to keep the design simple and the cost low.

---

[*] Polymethylmethacrylate cuvettes are useable between 250 and 1100 nm.

Phototubes are vacuum tubes (sometimes with quartz or fused silica windows) with a half-cylindrical photocathode coated on its inside surface. The coating may be a mixture of cesium, cesium oxide, silver, and silver oxide (however, other alkali metals may be used). An anode wire is mounted near the axis line of the cylinder. When photons impinge on the cathode photoelectrons are ejected and are attracted to the anode where they produce a photocurrent. This small current is readily amplified in an external circuit. It is necessary, however, for the instrument manufacturer to calibrate the output response to compensate for nonlinearity in the relationship between incident radiant power and the evoked photocurrent.

Photomultiplier tubes are cascade effect phototubes consisting of a photocathode and a series of dynode electron multipliers. Photons impinging on the cathode produce photoelectrons which are accelerated by a potential field toward the first dynode stage, where each striking electron ejects 2–5 additional electrons. These are, in turn, accelerated to the second dynode stage where the process is repeated. Commercial photomultipliers may have 9-16 dynode stages for an overall gain of 10^6–10^9 electrons at the anode for each impinging photon at the cathode. Each successive dynode stage is typically 75–100 V more positive than the preceding one, and a high stability power supply is essential. The photomultiplier output current can then be further amplified in an external circuit. Photomultiplier tube manufacturers use a standard code ("S" followed by a number) to designate the spectral response of these devices.

Both phototubes and photomultipliers are limited by their dark current and shot noise, which together represent the sum of the noise generated by the detector and its associated circuitry. Dark current is caused by the temperature dependent emission of thermal electrons. Shot noise is a statistically distributed background emission of photoelectrons. Design efforts to reduce both will result in improved photon detection limits.

Photodiode arrays (PDAs) and charge transfer devices (CTDs) are detectors which allow the simultaneous measurement of an entire wavelength region. PDAs are still relatively new in absorption spectrophotometry, however they have proven their value in rapid scanning and in multicomponent quantitative work. CTDs have great potential in this area as well. Both classes of detectors will be discussed further in Chapter Six since their application in optical emission is becoming commonplace.

3.4.4 Miscellaneous Equipment

Among the equipment that is sometimes useful in spectrophotometric measurement are flow-through or "sipper" cells for filling, rinsing, and refilling from an automatic sampler controlled by a personal computer. Accessories for photometric titration, specular, diffuse, and total reflection measurements, and fiber optic probes for remote absorbance measurement may prove valuable to certain laboratories for specialized applications.

Software has now become a major feature of nearly all analytical instruments and UV/visible spectrophotometers are no exception. The ability to control all instrument functions from a keyboard is common in mid- and high-priced designs. Often quantitative calculation programs specific to environmental or clinical standard methods are offered. Somewhat less common are multicomponent calculation programs for rapid scanning instruments capable of handling 10 or more components.

3.5 Instrumentation for Related Techniques

Vacuum ultraviolet spectrophotometry [1] is currently a rather specialized field because routine-use instrumentation of the sort that lends itself to process control simply does not exist. The vacuum ultraviolet is usually defined as the range 0.1–200 nm, which stretches from the low end of conventional UV/visible instruments down to the edge of the X-ray region. For most molecules electron transitions in this region are energetic enough to produce molecular ions (Rydberg transitions). As the name of this field implies, air must be scrupulously excluded from the optical path since oxygen begins to absorb at 195 nm, water vapor begins to absorb at 178 nm, and nitrogen begins at 145 nm.

Lenses and prisms for vacuum UV instruments are often manufactured from calcium or lithium fluoride since even fused quartz absorbs strongly in the vacuum UV. Light sources for this region are a series of gas-filled capillary discharge tubes, which each show short regions of a useable continuum. Thus, hydrogen extends down to 165 nm before producing a line spectrum; xenon extends to 147 nm; krypton to 124 nm, and argon to 107 nm.

There are also special design problems for the monochromator because scattered light may actually be more intense than the dispersed light. Special materials are required for reflection gratings because of a decrease in reflectivity at low wavelengths. Detectors range from fluorescent material coated photomultipliers to X-ray type radiation counters. One principal manufacturer of vacuum ultraviolet instruments is the McPherson Instrument Corporation. Potential applications of vacuum ultraviolet absorption spectrophotometry include the analysis of saturated hydrocarbons and metallic vapors.

Photoacoustic spectrophotometry [2] is based on a discovery published by Alexander Graham Bell in 1880. The sample (which may be opaque) is enclosed in a sealed chamber with a light permeable window. The light source is modulated by an audiofrequency signal. The sample absorbs these light pulses, producing pulses of heat. The heat pulses produce a corresponding expansion and contraction of the sealed chamber atmosphere. These acoustic waves are detected by a microphone, which converts them to an electrical signal (the absorbance). UV/visible scanning spectra produced in this way are often much higher resolution than the equivalent diffuse reflectance spectra.

Fourier transform UV/visible spectrophotometry, compared to its infrared region equivalent, is still at an early stage of development. Currently, only custom-built research grade instruments exist and are largely engaged in hyperfine absorbance line studies. The future of this technology is presently uncertain since formidable design constraints must be overcome to produce a commercial routine use instrument while the practical rewards of such an effort remain dubious.

3.6 Instrument Maintenance

It is important that every laboratory that utilizes a UV/visible spectrophotometer establish and follow a regular plan for the wavelength and absorbance accuracy monitoring of these instruments. The wavelength accuracy can be checked by the use of holmium or didymium oxide glass standards with published certificates that include absorbance spectra. A simple signed and dated notebook entry with the measured absorbance readings near and at two or three absorbance maxima will usually suffice to establish and document wavelength accuracy.

Response accuracy is a little more difficult to monitor. In theory one can carefully prepare a chromophore solution, measure its absorbance, then calculate its molar absorptivity and compare it to the literature value. In practice this approach may lead to frustration. Except in a highly controlled research environment or a metrology laboratory a close match with published values of ε may be elusive. There are a great many empirical factors involved in the instrumental measurement and the chemistry is subject to all manner of chromogenic effects. That is why all good methods stress careful attention to sample and reagent blanks. It is probably fair to say that absorbance spectrophotometry started as a comparative methodology and largely remains so today.

The best approach to an absorbance check is to utilize a neutral density filter standard of a form designed to be conveniently fitted into the light path. The expected absorbance values listed on the accompanying certificate should list the bandpass or spectral slit employed for each wavelength value.

If suitable absorbance standards are not available in a form that can be accommodated by a particular instrument it is possible to establish traceability by preparing a chromophore solution and measuring it on both an absorbance verified instrument and on the instrument for which neutral density

---

Didymium refers to a mixture of praseodymium and neodymium.

filter standards are lacking. Closely matching absorbance readings on the two instruments may be taken as proof of absorbance accuracy on the subject instrument. As with wavelength accuracy monitoring, appropriate notebook entries should be made.

In the event of evidence of an important discrepancy in either wavelength or absorbance readings some steps must be immediately taken before the instrument is used again for analytical work. Usually this involves a service call from the instrument manufacturer for remedial action.

4. METHOD CALIBRATION AND MEASUREMENT [3,4,10–13]

The conventional approach to the calibration of a UV/visible single-beam spectrophotometer for quantitative work involves adjusting the dark current with the light shutter closed so that the instrument reads out zero percent transmittance (or infinite absorbance). Reference solvent is then placed into position and the shutter is opened. The slit width, gain (or sensitivity) control is then set to read 100% transmittance (or zero absorbance). With the instrument adjusted in this way a series of graded concentration standards are measured against the reference solution, and then these absorbance values are plotted versus concentration. This widely used approach is far from the only calibration technique available, however. Some alternative protocols offer distinct advantages in certain situations.

The "transmittance-ratio" method involves setting the dark current to zero transmittance (infinite absorbance) with the shutter closed, as in the conventional approach, but then adjusting the 100% transmittance (zero absorbance) using a standard solution slightly lower in analyte than the lowest levels expected in the unknown sample solutions. A calibration curve is then prepared from standard solutions in the conventional manner. This technique, which is an example of *differential spectrophotometry*, amounts to an expansion of the measurement scale of the instrument. In effect this means that the absorbance difference between any unknown solution and the analyte concentration of the adjustment standard has been greatly magnified. If we neglect any errors in standard solution preparation we can presume a proportional increase in accuracy.

This technique can be utilized to make spectrophotometric measurements on high concentration analytes with an error equivalent to that obtained by volumetric and gravimetric methods. For example, suppose that we can measure cobalt at 5 mg/mL using a conventional spectrophotometric calibration with an error of 0.5%. We can set up the transmittance-ratio method with a 40 mg/mL standard cobalt solution set equal to 100% T and measure a 45 mg/mL unknown solution against it with an error of only 0.056% [.025(100)/

45]. This method ordinarily requires an instrument with sufficient reserve amplifier gain to adjust 100% T with an absorbing solution in the beam, but in some cases widening the slits is all that is necessary.

C.N. Reilley and C.M. Crawford [10] published an important paper in 1955 which examined the influence of spectrophotometric calibration modes on measurement precision. This study not only treated ordinary and transmittance-ratio methods, but also introduced two new methods. The first of these, termed by the authors the "trace analysis" method, involves setting the readout to 100% transmittance (zero absorbance) when the light beam passes through the reference solution, and then setting the readout to 0% T (infinite absorbance) on a standard solution slightly more concentrated in analyte than the unknown sample solution. This approach results in a significant gain in measurement precision for low absorbance samples. However, it requires the use of a spectrophotometer with a dark current control circuit with sufficient reserve "bucking voltage" capacity to offset currents much higher than those normally encountered with the shutter closed.

The same paper also described "the method of ultimate precision." In this approach a standard solution slightly lower in analyte concentration than the unknown is used to set 100% T (zero absorbance) and a standard solution slightly higher in analyte concentration than the unknown is used to set 0% T (infinite absorbance). This approach results in the greatest precision of the four calibration modes. In order to achieve these advances in measurement precision a high quality instrument is essential. Gain and dark current controls must be adequate; and in wavelength regions where it becomes necessary to widen the slits in order to accomplish the desired adjustment, stray light error must be low.

In some of the older literature on spectrophotometry there is much discussion regarding the delineation of an absorbance reading region of maximum precision, applicable to any calibration mode. Clearly, it seems intuitive that the best precision would not be obtained at extremely low absorbance readings or at extremely high ones.

From Eq.(7):

$$A = \log(P_o/P)$$

$$\text{or: } A = (1/2.3)\ln(P_o/P) \qquad (12)$$

It is possible to use differential calculus to locate the absorbance at which the greatest measurement precision is to be expected:

$$dA = (1/2.3)d\ln(P_o/P)$$

$$dA = [(-P_o/P^2)dP]/[2.3(P_o/P)]$$

$$dA = -dP/2.3P \qquad (13)$$

To obtain an expression for the relative error in absorbance (dA/A) we can divide both sides of (13) by A:

$$dA/A = -dP/2.3PA \tag{14}$$

Since equation (7) can be rearranged to $P = P_o \times 10^{-A}$ we can substitute this value in (14):

$$dA/A = dP/2.3AP_o \times 10^{-A} \tag{15}$$

If we set the spectrophotometer to read 100% T (zero absorbance), then we can normalize (15) by setting $P_o = 1$.

$$dA/A = dP/2.3A \times 10^{-A} \tag{16}$$

Obviously, the minimum in Eq.(16) occurs when ($2.3A \times 10^{-A}$) is at a maximum value. We, therefore, differentiate this term and set the derivative equal to zero:

$$d(2.3A \times 10^{-A})/dA = 10^{-A} - 2.3A \times 10^{-A} = 0$$

$$10^{-A}(1 - 2.3A) = 0$$

If $10^{-A} = 0$, then the absorbance and the error are both infinity; but the other factored term yields a practical result:

$$1 - 2.3A = 0$$

$$A = 0.435$$

This is the absorbance value that should have the lowest reading error. Plotting an error curve from a related treatment of concentration error it can be shown that the precision is fairly close to optimal over the range 0.2–0.7 absorbance units.

It must be stated, however, that these mathematical treatments apply primarily to old d'Arsonoval-type panel meter direct reading instruments, such as the Beckman model *B*™. In modern digital readout spectrophotometers the error limits are much more difficult to define, since they are largely determined by shot noise in the photomultiplier. With the better quality modern digital readout spectrophotometers it is probably safe to assume that the region of optimum precision extends beyond the 0.2 to 0.7 absorbance unit range.

The analysis of multicomponent analyte systems is commonplace in infrared organic work but also not uncommon in UV/visible trace metal analysis. In UV/visible work it is quite common that the absorbance peaks of the analytes are overlapping to some degree. Sometimes in a two component system, A and B, A contributes to the absorbance of B at its maximum absorbance wavelength, λ_B, but is, itself, free of interference at its own maximum, λ_A. In this case we prepare three calibration curves: A at λ_A, A at λ_B, and B at λ_B. The concentration of A found at λ_A is then used to determine a correction factor for the observed absorbance at λ_B.

Even more commonly in a two component system, A and B, A contributes somewhat to the absorbance of B at its maximum, λ_B, while B contributes to the absorbance of A at its maximum, λ_A. To determine both analytes we must prepare four calibration curves: A at λ_A, A at λ_B, B at λ_A, and B at λ_B. The unknown sample solution is then measured at λ_A and λ_B. For simplicity we shall assume that all absorbances have been properly blank corrected and, further, that all calibration curves obey Beer's law and pass through the origin. We can formulate a set of simultaneous equations:

$$A_A = c_A s_{AA} + c_B s_{BA}$$
$$A_B = c_A s_{AB} + c_B s_{BB}$$

where, A_A = the absorbance of the unknown solution at λ_A
A_B = the absorbance of the unknown solution at λ_B
c_A = the concentration of analyte A in the unknown
c_B = the concentration of analyte B in the unknown
s_{AA} = the slope of analyte A's calibration curve at λ_A,
s_{BA} = the slope of analyte B's calibration curve at λ_A,
s_{AB} = the slope of analyte A's calibration curve at λ_B, and
s_{BB} = the slope of analyte B's calibration curve at λ_B.

Since all of the quantities except c_A and c_B are known this set of simultaneous equations is easily solved by, for example, solving one equation for c_A and substituting that expression into the other equation, solving for c_B, then using that value to solve for c_A. This technique can sometimes be successfully extended to three components that show overlapping maxima at three wavelengths. For systems with more components the analysis must be *overdetermined*, that is, measurements must be made at more wavelengths than the number of desired analytes. Such procedures are generally performed with a rapid scanning instrument. It is also necessary to obtain individual wavelength scans of standard solutions of each analyte.

Data reduction in such cases almost always involves computer-based iterative techniques. Sustek [12] reported on studies to optimize both the degree of overdetermination and the selection of wavelength positions. Connors and Eboka [13] describe alternative graphical methods for calculating the results for two-, three- and four-analyte multicomponent systems which are overdetermined at many different wavelengths. Overdetermined multicomponent analyses are, predictably, more accurate than methods in which only a small number of wavelengths are utilized.

5. THE USES OF UV/VISIBLE MOLECULAR ABSORPTION SPECTROPHOTOMETRY [3]

It would be well to pause here and briefly summarize some of the practical uses to which this technique can be applied before we proceed to narrow our focus to fine detail. The organic chemist generally considers UV/visible spectrophotometry as a supplemental qualitative tool. The spectra are lean in information but can sometimes be used cleverly. For example, a weak peak at 280–290 nm that moves to a shorter wavelength when the solvent is switched to one of greater polarity suggests the carbonyl group. In addition, any organic compound containing at least one chromophoric group may be able to be determined quantitatively in neat solution or in simple mixtures. The same statement applies to certain inorganic species (e.g., chromate, nitrate, nitrite).

However, the majority of applications and the great strength and utility of this methodology involve the quantitative determination of nonabsorbing analytes (or weakly absorbing analytes) by forming derivatives that are strongly absorbing. Sometimes the formation of a strongly absorbing moiety is simultaneous with an isolation step (typically, solvent extraction), sometimes it follows a sequence of separations, and sometimes it occurs immersed in a sample background, which must be separately measured and subtracted out.

Most methods directly measure the absorbance of an analyte complex, but there are indirect methods, catalytic methods, and even turbidometric methods that are occasionally employed in trace metal analysis. Thus, the molybdenum of a phosphomolybdate precipitation separation may be reacted with thiocyanate to obtain an analytical result for phosphorus. The catalytic influence of osmium on ceric/cerous equilibrium may be exploited spectrophotometrically to measure osmium. And silver may be measured as a suspended silver chloride precipitate. There are also bleaching reaction methods in which increasing amounts of analyte produce decreasing amounts of absorbance of some complex that is destroyed by the analyte. Fluoride ion is commonly determined by this means.

6. PRACTICAL DETAILS

It is very important that sample cuvettes be cleaned and handled properly. An optically matched set of such cells is an important starting point, but the analyst must never assume the absence of a cell blank. Cell blanks should be checked using the reference solution each time the cells are used and at each wavelength measured. A zero cell blank at one wavelength is not a guarantee that a cell blank does not exist at higher or lower energy. Cuvettes should never be touched or picked up by the optical surfaces. They should be carefully wiped with lint-free wipers. In the case of those methods that utilize concentrated sulfuric acid as the reference solvent and the sample matrix solvent it is advisable to keep a cloth soaked in water close at hand. The outside of the filled cell is first wiped with the water-soaked cloth, then dried with lint-free wipers.

Cuvettes should be rinsed 3–4 times with the sample solution before they are filled and wiped for measurement. An effort should be made to place them in position in the light beam as reproducibly as possible. It is advisable to measure a graded series of analyte concentrations in order of *increasing* analyte concentration. If it is necessary to go from a high to a low analyte concentration a special effort at cell rinsing should be expended. In such cases some analysts will rinse the cell with water or other reference solvent before rinsing and filling with the next sample.

Filled cuvettes should always be examined before placing them in the spectrophotometer for measurement. They should be free of lint or dried residues on their outer optical surfaces, and the liquid within should be free of air bubbles, floating particulates, or refracting concentration gradients. When working with volatile solvents cuvette lids or stoppers should be used to minimize evaporation (which can lead to serious errors, harm the instrument, and pose safety problems).

Cuvettes must be cleaned promptly after use. Ordinarily, this can be accomplished by thoroughly rinsing with distilled water and allowing them to dry at room temperature. Stubborn residues can often be removed by filling the cells with a 1:1 ethanol:hydrochloric acid mixture, allowing them to stand for several minutes, then thoroughly rinsing with distilled water. Etched or badly contaminated cells should be discarded.

When developing or adopting a new spectrophotometric procedure it is always wise to obtain a wavelength scan even if it must be obtained laboriously on a single-beam instrument (by setting 0 and 100% T and measuring the analyte solution at each incremental wavelength, and manually plotting the results). The curve provides valuable information for the placement of the analytical wavelength.

For peaks with a broad flat top the absorbance maximum is the logical choice as the place to measure analyte absorbance. But for convoluted or sharply peaked absorbance/wavelength curves it may be prudent to select a flat shoulder region near the maximum rather than the maximum itself. Such a region (if it exists) may be less subject to subtle change with variations in the source, the detector, or the chemistry of the color reaction.

Unless one is dealing with a well documented literature method it is also wise to conduct color stability studies to determine the optimum time between the initial development of the color and its measurement. This work can take the form of a plot of the analyte complex absorbance at the selected wavelength versus time. In many methods a steady-state condition is reached in 0–10 min and persists for 1–24 hr; however, there are numerous exceptions.

Interference studies are also often appropriate, especially when a method is being adapted to a new sample alloy matrix. The effect of matrix elements at or slightly above typical matrix concentrations should be studied both individually and in synthetic mixtures which approximate real alloys. Other important details whose relative importance must be studied are the sample blank (sample minus the color reagent) and the reagent blank (all reagents minus the sample). In addition, if a masking agent is suggested in the literature its effect in a new matrix must be very carefully evaluated.

While many spectrophotometric calibration curves are stable with time one must not assume that they will hold indefinitely. As sources, detectors, and amplifier circuits age, instrument response will shift enough to invalidate calibrations. The unfaltering use of appropriate validation standards each time the method is applied is the best check on calibration. In some cases a control chart approach is feasible, enabling the analyst to spot a trend before any rejectable data is generated. Whatever the validation means, at the first sign of a problem the calibration should be redone and any questionable sample work repeated with appropriate validation standards.

Useful general references for spectrophotometric measurement include the *Manual on Recommended Practices in Spectrophotometry* and the standard practices under the jurisdiction of ASTM Committee E-13 on Molecular Spectroscopy (which appear in volume 03.06 of the *Annual Book of ASTM Standards*). Both are published by the American Society for Testing and Materials, West Conshohocken, PA [14,15]. Also useful is *Visible and Ultraviolet Spectrophotometry* by D.F. Boltz and G.F. Schenk, which appears in the *Handbook of Analytical Chemistry* (L. Meites, ed.), McGraw-Hill, NY, 1963 [16].

7. APPLICATIONS TO THE TRACE ANALYSIS OF METALS

7.1 General

Unlike most of the other measurement techniques applicable to trace metal analysis, UV/visible spectrophotometry shows relatively few applications that directly pertain to groups of elements or to broad categories of alloy matrices. For this reason it was decided that the field is most appropriately surveyed analyte-by-analyte with accompanying notes related to alloy types.

In the sections that follow no attempt has been made to comprehensively summarize the extant literature. Rather, the goal has been to focus on methodology that has found practical utility in commerce and industry.

Admittedly, this modest treatment will not uncover hidden treasure, but it will highlight the tried and true, while keeping the length of the discussion manageable. It is impractical and inappropriate to survey the periodic chart exhaustively, so we have culled our list of topics to those elements for which a practical spectrophotometric approach is at least feasible.

7.2 Alkali Metals

Spectrophotometric methods are not now and have never been the method of choice for these analytes. Today, the alkali metals are most often determined by atomic emission or atomic absorption methods. There *are* molecular absorption methods, however, that qualify by the above enjoiner, "at least feasible."

Lithium can be determined in a dilute potassium hydroxide solution containing acetone, using thoron [o-(2-hydroxy-3,6-disulfo-1-naphthylazo)-benzenearsonic acid]. The complex is measured at 486 nm. Under such conditions a 50-fold excess of sodium and a 10-fold excess of calcium and magnesium can be tolerated without effect. Rubidium and cesium can be tolerated at greater than 10-fold excess [17,18].

Sandell [18] recommends precipitating *potassium* as the chloro-platinate in an alcoholic medium, then dissolving the precipitate in water. Stannous chloride is then added and the color due to platinum is measured. The same reference also describes a method in which potassium is precipitated with sodium dipicrylaminate, then dissolved in hot water. A color is developed in dilute sodium hydroxide. In 1980 Nakamura et al. published a solvent extraction method for potassium using an NO_2^- derivative of 4'-picryl-aminobenzo-18-crown-6 in chloroform [19]. The extract was measured at 470 nm. The method has been successfully applied to the determination of potassium in Portland cement.

7.3 Alkaline Earths*

For the most part there is little to recommend spectrophotometric approaches for these elements since they can be conveniently determined by atomic absorption or optical emission. *Magnesium* has been measured as the Titan Yellow complex. This is one of a series of thiazole dyes that have been used for this determination. The presence of calcium increases the intensity of the color, but calcium can be "blanked out." Titan Yellow has been used to determine

* Beryllium is a group IIA element, as well, but is treated here as a separate category since it has unique properties and can be readily determined spectrophotometrically.

trace magnesium in titanium alloys, as has the dye Solachrome Cyanine R200 [20]. Magnesium has also been measured in electronic nickel as an 8-hydroxyquinoline complex [21].

Calcium, *strontium*, and *barium* have been determined at 575 nm with o-cresolphthalein complexone in weakly alkaline media, but the analyte must be completely separated from the remainder of the group IIA elements [18].

7.4 Aluminum

Spectrophotometric approaches for trace aluminum are common and useful for a large array of alloy types. A high percentage of the routinely applied procedures are based on one of two reagents: aluminon and 8-hydroxyquinoline. Aluminon is ammonium aurintricarboxylate, a dye which forms a bright red precipitate with high concentrations of aluminum, but forms a stable colored solution with low concentrations of aluminum. Iron interferes seriously, producing nearly half the absorbance of an equal concentration of aluminum. In steel analysis [22] this is usually handled by a preliminary mercury cathode electrolysis or by an isopropyl ether extraction from 8M HCl. Chromium (III) forms a red lake similar to that of aluminum, but chromium (VI) does not react with the dye. Steel samples are, therefore, usually fumed in perchloric acid.

In some versions of the color development procedure a 0.1% (w/v) solution of aluminon is prepared by dissolving 0.100 g of reagent in water, adding 10 mL of a 10% (w/v) benzoic acid in methanol solution, then diluting to 100 mL. The solution is then allowed to stabilize for 3 days. 1 g of gelatin is dissolved in hot water, then 10 mL of 10%(w/v) benzoic acid in methanol is added, and this second solution is diluted to 100 mL. An acetic acid/ammonium acetate buffer (pH 5.3) is prepared by very cautiously combining 470 mL of 15M ammonium hydroxide and 430 mL of glacial acetic acid, cooling, and diluting to 1 liter. A composite color reagent is then prepared by combining equal volumes of these three reagents.

A 5-mL sample aliquot containing 0.01–0.04 mg aluminum is combined with 15 mL of the composite reagent and heated in a water bath at 90–100°C for 10 min, allowed to stand at room temperature for 10 min, then diluted to 100 mL and measured at 525 nm versus a water reference. A sample blank must also be measured.

The aluminum/aluminon color is not a true dissolved species but a hydrosol (a dispersed colloid in water). The gelatin—itself a lyophillic colloidal suspension—acts to stabilize the color phase. Besides steel [22], the approach has been applied to ferroboron [23], electronic nickel [21], copper and its alloys [24] (including copper/beryllium [25]); and to titanium [20], and zirconium [18], and their alloys.

The second widely used aluminum reagent is 8-hydroxyquinoline (also known as 8-quinolinol). It can be extended to concentrations as low as 10 ppm in some cases. As with aluminon, iron is a major interferent and must be carefully removed, usually by an initial mercury cathode separation, followed by an extraction with cupferron/chloroform, and then sometimes by a final sodium hydroxide separation. The related reagent, 8-hydroxyquinaldine, had originally been recommended for iron "clean-up" [26] since steric hindrance is known to prevent its *precipitation* of aluminum, and, by analogy, it was assumed that aluminum did not *extract* with the reagent. However, it has since been found that aluminum partially extracts as an ion pair in the presence of acetate [27].

The unreacted 8-hydroxyquinoline reagent absorbs below 370 nm while the aluminum/8-hydroxyquinoline complex shows an absorbance maximum at 395 nm [18]. This minimizes the size of the reagent blank.

Any aluminum reagent can be adapted to the determination of "acid-soluble," "acid-insoluble," and total aluminum, but 8-hydroxyquinoline has been most frequently applied to these analytical schemes. The designation of what constitutes "acid-soluble" and "acid-insoluble" aluminum (or any other element for that matter) is always an empirical convention. In steel, acid-insoluble aluminum is generally associated with Al_2O_3 particles of a filterable size. "Total aluminum" is not an empirical concept, of course, but it must be recognized that the acid-resistant character of aluminum inclusions usually requires the analyst to filter the acid dissolved sample, ignite and fuse the residue, and recombine the leached fusion with the filtrate.

For steel the sample is usually dissolved in *aqua regia* and then fumed in perchloric acid. The chromium is volatilized by cautiously adding hydrochloric acid dropwise to the fuming solution. The cooled solution is diluted and filtered. The residue is washed with hot 2%(v/v) perchloric acid, then very thoroughly with hot water to remove all traces of perchloric acid. The filter and residue are then ignited at 1000°C in a platinum crucible and cooled in a desiccator. Silica is volatilized by adding 1:1 sulfuric acid:water, then hydrofluoric acid, and evaporating to dryness. Sodium bisulfate (fused salt) and sulfuric acid are added to the crucible and it is heated over a burner until no undissolved residue remains. The cooled melt is leached in a beaker with water and the crucible is removed and rinsed. The leachate is then added to the previously obtained filtrate.

WARNING: Extreme caution is required. Cyanide compounds and mercury are extremely toxic. Cupferron and chloroform are known carcinogens. Use proper safety equipment and practices and perform all work in an efficient fume hood.

The combined solution is then electrolyzed on a mercury cathode and filtered on double Whatman No. 41 paper or equivalent. The papers are washed

8 times with water, then discarded (after properly disposing of the waste mercury). Dilute sulfuric acid (10 mL of 1:1 H_2SO_4:H_2O) and ammonium persulfate (10 mL of a 100g/L solution) are added and the solution is boiled down to approximately 75 mL, then cooled to room temperature, and then finally in an ice bath. It is then transferred to a 250-mL separatory funnel, 15 mL of 60g/L cupferron solution (prepare fresh weekly, filter, and store in a refrigerator) is added, the funnel is agitated, and the precipitate is permitted to settle. Then 20 mL of chloroform is added and the solution is extracted for 1 min. The (bottom) organic layer is drawn off and discarded. The chloroform extraction is repeated until the organic layer remains colorless. The aqueous layer is then transferred back to the original beaker, reduced to 35 mL by heating. Then 25 mL of nitric acid are added and the solution is heated to fumes of SO_3. The cooled salts are dissolved by boiling with 75 mL of water and the cooled solution is transferred to a Teflon beaker.

The solution is neutralized with 200g/L sodium hydroxide solution (litmus paper) plus 10 mL excess; 1 mL of 30% hydrogen peroxide is added and the solution is held near the boiling point for 6 min, then filtered through a hardened, medium porosity filter paper (e.g., Whatman No. 540). The filtrate is collected in a Teflon beaker. The residue is washed with hot water. The filtrate is neutralized with hydrochloric acid (litmus paper) plus a slight excess and transferred to a 200-mL volumetric flask, diluted to volume and mixed. An aliquot containing approximately 0.05 mg of aluminum is transferred to a 250-mL beaker and diluted to approximately 50 mL.

In a hood 1 mL of ammonium acetate solution (180g/L) and 10 mL of sodium cyanide solution (100g/L) are added. With a pH meter the solution is adjusted to pH 9.0 ± 0.2 with ammonium hydroxide and 1:1 hydrochloric acid:water. The solution is transferred to a 125-mL separatory funnel and 1 mL of 8-hydroxyquinoline solution (25 g of reagent dissolved in 60 mL of acetic acid, diluted to 300 mL with warm water, filtered, then diluted to 500 mL) is added. The funnel is swirled to mix the solution, then 10 mL of chloroform is added and the solution is extracted for 20 sec. The (lower) organic layer is transferred to a dry 50-mL beaker. The extraction is repeated with a second 10-mL portion of chloroform and the extracts are combined. Then 0.5 g of anhydrous sodium sulfate is added and the extracts are transferred to a 25-mL volumetric flask. The beaker is rinsed with chloroform, which is then added to the flask. Then the flask is diluted to the mark with chloroform and mixed. The absorbance is measured versus chloroform at 395 nm [28].

Besides low alloy steel 8-hydroxyquinoline has been applied to the spectrophotometric determination of aluminum in high purity iron [18], lead, antimony, and tin alloys [18], thorium metal [18], zirconium and zirconium alloys [29], as well as to stainless steels [30], and to silicon and ferrosilicon [31].

If there is a third place for the popularity of color reagents for aluminum it might go to Eriochrome™ Cyanine R. Structurally related to aluminon, it forms a red-violet complex with aluminum at pH 5.8 that absorbs at 535 nm. This reagent has been applied to steel [32], copper metal [33], and zinc alloys [18].

Other reagents have been recommended and used for the practical determination of trace and low levels of aluminum in iron and steel. For example, Alizarin Red S in the presence of calcium ions forms a complex with aluminum which absorbs at 490 nm [34]. Extreme sensitivity for low concentrations has been claimed for some reagents when used in the presence of a surfactant—for example, haematoxylin in the presence of cetyltrimethylammonium bromide (CTAB) [35].

7.5 Antimony

Antimony exhibits both a +III and a +V oxidation state in aqueous solution, as well as a marked tendency to hydrolyze as a hydrous oxide precipitate. It is, therefore, essential that solution conditions be carefully controlled throughout its determination. The most common color reagent is Rhodamine B, a cationic dye that forms an ion pair with an anionic Sb(V) complex, such as $SbCl_6^-$; such ion association systems are readily extractable with aromatic solvents or ethers.

Luke [36] dissolved lead samples by fusing with pyrosulfate salt and sulfuric acid in an erlenmeyer flask, then cooling, adding dilute hydrochloric acid, boiling, and filtering off the lead sulfate precipitate. Additional hydrochloric acid (10 mL), 5 mL of 6% sulfurous acid and about 50 mL of water were added and the solution was boiled down to 40 mL. The cooled solution was transferred to a 150-mL separatory funnel and diluted to 60 mL; 1 mL of ceric sulfate solution [0.5 g $H_4Ce(SO_4)_4$ dissolved in 100 mL of 3:97 H_2SO_4:H_2O] was added and the solution swirled. This is followed immediately by 3 mL of Rhodamine B solution [2%(w/v) in water]. Luke then added 15.0 mL of benzene and extracted for 1 min. The modern analyst would likely substitute a safer solvent since benzene is a known carcinogen, banned from many laboratories. The aqueous layer was discarded and the organic layer was drained through Pyrex wool into a 2-cm path length cell and measured at 565 nm.

In addition to lead alloys [18,36,37] Rhodamine B has been used for antimony in copper and copper alloys [18], germanium [18], iron and steel [18], and cadmium metal [38]. Two related cationic dyes have also been used in this application. Brilliant Green is used to determine 0.0002–0.030% antimony in steel by extracting with toluene (also a hazardous solvent) in the presence of tartaric acid and pyrophosphate ion and measuring the absorbance of the organic layer at 645 nm [28]. Crystal Violet is used to determine 0.0004–0.0075% antimony in copper metal by extracting with trichloroethylene (a known carcinogen) and measuring at 570 nm [33]. All three dyes suffer from positive interference caused by the presence of gold and thallium.

Antimony also forms a colored solution with iodide ion which can be measured near either 330 nm or 425 nm. While there is interference from, for example, bismuth [16], the technique has been applied to copper [39], lead [40], and aluminum alloys [41].

The chelating reagent, Pyrocatechol Violet, has been applied to the determination of traces of antimony in copper, lead, and lead alloys. First, antimony (III) is extracted from iodide or bromide/sulfuric acid media with toluene. Second, the toluene layer is extracted with an aqueous solution of Pyrocatechol Violet (PCV) to produce the antimony/PVC complex. Third, a solution of tri-n-octylamine in toluene is added, the solution is extracted, and the organic layer is measured at 555 nm [42].

7.6 Arsenic

Arsenic, like antimony, can occur in the +III and +V oxidation states in aqueous solution. The reduction of the arsenomolybdate complex to form the "heteropoly blue" color has long been the basis of one of the most dependable and sensitive methods for arsenic. Applied to complex alloy systems, however, the method can be somewhat dependent upon operator skills. For iron and steel samples arsenic must be isolated, usually by distillation as the chloride. For brasses, bronzes, and bearing metals precipitation as elemental arsenic using hypophosphorous acid is an alternative to distillation. For germanium metal the best isolation approach is extraction with diethylammonium diethyldithiocarbamate/chloroform [18].

Donaldson employed coprecipitation with iron hydroxide and a xanthate extraction to isolate arsenic from copper-base alloys for the molybdenum blue method. A 0.2–1.0 g test portion, containing up to 150 µg of arsenic, is dissolved in hydrochloric, nitric, and sulfuric acids, then heated to fumes of SO_3. The cooled solution is diluted to 100 mL and 5 mL of hydrochloric acid and 80 mg of iron (in the form of Fe^{3+}) are added and the solution is heated. Ammonium hydroxide is added to blue litmus, plus 5 mL excess. The solution is filtered, the residue is washed; then the precipitate is redissolved into the original beaker with hot 15%(v/v) hydrochloric acid. The iron precipitation, filtration, and precipitate dissolution are repeated.

The solution is then evaporated down to approximately 3 mL using low heat. The solution is cooled to room temperature; 20 mL of hydrochloric acid and 3 mL of water are added and the solution is allowed to stand until all salts have dissolved. Then 10 mL of freshly prepared 5%(w/v) iron (II) sulfate (5 g of reagent in 100 mL of HCl) are added, and the solution is transferred to a 125-mL separatory funnel, using hydrochloric acid to a final volume of 50 mL.

Next, 10 mL of chloroform and 1 mL of freshly prepared 20%(w/v) potassium ethyl xanthate (PEX) are added to the funnel and the contents are immediately extracted for 1 min. The (lower) organic layer is drained into a second 125-mL separatory funnel. The aqueous phase is washed twice more, using a 10-mL and a 5-mL portion of chloroform and 1 mL and 0.5 mL of the PEX solution, respectively, and finally with just 5 mL of chloroform. All chloroform layers are combined in the second funnel.

Then 5 mL of 20%(v/v) bromine in carbon tetrachloride is added to the combined extracts and the solution is thoroughly mixed. The solution is allowed to stand 5 min, 15 mL of water is added and it is extracted for 1 min. The (lower) organic layer is drained off and discarded. The aqueous phase is drained into a beaker, rinsing the funnel with water. The solution is evaporated to about 20 mL using low heat. The cooled solution is transferred to a 50-mL volumetric flask and diluted to approximately 40 mL.

Next 5 mL of 1%(w/v) ammonium molybdate in 2.3M sulfuric acid and 1 mL of 0.5%(w/v) hydrazine sulfate solution are added and the flask is diluted almost to the mark, mixed and heated in a boiling water bath for 30 min. After cooling to room temperature, the flask is diluted to the mark and mixed. The absorbance of the heteropoly blue complex is measured in a 2-cm cell at 845 nm versus a reagent blank [43].

The heteropoly blue method for arsenic has also been applied to pig lead [37], cadmium metal [38], silicon and ferrosilicon [31], ferromanganese [44], silicomanganese, and ferrosilicon manganese [45], chromium metal and ferrochromium [46], and ferrochromium silicon [47]. It is also possible to measure the yellow unreduced species of heteropoly acid. A combination method has been described for steel in which both arsenic and phosphorus are measured as ion pair extracts with Ethyl Violet, then phosphorus alone is measured on a separate aliquot using masking agents. Arsenic is then determined by difference [48].

The evolution of arsine gas (AsH_3) and its subsequent reaction with a pyridine solution of silver diethyldithiocarbamate (Ag-DDC) to form a pink colloidal suspension of silver metal has been the basis of several approaches. Steel samples dissolved in *aqua regia* can be fumed in sulfuric acid so long as the temperature remains below 200°C. Hydrochloric acid, potassium iodide, and stannous chloride are added to the cooled and diluted solution. After 15 min. the solution is transferred to a generator flask, fitted with a lead acetate-treated cotton trap for H_2S, and connected by means of a fine capillary to the absorbing solution (0.5% Ag-DDC in pyridine). 3 g of arsenic-free zinc are added to the sample solution and the apparatus is immediately sealed. After 1 hour the absorbing solution is measured versus pyridine at 545 nm [49].

The direct arsine method has been plagued with reproducibility problems that can often be eliminated with preliminary separations. Extraction of $AsCl_3$ from concentrated hydrochloric acid solution with p-xylene (a health risk solvent) has been suggested [27].

7.7 Beryllium

Beryllium is determined more commonly by fluorimetry than by absorption spectrophotometry. There are a few procedures, however, that are worthy of mention, especially since comparatively few metal analysis labs are equipped with a spectrofluorimeter. Beryllium can be determined in bronze or aluminum alloys using Chromazurol S (a registered trademark of Geigy Co.). Transfer an aliquot of the dissolved sample, containing 1–80 μg of beryllium to a 25-mL volumetric flask. With an identical aliquot determine the amount of 10%(w/v) sodium hydroxide solution required to exactly neutralize the free acid to a methyl orange endpoint. To the volumetric flask add 1.0 mL of 1%(w/v) ascorbic acid, 2.0 mL of 10%(w/v) EDTA solution, 5.0 mL of acetate buffer (pH 4.60; 238 g sodium acetate · $3H_2O$ and 102 mL glacial acetic acid diluted to 1 liter), and the appropriate volume of 10%(w/v) sodium hydroxide solution, then dilute to approximately 20 mL with water. Add 2.00 mL of a 0.165%(w/v) Chromazurol S solution (in water) and dilute to the mark and mix. Measure in a 1-cm cell versus a reagent blank at 569 nm [27,50].

Sandell [18] mentions 4-(p-nitrophenylazo)orcinol as suitable for beryllium in titanium alloys. The same reagent has been used for aluminum alloys [41], while p-nitrobenzene-azo-orcinol has been applied to copper alloys [24]. Acetylacetone has been used for copper alloys after extensive separations [51]. An interesting approach was published by Matsubara and Takamura [52]. A mixture of sodium hexafluorotitanate(IV), hydrogen peroxide, and sodium fluoride dissolved in hydrochloric acid shows no color, but in the presence of beryllium (which has a stronger affinity for fluoride than titanium) titanium (IV) ions are released from their soluble fluoride complex and form the characteristic yellow peroxytitanium complex which is measured spectrophotometrically. The approach was utilized to determine beryllium in copper-beryllium alloys. A reference solution was prepared by extracting the beryllium with benzoylacetone in chloroform at pH 7–8 in the presence of EDTA.

7.8 Bismuth

Today, bismuth is most frequently determined by atomic absorption or atomic emission techniques, however, very good spectrophotometric methods for alloys are available. Bismuth forms a yellow color with iodide ion which can be measured at either 460 or 337 nm. As with most spectrophotometric procedures for this element, extensive separations are necessary for most alloy types. Copper (I), lead, thallium (I), silver, and cadmium form iodide precipitates, while antimony, tin, platinum, and palladium form competing colored complexes with iodide. Many separation schemes have been applied, including cupferron and dithizone extractions, and copper precipitation [18]. Donaldson has employed a sodium diethyldithiocarbamate extraction with chloroform from a basic solution containing citric and tartaric acids, EDTA, and cyanide. This is followed by back extraction into 12M hydrochloric acid, then chloroform extraction of the ethyl xanthate complex from a 2.5M hydrochloric acid/tartaric acid/ammonium chloride medium. The organics are wet ashed, and the colored iodobismuthite ion is formed with potassium iodide in the presence of hypophosphorous acid in 1M sulfuric acid. The method has been successfully applied to copper metal, and copper-, tin-, and lead-base alloys [40,53].

Thiourea yields a yellow color with bismuth that absorbs at 322 nm. This reagent has been applied to lead and tin alloys [18], copper alloys [33], and aluminum alloys [41]. Of wider applicability are methods based on dithizone. The bismuth complex is extracted with chloroform from an isolate solution that is pH 11.5 in the presence of cyanide; it is measured at 500 nm. Dithizone methods have been applied to steel, copper metal and copper alloys, lead metal, pig lead, "corrosion lead," antimonial leads, tin metal and tin alloys, zinc metal and slab zinc [51], as well as to uranium and uranium alloys [18].

7.9 Boron

Boron is one of a comparatively small array of analytes that the modern trace analysis laboratory may choose to determine spectrophotometrically despite the availability of faster, more cost effective techniques. There are a number of reasons for this. First, boron has uniquely stringent analytical requirements because its metallurgical effects are synergistic—ultratrace amounts can have a profound influence on alloy properties. Second, it is relatively easy to isolate by distillation, ion exchange, or other chemical techniques, while solids-based techniques may be confounded by matrix effects. Finally, cross check of a highly critical result with a "wet chemical" boron result provides vital assurance that stringent specifications are being met.

Boron is commonly determined spectrophotometrically after its separation as the methyl borate ester. Numerous colorimetric reagents have been employed on the distillate, including carmine, carminic acid, curcumin, quinalizarin, and turmeric. Solvent extraction separation is also possible; this approach has been joined with a slightly different list of color-forming reagents that includes methylene blue, Brilliant Green, and curcumin. Curcumin and the anthraquinone derivative, 1,1-dianthrimide, are sometimes used for boron without any preliminary separation.

The methyl borate distillation/curcumin method is typical of "umpire" approaches for low alloy steels. Dissolve a 2.0 g test portion in a quartz flask under reflux, using dilute sulfuric acid and hydrogen peroxide. Cool and filter, washing the acid insolubles, then transfer the paper to a platinum crucible and ignite at 1000°C.* Fuse the residue with 1.0 g of sodium carbonate. Leach in dilute sulfuric acid and add to the reserved filtrate. Transfer the combined solution to a 100-mL volumetric flask, dilute to the mark, and mix.

Transfer a 5-mL aliquot to a 125-mL quartz flask, connect the flask to a condenser, and add 30 mL of methanol. Seal the flask and apply heat, collecting the distillate in a Teflon vessel containing 5 mL of a 20g/L sodium hydroxide solution. When approximately 5 mL remain in the quartz distilling flask, cool to room temperature, add an additional 10 mL of methanol, and resume the distillation until less than 5 mL remain in the flask.

To the solution in the collection vessel add 3 mL of a 1.25g/L curcumin in glacial acetic acid solution and swirl to mix, then add 3 mL of a 1:1 acetone:water mixture, swirl to mix, and transfer to a 100-mL volumetric flask. Rinse the collection vessel and dilute the volumetric flask to the mark with 1:1 acetone:water. Measure the absorbance versus the 1:1 acetone:water mixture in 5-cm cells at 555 nm. Carry a reagent blank through the entire procedure. Calibration solutions must be distilled in the same manner as the samples [28].

In addition to low alloy steels curcumin has been used for stainless steels, high speed steels [54], nickel and nickel alloys [55], copper and copper alloys [24]. Donaldson [56] employed curcumin for iron and steel following a 2-ethyl-1,3-hexanediol/chloroform extraction.

Turmeric, which has been used as a color reagent for boron, is a crude extract, while curcumin is a refined extract from the same plant source. The dye carmine has been applied to boron determination in aluminum alloys [57]. Carmine is the crude dyestuff, of which carminic acid is the essential component; both are natural product derivatives of the cochineal plant [58]. Carminic acid has been used for boron determination in aluminum alloys [41], titanium alloys[59,60], nickel and cobalt coatings [61], among other applications. Like many reagents for boron, carmine and carminic acid require measurement in a sulfuric acid medium, using sulfuric acid as a reference solution. The colored complexes with boron are violet/blue, while the unreacted reagents, as their name implies, are red.

The color change when 1,1-dianthrimide reacts with boron in concentrated sulfuric acid is from dark green to blue. Samples are heated around 100°C to develop the color, which is measured at 620 nm versus sulfuric acid.

---

*Many methods specify the addition of sodium carbonate or calcium hydroxide to the crucible to prevent volatile loss of boron.

As with all boron methods a careful reagent blank is essential. If the boron is not isolated from the matrix a careful sample blank is also required. Danielsson applied 1,1-dianthrimide in a direct procedure for iron and low alloy steels [62], and Burke and Albright extended the application to nickel-base alloys [63].

At one time quinalizarin was the most popular reagent for boron. However, the unreacted reagent absorbs strongly at the analytical wavelength. Early work was conducted in 98% sulfuric acid but later methods utilized reagent grade (96%) [64]. Tetrabromochrysazin forms a rose-colored complex with boron in sulfuric acid that absorbs at 540 nm. Chromotrope B in sulfuric acid has also been applied to boron; the complex is measured at 620 nm. New solvent extraction procedures based on synthesized organic reagents have been applied to steel analysis[65,66].

7.10 Cadmium

The principal color reagent for cadmium is dithizone. It is generally applied after thorough separation procedures. For example, trace cadmium in zinc metal can be determined by applying a sulfide separation prior to chloroform extraction of the cadmium dithizonate and spectrophotometric measurement at 518 nm [18]. For copper alloys copper and lead are removed by electroplating as metal and oxide, respectively, prior to the application of the colorimetric finish [51].

7.11 Chromium

Diphenylcarbazide is almost exclusively employed as a colorimetric reagent for chromium (VI). It forms a violet complex that absorbs at 540 nm. Interference by iron (III) is masked by oxalate, while interference by molybdenum (VI) is masked by phosphate. Color due to vanadium(V) fades after 15 min [18,27] Diphenylcarbazide is used to determine trace levels of chromium in aluminum alloys [57], titanium alloys [67], beryllium metal [68], copper and copper alloys [24], zirconium and zirconium alloys [59], as well as traces in certain low chromium irons and steels [18].

7.12 Cobalt

Nitroso R salt (1-nitroso-2-naphthol-3,6-disulfonic acid, disodium salt) is the most important, if not the most sensitive, colorimetric reagent for cobalt. In the determination of 0.01–0.30% cobalt in steel a 0.5 g test portion is dissolved in *aqua regia*, cooled, and diluted to 50 mL in a 100-mL volumetric flask. A slurry of zinc oxide (165g/L) is added in 5-mL portions until the mixed solution looks like coffee with cream in it. The flask is then diluted to the mark and mixed. When the precipitate has settled the solution is poured through a dry low-porosity filter paper and collected in a dry 150-mL beaker after discarding the first 10–20 mL. A 10 mL aliquot of the filtrate is transferred to a 50-mL volumetric flask. Then 5 mL of sodium acetate solution (500 g/L—filtered) and 10 mL of nitroso R salt solution (7.5 g/L—filtered and prepared fresh weekly) are added, swirling after each addition. The flask is placed in a boiling water bath for 8 min, then 5 mL of nitric acid are added, the solution is swirled, and the boiling is continued for 3 min more. The flask is removed from the bath, cooled to room temperature, diluted to the mark and mixed. The absorbance is measured at 520 nm versus a reagent blank [28].

Nitroso R salt is used for low alloy steels [28], cast irons [69], tool steels [70], stainless steels [30], and iron- and nickel-base high temperature alloys [71,72], for nickel metal [21,73], copper/nickel alloys and copper/nickel/zinc alloys [74], for nickel/copper alloys [75], copper/beryllium alloys [25], aluminum alloys [41], titanium alloys, and zirconium alloys [59].

Cobalt is also sometimes determined by chloroform extraction of the tetraphenylarsonium/cobalt thiocyanate complex, which is measured at 620 nm. The method is sensitive and has been recommended for stainless steels [60] and nickel/copper alloys [76], although it is not widely used. A compound related to nitroso R, 2-nitroso-4-dimethylaminophenol hydrochloride (nitroso-DMAP-HCl), forms an extractable cobalt complex that shows high molar absorptivity. As little as 1 ppm of cobalt can be measured in a 1 g steel sample [27].

7.13 Copper

There are three highly selective colorimetric reagents that react with copper (I): cuproine (2,2'-biquinoline) and neocuproine (2,9-dimethyl-1,10-phenanthroline), which are suitable for low levels, and bathocuproine (2,9-dimethyl-4,7-diphenyl-1,10-phenanthroline), which is suitable for trace levels[27,77].

Cuproine (prepared in isoamyl alcohol [18] or dimethylformamide [78]) has been recommended for iron and steel. The violet copper complex forms in aqueous solution and absorbs at 545 nm. Neocuproine (prepared in absolute ethanol) is slightly more sensitive for copper; it forms an orange complex that is extracted into n-hexyl alcohol and measured at 479 nm. In each approach copper must be reduced to the +I oxidation state prior to color development. Hydroxylamine hydrochloride or ascorbic acid are suitable reducing agents.

The application of neocuproine to determine copper in electronic nickel is illustrative of most examples of the use of that reagent. A 0.25 g test portion is dissolved in 10 mL of nitric acid and 5 mL of phosphoric acid and then fumed with 10 mL of perchloric acid. The sample is cooled and diluted to the mark and mixed in a 250-mL volumetric flask. A 25-mL aliquot is transferred to a 150-mL beaker. It is diluted to 50 mL, then 1 mL of phosphoric acid, 10 mL of sodium citrate solution (300g/L), and 5 mL of hydroxylamine hydrochloride solution (100g/L) are added. The mixture is stirred magnetically for 30 sec., then 10 mL of neocuproine solution (1g/L in absolute ethanol) are added and the pH is adjusted to 5 with dilute ammonium hydroxide. The solution is transferred to a 250-mL separatory funnel, 10 mL of chloroform are added, and the solution is extracted for 30 sec. The (lower) chloroform layer is drained into a 25-mL volumetric flask containing 4 mL of absolute ethanol. A second extraction with 5 mL of chloroform is drained into the same flask; then it is diluted to the mark with absolute methanol and mixed. The absorbance is measured in a 1-cm cuvette at 457 nm versus a reagent blank[21,73].

Similar methods using neocuproine have been described for magnesium alloys [79], low alloy steels [28], cast irons [69], tool steels [70], stainless steels [30], high temperature alloys[71,72], cadmium metal [38], titanium and zirconium alloys [59], tungsten alloys, aluminum alloys, and lead/tin solders [18]. Bathocuproine and the disodium salt of bathocuproine disulfonic acid are suitable for extremely low amounts of copper, the latter allowing ultratrace work in aqueous solution, eliminating the extraction step [27].

Many older reagents for copper are still in routine use. Hydrobromic acid forms a color with copper (II) that is employed for magnesium alloys [79] and for electronic nickel [21]. α-Benzoinoxime has been employed for years as a color reagent for copper. It forms a chloroform extractable green complex in the presence of sodium potassium tartrate at pH 11.3–12.3. It has been used for ferrous alloys in which nickel is $< 2\%$ and cobalt is $< 0.25\%$[80]. Sodium diethyldithiocarbamate/chloroform extraction has been frequently used for titanium alloys[20,67], and bearing metal [60]. Butyl acetate has been recommended as a solvent for the use of this reagent for iron, steel, and tantalum alloys [18]. The diethylammonium diethyldithiocarbamate/chloroform system has been used for lead alloys [18]. Cuprizone (biscyclohexanoneoxaldihydrazone) [27] is an extremely sensitive reagent for copper. It is soluble in hot ethanol and reacts with copper in aqueous solution to form a complex which absorbs at 600 nm. It has been used for titanium alloys [20,67]. A related compound, oxalyldihydrazide, has been used to check the spent electrolyte after copper electrogravimetric analysis [81].

7.14 Gallium

The principal reagent for gallium is 8-hydroxyquinoline, which has been used after the analyte's separation from aluminum alloys using diisopropyl ether extraction from 8M hydrochloric acid media [41]. Measurement of the extracted 8-hydroxyquinolate is made at 400 nm versus chloroform [60]. The red Rhodamine B complex can also be employed; it absorbs at 565 nm. The azo dyes known as Gallion and Lumogallion have been used for gallium. The former forms a complex at pH 6 that absorbs at 560 nm; the latter forms a complex at pH 3 that absorbs at 580 nm. Gallion also forms a ternary complex with 8-hydroxyquinoline and gallium that forms at pH 3 and absorbs at 640 nm [27].

7.15 Germanium

The most important spectrophotometric technique for germanium is the phenylfluorone method. Phenylfluorone is widely employed for both germanium and tin determinations. Germanium is best isolated by extracting it with carbon tetrachloride from 9M hydrochloric acid solution [82]. An aliquot of the organic layer can be combined with the reagent (in ethanol and hydrochloric acid), diluted to volume with ethanol, and measured at 508 nm [27].

In acidic solution, germanium reacts with molybdate to form yellow molybdigermanic acid. This complex can be reduced with ferrous ammonium sulfate or other reductants to a heteropoly blue color and measured spectrophotometrically. However, while the reaction has been shown to be more sensitive, it is also less reproducible than spectrophotometric measurement of the unreduced yellow color [83]. A method has been published based on the formation of a germanium (IV)-mandelic acid complex and its extraction into chlorobenzene as an ion pair with Malachite Green [84].

7.16 Gold

Trace levels of gold (after a suitable isolation, such as coprecipitation with a tellurium carrier using stannous chloride) can be determined spectrophotometrically at 500 nm with Rhodanine (p-diethylaminobenzylidenerhodanine). The presence of tellurium does not affect the formation and measurement of the violet-red colloidal suspension. Oxidizing acids, silver, palladium, mercury, and thallium must be absent or chemically removed. Gold is also determined with Rhodamine B, employing diisopropyl ether extraction of the complex from dilute hydrochloric acid/ammonium chloride solution and measurement versus the pure solvent at 550 nm. The orange-colored complex that forms with hydrobromic acid can be measured at 380 nm. The yellow o-tolidine complex with gold can be measured at 437 nm [18,60].

7.17 Hafnium (See Section 7.49, Zirconium and Hafnium)

7.18 Indium

Indium has been determined with dithizone at 510 nm, but it must be well isolated from its alloy matrix. For zinc metal an extraction with diisopropyl ether from 6M hydrobromic acid is required. The indium is returned to the aqueous phase by re-extraction with neutral water. The aqueous phase is then reacted with 50 mL of a solution prepared by titrating 150 mL of 10%(w/v) hydroxylamine hydrochloride to pH 9 with 1M ammonium hydroxide, then adding 40 mL of 5%(w/v) potassium cyanide solution. The mixture is diluted to 500 mL and purified by extraction with dithizone. After the sample reagent solution has been allowed to stand for 1 hr to ensure iron reduction, dithizone/chloroform is added and the solution is extracted for 2 min and the organic layer is measured [85].

Another approach for zinc and zinc alloys involves the same preliminary separation, followed by extraction of the Bromopyrogallol Red complex with benzyl alcohol and measurement of the organic phase at 540 nm [86]. A synthesized reagent, 1-(2-pyridylmethylideneamine)-3-(salicylideneamine) thiourea, has been applied to nickel alloys following a two-step isolation of the indium [87].

7.19 Iridium

Once iridium has been separated from other metals of the platinum group, the most sensitive reagent for its spectrophotometric determination for many years has been p-nitrosodimethylaniline. In this approach an acid solution containing 1.5–10 ppm of iridium is evaporated to dryness in the presence of 40 mg of sodium chloride. Then the residue is moistened with 4 mL of *aqua regia* and again evaporated to dryness. Then the evaporation is repeated three times with hydrochloric acid. The salts are rinsed with a small amount of water in a test tube containing 2 mL of a 7.2–7.3 pH buffer solution [(10 g disodium hydrogen phosphate + 4.1 g potassium dihydrogen phosphate)/100 mL] and 2 mL of p-nitrosodimethylaniline (150 g/100 mL 95% ethanol—filtered). The solution is diluted to 8 mL, mixed, and heated in a hot water bath at 70°C for 40 min. The solution is cooled to room temperature, diluted to 10 mL with 6M hydrochloric acid, and measured at 530 nm [88].

Earlier, less sensitive, approaches include the application of the purple color that develops upon heating iridium solutions with perchloric, phosphoric, and nitric acids [89], and the red color that is produced by heating iridium solutions with cerium (IV) sulfate [90]. A method of superior sensitivity and selectivity was developed by Ayres and Bolleter. It is based on the oxidation of Leuco-Crystal Violet by iridium (IV) and absorbance measurement at 590 nm. Careful sample solution preparation is required, and gold must be first removed by precipitation with hydroquinone [91].

7.20 Iron

The two most frequently used colorimetric reagents for iron in nonferrous metals and alloys are thiocyanate and 1,10-phenanthroline (also called o-phenanthroline). For example, thiocyanate can be used for copper/beryllium alloys. The reserved electrolyte from the electrogravimetric determination of copper is evaporated to fumes of SO_3 (the original 5 g test portion having been dissolved in a water/sulfuric acid/nitric acid mixture). The cooled solution is diluted to the mark in a 250-mL volumetric flask and mixed. A 5.0-mL aliquot is transferred to a 100-mL volumetric flask, 50 mL of water is added, then 10 mL of 10%(v/v) hydrochloric acid, 10 mL of ammonium persulfate solution (5g/L—freshly prepared), and then 10 mL of sodium thiocyanate solution (200g/L), swirling to mix after each addition. The flask is diluted to the mark, mixed, and the absorbance measured versus a reagent blank at 490 nm [25]. Thiocyanate is also used for many other alloy systems. Examples are nickel [21,73], nickel/copper [75], copper metal [33], zinc, and tin[51].

1,10-Phenanthroline is the most popular (but not the most sensitive) of a group of structurally related compounds sometimes called the "ferroin reagents." For the determination of iron in aluminum alloys a 0.5 g test portion (freed of adventitious iron contamination by treatment with dilute hydrochloric acid, then water, then acetone; then dried) is transferred to a nickel beaker. The alloy is dissolved with small additions (15 mL total volume) of sodium hydroxide solution (200g/L). The vessel walls are rinsed down with water and the solution is boiled, then cooled to room temperature. The solution is cautiously transferred to a borosilicate glass beaker containing 25 mL of 1:1 hydrochloric acid: water. The nickel beaker is then policed and rinsed into the glass vessel. The solution is boiled and 30% hydrogen peroxide is added dropwise until a clear solution is obtained. The boiling is continued until excess hydrogen peroxide, oxygen, and chlorine have been expelled. The solution is cooled, transferred to a 250-mL volumetric flask, diluted to the mark, and mixed.

An aliquot containing 0.02–0.20 mg of iron is transferred to a beaker and diluted to about 30 mL. Then 2 g of finely granulated high-purity lead metal are added and the solution is boiled for 8 min. The liquid is transferred to a 100-mL volumetric flask; 4 mL of hydroxylamine hydrochloride solution (100g/L), 10 mL of a buffer solution (272 g of sodium acetate · $3H_2O$ is dissolved in 500 mL of water; 240 mL of glacial acetic acid is added with stirring; after cooling the solution is diluted to 1 liter and mixed), and 10 mL of 1,10-phenanthroline solution (0.4 g 1,10-phenanthroline · H_2O in 200 mL of methanol) is added. The flask is diluted to volume and mixed, then allowed to stand for 5 min. The absorbance is then measured in a 2-cm cell versus a reagent blank [57].

1,10-Phenanthroline is used to determine low levels of iron in many other types of alloys, including magnesium [79], pig lead [37], titanium [67], molybdenum [92], beryllium [68], copper [93], and manganese/copper [94].

Bathophenanthroline (4,7-diphenyl-1,10-phenanthroline) is twice as sensitive as 1,10-phenanthroline and three times as sensitive as thiocyanate. Iron is reduced with hydroxylamine hydrochloride, the solution is buffered with acetate at pH 4, the reagent is added (dissolved in 1:1 ethanol:water), and the solution is extracted with isoamyl alcohol. The absorbance is measured at 533 nm. Bathophenanthroline has been used to measure trace and ultratrace levels of iron in tungsten, molybdenum, vanadium, chromium, titanium, niobium, tantalum, uranium, bismuth, gold, beryllium, copper, aluminum, gallium, arsenic, nickel, and plutonium [95].

Intermediate in sensitivity between thiocyanate and bathophenanthroline there are also a number of other iron reagents that are sometimes employed for metals analysis. Thioglycolic acid (also known as mercaptoacetic acid) reacts with both iron(II) and iron(III), although the analyte is generally reduced to the ferrous state prior to color development. This reagent has been used for copper metal [24] and titanium alloys [20], magnesium and magnesium alloys, brasses and bronzes, tin- and lead-base alloys, and precious metals [60]. Nickel metal has been analyzed for trace iron using 2,2'-bipyridine (also known as α,α'-bipyridyl), which is a "ferroin" class reagent. Sulfosalicylic acid has been suggested for iron in tin-base Babbitt alloys [60].

7.21 Lead

Dithizone remains the principal colorimetric reagent for lead. Early work with high-grade copper alloy matrices dates back to 1948 when lead levels below 0.02% were state-of-the-art. Lead is chloroform- or carbon tetrachloride-extractable as a dithizonate from basic, cyanide-containing solutions. Bismuth, thallium (I) and tin (II) extract under the same conditions. Tin can be oxidized to tin (IV), which does not interfere, while bismuth and thallium are often too low to be a problem.

The determination of lead in magnesium alloys is a straight-forward example of the application of dithizone. Weigh a sample that contains 0.1–0.7 mg of lead, add 30 mL of water and 20 mL of 1:1 hydrochloric acid:water for each gram of sample. Boil, cool, dilute to 200 mL, transfer to a 500-mL volumetric flask, dilute to volume, and mix. Transfer a 10.0-mL aliquot to a 125-mL separatory funnel and a second 10.0-mL aliquot to a beaker. Titrate the beaker sample with 1:9 ammonium hydroxide:water to a methyl red endpoint. To the separatory funnel add 15 mL of a solution (A) (435 mL of water + 30 mL of 50g/L potassium cyanide solution + 30 mL of 50g/L ammonium citrate solution + 5 mL of ammonium hydroxide), then add the titrated amount of 1:9 ammonium hydroxide:water.

Add dithizone solution (0.0025g/100 mL of chloroform) from a buret to a faint purple-green color. Add chloroform from a buret to make the dithizone/chloroform volume exactly 10.0 mL. Extract for 1 min and draw off the (lower) chloroform layer into a second 125-mL separatory funnel that contains 20 mL of a solution (B) (500 mL water + 10 mL of 50g/L potassium cyanide solution + 5 mL of ammonium hydroxide). Extract for 1 min, then drain the (lower) chloroform layer into a third 125-mL separatory funnel containing 20 mL of the same solution (B). Extract for 1 min and drain the (lower) chloroform layer through a cotton plug in the funnel stem into a 1-cm cell and measure the absorbance at 520 nm versus a reagent blank [79].

Luke employed a preliminary iodide/methylisobutylketone extraction to isolate lead prior to a dithizone finish for iron and steel, zinc, magnesium metal, nickel, cobalt, titanium, and titanium alloys [96]. Dithizone has been used for trace lead determination in ferromanganese [44], silicomanganese and ferrosilicon manganese [45], chromium metal and ferrochromium [46], ferromolybdenum [97], cadmium metal [38], bismuth metal, copper and copper alloys, tin metal, zinc and zinc alloys [51].

7.22 Manganese

The development of the permanganate color is the basis of nearly all spectrophotometric methods for manganese. Iodate is the most frequently used oxidant, although persulfate is also used. The determination of 0.0005–0.10% manganese in titanium alloys will serve to illustrate the technique, which with minor changes is applied to many metals.

Transfer a 1.00 g test portion to a beaker and add 45 mL of 1:4 sulfuric acid:water and 2 mL of fluoboric acid. When dissolved, add nitric acid dropwise until a colorless solution is obtained, plus one drop in excess. Boil for 3 min, cool, and dilute to 60 mL with water. Add 10 mL of nitric acid, heat to boiling, then add 15 mL of potassium periodate solution (50g/L in 1:4 HNO_3:H_2O; heat to dissolve). Boil gently for 30 min, cool to room temperature, transfer to a 100-mL volumetric flask, dilute to volume, and mix. Measure the absorbance of the test solution and a reagent blank versus water at 525 nm, using a 5-cm cell if manganese is in the range 5–250 μg or in a 1-cm cell if manganese is in the range 200–1000 μg. Add dropwise with swirling freshly prepared sodium nitrite solution (20g/L) to the remainder of the test solution in the volumetric flask until the permanganate color is destroyed, plus one drop in excess. Add an equal number of drops to the remainder of the reagent blank. Measure the background color on each at 525 nm. Subtract the background reagent blank from the background test portion readings and use the results to correct the test solution readings [67].

This general approach has been employed for aluminum alloys [57], magnesium alloys [79], nickel [73], copper [39], and copper alloys [74], nickel/copper alloys [75], low alloy steels [28], cast irons [69], tool steels [70], stainless steels [30], iron-, nickel-, and cobalt-base high temperature alloys[71,72], beryllium [68], and zirconium alloys [59].

Donaldson (nee Penner) and Inman [98] determined 0.0002–0.12% manganese in high-purity molybdenum and tungsten metals by measuring at 562 nm the extracted complex formed between PAN [1-92-pyridylazo)-2-naphthol] and manganese(II) ion. The analyte was initially isolated by solvent extraction.

7.23 Mercury

Mercury rarely survives metallurgical processing but most labs that deal with high performance alloys receive some requests for trace mercury determinations because its presence in certain materials, even in minute amounts, can adversely effect properties. Today, these are likely to be treated by a cold vapor atomic absorption approach, although applicable color methods exist.

Mercury extracts with dithizone/chloroform from dilute acid solution and can be measured at 490 nm. Mercury forms a colorless complex with thiocyanate that absorbs in the ultraviolet (281 nm) [60]. Thio-Michler's Ketone [4,4'-bis(dimethylamino)thiobenzophenone] is a highly sensitive color reagent for mercury (II). The complex forms at pH 3.2 and is measured at 560 nm. Silver, gold, and palladium are among the potential interferences [27].

7.24 Molybdenum

The most important spectrophotometric reagent for molybdenum is thiocyanate. The molybdenum (V) thiocyanate complex forms in the presence of stannous chloride in aqueous solution and is reasonably sensitive, but the color is unstable. Moreover, full color development requires a *minimum* presence of a gram-atom of iron for each gram-atom of molybdenum. In the presence of copper (II) less iron can be present. By working with a strict protocol (an exact time interval between color development and measurement, or, perhaps, a timed heating cycle) or by adding a stabilizing substance [such as Butyl Cellosolve™ (ethylene glycol monobutyl ether)] it is possible to do accurate work in aqueous media. Most umpire work, however, is performed using a solvent extraction approach. The molybdenum (V) thiocyanate complex requires an oxygenated solvent—ethers, alcohols, or esters [18].

For 0.0050–0.100% molybdenum in titanium alloys dissolve a 1.00 g test portion in 15 mL of 1:1 sulfuric acid:water, 40 mL of 1:1 hydrochloric acid:water, and 1 drop of hydrofluoric acid. Add 30% hydrogen peroxide dropwise until a permanent yellow color forms. If the molybdenum content is in the range 5–35 μg evaporate to fumes of SO_3, cool, and dilute to 60 mL; otherwise select a dilution and an aliquot to yield a molybdenum content in the indicated range, add 8–15 mL of 1:1 sulfuric acid:water, fume, cool, and dilute to 60 mL. Add 1 mL of ferric sulfate solution (50 g $Fe_2(SO_4)_3$-$9H_2O$/L) and 10 mL of citric acid solution (250g/L), heat for 5 min, and cool to room temperature.

Transfer the solution to a 250-mL separatory funnel and cool in running water; add 10 mL of sodium thiocyanate solution (100g/L—filtered and stored in a dark bottle) and mix well. Add 10 mL of stannous chloride solution (175 g of $SnCl_2 \cdot 2H_2O$ + 5 mg of $CuCl_2 \cdot 2H_2O$ + 100 mL of 1:1 HCl:H_2O; warm to dissolve and dilute to 500 mL) and shake for exactly 1 min. Add 20.0 mL of butyl acetate and extract for 30 sec. Cool in running water for 2–3 min. and extract for 30 sec. more. Drain off and discard the (lower) aqueous layer. Add 50 mL of 1:9 sulfuric acid:water, 10 mL of sodium thiocyanate solution, and 5 mL of stannous chloride solution. Extract for 30 sec; let stand 2–3 min., then extract for 30 sec more. Drain off the (lower) aqueous layer and discard it. Transfer the butyl acetate layer to a dry glass stoppered vessel and allow to stand until the solution is clear. Measure the test solution and the reagent blank in 2-cm cells at 470 nm versus pure butyl acetate [67].

Besides titanium alloys, the thiocyanate approach is used for low alloy [28,99], tool [70,99] and stainless steels [30,99], cast irons [69,99], high temperature alloys [71,72], zirconium and zirconium alloys [59], among other applications.

The toluene-3,4-dithiol (4-methyl-1,2-dimercaptobenzene) methods for molybdenum are less sensitive than the thiocyanate methods. The dithiol reagent forms a complex with molybdenum (VI) that is extractable by both polar and nonpolar solvents, but there is little to recommend it over thiocyanate except when the sample contains tungsten since it is possible to measure both analytes simultaneously. The tungsten complex absorbs at 630 nm and the molybdenum complex absorbs at 680 nm [18,60].

Molybdenum has been determined in plutonium alloys using chloranilic acid, following extraction with 4-methyl-2-pentanone from 6M hydrochloric acid/0.4M hydrofluoric acid [100]. The thiocyanate procedure has been applied following extraction of molybdenum from a ferrous alloy matrix using acetylacetone [101]. Phenylfluorone was used for the spectrophotometric measurement of molybdenum in low alloy steels following ion exchange separation [102]. Flow injection systems for molybdenum in steels using thiocyanate [103] and chloroform extraction of the 8-hydroxyquinolate [104] have been described.

7.25 Nickel

Dimethylglyoxime remains the most commonly used reagent for the colorimetric determination of nickel even though a related compound, α-furildioxime, is known to be more sensitive. Dimethylglyoxime reacts with nickel in an ammoniacal citrate medium in the presence of an oxidizing agent—iodine, bromine, or persulfate (in order of decreasing preference)—to form a red-colored solution that can be measured directly in the aqueous medium at 535 nm or extracted with chloroform then returned to an aqueous medium for measurement. While proof remains elusive, it is generally believed that the complex involves the rarely seen nickel (III) oxidation state.

For steels, a 0.5 g test portion is dissolved in 10 mL of 2:1:2 hydrochloric acid:nitric acid:water plus 10 mL of perchloric acid. The solution is then heated to fumes of $HClO_4$; fuming is continued for 3 min. The solution is cooled to room temperature; 50 mL of water is added and the solution is diluted to the mark in a 250-mL volumetric flask and mixed. The solution is then dry-filtered into a dry beaker, discarding the first 50 mL. A 25.0-mL aliquot is transferred to a 100-mL volumetric flask, 20 mL of ammonium citrate solution (dissolve 250 g of citric acid · H_2O cautiously with stirring in 250 mL of 1:1 ammonium hydroxide:water and dilute to 1 liter) is added and the solution is swirled to mix. Next, 3 mL of iodine solution (dissolve 25 g of KI + 12.7 g of I_2 in a minimum of water, then dilute to 1 liter) is added and the solution is swirled again. The solution is allowed to stand for 5 min. Then 20 mL of dimethylglyoxime solution (1 g of reagent dissolved in 500 mL of 1:1 ammonium hydroxide:water, then dilute to 1 liter) is added and the flask is diluted to the mark and mixed. After exactly 10 min. the solution is measured at 535 nm versus a reagent blank. A sample blank that includes all reagents (including ammonium hydroxide) but excluding dimethylglyoxime must be also prepared and measured [105].

Dimethylglyoxime has also been utilized for nickel determination in magnesium alloys [79], molybdenum metal [92], beryllium metal [68], copper metal [33], copper alloys [25,93], aluminum alloys [41], tungsten metal [18], bismuth metal [51], lead and lead alloys [51], tin metal [51], zirconium and zirconium alloys [59].

Nickel reacts with α-furildioxime in basic solution ($pH > 7.2$) to form a yellow, chloroform-extractable complex that absorbs at 438 nm. The lower limit of quantification is approximately four times lower than that for an aqueous dimethylglyoxime determination [27].

7.26 Niobium

There are many good reagents for niobium determination but none are broadly used. Thiocyanate forms a complex with niobium that is extractable from 1.2M hydrochloric acid with diethyl ether. Tartaric acid and stannous chloride must be present in the aqueous phase prior to extraction. The ether phase is diluted to volume with reagent-equilibrated ether and measured in a stoppered cell at 385 nm. Molybdenum, tungsten, and titanium interfere, requiring separation or correction by means of a series of interferent calibration curves at the analytical wavelength. The method has been applied to steels [106], titanium alloys, tantalum metal [60], and to uranium-fissium alloy (after N-benzoyl-N-phenylhydroxylamine extraction) [107].

Niobium forms a yellow peroxy-complex with hydrogen peroxide in concentrated sulfuric acid. The test solution is measured against concentrated sulfuric acid at 365 nm. Molybdenum, tungsten, and rhenium interfere. Vanadium does not interfere, provided it is first reduced to the +IV state. Tantalum forms a complex that absorbs at 290 nm and titanium forms one that absorbs at 450 nm. Thus, multicomponent spectrophotometry is a possibility [18,60].

Niobium can also be determined simultaneously with titanium in iron-, nickel-, and cobalt-base alloys as follows. Dissolve a 0.2 g test portion by adding (in this exact order, swirling after each addition) 25 mL of hydrochloric acid, 5 mL of nitric acid, 25 mL of water, and 25 mL of solvent mix (400 mL of H_3PO_4 dissolved in 300 mL of water with stirring and cooling; then very cautiously add 120 mL of H_2SO_4 with stirring and cooling; dilute to 1 liter and mix; cool to room temperature). Heat gently until the test portion is dissolved, then heat to light fumes. Fume for exactly 1 min, then remove from the heat and cool for 1 min. Then cautiously, with swirling add 25 mL of dilution mix (cautiously, while cooling and stirring, add 350 mL of H_2SO_4 to 600 mL of water, dilute to 1 liter, mix, and cool to room temperature).

Transfer the test solution to a 50-mL volumetric flask using 3.0 mL of hydrochloric acid and 10.0 mL of stannous chloride solution (88g of stannous chloride-$2H_2O$ + 60 mL HCl, warm to dissolve; in a separate vessel: 50 g tartaric acid + 100 mL of water, warm to dissolve; combine the two solutions and dilute to 250 mL and mix), cool and dilute to the mark with dilution mix. Transfer a 1.0-mL aliquot to each of two dry 50-mL beakers. To one beaker add 25.0 mL of concentrated sulfuric acid, to the other add 50 mL of hydroquinone solution (15 g hydroquinone dissolved in 500 mL of sulfuric acid; stored in the dark; prepared 24 ± 1 hour before use).

Swirl to dissolve the white precipitate that forms and let stand 10 min. Read the absorbance in 1-cm cells versus concentrated sulfuric acid at both 400 nm and 500 nm. Correct the test portion readings for sample-, reagent-, and cell-blanks at both wavelengths. From prepared calibration curves for titanium and niobium at both wavelengths calculate the concentrations for both analytes by solving two simultaneous equations in two unknowns, as described in section 4 of this chapter. The same approach can be applied to niobium–tungsten combinations [108,109].

In addition to high temperature alloys the hydroquinone approach for niobium determination has been applied to steels [108,109], titanium alloys [59], and uranium alloys [110]. Other methods for niobium include procedures that utilize 4-(2-pyridylazo)resorcinol, which forms a colored complex that absorbs at 560 nm [111]. This approach has been applied to steel [112] and zirconium metal [113]. Sulfochlorophenol S forms a complex that absorbs at 650 nm; it has been applied to steels [114,115], among other metals. The Bromopyrogallol Red-niobium complex absorbs at 610 nm and has, likewise, been applied to steel analysis [116].

7.27 Nitrogen

Except for special studies, nitrogen is seldom determined these days by a wet chemical/spectrophotometric approach. In the classical procedure the metal or alloy is dissolved in dilute sulfuric acid (sometimes with added potassium sulfate) or 1:1 hydrochloric acid:water (with or without added hydrofluoric acid) in a round-bottomed, long-necked flask which constitutes part of a Kjeldahl or micro-Kjeldahl distillation apparatus. For total nitrogen hydrogen peroxide may also be added during the dissolution. Various schemes have been devised for determining empirical quantities, such as "soluble nitrogen," insoluble nitrogen," and "ester-halogen nitrogen." The sample flask is sealed into the distillation apparatus after the addition of 20%(w/v) sodium hydroxide solution and ammonia is then steam distilled out of the solution [60,117].

Nessler's reagent (first proposed as a reagent for ammonia in 1856) is still the preferred colorimetric reagent. It can be prepared in several ways. For example, dissolve 100 g of mercuric iodide and 70 g of potassium iodide in 100 mL of (ammonia-free*) water. Add this solution, slowly with stirring, to a solution of 160 g of sodium hydroxide in 700 mL of (ammonia free) water. Dilute to 1 liter with ammonia-free water. Allow to stand for several days, allowing the precipitate to settle. Decant the clear supernatent solution for use. Generally 1 mL of reagent is added to 50 mL of distillate. The absorbance is measured at 420 nm after a fixed development time (often 20 min) [117].

An alternative approach that has been applied to vanadium, titanium, and uranium samples is a potassium hydroxide fusion in a nickel boat in a 550°C tube furnace. A nitrogen stream carries the evolved ammonia to a dilute hydrochloric acid solution. After 20 min the absorbing solution is buffered at pH 10 and reacted with sodium hypochlorite to form monochloramine. Thymol dissolved in dilute caustic and acetone is added, the solution is adjusted to pH 11.7 and allowed to stand in the dark for 1 hr, then measured at 660 nm [118].

---

*Ammonia-free water is prepared by ion exchange of distilled water using a mixed bed resin consisting of 2 parts Amberlite IRA-400 to 1 part of IR-120-H.

7.28 Osmium

Thiourea yields a red color with osmium that absorbs at 480 nm. Several members of the platinum group interfere, including rhodium, ruthenium, platinum, and palladium. Osmium can be isolated by distillation prior to color development [18,60]. Osmium(VI)-naphthylaminesulfonic acid complexes have been shown to be more sensitive than thiourea, but these approaches still require distillation of OsO_4 prior to their application [119].

7.29 Palladium

There are several important reagents for palladium. Both 1-nitroso-2-naphthol (complex measured at 430 nm) and 2-nitroso-2-naphthol (complex read at 370 nm) are utilized (the former for titanium alloys [67]). p-Nitrosodiphenylamine forms a red complex that absorbs at 520 nm, however, Sandell [18] suggests o-nitrosodiphenylaniline (red; with an absorbance maximum at 525 nm) for faster development and greater color stability. α-Furildioxime forms a complex that can be extracted with chloroform and measured at 380 nm.

Palladium forms an orange/red complex with bromide that absorbs at 505 nm. This finish requires the isolation of the palladium. Precipitation with dithiooxamide, extraction with phenylthiourea/ethyl acetate, and removal of gold (III) by amyl acetate extraction from 1:1 hydrochloric acid:water represents one approach. Palladium (II) forms a red color with tin (II) phosphate in perchloric acid/phosphoric acid solution that absorbs at 487 nm. Palladium (II) forms a green color with tin (II) chloride in hydrochloric acid/perchloric acid solution that absorbs at 635 nm [120]. Other reagents include 3-hydroxy-1-p-sulfonatophenyl-3-phenyltriazine [121], phenyl-alpha-pyridyl ketoxime [122], and N,N'-bis(3-dimethylaminopropyl)dithio-oxamide [123].

7.30 Phosphorus

There are two basic trace methods for phosphorus in widespread use: one is based on the phosphomolybdate complex and the other on the phosphovanadomolybdate complex. The first involves development of the yellow phosphomolybdate complex in aqueous solution. It can be measured directly at 389 nm, but, more typically, is reduced to the heteropoly blue complex and measured at 650 or 830 nm. The reduced blue form shows ten times the sensitivity of the yellow form. Sometimes the yellow phosphomolybdate complex is extracted with isobutanol to isolate phosphorus from most of the sample matrix. The complex is then reduced and measured at either 625 or 725 nm.

The second basic method is based on the formation (also in aqueous solution) of the yellow phosphovanadomolybdate complex, which can be measured directly at 460 nm or extracted with methylisobutyl ketone and measured at 355 nm. Unlike the heteropoly blue complex, which shows features of a colloidal suspension, the phosphovanadomolybdate species is in true solution. Although considerably less sensitive than the heteropoly blue procedure, it is free of some interferences that can trouble the other method. However, both methods require special provision for arsenic, silicon, chromium, and elements that can readily hydrolyze in acid solution (e.g., tungsten, titanium, niobium, and tantalum) [60,124].

The following is a procedure suggested for nickel and nickel alloys that involves extraction of the phosphovanadomolybdate complex. It is reasonably tolerant of common interferences. Dissolve a sample estimated to contain between 5 and 100 mg of phosphorus in 25 mL of 1:4 nitric acid:water and 4 mL of hydrofluoric acid in a Teflon beaker. Add 10 mL of perchloric acid and heat to fumes of $HClO_4$. Fume for 3 min, then cautiously add hydrochloric acid dropwise to volatilize the chromium. Cool to room temperature. Add 10 mL of sodium nitrite solution (50g $NaNO_2$/L—freshly prepared) and boil for 10 min. Add 40 mL of fluoboric acid (75g H_3BO_3 + 600 mL hot water + 50 mL HF in a Teflon beaker; stir cautiously and warm on a sandbath on a hotplate until dissolved) and cool rapidly in a water bath to 25°C. Add 10 mL of ammonium metavanadate solution (25g NH_4VO_3/L) and 15 mL of ammonium molybdate solution [15g $(NH_4)_6Mo_7O_{24} \cdot 4H_2O$/100mL—prepare fresh]. Mix and let stand 11 ± 4 min., then transfer to a 250-mL separatory funnel and dilute to 100 mL. Add 10 mL of citric acid solution (500g/L), mix; add 40.0 mL of methylisobutyl ketone and extract for 30 sec. Discard the (lower) aqueous layer. Filter the organic phase into a small, dry beaker. Measure the absorbance at 355 nm versus methylisobutyl ketone in 1-cm cells [125].

The phosphovanadomolybdate method has also been applied to steels and cast irons [126], copper alloys [39], including manganese copper alloys [94], to tin metal [51], and to uranium metal [60]. The more widely used heteropoly blue method has been extensively studied, in particular with regard to the reducing agent utilized. Different reducing reagents, as well as different acidities, result in complexes that show different absorption maxima. Most standard heteropoly blue methods utilize hydrazine sulfate as the reducing agent, although stannous chloride, ferrous sulfate, ascorbic acid, and many others have also been employed [60]. Other work has centered around the problem of interferences [127]. The heteropoly blue method is routinely applied to all types of steels and ferrous alloys [28,30,69,70], to aluminum alloys [41], titanium alloys, and zirconium alloys [59], as well as nickel alloys [128].

7.31 Platinum

The orange/yellow color produced by adding stannous chloride to a solution of platinum (IV) chloride remains one of the most useful approaches. It is probably true to say that the reaction is imperfectly understood. Tin is known to be present in the chromophore and the color is affected by pH, temperature, and the presence of any tin (IV). In aqueous media the absorbance is measured at 403 nm, but if the color is extracted with amyl acetate it is measured at 390 nm.

The most sensitive reagent appears to be p-nitrosodimethylaniline, which forms an orange/red complex that absorbs at 525 nm. Palladium is a major interferent that can be simultaneously determined by a temperature-based scheme. A solution incubated at 100°C for 20 min yields an absorbance (at 525 nm) that corresponds to the sum of platinum plus palladium, while a solution that is not heated yields an absorbance at the same wavelength that corresponds to palladium alone. The difference yields a value for platinum [18,60].

7.32 Rare Earths (including Scandium and Yttrium)

Some rare earth ions exhibit uncommonly sharp absorption lines in the visible, but these have little utility for the element's trace determination in alloys. The chemical similarity of these elements makes it unlikely that selective reagents for their individual determination will ever be found. Cerium is often measured at 505 nm as the red/brown 8-hydroxyquinolate. Alizarin Red S forms red complexes with all the rare earths. They absorb at 550 nm, and are at least two orders of magnitude more sensitive than direct measurement of the aqueous elemental ions in the visible and near infrared [60]. Many other reagents have been applied: Arsenazo-III [129,130], Pyrocatechol Violet [131], p-nitrochlorophosphonazo [132], N-benzoyl-N-phenylhydroxamine [133], and others.

7.33 Rhenium

Rhenium forms a brown/yellow color with stannous chloride and thiocyanate. It can be measured in aqueous solution at 400 nm, or extracted with ether and measured at 420 nm. In either case, molybdenum must first be separated. This is usually accomplished by means of an α-benzoinoxime precipitation of the interferent [60,76]. Rhenium also forms a colored complex with α-furildioxime in the presence of stannous chloride. After 45 min. of color stabilization the solution is measured at 530 or 540 nm. Molybdenum must be removed here, as well. Ion exchange [60] or extraction of the rhenium with tetraoctylammonium chloride in chloroform [134] have been used.

7.34 Rhodium

Rhodium reacts with stannous chloride at boiling water temperatures to form a light red color that is measured at 475 nm. Platinum can be simultaneously measured by also reading the absorbance at 403 nm and solving a pair of simultaneous equations, as previously described. Sandell [18] describes an alternative use of stannous chloride and potassium iodide that produces a stronger color for rhodium, showing a maximum absorbance at 435 nm. A complex with 2-mercapto-4,5-dimethylthiozole exhibits an amber to red color that absorbs at 430 nm. Other color reagents include sodium hypochlorite, 2-mercaptobenzoxazole, and N,N'-bis(3-dimethylaminopropyl)dithio-oxamide [135].

7.35 Ruthenium

From its position in the periodic chart below iron it is reasonable to expect ruthenium to undergo many analytically useful reactions. Thiourea is a good reagent for ruthenium, forming a blue/violet color that is measured at 620 nm. Rubeanic acid (also called dithiooxamide) reacts with ruthenium in 1:1 sulfuric acid:ethanol at 85°C to form a blue complex that absorbs at 650 nm. 1,10-Phenanthroline forms a yellow complex that absorbs at 448 nm; and p-nitrosodimethylaniline forms a green complex that absorbs at 610 nm [18,60].

For most methods and samples it is necessary to separate ruthenium from its matrix by distillation from either perchloric acid/phosphoric acid/sodium bismuthate or from 6M sulfuric acid/sodium bismuthate, collecting the distillate in strong caustic. Osmium must be distilled out first [60]. If a trace of chloride is added to the sample solution, or if hypochlorite is added to the collection solution the perruthenate color will develop and can be measured at 380 nm [136]. An alternative approach involves oxidation to the tetroxide, RuO4, and extraction with carbon tetrachloride. Back extraction with an aqueous solution of sodium thiocyanate produces the deep blue $RuCNS^{2-}$ ion, which can be measured at 590 nm [137].

7.36 Selenium

The principal colorimetric reagent for selenium is now 3,3'-diaminobenzidine with which selenium(IV) forms a yellow diphenylpiazselenol compound. It was originally read in aqueous solution at 340 nm [138], but now is more commonly measured in a toluene extract at 420 nm. Interferences are masked by EDTA. The colored selenium complex forms below pH 4 but is extracted at pH 5 [139].

Selenium has been determined with 3,3'-diaminobenzidine in several different alloy systems, including copper [24,140,141], stainless steels [60,141], arsenic metal [60,139], and gallium metal [142]. A related compound, 4-methyl-o-phenylenediamine, has been applied to the determination of selenium in iron and stainless steels, following extraction and removal of interferences with capric acid/chloroform and with 8-hydroxyquinoline/chloroform. The colored complex is fully formed at pH 1.5–2.0 in the presence of EDTA and hydrogen peroxide after 2 hours, then measured at 337 nm [143].

Older methods are still sometimes employed. Selenium in boiling neutral solution will form a colloid if hydrazine hydrate solution is rapidly added. The absorbance is measured in the range 250–300 nm. Selenium (IV) also reacts with cadmium iodide to form iodine, which can be measured as the triiodide at 352 nm, or as the blue starch complex at 615 nm [60,144].

7.37 Silicon

The only silicon color methods of broad utility are based on the reduced and unreduced forms of the silicomolybdate complex. Of these, the reduced heteropoly blue complex is the most frequently employed. However, the unreduced yellow silicomolybdate is the basis of methods used for aluminum alloys [41,57], magnesium alloys [79], and copper alloys [39,51,145].

An example of the use of the heteropoly blue approach is its application to iron and steel. In this procedure only 10 megohm · cm (or better), membrane-filtered water should be employed for all reagent preparation and dilutions. A test portion containing 0.1–1.0 mg of silicon is transferred to a Teflon beaker, then 10.0 mL of hydrochloric acid and 5.0 mL of nitric acid are added by pipet. The sample is allowed to react over low heat (covered with a Teflon watchglass) until dissolved. The sample is cooled to room temperature, 6 drops of hydrofluoric acid are added, and the sample is warmed over low heat for 10 min. Then 25 mL of a 5%(w/v) boric acid solution is added and the beaker is warmed again for 10 min. The solution is cooled to room temperature, transferred to a 100-mL polyethylene volumetric flask, 20 mL of 5%(w/v) boric acid solution is added, and the flask is diluted to the mark and mixed.

A 10.0-mL aliquot is transferred to a 250-mL separatory funnel, 1 g of sodium sulfate is added and the solution is diluted to 80 mL and mixed. Three drops of 1:1 sulfuric acid:water are added to all samples except the blanks. Then 10.0 mL of 5%(w/v) ammonium molybdate is added, the solution is diluted to 100 mL, mixed, and allowed to stand for at least 10 min. Next, 2.0 mL of 10%(w/v) oxalic acid, then 5.0 mL of 1:1 sulfuric acid:water are added, mixing after each addition. Then 25 mL of N-butanol is added and the solution is extracted for 30 sec. The (lower) aqueous layer is drained out and discarded. Then 2 mL of reducing mix [0.5g 1-amino-2-naphthol-4-sulfonic acid + 1.0 g sodium sulfite + 30 g sodium bisulfite +200 mL of hot water—filter; prepare fresh weekly] is added, followed by 10 mL of water. The solution is extracted for 15 sec and the (lower) aqueous layer is drained into a 50-mL volumetric flask, diluted to the mark and mixed. The absorbance is measured at 815 nm versus water in 1-cm cells. At least two reagent blanks should be taken through the entire procedure.

The heteropoly blue method for the determination of silicon is used, in one form or another, for nickel metal [21], molybdenum metal [92], titanium and titanium alloys [59], zirconium and zirconium alloys [59], as well as for iron and steel [28,146]. The analytical wavelength varies among the different procedures largely because different reducing agents produce slightly different effects.

7.38 Sulfur

Except for highly specialized applications, colorimetric methods are no longer used for sulfur determination. Today, these exceptions consist primarily of certification, umpire, or cross-check work at very low concentrations. Methylene blue methods for trace sulfur are probably the most frequently employed color procedures. A reducing agent is added to the dissolved test portion and hydrogen sulfide gas is evolved in a special all-glass apparatus and collected usually in a basic zinc acetate/ammonium chloride solution. Then iron (III) chloride and N,N-dimethyl-p-phenylenediamine solution (in dilute hydrochloric acid) are added to the absorbing solution. The solution is diluted to a fixed volume, mixed, and allowed to stand for 15 min., then measured at 665 nm. To utilize this approach successfully at the trace and ultratrace level it is essential that all glassware be scrupulously clean.

The hydrogen sulfide evolution/methylene blue colorimetric approach has been utilized for iron and steel [147,148]. One group of authors described the use of an FEP Teflon waveguide capillary cell 2 meters in length and a 632.8 nm helium/neon laser light source for the measurement of the methylene blue absorbance. They were able to measure as little as 0.2 ppm of sulfur in steel by this means[149]. The method has also found use for refined nickel [150].

An induction furnace from a carbon/sulfur combustion apparatus can be set up so that the effluent gas stream bubbles through a solution of p-rosaniline (fuchsine) and formaldehyde (which is a known carcinogen) in a water/sulfuric acid/ethanol solvent. The resultant solution is diluted to a fixed volume and measured at 580 nm [60].

Turbidimetric methods were once common; lead sulfide was probably the most widely used. Luke [151] applied the lead sulfide approach to the determination of trace sulfur in elemental arsenic, antimony, tin, germanium, tellurium, selenium, lead alloys, silver alloys, mercury, molybdenum, aluminum, and magnesium. Barium sulfate methods were also widely employed [152].

7.39 Tantalum

Pyrogallol is by far the most important reagent for tantalum. Tantalum in low concentrations is often first isolated by ion exchange and then by a cupferron coprecipitation (e.g., with hafnium). The combined cupferrates are ignited to oxides, fused with potassium pyrosulfate, and leached in an ammonium oxalate solution (which serves to retard the hydrolysis of Ta_2O_5). Pyrogallol/stannous chloride/hydrochloric acid reagent is added and the solution is diluted to a fixed volume, mixed, and measured at 325 nm [60]. Pyrogallol has been applied to ferroniobium [15]3, high temperature alloys [72], titanium and titanium alloys [59], steels, and stainless steels [60]. A new reagent for tantalum, 4,4-dibromo-o-nitrophenylfluorone, has been applied to ferroniobium, and to nickel-base alloys [154].

7.40 Tellurium

Reasonably reliable methods are based on the formation of an elemental hydrosol by reduction of tellurium (IV) with either stannous chloride or hypophosphorous acid. A gum arabic solution is added to stabilize the colloid. If stannous chloride is used absorbance measurements are made at 420 nm. If hypophosphorous acid is used the absorbance maximum in the visible is dependent on the reductant concentration, which relates to the particle size of the resultant sol. For this reason absorbance readings are generally taken in the ultraviolet in the 240–290 nm region. For both reductants all reaction and measurement conditions must be rigorously standardized. Selenium interferes, as it also does in methods based on colored complexes of tellurium with iodide (measured at 335 nm) and thiourea (measured at 366 nm) [60,144].

Donaldson isolated tellurium from copper alloys by coprecipitation with iron (III) hydroxide, then separated the tellurium from the iron by an ethyl xanthate/chloroform extraction. Finally, tellurium (IV) is reacted with diantipyrylmethane in the presence of potassium bromide to form the colored complex, which is extracted with chloroform and measured at 336 nm. Up to 500 μg of selenium do not interfere [155].

7.41 Thallium

Thallium forms a red/violet complex with Rhodamine B which can be extracted and measured at 560 nm [18]. This approach has been used for thallium in cadmium metal [38]. Thallium (III) will oxidize iodide to free iodine, which can be extracted with carbon tetrachloride and measured at 352 nm, or reacted with starch and measured at 610 nm [60].

7.42 Thorium

Thorin [2-(2-hydroxy-3,6-disulfo-1-naphthylazo)-benzenearsonic acid] yields a red complex with thorium (IV) which absorbs in aqueous solution at 545 nm. Carminic acid forms a blue color at pH 4.2 that is measured at 575 or 580 nm. Morin reacts to form a yellow complex at pH 2; it is measured at 410 nm. All of these procedures require some degree of analyte separation from most matrices [18,27,60].

7.43 Tin

Based on early work by C.L. Luke the phenylfluorone procedure for tin became popular. It proves necessary to isolate the analyte, however. This can sometimes be accomplished by acid sulfide precipitation, by extraction with cupferron/chloroform, and/or by extraction with diethylammonium diethyldithiocarbamate/chloroform. The color is developed at pH 5 in the presence of hydrogen peroxide and oxalate ion. The method has been applied to lead and 1% antimony/lead alloys [156], and to steel [157].

Toluene-3,4-dithiol has been suggested as a reagent for the determination of tin in iron and steel. A magenta-red precipitate forms with the reagent in the presence of thioglycollic acid and dilute sulfuric acid. Formation of a stable colloid requires the use of a dispersing agent. The absorbance is measured at 530 nm. Since many elements interfere a preliminary distillation of the tin is necessary [16,60]. Tin in cadmium metal has been determined by extraction of the 8-hydroxyquinolate with chloroform [38]. Quercetin forms a complex with tin in dilute hydrochloric acid solution. The measurement is made at 420 nm. The method has been applied to copper metal samples [33].

Pyrocatechol Violet (also sometimes called Catechol Violet) has been employed as an effective reagent for tin. Ross and White used it as the spectrophotometric finish for their extraction studies of tin with tris(2-ethylhexyl) phosphine oxide (TEHPO) (see the previous chapter). The color is developed in an aliquot of the cyclohexane extract in the presence of pyridine and ethanol after heating at 55–60°C for 20 min. Absorbance is measured at 575 nm versus ethanol. The procedure was applied to lead-base bearing metal, sheet brass, steel, zinc-base die casting alloy and Zircaloy [158]. Ashton et al. [159] employed a toluene extraction of tin (IV) iodide to isolate tin from steels, then developed the Pyrocatechol Violet complex in the presence of cetyltrimethylammonium bromide as a sensitizing surfactant. Lactic acid was utilized as a masking agent for many interferences. The color was measured at 662 nm [159]. Other workers applied a similar procedure to copper-base alloys without separation, using HEDTA as a masking agent [160].

Donaldson employed the red complex with gallein to determine tin in iron, steel, copper-, zirconium-, aluminum-, and zinc-base alloys over the range 0.0005–1%. Tin was first isolated by toluene extraction of tin (IV) iodide [40].

7.44 Titanium

Today, the reagent of choice is diantipyrylmethane, which is much more selective and sensitive than hydrogen peroxide [27]. Steels can be analyzed as follows. Weigh a test portion estimated to contain 0.2–0.7 mg of titanium and dissolve it in hydrochloric and nitric acids and evaporate the solution to dryness. Dissolve the residue in 5 mL of hydrochloric acid, then add 15 mL of water. Filter into a 100-mL volumetric flask. If insoluble titanium compounds are known or suspected to be present ignite the washed paper at 700°C in a platinum crucible, volatilize silica with 1:1 sulfuric acid:water and 10 drops of hydrofluoric acid, heating to dryness on a sand bath. Fuse the residue with 2 g of potassium pyrosulfate. Dissolve the cooled melt with 10 mL of tartaric acid solution (100g/L). Rinse the solution into the reserved filtrate. Dilute to volume, mix, and then transfer a 10-mL aliquot into each of two 50-mL volumetric flasks. Add 1 mL of 1:1 hydrochloric acid:water into each, then 5 mL of ascorbic acid solution (100g/L). Allow to stand for 10 min, then add 10 mL of diantipyrylmethane solution (dissolve 5g of reagent in 250 mL of 1:9 hydrochloric acid:water—prepare fresh) to one of the two flasks and dilute both to the mark and mix. After at least 90 min measure the absorbance of the test solution and the sample blank solution versus water at 390 nm using 2-cm cells. Also prepare and measure a reagent blank using 10 mL of water instead of the sample aliquot [28].

Diantipyrylmethane can be used for aluminum alloys [57], low alloy steel [28], cast irons [69], stainless steels [30], high-purity molybdenum and tungsten [161], and nickel-base alloys [40].

Hydrogen peroxide is certainly second in importance as a reagent for colorimetric titanium. The best approach for many alloy types is to use the mercury cathode after fuming the dissolved test portion with sulfuric acid. Vanadium will accompany titanium and interfere. It can be corrected for by preparing a vanadium correction curve at the titanium wavelength (390 nm) or by measuring the solution at both 400 nm (where titanium and vanadium both absorb) and at 460 nm (where vanadium alone absorbs) and solving for both analytes, using calibration curves for both analytes at 400 nm and for vanadium at 460 nm [18,60]. A third alternative is to remove vanadium in the filtrate of a sodium hydroxide or sodium hydroxide/sodium peroxide precipitation. The precipitated titanium can be washed free of the interferent, then dissolved and measured. Hydrogen peroxide has been used for titanium in steel [60], aluminum alloys [41, zirconium alloys [59], and ferroniobium [153].

The disodium salt of chromotropic acid (disodium 1,8-dihydroxy-3,6-naphthalenedisulfonate) forms a yellow complex with titanium(IV). The reagent is sensitive but subject to interferences, and pH conditions must be carefully controlled. It has found use in aluminum alloy analysis [57]. Tiron (disodium 1,2-dihydroxybenzene-3,5-disulfonate) also forms a yellow complex with titanium. EDTA and sodium dithionite are used to minimize certain interferences. It has been used to determine titanium in electronic nickel [21]. Steels have been analyzed for titanium by extraction of the N-benzoyl-N-phenylhydroxamine complex from 10M hydrochloric acid in the presence of stannous chloride, followed by its spectrophotometric measurement [162].

7.45 Tungsten

The methods based on thiocyanate remain very important. A strong reducing agent—usually stannous chloride—reduces tungsten to the +V state which forms the yellow complex with thiocyanate. At the same time molybdenum is reduced to the +III state which forms only a weakly colored complex with thiocyanate. Many versions of the procedure are subject to vanadium interference; however, it is possible to extract a ternary complex, such as the tetraphenylarsonium-tungsten (V)-thiocyanate ion association complex, which eliminates the problem. Titanium and niobium, which also form colored complexes with thiocyanate, can be handled by the addition of ammonium bifluoride prior to the color development [60,163–165].

The thiocyanate approach has been applied to steel [60,164], high temperature alloys [60], zirconium alloys [59], and titanium alloys [20]. The tungsten (V)-thiocyanate complex in aqueous solution or chloroform extract is measured at 400 nm.

For over 100 years it has been known that tungsten forms a red color with hydroquinone in concentrated sulfuric acid [166]. In 1946 C.M. Johnson published a method for tungsten steels that became a standard of the steel industry [167]. As discussed under the heading for niobium (Section 7.26), hydroquinone is the basis for an important simultaneous method for niobium and titanium or niobium and tungsten determinations [108,109]. Titanium is a major interference for the tungsten determination and must be separated (e.g., by precipitation with sodium hydroxide). It is possible to obtain values

---

It is molybdenum(+V) that forms an intensely colored thiocyanate complex.

for tungsten–titanium–niobium combinations by coprecipitating tungsten with molybdenum using α-benzoinoxime, measuring niobium and titanium in the processed filtrate simultaneously with hydroquinone, then measuring tungsten in the wet ashed precipitate with hydroquinone. Tungsten alone (or with the molybdenum carrier) is measured at 530 nm. Tungsten–niobium combinations (like titanium–niobium combinations) are measured at 400 and 500 nm. Besides steel, hydroquinone has been applied to tungsten determination in uranium/tantalum/tungsten alloys [168]. Hydroquinone procedures have been used more frequently for major and minor levels of tungsten, but they can be extended to near trace levels.

Toluene-3,4-dithiol (often called, simply "dithiol") might be regarded as the principal reagent for trace tungsten determination. Machlan and Hague published a procedure for steel and titanium alloys that involves a preliminary dithiol/chloroform extraction of molybdenum to remove its interference. Insoluble dithiolates, such as copper, are removed by filtration. Then the tungsten-dithiol complex is formed in the presence of stannous chloride and hydrochloric acid and extracted with butyl acetate. The absorbance is measured at 635 nm [169]. Besides steel and titanium [20,67], the method has been applied to tantalum and niobium metal [170].

7.46 Uranium

The book, *Analytical Chemistry of the Manhatten Project* (C.J. Rodden, editor-in-chief; McGraw-Hill, NY, 1950), in the Chapter titled *Uranium* by C.J. Rodden and J.C. Warf lists 67 reagents for the colorimetric determination of uranium. Probably none of these would be used today. Arsenazo-III has been suggested for trace uranium, as follows. To a test solution aliquot (1–5 mL) containing 5 to 50 μg of uranium add 25 mL of 60%(w/v) ammonium nitrate/ 0.25%(w/v) disodium EDTA solution. Adjust to pH 2.5–3.0 with dilute ammonium hydroxide or dilute nitric acid. Add 15 mL 1:4 tributyl phosphate:carbon tetrachloride and extract for 2 min. Filter the organic phase and retain it. Repeat the extraction of the aqueous phase with 10 mL of the mixed solvent. Filter and combine the organic phase with the previous extract. Add 15.0 mL of 0.006%(w/v) Arsenazo-III to the combined organic phases and extract for 1 min. Measure the absorbance of the aqueous phase in 2-cm cells at 655 nm versus a reagent blank [27].

Hydrogen peroxide produces a yellow to orange color with uranium in sodium carbonate solution that can be measured at 425 nm, but the response is insensitive. Thiocyanate produces a yellow complex in strong hydrochloric or sulfuric acid solutions in the presence of stannous chloride that absorbs at 365 nm. The β-diketone, dibenzoylmethane, is 5–6 times more sensitive than thiocyanate. At neutral pH a complex is produced that absorbs at 400 nm.

Extraction with ethyl acetate from pH 7.5 solution in the presence of EDTA minimizes interferences [60]. This reagent has been used for zirconium alloys [20].

7.47 Vanadium

Hydrogen peroxide remains an important reagent, showing absorbance maxima in both the visible (450 nm) and the UV (290 nm). The sample solution should contain about 10 mL of perchloric acid in 50 mL of final solution volume. The mercury cathode removes many interferences, however, titanium, tungsten, and some of the molybdenum will remain to create difficulties. The best approach is a compensating blank [60].

N-Benzoyl-N-phenylhydroxamine (BPHA) is probably the reagent of choice today for many alloys. The determination of vanadium in aluminum alloys will illustrate its use. Weigh a test portion containing 0.02–0.25 mg of vanadium and dissolve it in 25 mL of water, 7 mL of 1:1 sulfuric acid:water, and 2 mL of hydrochloric acid. Add 20 mL of 1:1 sulfuric acid:water and heat to fumes of SO_3. Fume for 15 min, cool, and transfer to a 250-mL separatory funnel. Dilute to approximately 100 mL. Add potassium permanganate solution (2g/L) dropwise until a faint pink persists for 10 min. Add 30 mL of BPHA solution (1g/L in chloroform—store in the dark) and 34 mL of hydrochloric acid. Extract for 1 min. and drain the (lower) organic layer into a 50-mL volumetric flask. Repeat the extraction with an additional 10 mL of BPHA solution and drain the organic layer into the same flask. Dilute to volume with chloroform, mix, and let stand for at least 1 hour. Measure the absorbance in 1-cm cells versus a reagent blank at 530 nm. Titanium is an important interference [57]. BPHA has also been applied to steels [171], and nickel-base alloys [40]. A related compound, N-benzoyl-N-o-tolylhydroxylamine, is more specific for vanadium [27].

The phospho-tungsto-vanadate complex, which absorbs at 400 nm, was once the basis for an important procedure for vanadium. Benzohydroxamic acid forms a complex that can be extracted from pH 2 solution with 1-hexanol and measured at 450 nm. Alloy steels have been analyzed for vanadium using 3-methylcatechol [172]. Acetylacetone has also been utilized for steel [173].

7.48 Zinc

Zinc has been often determined with either dithizone or zincon. Dithizone forms a complex with zinc at pH 5.5 that is extracted in the presence of cyanide ion using a carbon tetrachloride or chloroform solution of dithizone. Absorbance measurement is in the vicinity of 525 nm. The extraction of zinc dithizonate is the basis of methods for chromium/nickel steels [18], aluminum alloys [18], cadmium metal [18], copper metal [33], and copper alloys [51], bismuth metal [51], lead and lead alloys [51], tin [51], tungsten and tungsten alloys [59].

Zincon (2-carboxy-2'-hydroxy-5'-sulfoformazylbenzene) forms an intense blue complex with zinc at pH 8.5–9.5. There are many interferences, but zinc can be adequately separated by ion exchange. At pH 9.0 both zinc and copper complexes absorb at 600 nm. At pH 5.2 the copper complex absorbs, but the zinc complex does not. Thus, both elements can be determined, calculating zinc by difference. Zinc in the absence of copper is read at 620 nm [27,60].

7.49 Zirconium (and Hafnium)

There is a fairly rich literature on the spectrophotometric determination of zirconium and hafnium. The high degree of similar reactivity of these two elements has led to a few published methods for exploiting minor differences to determine them simultaneously [174–178]. However, most routine procedures are not designed to distinguish these two elements, and so it is fortunate that they occur together in comparatively few alloys. Many different color reagents have been used for both elements.

Older reagents employed for zirconium (and hafnium) include Alizarin Red S, which forms a red lake at pH 0.7 that absorbs at 540 nm, and quercetin, which forms a yellow color at pH 0.5 that is measured at 440 nm [60,79]. A newer reagent, Arsenazo-III, has been applied to aluminum alloys as follows. Dissolve a 0.25 g test portion in 1:1 hydrochloric acid:water. Transfer the solution to a 200-mL volumetric flask and dilute to volume with 1:1 hydrochloric acid:water and mix. Transfer a 20.0-mL aliquot to a 50-mL volumetric flask. Add 2 mL of aluminum solution (25g/L in 1:1 $HCl:H_2O$) and 1.0 mL of Arsenazo-III solution (dissolve 0.250g of reagent in 90 mL of water containing 0.300g of sodium carbonate, heating gently; adjust to pH 4.0 ± 0.1 with 1:1 $HCl:H_2O$ and cool; transfer to a 100-mL volumetric flask, dilute to volume and mix). Dilute to volume with 1:1 hydrochloric acid:water and mix. Measure versus a reagent blank at the maximum absorbance found between 630 and 670 nm (the exact absorbance maximum varies with reagent lot) [57]. Arsenazo-III has also been extensively developed as a reagent for zirconium in steels [179,180], nickel-base alloys [181], and molybdenum-base alloys [182].

Another reagent that has evoked interest is Pyrocatechol Violet. Young and White [183] utilized a preliminary isolation of the zirconium (and hafnium) by extraction from 7M nitric or hydrochloric acid solution with trioctylphosphine oxide (TOPO)/cyclohexane. The procedure was developed further by Wood and Jones [184], who applied it to niobium and nickel-base alloys, and by Ratcliffe and Byford [185], who applied it to steels. The color is developed in an aliquot of the cyclohexane extract with added pyridine and ethanol. The absorbance is measured at 655 nm.

K.L. Cheng extensively studied Xylenol Orange and the structurally related, Methylthymol Blue, for this application [174,176,186,187]. Xylenol Orange complexes show absorbance maxima at 530 nm (Hf) and 535 nm (Zr), while Methylthymol Blue complexes show absorbance peaks at 570 nm (Hf) and 580 nm (Zr). Zirconium in aluminum-magnesium alloys was determined in a procedure published by R.F. Rolf [188] that is based on extraction with di-n-butyl phosphate/chloroform and development of a red color with 1-(2-pyridylazo)-2-naphthol. The complex absorbs at 555 nm.

REFERENCES

1. Olsen, E.D. Modern Optical Methods of Analysis, McGraw-Hill, NY, 1975, pp.83–154.
2. Cheng, K.L.; Young, V.Y. "Ultraviolet and Visible Absorption Spectroscopy" Instrumental Analysis 2nd ed. (G.D. Christian and J.E. O'Reilly, eds.) Allyn and Bacon, Boston, 1986, pp.161–208.
3. Skoog, D.A.; West, D.M. Principles of Instrumental Analysis, 2nd ed., Saunders, Philadelphia, 1980, pp.169–197.
4. Day Jr., R.A.; Underwood, A.L. Quantitative Analysis, 2nd ed., Prentice-Hall, Englewood Cliffs, NJ, 1967, pp.290–331.
5. Ciurczak, E.M.; Workman Jr., J. "Getting Started with UV/Visible Spectroscopy, Spectroscopy (suppl.)" September, 1966.
6. Van Nostrand's Scientific Encyclopedia, 4th ed., D. Van Nostrand, Princeton, NJ, 1968.
7. Hughes, H.K. Analytical Chemistry, 24, 1349 (1952).
8. Annual Book of ASTM Standards, Vol. 03.06, American Society for Testing and Materials, West Conshohocken, PA, 1996, Designation E131.
9. Snell, F.D.; Snell, C.T. Colorimetric Methods of Analysis D. Van Nostrand, NY, 1948.
10. Reilley, C.N.; Crawford, C.M. Analytical Chemistry, 27, 716 (1955).
11. Annual Book of ASTM Standards, Vol. 03.06, American Society for Testing and Materials, West Conshohocken, PA, 1996, Designation E169.
12. Sustek, J. Analytical Chemistry, 46, 1676 (1974).
13. Connors, K.A.; Eboka, C.J. Analytical Chemistry, 51, 1262 (1979).
14. Manual on Recommended Practices in Spectrophotometry, American Society for Testing and Materials, West Conshohocken, PA, 1966.
15. Annual Book of ASTM Standards, Vol.03.06, American Society for Testing and Materials, West Conshohocken, PA, 1996.
16. Boltz, D.F.; Schenk, G.H. "Visible and Ultraviolet Spectrophotometry" Handbook of Analytical Chemistry, 1st ed. (L. Meites, ed.)., McGraw-Hill, NY, 1963, pp.6-6–6-102.

17. Thomason, P.F. Analytical Chemistry, 28, 1527 (1956).
18. Sandell, E.B. Colorimetric Determination of Traces of Metals, 3rd ed., Interscience, NY, 1959.
19. Nakamura, H.; Takagi, M.; Ueno, K. Analytical Chemistry, 52, 1668 (1980).
20. Elwell, W.T.; Wood, D.F. Analysis of the New Metals: Titanium, Zirconium, Hafnium, Niobium, Tantalum, Tungsten and Their Alloys, Pergamon, Oxford, 1966.
21. Annual Book of ASTM Standards, Vol. 03.05, American Society for Testing and Materials, West Conshohocken, PA, 1996, Designation E107.
22. Craft, C.H.; Makepeace, G.R. Industrial and Engineering Chemistry, Analytical Edition, 17, 206 (1945).
23. Annual Book of ASTM Standards, Vol. 03.05, American Society for Testing and Materials, West Conshohocken, PA, 1996, Designation E31.
24. Elwell, W.T.; Scholes, I.R. Analysis of Copper and Its Alloys, Pergamon, Oxford, 1967.
25. Annual Book of ASTM Standards, Vol. 03.05, American Society for Testing and Materials, West Conshohocken, PA, 1996, Designation E106.
26. Hynek, R.J.; Wrangell, L.J. Analytical Chemistry, 28, 1520 (1956).
27. Cheng, K.L.; Ueno, K.; Imamura, T. Handbook of Organic Analytical Reagents, CRC Press, Boca Raton, FL, 1982.
28. Annual Book of ASTM Standards, Vol. 03.05, American Society for Testing and Materials, West Conshohocken, PA , 1996, Designation E350.
29. Codell, M. "Methods for The Analysis of Titanium, Tungsten, and Zirconium" Handbook of Analytical Chemistry (L. Meites, ed.)., McGraw-Hill, NY, 1963, pp.13–70.
30. Annual Book of ASTM Standards, Vol.03.05, American Society for Testing and Materials, West Conshohocken, PA, 1996, Designation E353.
31. Annual Book of ASTM Standards, Vol. 03.06 , American Society for Testing and Materials, West Conshohocken, PA, 1996, Designation E360.
32. Hill, U.T. Analytical Chemistry, 31, 429 (1959).
33. Studlar, K.; Janousek, I. Chemist-Analyst, 50, 36 (1961).
34. Corbett, J.A.; Guerin, B.D. Analyst, 91, 490 (1966).
35. Zaki, M.T.M.; El-Didamony, A.M. Analyst, 113, 577 (1988).
36. Luke, C.L. Analytical Chemistry, 25, 674 (1953).
37. Annual Book of ASTM Standards, Vol.03.05, American Society for Testing and Materials, West Conshohocken, PA, 1996, Designation E37.
38. Annual Book of ASTM Standards, Vol. 03.06, American Society for Testing and Materials, West Conshohocken, PA, 1996, Designation E396.
39. Annual Book of ASTM Standards, Vol. 03.05, American Society for Testing and Materials, West Conshohocken, PA, 1996, Designation E62.
40. Donaldson, E.M. Some Instrumental Methods for the Determination of Minor and Trace Elements in Iron, Steel, and Non-ferrous Metals and Alloys, Monograph 884, Energy, Mines, & Resources Canada, Ottawa, Canada, 1982.
41. Oliver, R.T.; Moss, M.L. "Methods for the Analysis of Aluminum-Base Alloys" Handbook of Analytical Chemistry, 1st ed. (L. Meites, ed.)., McGraw-Hill, NY, 1963, pp.13-14–13-29.

42. Tsukahara, I.; Sakakibara, M.; Tanaka, M. Analytica Chimica Acta, 92, 379 (1977).
43. Donaldson, E.M. Talanta, 24, 105 (1977).
44. Annual Book of ASTM Standards, Vol.03.06, American Society for Testing and Materials, West Conshohocken, PA, 1996, Designation E361.
45. Annual Book of ASTM Standards, Vol. 03.06, American Society for Testing and Materials, West Conshohocken, PA, 1996, Designation E362.
46. Annual Book of ASTM Standards, Vol. 03.06, American Society for Testing and Materials, West Conshohocken, PA, 1996, Designation E363.
47. Annual Book of ASTM Standards, Vol. 03.06, American Society for Testing and Materials, West Conshohocken, PA, 1996, Designation E364.
48. Motomizu, S.; Wakimoto, T.; Toei, K. Analyst, 108, 944 (1983).
49. Bhargava, O.P.; Donovan, J.F.; Hines, W.G. Analytical Chemistry, 44, 2402 (1972).
50. Pakalns, P. Analytica Chimica Acta, 31, 576 (1964).
51. Silverman, L. "Methods for the Analysis of Other Non-ferrous Metals and Alloys" Handbook of Analytical Chemistry, 1st ed. (L. Meites, ed.)., McGraw-Hill, NY, 1963, pp.13-29–13-51.
52. Matsubara, C.; Takamura, K. Analytica Chimica Acta, 77, 255 (1975).
53. Donaldson, E.M. Talanta, 25, 131 (1978).
54. "Steel—Determination of Boron Content—Curcumin Spectrophotometric Method" ISO 10153:1997 International Organization for Standardization (ISO).
55. "Nickel and Nickel Alloys—Determination of Total Boron Content—Curcumin Molecular Absorption Spectrometric Method" ISO 11436:1993 International Organization for Standardization (ISO).
56. Donaldson, E.M. Talanta, 28, 825 (1981).
57. Annual Book of ASTM Standards, Vol. 03.05, American Society for Testing and Materials, West Conshohocken, PA, 1996, Designation E34.
58. The Merck Index, 7th ed., Merck & Co.,Rahway, NJ, 1960.
59. Codell, M. "Methods for the Analysis of Titanium, Tungsten, and Zirconium" Handbook of Analytical Chemistry, 1st ed. (L. Meites, ed.)., McGraw-Hill, NY, 1963, pp. 13-65–13-72.
60. Snell, F.D.; Snell, C.T.; Snell, C.A. Colorimetric Methods of Analysis, Vol. IIA, D. Van Nostrand, Princeton, NJ, 1959.
61. Norwitz, G.; Gordon, H. Analytica Chimica Acta, 94, 175 (1977).
62. Danielsson, L. Talanta, 3, 203 (1959).
63. Burke, K.E.; Albright, C.H. Talanta, 13, 49 (1966).
64. Porter, G.; Shubert, R.C.; "Boron" Colorimetric Determination of Nonmetals (D.F. Boltz, ed.)., Interscience, NY, 1958, pp. 339–353.
65. Toei, K.; Motomizu, S.; Oshima, M.; Watari, H. Analyst, 106, 776 (1981).
66. Oshima, M.; Fujimoto, K.; Motomizu, S.; Toei, K. Analytica Chimica Acta, 134, 73, (1982).
67. Annual Book of ASTM Standards, Vol. 03.05, American Society for Testing and Materials, West Conshohocken, PA, 1996, Designation E120.
68. Annual Book of ASTM Standards, Vol 03.06, American Society for Testing and Materials , West Conshohocken, PA, 1996, Designation E439.

69. Annual Book of ASTM Standards, Vol 03.05. American Society for Testing and Materials, West Conshohocken, PA, 1996, Designation E351.
70. Annual Book of ASTM Standards, Vol. 03.05. American Society for Testing and Materials, West Conshohocken, PA, 1996, Designation E352.
71. Annual Book of ASTM Standards, Vol. 03.05. American Society for Testing and Materials, West Conshohocken, PA, 1996, Designation E354.
72. Annual Book of ASTM Standards, Vol. 03.06. American Society for Testing and Materials, West Conshohocken, PA, 1996, Designation E1473.
73. Annual Book of ASTM Standards, Vol. 03.05. American Society for Testing and Materials, West Conshohocken, Pa, 1995, Designation E39.
74. Annual Book of ASTM Standards, Vol. 03.05. American Society for Testing and Materials, West Conshohocken, PA, 1996, Designation E75.
75. Annual Book of ASTM Standards, Vol. 03.05. American Society for Testing and Materials, West Conshohocken, PA, 1996, Designation E76.
76. Wagner, W.; Hull, C.J.; Markle, G.E. Advanced Analytical Chemistry, Reinhold, NY, 1956.
77. Diehl, H.; Smith, G.F. The Copper Reagents: Cuproine, Neocuproine, Bathocuproine, 1st ed., 1958; Schilt, A.A.; McBride, L., 2nd ed., 1972, G.F. Smith Chemical Co., Columbus, OH.
78. "Steel and Cast Iron—Determination of Copper Content—2,2'-Diquinolyl Spectrophotometric Method," ISO 4946:1984, International Organization for Standardization (ISO).
79. Annual Book of ASTM Standards, Vol. 03.05. American Society for Testing and Materials, West Conshohocken, PA, 1996, Designation E35.
80. Dunleavy, R.A.; Wiberley, S.E.; Harley, J.H. Analytical Chemistry, 22, 170 (1950).
81. Annual Book of ASTM Standards, Vol. 03.06. American Society for Testing and Materials, West Conshohocken, PA, 1996, Designation E478.
82. Luke, C.L.; Campbell, M.E. Analytical Chemistry, 28, 1273 (1956).
83. Boltz, D.F.; Mellon, M.G. Industrial and Engineering Chemistry, Analytical Edition, 19, 873 (1947).
84. Sato, S.; Tanaka, H. Talanta, 36, 391 (1989).
85. Collins Jr., T.A.; Kanzelmeyer, J.H. Analytical Chemistry, 33, 245 (1961).
86. Jadhav, S.G.; Murugaiyan, P.; Venkateswarlu, Ch. Analytica Chimica Acta, 82, 391 (1976).
87. Rosales, D.; Millan, I.; Gomez Ariza, J.L. Talanta, 33, 607 (1986).
88. Westland, A.D.; Beamish, F.E. Analytical Chemistry, 27, 1776 (1955).
89. Ayres, G.H.; Quick, Q. Analytical Chemistry, 22, 1403 (1950).
90. Maynes, A.D.; McBryde, A.E. Analyst, 79, 230 (1954).
91. Ayres, G.H.; Bolleter, W.T. Analytical Chemistry, 29, 72 (1957).
92. Annual Book of ASTM Standards, Vol. 03.05. American Society for Testing and Materials, West Conshohocken, PA, 1996, Designation E315.
93. Annual Book of ASTM Standards, Vol. 03.06. American Society for Testing and Materials, West Conshohocken, PA, 1996, Designation E478.
94. Annual Book of ASTM Standards, Vol. 03.06. American Society for Testing and Materials, West Conshohocken, PA, 1996, Designation E581.

95. The Iron Reagents. 3rd ed., G. F. Smith Chemical Co., Columbus, OH, 1980, pp.7–21.
96. Luke, C.L. Analytica Chimica Acta, 39, 447 (1967).
97. Annual Book of ASTM Standards, Vol. 03.06, American Society for Testing and Materials, West Conshohocken, PA, 1996, Designation E368.
98. Penner (Donaldson., E.M.; Inman, W.R. Talanta, 13, 489 (1966).
99. "Steels and Cast Irons—Determination of Molybdenum," ISO 4941:1994, International Organization for Standardization (ISO).
100. Waterbury, G.R.; Bricker, C.E. Analytical Chemistry, 29, 129 (1957).
101. McKaveney, J.P.; Freiser, H. Analytical Chemistry, 29, 290 (1957).
102. Black, A.H.; Bonfiglio, J.D. Analytical Chemistry, 33, 431 (1961).
103. Bergamin, H.; Krug, F.J.; Reis, B.F.; Nobrega, J.A.; Mesquita, M.; Souza, I.G. Analytica Chimica Acta, 214, 397 (1988).
104. Burns, D.T.; Harriott, M.; Porsinlapatip, P. Analytica Chimica Acta, 281, 607 (1993).
105. "Steel and Cast Iron—Determination of Nickel Content —Dimethylglyoxime Spectrophotometric Method," ISO 4939:1984, International Organization for Standardization (ISO).
106. Iyer, C.S.P.; Kamath, V.A. Talanta, 27, 537 (1980).
107. Villarreal, R.; Barker, S.A. Analytical Chemistry, 41, 611 (1969).
108. Ikenberry, L.; Martin, J.L. Boyer, W.J. Analytical Chemistry, 25, 1340 (1953).
109. McKaveney, J.P. Analytical Chemistry, 33, 744 (1961).
110. Waterbury, G.R.; Bricker, C.E. Analytical Chemistry, 30, 1007 (1958).
111. Siroki, M.; Maric, L.; Herak, M.J. Analytical Chemistry, 48, 55 (1976).
112. "Steel—Determination of Niobium Content—PAR Spectrophotometric Method," ISO 9441:1998 International Organization for Standardization (ISO).
113. Pakalns, P.; Ivanfy, A.B. Analytica Chimica Acta, 41, 139 (1968).
114. Cizek, Z.; Dolezal, J. Analytica Chimica Acta, 109, 381 (1979).
115. Cizek, Z.; Studlarova, V. Analyst, 108, 524 (1983).
116. Ramakrishna, T.V.; Rahim, S.A.; West, T.S. Talanta, 16, 847 (1969).
117. Taras, M.J.; "Nitrogen" Colorimetric Determination of Nonmetals (D.F. Boltz, ed.)., Interscience, NY, 1958, pp. 75–124.
118. Hashitani, H.; Yoshida, H.; Adachi, T. Analytica Chimica Acta, 76, 85 (1975).
119. Steele, E.L.; Yoe, J.H. Analytical Chemistry, 29, 1622 (1957).
120. Ayres, G.H.; Alsop III, J.H. Analytical Chemistry, 31, 1135 (1959).
121. Sogani, N.C.; Bhattacharyya, S.C. Analytical Chemistry, 29, 397 (1957).
122. Sen, B. Analytical Chemistry, 31, 881 (1959).
123. Jacobs, W.D. Analytical Chemistry, 32, 512 (1960).
124. Boltz, D.F.; Lueck, C.H. "Phosphorus" Colorimetric Determination of Nonmetals (D.F. Boltz, ed). Interscience, Ny, 1958, pp. 29–38.
125. Nickel, Ferronickel, and Nickel Alloys—Determination of Phosphorus Content—Phosphovanadomolybdate Molecular Absorption Method, ISO 11400:1992, International Organization for Standardization (ISO).
126. "Steel and Cast Iron—Determination of Phosphorus Content—Phosphovanadomolybdate Spectrophotometric Method" ISO 10714:1992 International Organization for Standardization (ISO).

127. Zatka, V.J.; Zelding, N. Analytical Chemistry, 56, 1734 (1984).
128. "Nickel Alloys—Determination of Phosphorus Content—Molybdenum Blue Molecular Absorption Spectrometric Method," ISO 9388:1992, International Organization for Standardization (ISO).
129. Musil, J.; Dolezal, J. Analytica Chimica Acta, 87, 239 (1976).
130. Huro, K.; Russell, D.S.; Berman, S.S. Analytica Chimica Acta, 37, 209 (1967).
131. Young, J.P.; White, J.C.; Ball, R.G. Analytical Chemistry, 32, 928 (1960).
132. Hsu, C-G; Pan, J-M Analyst, 110, 1245 (1985).
133. Murugaiyan, P.; Sankar Das, M. Analytica Chimica Acta, 48, 155 (1969).
134. Budesinsky, B.W. Analyst, 105, 278 (1980).
135. Jacobs, W.D. Analytical Chemistry, 32, 514 (1960).
136. Larsen, R.P.; Ross; L.E. Analytical Chemistry, 31, 176 (1959).
137. Belew, W.L.; Wilson, G.R.; Corbin, L.T. Analytical Chemistry, 33, 886 (1961).
138. Hoste, J.; Gillis, J. Analytica Chimica Acta, 12, 158 (1955).
139. Cheng, K.L. Analytical Chemistry, 28, 1738 (1956).
140. Donaldson, E.M. Talanta, 24, 441 (1977).
141. Cheng, K.L. Chemist-Analyst, 45, 67 (1956).
142. Danchik, R.S.; Thompson, D.E.; Hillegass, H.F. Analytical Letters, 9, 687 (1976).
143. Kawashima, T.; Ueno, A. Analytica Chimica Acta, 58, 219 (1972).
144. Johnson, R.A. "Tellurium and Selenium" Colorimetric Determination of Nonmetals (D.F. Boltz, ed.), Interscience, NY, 1958, pp. 309–338.
145. Case, O.P. Industrial and Engineering Chemistry, Analytical Edition, 16, 309 (1944).
146. "Steel and Cast Iron—Determination of Total Silicon Content—Reduced Molybdosilicate Spectrophotometric Method" Parts 1 and 2, ISO 4829-1:1986 and ISO 4829:2:1988, International Organization for Standardization (ISO).
147. Kriege, O.H.; Wolfe, A.L. Talanta, 9, 673 (1962).
148. "Steel and Iron—Determination of Sulfur Content—Methylene Blue Spectrophotometric Method" ISO 10701:1994 International Organization for Standardization (ISO).
149. Chiba, K.; Inamoto, I. ; Tsunoda, K.; Akaiwa, H. Analyst, 119, 709 (1994).
150. Annual Book of ASTM Standards, Vol.03.06, American Society for Testing and Materials, West Conshohocken, PA, 1996, Designation E1587.
151. Luke, C.L. Analytical Chemistry, 21, 1369 (1949).
152. Patterson Jr., G.D. "Sulfur" Colorimetric Determination of Nonmetals (D.F. Boltz, ed.), Interscience, NY, 1958, pp. 261–285.
153. Annual Book of ASTM Standards, Vol. 03.06, American Society for Testing and Materials, West Conshohocken, PA, 1996, Designation E367.
154. Wu, Z.; Hu, Z.; Jia, X. Analytica Chimica Acta, 231, 101 (1990).
155. Donaldson, E.M. Talanta, 23, 823 (1976).
156. Luke, C.L. Analytical Chemistry, 28, 1276 (1956).
157. Luke, C.L. Analytica Chimica Acta, 37, 97 (1967).
158. Ross, W.J.; White, J.C. Analytical Chemistry, 33, 424 (1961).
159. Ashton, A.; Fogg, A.G.; Burns, D.T. Analyst, 98, 202 (1973).
160. Spinola Costa, A.C.; Teixeira, L.S.G.; Ferreira, L.C. Talanta, 42, 1973 (1995).

161. Donaldson, E.M. Talanta, 16, 1505 (1969).
162. Afghan, B.K.; Marryatt, R.G.; Ryan, D.E. Analytica Chimica Acta, 41, 131 (1968).
163. Fogg, A.G.; Marriott, D.R.; Burns, D.T. Analyst, 95, 848 (1970).
164. Fogg, A.G.; Marriott, D.R.; Burns, D.T. Analyst, 95, 854 (1970).
165. Donaldson, E.M. Talanta, 22, 837 (1975).
166. Norwitz, G.; Gordon, H. Analytica Chimica Acta, 69, 59 (1974).
167. Johnson, C.M. Iron Age, 157, 66 (1946).
168. Bricker, C.E.; Waterbury, G.R. Analytical Chemistry, 29, 1093 (1957).
169. Machlan, L.A.; Hague, J.L. Journal of Research of the National Bureau of Standards, 59, 415 (1957).
170. Topping, J.J. Talanta, 25, 61 (1978).
171. Ryan, D.E. Analyst, 85, 569 (1960).
172. Nardillo, A.M.; Catoggio, J.A. Analytica Chimica Acta, 86, 299 (1976).
173. McKaveney, J.P.; Freiser, H. Analytical Chemistry, 30, 526 (1958).
174. Cheng, K.L. Talanta, 3, 81 (1959).
175. Cerrai, E.; Testa, C. Energia Nucleare, 7, 477 (1960).
176. Cheng, K.L. Analytica Chimica Acta, 28, 41 (1963).
177. Challis, H.J.G. Analyst, 94, 94 (1969).
178. Dulski, T.R. Talanta, 29, 467 (1982).
179. Pakalns, P. Analytica Chimica Acta, 57, 51 (1971).
180. Ashton, A; Fogg, A.G.; Burns, D.T. Analyst, 99, 108 (1974).
181. Sekine, K.; Onishi, H. Analytica Chimica Acta, 62, 204 (1972).
182. Dupraw, W.A. Talanta, 19, 807 (1972).
183. Young, J.P.; White, J.C. Talanta, 1, 263 (1958).
184. Wood, D.F.; Jones, J.T. Analyst, 90, 125 (1965).
185. Ratcliffe, D.B.; Byford, C.S. Analytica Chimica Acta, 58, 223 (1972).
186. Cheng, K.L. Talanta, 2, 61 (1959).
187. Cheng, K.L. Talanta, 2, 266 (1959).
188. Rolf, R.F. Analytical Chemistry, 33, 125 (1961).

5
Atomic Absorption Spectrophotometry

1. INTRODUCTION

As with many ideas in science and technology, atomic absorption experienced a long latency between its original conception and its practical realization as a laboratory technique. Arguably, the first atomic absorption measurement was the observation of the Fraunhofer lines in the spectrum of the sun made by Wollastin in 1802 [1]. In 1860 Kirchhoff and Bunsen demonstrated that when light from a continuum source was passed through a flame, then was dispersed by a prism, a black line appeared in the continuous spectrum [2].

By 1900 no one doubted Kirchhoff's Law: that elements which can be made to emit light of a given wavelength will also absorb light of the same wavelength. However, for half a century more little practical use would be made of atomic absorption phenomena, except in astronomy where dark lines in the spectra of stars were used to infer elemental composition. An important exception was the laboratory measurement of mercury as a cold vapor in 1939 by Woodson [3].

The landmark year for atomic absorption was 1955. In that year independent work by Walsh [4] and by Alkemade and Milatz [5,6] suggested that a new and revolutionary analytical tool was at hand. It was Walsh and his coworkers at the Commonwealth Scientific and Industrial Research Organization (CISRO) in Australia who ultimately developed and refined both the theory and practice of atomic absorption as a laboratory technique.*

One key insight was that since absorption resonance lines are very sharp, no monochromator could isolate a corresponding line source of sufficient

* For an interesting historical account by Sir Alan Walsh see *Analytical Chemistry 46*, 698A (1974).

intensity to allow its absorbance to be accurately measured. The answer was the hollow cathode—a lamp containing a discharge source of the element of interest that emits intense and discrete lines, including the needed resonance lines.

The first commercial atomic absorption spectrophotometer was the Aztec Techtron AA3, based on Walsh's original design, marketed in the early 1960s. Other instruments followed shortly, including atomic absorption adaptations to flame emission spectrometers. Refinements followed. Double-beam systems corrected for variations in source intensity and detector response. Nitrous oxide/acetylene became a useful, if initially somewhat dangerous, alternative to air/acetylene flames. Delves cup, hydride techniques, and various electrothermal atomizer approaches followed shortly, each promising to extend detection limits into new realms.

The period 1965–1975 might be regarded as the golden age for atomic absorption. It seemed that there was no inorganic analysis problem that could not be tackled by this upstart technique. And no self-respecting metals analysis lab could be without some kind of AA instrument. There were even attempts to control rapid metallurgical processes by this means, especially by smaller companies attempting to save the cost of an optical emission direct reader.

There were challengers, however. The interval mentioned begins with the introduction of the first commercial DC plasma optical emission instrument (1965) and ends with the introduction of the first commercial inductively coupled plasma OES (1975). These two techniques, and especially the latter, would ultimately raise questions about the future of atomic absorption. In particular: why did a laboratory need to retain a one-element-at-a-time, high operating cost instrument when an equivalent multi-element, low operating cost device could be installed (albeit with a considerable initial investment)?

The answer came in the decade of the 1990s. Electrothermal AA techniques had become the cheapest and most direct route to newly required levels of extreme sensitivity, while line interference problems at higher concentrations of selected elements continued to thwart ICP-OES. A picture emerged in which a good lab with a diverse workload needed both methodologies to function effectively.

Today, after nearly 50 years, atomic absorption spectrophotometry is alive and well. If you can't yet justify the cost of an ICP-MS you determine your tramps this way. And if your OES cobalt line exhibits serious interferences you check your results this way. If you can't tie-up the ICP-OES to run 75 samples for nickel you divert the work to AA. And if somebody wants an independent check by another technique, what could be more cost effective than AA?

In this chapter we will first briefly examine the underlying theory of atomic absorption and attempt to place it in perspective with two related phenomena: atomic emission and atomic fluorescence. The normally experienced deviations from ideal behavior will also be described. The current commercial AA methodologies will then each be outlined: flame, electrothermal, hydride, cold vapor, glow discharge, and related techniques. Approaches to calibration will be described next, followed by a discussion of interferences —chemical, ionization, spectral, and physical—and how each is dealt with. Some details about the lower working limits in the utilization of this methodology follows next. Finally, a large section is devoted to specific applications of atomic absorption for the trace analysis of metals.

Unlike the previous chapter on molecular absorption where applications were listed by analyte element headings, here it becomes more useful to list applications by commodity groups. While there are many analyte-specific atomic absorption procedures, there are far more methods which can be ascribed to groups of analytes in an alloy class. Thus, the final section contains subheadings for ferrous alloys, nickel- and cobalt-base alloys, copper-base alloys, aluminum alloys, lead, tin, and zinc alloys, refractory metals and alloys, precious metals, and the inevitable miscellaneous alloys.

2. PRINCIPLES

2.1 Atomic Vapor and Atomic Spectra [7–11]

Atomic absorption must be ultimately regarded as a low temperature phenomenon despite the routine use of temperatures of 2800°C (as in a nitrous oxide/acetylene flame). The rationale for such a statement lies in the needed intent to select conditions that minimize the extent of analyte ionization. However, concurrent with this requirement is the need to generate a representative, reproducible ground state cloud of analyte atoms, either continuously in a flame or discretely in a furnace.

While molecular absorption is readily observed in liquid solutions, the only practical way to study atomic absorption is with the analyte element as a monoatomic gas. With few exceptions, the only practical way to produce an atomic vapor of measurable concentration is by applying heat—by most standards, a lot of heat. Fortunately, thermal sources (as contrasted with electrical discharge sources) can be selected to maximize the abundance of ground state atoms. Below 3000°C the number of excited atoms and ions is usually minuscule. This is important because atomic absorption occurs when a ground state atom absorbs a photon of characteristic wavelength and is elevated to one of several excited levels. Each of these so-called *resonance* wavelength absorptions results in a valence electron transition. Resonance absorption spectra are simple compared to emission spectra since all electron transitions begin at the lowest state. At these "low" temperatures electron transitions from excited states to higher excited states occur to a negligible degree.[1]

Figure 5.1 is a simplified Grotian diagram that illustrates the energy levels for the single outer shell electron of the sodium atom. The splitting of the *p* orbitals results from the two states in which the electron spin direction opposes or complements its orbital motion. Thus the absorbance of light at 589.0 and 589.6 nm results in the transition of 3s electrons to the 3p level, while weaker absorbance at 330.2 and 330.3 nm results in the transition of 3s electrons to the 4p level. For atoms with two outer shell electrons (the alkaline earths, for example) the spin of the two valence electrons may be opposed (or "paired"), which produces singlet states, or it may be parallel, which produces three levels (triplet states) for each orbital. The combined effect of two unpaired electron spin moments and the orbital magnetic moment produces the triplet. For atoms with three outer shell electrons doublets and quadruplets occur. With four outer shell electrons singlets, triplets, and quintet states are observed. Thus, even just considering resonance from the ground state, predicting spectral behavior can be daunting, and often impossible.

It seems intuitive that the most sensitive absorption resonance line should be the so-called "last line," representing the jump from the ground state to the first excited state. This should be the longest wavelength (lowest energy) resonance in the absorption spectrum of an element. In fact, except for simple examples such as the alkali metals and the alkaline earths, this is rarely the case. It is common that the most intense absorbance occurs at a resonance line representing a transition from the ground state to some excited state far above the first excited state. This is the first of many examples we will encounter in this chapter illustrating how practice eludes theory in this methodology. In fact, all the most sensitive resonance lines for the elements were determined by empirical measurement.

The production of an analytically useful ground state atom cloud is one of the key features of this methodology. In flame methods the sample solution is nebulized to a fine spray aerosol with a droplet size between 1 and 10 μm diameter. Since most AA nebulizers are 3–15% efficient, between 97 and 85% of the sample never reaches the flame. More sample is lost to flow spoilers or baffles that intercept all but the finest droplets. Once exposed to the rapidly expanding hot gases of the flame the tiny droplets are first dried (desolvated) to salt particles. These are then vaporized, producing some equilibrium condition of molecular vapor, ground state atomic vapor, excited

---

[1] In stellar absorbance phenomena, such as the Fraunhoffer lines in the solar spectrum, such extreme temperature transitions *are* observed.

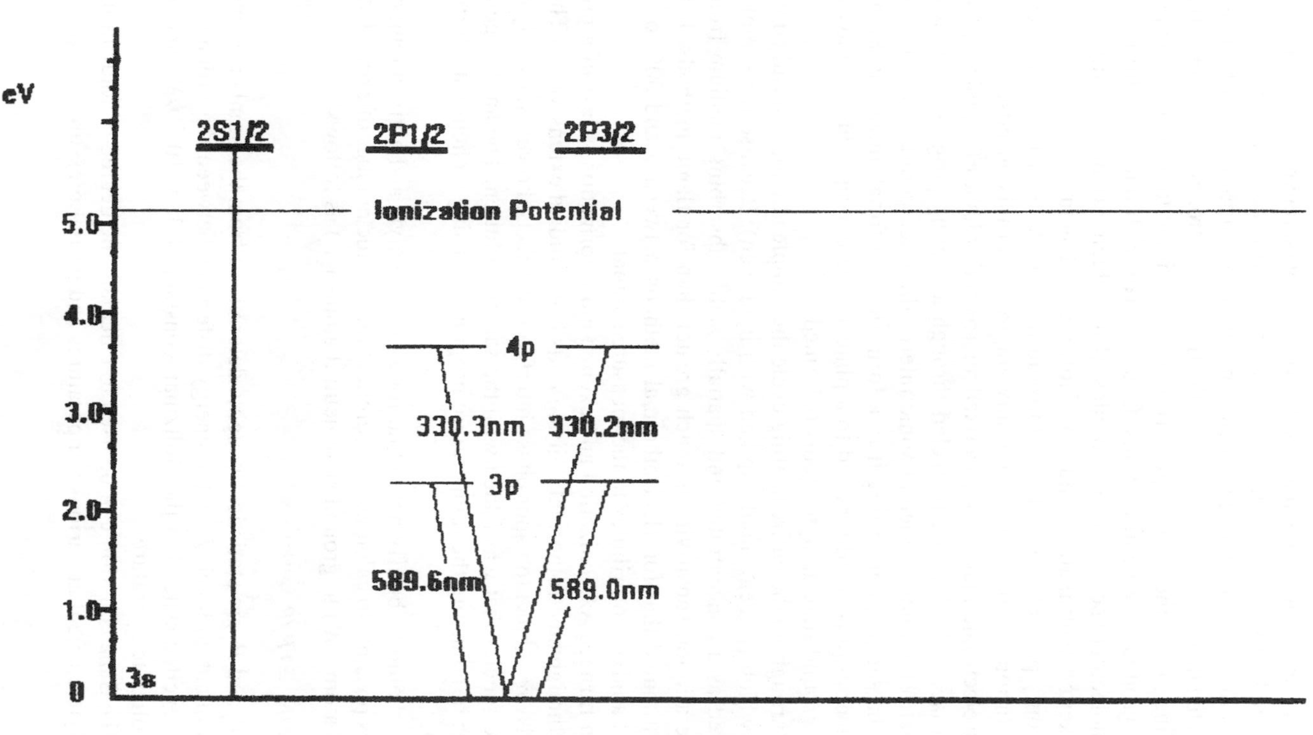

Figure 5.1 Simplified Grotian diagram for sodium.

atomic vapor, and ions. Each aerosol particle may be regarded as following a high speed ballistic path during which these events occur. Since ground state neutral atoms are the desired species in atomic absorption it can be readily appreciated that observation height in the flame will strongly affect sensitivity.

Fine tuning the vapor equilibrium is also critical to maximize the presence of ground state neutral atoms. Cool, oxidant-rich flame conditions will favor molecular species such as oxides. Hot, reducing flame conditions will favor excited neutral atoms and ions. However, balancing flame conditions is not the only possible control over this equilibrium. As we will see in Section 6.1.2 solution additives are commonly employed for this purpose.

In electrothermal methods a fixed microsized volume of sample is placed in a furnace. The furnace is cycled through a heating program to first dry (desolvate) the aliquot, then at some intermediate temperature to "char" (or "ash") the sample, converting it to a form suitable for atomization. Finally, the temperature is rapidly raised to a plateau where a reproducible cloud of neutral ground state analyte atoms is formed.

Throughout the entire heating cycle the sample aliquot is sheathed with argon (which is occasionally spiked with hydrogen). Because the sample is measured as a concentrated and spatially stable absorbing medium from a discrete aliquot, sensitivity is much greater than for flame methods. It has been estimated that for electrothermal methods between 5 and 50% of the injected analyte contributes to the measured signal.

In furnace AA ionization and excited atom formation is less of a problem (although electrothermal emission *has* been studied extensively.[*]) These methods *are* subject to vapor phase interference effects, however, usually due to some species that was added with the sample solution. The answer generally lies in adjusting the temperature program, or using a chemical additive, or both.

It should be briefly noted that for a given analyte and a given temperature it is possible to calculate a theoretical ratio of the number of excited state neutral atoms (N_l) to ground state neutral atoms (N_0) as follows:

$$N_l/N_0 = (P_l/P_0)e^{-E_l/kT}$$

where P_l and P_0 are statistical weights determined from the number of states at each quantum level, E_l is the energy difference between the ground state and the excited state, k is the Boltzman constant ($1.38 \times 10^{-23} J/K$), and T is the absolute temperature.

The Saha equation can be used to calculate the degree of ionization of an analyte element at various temperatures and partial pressures:

---

[*] See Section 3.6.1.

$$\log K = -(U/4.573T) + (5/2)\log T - 6.49 + \log (g_0\, g_e/g_i)$$

where K is the equilibrium constant for the ionization reaction, U is the ionization potential (in cal/mole), and the g terms are statistical weights (g_0 = unionized element; g_e = electron; g_i = ionized element).

From such first principles it is possible to predict the temperature of maximum ground state atom population. The Saha equation can also be used to predict the effect of those additives which act as ionization buffers. In practice, however, optimum flame conditions and the effect of additives are always investigated empirically.

2.2 Line Width [7,9,10]

Walsh demonstrated that the probability distribution over electron energy levels and the Heisenberg uncertainty principle combine to yield a "natural width" for an atomic absorption resonance line of about 10^{-5} nm (10^{-4} Å). Observed line widths are considerably broader [of the order of 0.002–0.005 nm (0.02–0.05 Å)] but still much narrower than the bandpass of all but the most sophisticated monochromators.

The discrepancy of two orders of magnitude between the theoretical and observed line widths is accounted for by the combined influence of Doppler, electric field, and pressure effects. The *Doppler width* depends upon the analyte element, the resonance line, and the temperature. It is produced by the thermal motion of atoms in the atom cloud. The Doppler width, D, is defined as:

$$D = (\nu/c)(2RT/M)^{-1}$$

where ν is the frequency of the resonance line, c is the speed of light, R is the gas law constant, T is the temperature, and M is the atomic weight.

Collisional broadening refers to the slight differences in energy of quanta absorbed due to the collision of atoms. The atomic collisions may involve neutral or charged particles, and they may involve collisions between analyte atoms, or between analyte atoms and other particles. This effect is often called Lorenz broadening.

Moreover, in some cases nuclear spin produces a hyperfine structure that leads to line broadening. Also, when an analyte element has several stable natural abundance isotopes, an isotope shift produces a resonance line for each isotope. Except for very heavy or very light elements these isotopic lines will not be resolved by an ordinary AA monochromator. The result will be further broadening of the resonance line.

Despite all these broadening effects absorption resonance lines are much too narrow to allow the use of an ordinary white light source monochromator as the line source. As we saw in the last chapter Beer's Law is obeyed only for monochromatic light. Another means of stating this is that a linear correspondence between analyte concentration and observed absorbance requires the line source to be narrower than the absorbance line. If the line source is broader than the resonance line we will be measuring only a fraction of the total absorbance, since the detector will be responding to the "wings" of the line source.

One of Walsh's great insights was to recommend the use of a sharp line emission source produced by the element of interest as the line source. If the source lamp is designed to operate at a lower temperature than that of the analyte atoms in the atom cloud it will be less broadened than the absorption line. A monochromator is then placed *after* the atom cloud to isolate the resonance line from other source lines and from emissions from the atom cloud. It need not be a high resolution device.

If one were to attempt to substitute a high intensity continuum lamp and a research grade narrow bandpass monochromator for the hollow cathode or electrodeless discharge lamps that have been the mainstay line sources for atomic absorption, it would soon be found that insufficient light throughput would thwart the effort. At some future time, perhaps, a tuneable dye laser device may be developed that could serve this purpose. At present, some workers have had success building single purpose dedicated instruments using a laser line source that matches a specific element's absorption resonance.

2.3 Beer's Law and the Absorption Coefficient [7–11]

The proportion of the incident line source photons that are absorbed by ground state analyte atoms in the light path is directly related to the total number of such atoms and to their effective cross section, termed the *absorption coefficient*, χ. This term is a parameter for the quantity of radiation at a resonance frequency, ν, that can be absorbed by an atom for the transition $j \rightarrow k$, as follows:

$$\chi_{jk} = B_{jk}(h\nu)/c$$

where B_{jk} is the Einstein probability coefficient of absorption, h is Planck's constant, and c is the speed of light. As χ is termed a cross section it has the dimensions of an area.

Another way to regard an atom is as a harmonic oscillator. The absorption of light causes higher frequency oscillations. The absorption coefficient, expressed in terms of electrodynamics is given by:

$$\chi_{jk} = f_{jk}(\pi e^2/m_e c)$$

where f_{jk} is the *oscillator strength*, which represents the number of classical electron oscillators (average number of electrons) involved in the $j \rightarrow k$ transition, for each atom; e is the charge of an electron; and m_e is the mass of an electron.

More commonly, a related means to express the absorption coefficient, K, relates to an effective volume of analyte atoms:

$$K_{jk} = N\chi_{jk}$$

where N is the number of free ground state atoms in the light path. Thus, we can write:

$$K_{jk} = Nf_{jk}(\pi e^2/m_e c)$$

Unlike χ_{jk}, K_{jk} is an experimentally determinable quantity; it is the probability of light absorption per unit volume. If N can also be measured then f_{jk} can be calculated. The strategy was proposed by Walsh in 1955 [4] and since that time has been used by many investigators to generate tables of oscillator strengths.

Oscillator strength is thus a measure of the sensitivity of an atomic absorption resonance line. It is always a good indicator of the relative sensitivity among several lines for the same element.* Difficulties arise in comparing sensitivities between elements by this means because of differences in molecular dissociation and atom vapor formation between analytes.

As we saw in the previous Chapter, Beer's Law is commonly expressed as:

$$A = \log (P_0/P) = abc$$

where A is the absorbance, P_0 is the incident radiant power, P is the transmitted radiant power, a is the absorptivity constant, b is the pathlength (in cm) and c is the analyte concentration (in g/L). This derived from a more general expression:

$$P = P_0 e^{-kbc}$$

Applying the atomic absorption coefficient, defined above, this becomes:

$$P = P_0\, e^{-\chi Nb} = P_0 e^{-Kb}$$

or: $$A = \log (P_0/P) = 2.3Kb$$

While Beer's Law is the applicable treatment for describing in general terms the linear portion of an atomic absorption calibration, the dynamic nature of the absorption region makes it impossible to apply the formula rigorously. In practical systems neither the path length nor the number of analyte

* Provided that correction is made for differences in line shape.

atoms in the lightpath can be precisely defined. Moreover, atomic absorption is subject to a number of spectral influences whose combined effect, despite the use of clever correction schemes, severely limits the linearity of the response.

The use of a sharp line source, such as a hollow cathode lamp, is not an absolute guarantee that a nonabsorbing line or a second absorbing line (from the source element or the fill gas) will not make its way to the detector. Attentive care in setting the monochromator slit width is always important to avoid this problem. Emission from the atom cloud is largely compensated for by modulating the line source. A principle source of this emission is usually the analyte itself (although there are exceptional cases, such as zinc, which does not emit at its most sensitive resonance absorbance line). The detection system is designed to respond to the AC line source signal. The atom cloud emission is a continuous background which is not measured. In the case of a very strong emission, such as an extremely luminous flame, for example, the signal to noise ratio will deteriorate.

More important and sometimes more difficult to control are background absorbance losses and scattering losses due to the sample solution or alloy matrix components. Simply measuring a reagent blank and adjusting the response to zero absorbance will effectively compensate for the interaction of reagent solution components and the flame. However, the alloy components of the test portion may have a more profound effect of scattering and nonspecific absorption. It is usually not practical, or even possible to match the sample matrix with a synthetic blank that is free of the analyte element. Today, there are several effective approaches to solving this problem, notably, the continuum-source, Zeeman effect, and Smith-Hieftje correction schemes. These will be discussed in some detail in Section 6.2.

Despite all these efforts to control extraneous effects, all atomic absorption calibration curves have a restricted linear range, commonly curving toward the concentration axis at high absorbance. This intractable curvature is traceable to several causes—notably stray light and background radiation from the line source, but also in the case of very high or very low Z-number elements to hyperfine structure caused by isotopic effects. Background radiation from the line source is attributable to self-absorption or self-reversal within the lamp itself. This is caused by a neutral atom cloud of the lamp element absorbing radiation from the emission line maximum. It results in a bifurcated line profile whose "wings" will not be absorbed by the analyte.

It has been generally accepted that an important advantage of atomic absorption over atomic emission is the fact that the atomization temperature does not exert a sensitive influence on the results. This is because at the temperatures utilized in AA (<3000°C) nearly all atoms are free and in the ground state. However, with increasing analyte concentration the degree of dissociation of molecular species will drop off to some extent, producing calibration curvature toward the concentration axis. This effect is strictly thermal and can sometimes be eliminated by increasing the flame temperature. Some elements (e.g., barium in a N_2O/C_2H_2 flame) are highly ionized when present at low concentrations but proportionately less ionized when present at high concentrations. The result is calibration curvature *away* from the concentration axis. In practice the additives known as ionization suppressors would be employed to eliminate this effect.

2.4 Element Sensitivities in Atomic Absorption

Before closing a discussion on basic principles it is appropriate here to point out that a rigorous understanding of the relative sensitivities of the elements in atomic absorption measurement remains elusive. We know that lower atomic weight elements tend to form higher ground state atom populations under typical AA conditions than do higher atomic weight elements. Thus, the lighter elements tend to be more sensitive under identical conditions. Some elements have many low-lying energy states just above the ground state. Under typical AA conditions so many of these states are occupied that the ground state population is severely depleted. Many of the rare earth elements are insensitive due to this effect. Such problems would be revealed by the oscillator strength of the resonance line, but many other effects would not.

For example, if an element's only good ground state resonance line is obscured by a molecular band (from the flame's fuel gases, for example) the element would be regarded as insensitive. Elements that react to form refractory oxides, nitrides, or carbides (in either flame or furnace) (e.g., B, Hf, Nb, Ta, W, Zr) are insensitive because typical AA conditions are not sufficiently energetic to dissociate such compounds. In glow discharge AA such analytes can be determined because the sources of oxygen, nitrogen, and carbon required to form such compounds are absent.

3. ATOMIC ABSORPTION METHODOLOGIES

3.1 Flame [9,10]

The use of a fuel/oxidant gas flame to produce the analyte atom cloud was a key development for the evolution of atomic absorption as a generally applicable laboratory tool. The direct injection burner (also known as the turbulent flow or total consumption burner), which had been widely used in flame emission spectroscopy, was employed in some early work. However, the premix (laminar flow) burner, which soon became standard, had been under development since 1960.

The direct injection burner consists of two concentric metal tubes with a sample capillary in the center of the innermost tube. All three orifices terminate at the burner tip. Fuel gas (generally, acetylene) is fed between the two tubes while oxidant gas (generally, oxygen) is fed up the innermost tube. A Venturi pressure drop caused by the flowing gas draws sample solution up the central capillary. The sample solution is thus nebulized into droplets at the burner tip and instantly vaporized by the flame. The premix burner, on the other hand, uses a special nebulizer device that employs a stream of oxidant (air or nitrous oxide) to produce a spray of sample droplets. This aerosol then passes into a premix chamber, where it is blended with fuel and additional oxidant and where baffles or "spoilers" remove large droplets. The mixture of fuel/oxidant/fine droplets then passes through a slot and creates the flame and atom cloud.

The direct injection burners were inexpensive and inherently safe since a flashback explosion could not occur. They were designed to allow all of the sample which was aspirated to be measured. And they could be designed to handle any combination of fuel and oxidant. Unfortunately, they were optically noisy and presented a very small pathlength for the line source. They were also loud, producing a piercing whistle. The premix burner produces a comparatively quiet flame (optically and acoustically) and allows for a significantly longer pathlength for the line source (10 cm for air/acetylene and 5 cm for nitrous oxide/acetylene are common). While the possibility of an explosive flashback to the premix chamber is present, modern designs make this event unlikely. Oxygen, however, cannot be used since its flame propagation rate with acetylene is too high for this design and a flashback *would* occur. The baffles and flow spoilers in the premix chamber result in a significant proportion of the aspirated sample being thrown away down a drain tube, but the advantages of a long, quiet optical path through the atom cloud compensate for this.

In the simplest, single-beam optical designs chopped light from a hollow cathode or electrodeless discharge lamp passes through the flame, then into a monochromator and ultimately to a photomultiplier detector, which, with its associated electronics, responds only to the modulated signal. More sophisticated and more commonplace today are double-beam designs in which the pulsed light is split into two paths—one carrying it through the flame, the other tracing a separate path that avoids the flame. The two beams then pass into a monochromator and, ultimately are measured by a photomultiplier detection system that produces a signal proportional to the ratio of the intensities of the two beams. Such a system automatically compensates for variations in the lamp intensity and in the response sensitivity of the detection system, but does not compensate for variations in the aspiration or atomization processes.

At various times dual channel atomic absorption instruments have been marketed. These consist of designs with completely separate monochromators and detectors for two element channels. With a multi-element hollow cathode lamp or with two hollow cathode lamps they can be used to measure two elements simultaneously. Also, they can be used in a mode in which one element channel is used to measure an internal standard element (either added or present as part of the sample matrix). The measured ratio of analyte to internal standard and the fixed amount of the internal standard element in the calibration standards, blanks, and samples allows compensation for variation in the mass transport process (aspiration) and the flame. Dual channel instruments can also be used to measure and correct for background near the resonance line.

The advantage of such background correction is less important today with the commercial availability of single channel designs incorporating Zeeman, deuterium arc, or Smith-Hieftje correction features.

There have been various adaptations to the standard flame technique over the years in an ongoing effort to extend atomic absorption to lower analyte concentrations. Many of these strategies involved some means of inserting a small discrete-volume test portion into the flame and measuring the pulse signal that results either by peak height or peak area. Besides the natural advantage that accrued from measuring a significant portion of the total analyte introduced, preconcentration techniques could be applied to great advantage without concern about a flashback. In one version of this approach miniature tantalum boats were employed. A mechanism was provided for drying and ashing the contained sample near the flame, then inserting the boat into the flame for atomization and measurement. The Delves cup approach utilized a nickel cup that was inserted into the flame just below the vertical hole in a horizontal ceramic tube through which the line source beam passed. The tube functioned to hold and concentrate the atom cloud.

Nonspecific absorbance due to smoke was a significant problem with these approaches, but it could be readily solved by the use of background correction. Today, these techniques have been supplanted by electrothermal atomization.

Originally, hydride generation techniques for antimony, arsenic, bismuth, germanium, selenium, tellurium, and tin involved passing the gaseous compounds either directly, or after trapping, into a flame (often argon-sheathed hydrogen). Later developments involved passing the hydrides into a flame-heated quartz tube. Today an electrically heated quartz tube has become dominant. This subject will be discussed further in Section 3.3.

3.2 Electrothermal [9–12]

It wasn't long after the success of discrete techniques, like the Delves cup, that resistance heated furnace designs began to be commercialized. The idea was to produce a configuration that reproduced the atom cloud and optimized its confinement for measurement. The marketed instruments were modifications of designs selected from among numerous experimental configurations in the published literature. The three approaches that found the largest commercial success were the carbon rod atomizer first sold by Varian Techtron in 1970 and by Shandon Southern in 1971 (although in different forms—both based on a design by Williams and West); the tantalum filament atomizer first sold by Instrumentation Laboratory in 1971 (based on a design by Donega and Burgess); and the graphite tube design by Massman first marketed by Perkin Elmer in 1970.

It was the Massman furnace that achieved the widest acceptance and its incremental refinement has led to its present role as a standard of the industry. The Massman furnace is a graphite cylinder supported at the ends by graphite cones in a water-cooled copper housing. The tube is heated resistively by the application of low voltage/high current.

In all versions of the modern instrument the graphite furnace is programmed through a series of heating cycles to first dry the liquid aliquot to a salt deposit, then char out sample components that would otherwise produce smoke during the atomization. The "char" step also is regarded as preparing the sample for atomization by converting the analyte to an optimal compound. The final step is the high temperature atomization during which the analyte measurement is made. This is followed by a maximum temperature burnout step that removes the last removeable vestiges of the sample matrix and otherwise prepares the graphite tube for the next sample. Most programs allow the analyst to set not only the temperatures and hold times for each step but also to specify the heating rate (ramp angle) between each step.

Throughout the heating program the tube is flushed with argon gas, also under program control. A gas stop may be programmed to occur during part of the heating regimen and also always occurs at or immediately before the beginning of the atomize cycle. Today, most furnace designs have end windows of optical quartz, effectively creating a sealed chamber in which to contain the atom cloud. Argon enters at the tube ends and exits out the central sample injection hole.

As with certain other types of matrix, for most metal samples even the most rigorous "char" will still leave components that will produce nonspecific absorbance (smoke) during the atomize cycle. To compensate for this serious background interference all graphite furnace atomic absorption units incorporate some form of background correction. The three approaches—deuterium arc, Zeeman, and Smith-Hieftje—each have their own strengths and weaknesses. For a detailed discussion of this topic see Section 6.2.

Early improvements included the use of pyrolytic graphite-coated tubes which showed reduced permeability to metal ions, longer tube life, and improved sensitivity for carbide-forming analytes. Autosampler carousel devices have eliminated the precision problems associated with manual micropipetting and made unattended operation feasible. Recent advances in commercial electrothermal instrumentation include the transverse-heated furnace, the L'vov platform insert, and instruments designed for the simultaneous determination of four to six analytes.

A transverse-heated furnace was developed by Perkin Elmer. The new design is based on the flow of resistance current at a right angle to the conventional direction. Its principal advantage is even heating down the length of the graphite tube so that atomized elements no longer recondense at the cooler ends of the tube.

The stabilized temperature platform furnace is based on work by B.V. L'vov. A graphite platform with a recession for holding the sample is centered in a pyrolytically coated graphite tube. There is minimal surface area contact between the platform and the tube wall so that the platform and its sample lag significantly in temperature as the tube wall and the argon are rapidly heated during the atomize cycle. When the platform begins to heat rapidly the atom cloud is evolved into a thermal environment that is at virtual equilibrium. The resultant peak area measurement is generally characterized by reduced chemical interference effects. It should be noted, however, that in atomic absorption few innovations have proven to be a panacea for all analytes. Thus, bismuth is sometimes measured by atomization from the furnace wall (for improved peak height sensitivity when the less sensitive 306.8 nm line must be used); and phosphorus shows better performance with porous graphite than it does with "pyro-coated" tubes.

Simultaneous graphite furnace instruments are designed to improve productivity. But this advantage comes with some necessary compromise in performance. Four to six analytes are typical, each line source traversing through the same graphite tube. The furnace program must be carefully chosen to optimize conditions for each element without severely degrading the conditions for all of the others. Instruments have been marketed by Hitachi, Perkin Elmer, and Leeman Labs.

3.3 Hydride Generation [9,10]

Antimony, arsenic, bismuth, germanium, lead, selenium, tellurium, and tin have been determined by chemically forming their gaseous covalent hydrides, then thermally dissociating them to form an analyte atom cloud in the line source light path. Early work utilized granular zinc metal to form nascent hydrogen in the acid sample solution which reacted to form hydrides. But yields were poor and the method was limited to antimony, arsenic, and selenium. Blanks were sometimes erratic due to impurities present in the zinc metal.

When sodium borohydride was substituted as a reductant all eight elements could be determined, although even today with varying degrees of success. The weakest of these applications is the determination of lead, which requires critical solution concentrations and is relatively insensitive. Even with optimized conditions less than 5% of the lead present is converted to the hydride [13].

Today, sodium borohydride is nearly always added as a solution [typically 1% (w/v) $NaBH_4$] to the dissolved test portion in dilute acid solution (often 1M HCl). Sometimes the sodium borohydride solution is stabilized with caustic and sometimes vacuum filtration through a 0.45 μm membrane filter is necessary. Reagent prepared in this manner remains useable for three weeks. Without these precautions sodium borohydride solutions should be prepared immediately before use.

The reaction with the sample (usually prereduced with potassium iodide and ascorbic acid) takes place in a sealed glass vessel which is continuously swept by a gentle stream of argon gas while being stirred magnetically. Or else the argon stream is bubbled through the sample solution to provide agitation.

In early work the evolved hydrogen and the covalent hydrides were swept by the argon into an argon-sheathed hydrogen flame in the light path. Later the hydrogen/argon flame was replaced by an air/acetylene-heated quartz tube. Streams of nitrogen applied at the tube ends displaced the ignition of sample/reagent hydrogen from the light path, preventing the occurrence of a serious amount of nonspecific absorbance. Today the quartz tube is typically heated electrically (to 1100°C). The quartz tube surface appears to play an active role in the atomization process. Welz [9] reports that treating the quartz tube in a hydrofluoric acid bath for a few minutes prior to use produces optimal performance and that similar treatment restores used tubes to their original performance.

3.4 Cold Vapor

The cold vapor technique is specific to one analyte—mercury. This is because at the ambient temperature utilized mercury is the only metallic element with a significant partial pressure. With the cold vapor approach, mercury in the sample solution is reduced with stannous chloride and passes into a pumped recirculating argon stream which passes through an optical cell in the line source light path. The free atom concentration of mercury increases in the recirculating gas loop, reaching a maximum in less than 1 min. It is then measured at 253.7 nm. The apparatus is then vented to a hood and the reaction vessel is removed. When the mercury signal returns to zero the next sample solution in a clean reaction vessel is reduced and connected to the recirculating system.

The optical cell may be up to 15 cm long with a diameter of 0.75 cm. It is conveniently supported by a burner head assembly in the optical path. Early publications raised concerns about spectral interference from water vapor and significant attention was paid to the use of suitable drying agents in the recirculating argon loop (magnesium perchlorate or sulfuric acid are suitable, while calcium chloride is unsuitable). It is now certain that water vapor only causes problems if it condenses on the optical windows of the cell. Warming the cell may be sufficient to prevent this from occurring.

Analyte loses are often related to materials that contact the recirculating mercury vapor. Both red rubber and polyvinylchloride tubing have been implicated. The presence of any organic matter in the sample solution must be dealt with by prior oxidation with sulfuric acid and permanganate, followed by removal of excess permanganate by reaction with hydroxylamine hydrochloride.

In place of stannous chloride as a reducing agent for mercury, sodium borohydride has become widely accepted. In this case the closed recirculating system is replaced by an open, single-pass design. In the first place, with sodium borohydride so much hydrogen is evolved that a closed system would become dangerously pressurized. And in the second place, the higher rate kinetics are favorable for sensitive peak height measurement. Also, the shorter contact time between the mercury vapor and the system components results in less losses due to adsorption and diminished memory effects, as well.

Mercury determination by the cold vapor AA approach is a very sensitive technique. This must be qualified to some extent because mercury vapor in the gaseous atomic state is in equilibrium with mercury in aqueous solution. An optimal sensitivity design, therefore, minimizes both liquid volume and the total volume of the apparatus. Extreme sensitivity can be achieved by amalgamating ultratrace amounts of mercury with tin, silver, or gold, then heating the amalgam rapidly to 500–700°C and measuring the evolved mercury in a cold vapor cell.

3.5 Glow Discharge [14,15]

Cathodic sputtering in an argon glow discharge can be used to produce a ground state atom cloud from a solid metal sample. Some interesting work can be performed in this manner because the sputtering process is essentially nonthermal and is conducted in a partial vacuum. Elements such as boron, niobium, rhenium, tantalum, tungsten, and zirconium, which are very insensitive by other atomic absorption approaches, show good response with this technique. These elements form refractory oxides or carbides in graphite furnace AA—compounds which are not significantly dissociated at the accessible temperatures. With glow discharge AA, however, oxygen and carbon are absent and so the instrument response for such analytes is more closely related to their oscillator strengths. The technique requires a conductive sample and is, therefore, ideally suited to metals and alloys.

In practice the sample in the form of a flat metal solid is sealed with an O-ring to an annulus in the sputtering chamber. The chamber is fitted with optical windows and sits in the light path of the line source. High velocity argon jets impinge upon the metal surface in the chamber, which is pumped to 0.01 atmosphere (7.6 Torr or 1013 Pa). A high voltage/low current power supply creates a potential between an anode and the sample which ionizes the argon. The impinging jets of argon plasma kinetically dislodge atoms from the sample surface and the vacuum pulls them away from the sample to create an atom cloud.

Background effects appear to be due to agglomerations of atoms which produce broad band absorbances. Compensation for this background is accomplished by modulating both the line source and the sputtering process. This is possible since the sputtering can be started and stopped within milliseconds.

Adjustment of the sputtering current and the argon flow rate affords a degree of control over the sensitivity of the analysis so that the useful concentration range for a given resonance line can be extended. The methodology has been applied to iron-, copper-, aluminum-, zinc-, nickel-, cobalt-, and titanium-base alloys from trace levels to major concentrations. Intensity-ratio and normalization approaches to calibration have proven useful with this technology. A commercial instrument was first sold by Analyte Corporation and is now marketed by Leeman Labs.

3.6 Related Techniques

It is appropriate to pause here and spend a brief time describing a few associated methodologies in order to place them in perspective with atomic absorption. These fall into two categories: atomic emission and atomic fluorescence.

3.6.1 Atomic Emission [7–10]

This is a very broad topic that will be treated in detail in Chapter 6. Work with metals and alloys is largely connected with high energy excitation sources—arcs, sparks, and plasmas. However, valuable work in atomic emission can be accomplished in and just above the temperature regimes of atomic absorption. Alkali metal and alkaline earth determinations, for example, are still a *forte* of flame emission with detection limits in the fractional to low part-per-million range (in pure solution). Other elements with good sensitivity include aluminum, chromium, copper, indium, manganese and vanadium. However, it is always necessary to include the proviso that an interference-free analyte emission line can be resolved by the monochromator (or that interferences can be corrected mathematically). For most metal matrices *some* interferences will need to be dealt with. Some particularly *insensitive* analytes at flame temperatures include arsenic, boron, beryllium, bismuth, iridium, tantalum, uranium, and zinc.

In the past much flame emission work was done utilizing an oxygen/ acetylene turbulent flow total consumption burner. Today, when the technique is used, typically a nitrous oxide/acetylene laminar flow burner is used instead. As remarked earlier, the number of excited and ionized atoms at flame temperatures is minute and so, in general, the technique is measuring only a small fraction of the total analyte population. And the response measured is highly sensitive to temperature changes in the flame. However, such changes do not normally occur unless one alters the fuel/oxidant ratio or begins to aspirate samples in a different solvent.

Background emission from the flame, the reagents, and the metal matrix must be dealt with in some manner. Manual measurements may be made on one or both sides of the analytical line, or a vibrating mirror device may allow "on-the-fly" background measurement. Unlike instruments exclusively designed for atomic absorption the unit employed for atomic emission must be equipped with a monochromator of good resolution. Dual or multichannel designs allow the use of an internal standard element inoculated in identical amounts in the blank, calibration solutions and test solutions. This approach will correct for variations in sample uptake rate due, for example, to slight viscosity differences between solutions.

Self-absorption and self-reversal phenomena tend to bend calibration curves toward the concentration axis at high levels of analyte. Some elements (e.g., potassium) also show curvature toward the emitted intensity (y-axis in conventional presentation) at low concentrations. This is caused by increased ionization by highly dilute samples. The result is an "S"-shaped calibration curve.

Graphite furnace emission has been studied by a number of investigators but remains more of a research curiosity than a practical tool at present. The rationale for this work is to extend the advantages of a discretely atomized and concentrated analyte measurement to atomic emission so that trace quantities of elements like cerium, thorium, boron, and others, which are weak or impossible by AA, can be measured conveniently. It could be argued that acceptance of the approach has been retarded by the absence of a commercial graphite furnace spectrophotometer with a high-resolution monochromator onboard. However, work with optimized assemblies (including devices for automatic background correction) confirm that the graphite furnace is a unique and useful emission source. The comparatively low excitation temperature and the completely thermal character of the excitation mechanism results in minimal ionization (excited neutral atoms yield a simpler spectrum), low background, and reduced interferences, particularly as compared to emission in the regime of arcs, sparks, and plasmas.

3.6.2 Atomic Fluorescence [7–10]

Atomic fluorescence spectrometry is a complementary technique to atomic absorption in that it measures the light that is re-emitted after absorption. There are a number of different mechanisms by which electronic transitions in atoms can lead to fluorescence. The most common of these are summarized in Table 5.1.

The fluorescence signal is normally measured at a 90° angle to the exciting line source to keep transmitted line source radiation out of the detector. An air/acetylene flame is the most common atom source, although Baird Corporation has marketed an instrument that employs an inductively-coupled plasma.

The intensity of the fluorescence is directly related to the intensity of the line source, so high intensity sources are sought and used. These include electrodeless discharge lamps, tuneable dye lasers, and pulsed hollow cathode lamps. For extremely intense sources, such as dye lasers, the fluorescence response becomes saturated. This is a desirable, in fact, ideal, condition since it renders the fluorescence insensitive to minor source fluctuations. As with atomic absorption, the line source and detection system are modulated to distinguish the fluorescence signal from flame emission.

Table 5.1 Summary of Some Electronic Transitions in Atomic Fluorescence Spectrometry

Type of Fluorescence	Description
Resonance	Absorption and fluorescence have the same upper and lower electron states. Absorption and fluorescence occur at the same wavelength.
Stokes Direct line	Absorption and fluorescence have the same upper energy level, but absorbance occurs at a lower wavelength than fluorescence.
Anti-Stokes Direct Line	Absorption and fluorescence have the same upper energy level, but absorbance occurs at a higher wavelength than fluorescence.
Stokes Stepwise Line	Absorption and fluorescence have different upper energy levels due to either collisional deactivation or thermal excitation/radiational deactivation. Absorbance occurs at a lower wavelength than fluorescence.
Anti-Stokes Stepwise Line	Same as above, except absorbance occurs at a higher wavelength than fluorescence.
Sensitized	The analyte atom fluoresces when collisionally activated by a different atom that has been excited by the absorption of resonance wavelength light.

Light scattering from particles in the atom cloud can be a serious concern. Rains, et al. [16] suggested measuring the difference in the fluorescence response between a xenon arc lamp continuum source and a line source when each is alternately passed through the flame. Thus, in a manner analogous to the deuterium arc background correction in atomic absorption, light scatter in atomic fluorescence is automatically corrected. Quenching of the fluorescence signal can also occur due to nonradiative collisions of excited atoms, especially in very hot and nitrogen-bearing flames. Self-absorption can occur as analyte ground state atoms in the flame absorb their own resonance fluorescence. For this reason nonresonance lines are sometimes sought and used (see Table 5.1).

Atomic fluorescence shows the element selectivity of atomic absorption with significantly better sensitivity. Calibration curves are linear over at least twice the concentration range of atomic absorption (largely because the detector does not "see" direct radiation from the line source). It has even been demonstrated that atomic fluorescence can be made to function with a high power continuum source instead of an array of line sources, and with a filter photometric readout instead of a monochromator.

Despite these apparent advantages atomic fluorescence is little used outside of the research community. Commercial instrument manufacturers have by-and-large shunned the technique. Perhaps part of the reason is that it is less robust with respect to sample matrices than atomic absorption. But in large part the reason is that electrothermal atomic absorption, in which a great development effort was invested, is inherently more sensitive than flame- or plasma-based atomic fluorescence. Possibly one day it may return as a furnace-based discrete sampling device where the technique's unique advantages may show it to compare favorably with GFAA.

4. INSTRUMENTATION

4.1 Optics [9,10]

The optical design of an atomic absorption spectrophotometer makes no unusually stringent demands of the instrument maker. The line source ideally should be a focused beam that transects a near-cylindrical volume of the analyte atom cloud and then is refocused on the monochromator entrance slit. A monochromator with an f-stop aperture of *f*.10 is considered adequate for atomic absorption work.

Most atomic absorption instruments utilize a grating monochromator, although a quartz prism system would present some advantage for analytes with a principal resonance line below 240 nm. However, gratings show uniform dispersion and bandpass for a fixed slit width over the entire UV-visible wavelength range. This feature makes them indispensable for the monochromator of a versatile commercial instrument. Common resonance lines range from 193.7 nm (arsenic) to 852.1 nm (cesium), and so it is highly desirable to take advantage of a grating monochromator's wavelength independent resolution.

The monochromator of an atomic absorption spectrophotometer is always positioned to receive light *after* it has been partially absorbed by the analyte. This is in marked contrast to the optical plan of molecular absorption UV/visible spectrophotometers which position the analyte between the monochromator and the detector. The need for this arrangement in atomic absorption derives from the fact that most atomic vapor clouds are light emitters as well as light absorbers (the mercury cold vapor cell is a notable exception). And so it is necessary that the monochromator be able to isolate the resonance line from *both* line source interferences and atomizer interferences. The line source may emit nearby lines from the lamp cathode element or from the filler gas. Multi-element hollow cathode lamps compound the problem. The flame or furnace may emit nearby molecular bands or sample matrix emission lines.

It is generally regarded that an atomic absorption monochromator should have a *spectral bandpass* no wider than 0.2 nm in order to resolve the resonance line of most elements. At the same time the amount of radiant power passing through the entrance slit of the monochromator should be maximized. Maximizing the light throughput optimizes the signal-to-noise ratio at the detector and the associated readout circuitry. This leads to optimal detection limits and precision in the analyte measurement. Thus, the mechanical slit width should be as wide as possible while yielding the needed resolution.

As we noted in the last chapter, for a given focal length the mechanical slit width and the spectral bandpass are related by the dispersion of the grating:

$$\Delta\lambda = s_m d_r$$

where: $\Delta\lambda$ is the spectral bandpass (in nm); s_m is the mechanical slit width (in mm); and d_r is the reciprocal linear dispersion of the grating (in nm/mm). Thus, if we must hold λ at 0.2 nm or lower, but we want s_m to be as large as possible, we require a grating with as small a value of d_r as possible. Good quality instruments today employ gratings with values of d_r that are 1 nm/mm or less.

Several different monochromator optical designs have been incorporated in atomic absorption spectrophotometers, the Littrow and Czerny-Turner plans being the most common. Similarly, the "preslit" optics plan of a commercial instrument will depend upon whether it is a single- or double-beam design, and on which of several background correction systems have been incorporated.

4.2 Light Sources [9–10]

4.2.1 Line Sources

The feature most closely identified with the early development of atomic absorption remains one of its most useful and versatile accessories—the *hollow cathode lamp*. While some fundamental studies may require a demountable version of the hollow cathode lamp in which the cathode material is changed by the analyst, most practical work is performed with purchased, permanently sealed lamps. These consist of a cylindrical glass envelope, evacuated, then back-filled with low pressure [4–10 torr (0.005–0.013 atm. or 533–1333 Pa)] argon or neon. The cathode is a small cylinder either made of the analyte element or containing it in some form. The anode is a tungsten or nickel wire or pin. Both electrodes are sealed through the glass for external electrical connection. A glass or quartz window permits exit of the emitted resonance lines.

When a potential of several hundred volts is applied to the circuit positive inert gas ions are formed and impact the surface of the cathode, dislodging the analyte element atoms. These atoms diffuse toward a region of a sustained glow discharge, consisting of inert gas ions and excited inert gas atoms. Here they attain sufficient energy to emit their characteristic spectrum. Under appropriate low current conditions the discharge and emission occur only from within the cathode cylinder, which in modern high-intensity lamps may be only 2 mm in diameter. A mica shield, or in some designs a ring-shaped anode, aids in confining the discharge within the cylinder.

Hollow cathode lamps can be operated over a range of milliamperes. The intensity of the emitted resonance lines is directly related to current until line broadening and self-absorption begin to occur. Some lamps are supplied with information on a maximum current beyond which they cannot be operated without sustaining permanent damage. It is prudent to operate hollow cathode lamps at the lowest current that results in a stable output. This produces sharp resonance lines and extends the life of the lamp.

Hollow cathodes can be operated in a pulsed mode to achieve a higher effective intensity and improved signal-to-noise without significantly accelerating deterioration. In the Smith-Hieftje background correction mode hollow cathodes are pulsed so intensely that they produce bursts of self-absorbed, line-broadened continuum radiation from the cold atom cloud that forms in front of the cathode cylinder. This continuum is then used to measure and automatically correct for nonspecific absorbance interferences (see Section 6.2.3). So-called "source-shifted" Zeeman background correction, however, is not particularly applicable to hollow cathode lamps because glow discharges are destabilized in a strong magnetic field (see Section 6.2.2).

The multi-element hollow cathode lamp is commonly assembled with a hollow cathode formed from a sintered mixture of pure powdered metals.

These devices can be a time and cost savings for labs that need to run certain element combinations but lack an instrument with a lamp turret. Lower intensities and the possibility of line interferences limit their usefulness, however. Some laboratories never use them, preferring to work exclusively with single-element hollow cathode lamps.

The *electrodeless discharge lamp*, or EDL, is a more intense and sharper line source than the hollow cathode, but, for the most part, practical EDL's are available for only the volatile elements. A small sealed quartz tube or bulb contains a small quantity of the analyte metal or one of its volatile salts. The tube contains a few torr of inert gas that initiates and sustains the plasma discharge. The optimum size of the quartz tube is analyte-specific, based on the vapor pressure of the analyte form employed and the nature of the excitation cavity. The object is to achieve a uniform discharge. Microwave energy was utilized in most early designs, but today radio frequency energy (typically 27.12 MHz) is more often employed. In some designs the lamp and the RF source are separate interchangeable devices; thus one RF source can serve for an array of lamps.

EDLs' enhanced intensity is of special advantage for those elements with prominent resonance lines in the ultraviolet because it tends to compensate for the light losses common to that spectral region. There is also a distinct signal-to-noise ratio advantage in utilizing a more intense source in any part of the spectrum. This leads to improved detection limits and better precision.

Electrodeless discharge lamps and hollow cathode lamps are complementary since EDLs are strong, stable, long-lived sources for the volatile elements, while hollow cathodes for these analytes can be unstable and short-lived. EDLs are available for antimony, arsenic, bismuth, cadmium, germanium, lead, mercury, phosphorus, selenium, tellurium, thallium, tin, titanium, and zinc, as well as for certain alkali metals.

4.2.2 Continuum Sources

It would require a monochromator of truly exceptional resolution combined with an extremely intense continuum source to serve for analyte measurement by atomic absorption. Such an instrumental arrangement would save the cost and inconvenience of changing line source lamps. Several different high intensity/high stability continuum sources are available, including high pressure xenon lamps, halogen lamps, hydrogen arc, and deuterium arc lamps. However, even when very bright sources, such as a 200 watt xenon arc lamp have been utilized, the transmitted intensity over the half-width of the average resonance line (0.002–0.005 nm) is minute compared to that from a line source. Such instruments have been built utilizing an echelle grating to achieve the required resolution, but these designs are not yet commercially viable, and, indeed, may never prove practicable.

However, continuum sources, such as the deuterium arc, can and do serve a useful purpose in atomic absorption as part of one important means of background correction (see Section 6.2.1).

4.3 Atomization Sources

4.3.1 Flame [7,9,10]

Today most work in flame atomic absorption is conducted with the premix, laminar flow, single-slot burner. *Air/acetylene* is still the most commonly used oxidant/fuel mixture. A typical burner for air/acetylene has a single 10-cm slot, although a three-slot (Boling) burner is available for samples with a high solids loading. This flame can achieve a maximum temperature of 2250°C. It has been found suitable for the determination of about 30 elements, including the alkali metals, which partially ionize. Background absorbance is very low—down to about 230 nm. Below this region absorbance climbs steeply. Air/acetylene has only a simple emission spectrum when pure water is atomized, showing peaks for C_2, CH, and OH.

The air/acetylene flame is effective for all elements that do not form refractory oxides. These include the alkali metals, the alkaline earths, bismuth, cadmium, cobalt, copper, indium, iron, lead, manganese, mercury, nickel, selenium, silver, tellurium, thallium, and zinc. While the mixture is commonly held to stoichiometric or weakly oxidizing proportions, certain noble metals (e.g., gold, iridium, palladium, platinum, and rhodium) show the highest sensitivity in a very fuel-lean (strongly oxidizing) air/acetylene flame. In contrast, the alkaline earth elements show the highest sensitivity in a slightly fuel-rich (slightly reducing) flame. Such a flame may be 150°C cooler than a stoichiometric flame.

The chemistry occurring in the flame is too complex to define a temperature limit for a given analyte based strictly on the dissociation energies of metal-oxide species. But it is clear that elements like chromium, molybdenum, and tin are not completely dissociated to free atoms by the air/acetylene flame. Other elements, like aluminum, titanium, zirconium, niobium, and tantalum show extremely low free atom populations in the air/acetylene flame.

The accepted solution to this problem came in 1965 when J.B. Willis introduced the *nitrous oxide/acetylene* flame for atomic absorption [17]. This mixture can achieve a maximum temperature (2955°C) about 700°C hotter than that attainable by air/acetylene. This temperature is sufficient to dissociate many metal-oxide bonds that remain intact in an air/acetylene flame. Just as importantly, the nitrous oxide/acetylene flame has a low burning velocity, which is conducive to the production of an optimal atom cloud. Hotter mixtures, such as oxygen/acetylene (3060°C), or even oxygen/cyanogen (4500°C), have been investigated. In most cases the propagation rate was too high for safe operation, or ionization effects came into play, decreasing sensitivity.

Typically, nitrous oxide/acetylene mixtures are slightly fuel-rich and exhibit three rather distinct reaction zones: a blue-white region extending about 3 mm above the burner slot, then a 0.5–5.0 cm red reduction zone (sometimes referred to as the "red feather"), and finally an upper pale blue region. The red feather is the area of optimal atom cloud formation. The nitrous oxide:acetylene ratio is stoichiometric at 3:1; however, the optimal ratio for a given analyte, as well as the optimal observation height in the red feather are method-specific.

Nitrous oxide/acetylene mixtures are recommended for aluminum, beryllium, chromium, gallium, germanium, molybdenum, silicon, tin, titanium, and vanadium. They are also recommended for the rare earth elements—both those with good AA sensitivity (dysprosium, erbium, europium, thullium, and ytterbium) and those with poor AA sensitivity (gadolinium, holmium, lanthanum, lutetium, neodymium, praseodymium, samarium, scandium, terbium, and yttrium).* Only cerium among this group shows no measurable response. Other elements with poor response due to refractory oxide formation are also typically determined with nitrous oxide/acetylene, since cooler gas mixtures are likely to yield no response whatever. These analytes include boron, hafnium, niobium, osmium, phosphorus, rhenium, tantalum, tungsten, uranium, and zirconium. In some cases the response in the nitrous oxide/acetylene flame can be enhanced by the use of a *releasing agent*. For example, Bond found an 8- to 10-fold increase in sensitivity for zirconium by preparing test solutions that were 0.1 molar in ammonium fluoride. Similar enhancements were observed for hafnium, tantalum, titanium, and uranium [18]. Urbain and Chambosse saw an improvement in zirconium response in the presence of hydrochloric acid, ammonium chloride, and aluminum chloride [19].

In the early days of atomic absorption flashback explosions were commonplace, albeit always unnerving. The flame front was drawn back through the burner slot and detonated the gases in the mixing chamber. This could occur whenever the burner slot became clogged with soot or salts, or whenever the gas flow rates or ratios were incorrect. In early instrument designs there was some danger from flying metal, but soon "blowout plugs" were incorporated and the principal damage from a flashback was to the analyst's peace of mind. The greatest concern was in the use of nitrous oxide/acetylene mixtures.

The nitrous oxide/acetylene burner is typically constructed of titanium. It incorporates a 5-cm slot, about 0.5 mm wide. Some designs are flat, while

---

* Here we continue our practice of including scandium and yttrium with the lanthanides under the category "rare earths."

others have grooves to stabilize the flame with entrained air and to prevent heat dissipation from the slot, minimizing carbon buildup. The flame is ignited as air/acetylene, then the acetylene flow is increased and the air flow is switched to nitrous oxide. The process is reversed for flame shutdown. During the measurement the slot must be kept free of residue by cautious wiping with a thin metal wand from time to time. In the case of excessive buildup the flame must be shut down, the burner allowed to cool and the slot cleaned. In modern instruments the lighting and shutdown is under programmed control and flashbacks are extremely unlikely [20].

When pure water is atomized the nitrous oxide/acetylene flame shows low absorbance down to 215 nm, then absorbance rises precipitously. It shows the same molecular emission features as the air/acetylene flame (from C_2, CH, and OH) and, in addition, shows three strong band regions from CN and one from NH. If such bands correspond to the wavelength of the atomic absorption resonance line of the analyte they may contribute emission noise that finds its way through the pulsed lamp modulation, which is supposed to correct for atomizer variations. The result is a decrease in the precision of the measurement. Another potential problem is due to the fact that many elements are extensively ionized by the nitrous oxide/acetylene flame.

The experience of the preceding three decades suggests that most analytical problems in flame atomic absorption can be solved by the use of either air/acetylene or nitrous oxide acetylene.

There *are* a few other mixtures that prove useful in selected niches, however. Tin, for example, shows the highest sensitivity in an air/hydrogen flame (even though there are less chemical interferences in a nitrous oxide/acetylene flame). An argon/hydrogen flame utilizes argon to nebulize the sample and the argon-diluted hydrogen burns in the air which surrounds the burner. Such a flame is hot at the outside edges and cool in the middle. It is subject to many interferences, both chemical and spectral, but it remains transparent down to very low wavelengths. This low absorbance at low wavelengths results in good sensitivity and low detection limits for elements with UV resonance lines, such as arsenic (193.7 nm) and selenium (196.0 nm). Specialized burner designs are used for such work.

The nebulizer/spray chamber/burner head assembly must be designed to achieve optimum performance in flame atomic absorption. Today, commercial designs have stabilized on just a few variant configurations. Unlike the newer field of plasma emission spectrometry where a range of designs still prevail, atomic absorption has settled upon pneumatic nebulizers. These are Venturi effect designs that produce a range of droplet sizes from <5 µm to >25 µm. The nebulization is adjusted by means of a screw that adjusts the position of the capillary end with respect to gas flow and the spray orifice. The materials of construction should be noble metals and fluorocarbon polymers for instruments used for metals analysis, since acid solutions will readily degrade a nebulizer constructed of stainless steel.

The spray chamber serves to both mix the fuel and oxidant gases and to prevent sample droplets larger than about 10 μm from reaching the burner. It is generally a metal cylinder coated on the inside with an easily wetted polymer. There is a blowout plug, a drain for the run-off of large droplets, and direct connection to the burner assembly. A flow spoiler is a fan- or propeller-shaped device in the chamber that serves to mix the gases and intercept large droplets. It is also wettable polymer coated. There may also be an impact bead directly in the nebulizer spray path to fragment droplets into smaller sizes. The drain tube must lead immediately to a liquid trap (sometimes just a loop of clear polyvinylchloride tubing filled with liquid) to provide back-pressure for the spray chamber; otherwise, the flame may be drawn back into the spray chamber and a flashback explosion will occur.

Burners are connected by a pressure fitting to the top of the spray chamber. They are also connected by means of loose wires to allow the release of pressure without their becoming a flying object in the event of a flashback. Most modern instruments are equipped with an automatic spark ignitor to light the flame.

4.3.2 Furnace [9,10]

Normal graphite was exclusively employed for furnace tubes in the early days of the commercialization of the Massman design. But normal graphite is porous and becomes more so after repeated heating cycles. In fact, during normal operation some analytes diffuse more readily through the walls of normal graphite tubes than they diffuse out open tube ends. L'vov's early work clearly showed that nonporous pyrolytic graphite, either as a material of manufacture or as a surface coating, produced a superior tube, free of wall-diffusion. Pyrolytic graphite shows less tendency to form stable carbides with analytes like titanium, vanadium, tungsten, and others. And these tubes are not readily eroded by oxidizing acids. In some cases switching to pyrolytic graphite tubes increases the sensitivity for an analyte by 100%.

Porous graphite tubes can be "pyro-coated" right in the furnace by heating above 2000°C with a methane-inert gas mixture passing through the tube. The coating can be continuously healed and renewed by substituting a 10% methane/balance argon ("P10") mixture for the inert gas during normal graphite furnace operation. However, "homemade" pyro-coated tubes typically lack uniformity and are unlikely to result in performance as satisfactory as that from the purchased item (on the other hand, the cost savings may be significant). Certain elements—typically, those that do not form refractory carbides—experience no benefit from the use of pyro-coated tubes. Phosphorus actually yields superior data using the normal porous graphite tubes.

The use of the stabilized temperature platform (or L'vov platform) technique is facilitated by pyrolytic graphite since this material transmits heat poorly at right angles to the layer planes. In the atomize cycle the platform containing the dried and charred sample thus lags significantly during the rapid heating of the tube wall and gas. When equilibrium is reached, the hot gas and wall radiation rapidly heat the platform, atomizing the sample into a thermally stable environment.

Temperature control is extremely important to the successful use of all electrothermal atomic absorption instruments. Closed loop feedback control usually involves thermistors or similar devices in the low temperature (<800°C) range and optical radiation measurement above 800°C. Experience has shown that for most analytes the atomize jump (from "char" to "atomize") should be less than 1000°C to ensure that the analyte is atomized into a region that is in thermal equilibrium, and that the rate of increase should be as fast as possible. Modern instruments are capable of heating rates in excess of 2000°C/sec. Fast heating rates generally achieve the most analyte response at the lowest atomize temperature. A maximum temperature cleanout cycle always follows. This may be followed by a blank to check on how effectively memory effects have been eliminated.

The inert gas flow may be programmed to start and stop several times during the programmed heating cycle. It is always stopped at or just before "atomize" to allow the analyte atom cloud to expand into a region that is in thermal equilibrium without cooling or dilution by flowing argon.

Automated sample introduction has become an essential feature of all modern graphite furnace instruments. A robot sampler can introduce microliter sample solutions much more accurately and reproducibly than a human analyst with a micropipet. Modern samplers can be programmed to automatically add matrix modifiers or standard spikes, and to calibrate and validate the analysis. Unattended overnight operation is, thus, a feasible, if somewhat risky, option.

4.3.3 Other Atomization Sources [9]

Hydride generation can be accomplished by the addition of a basic sodium borohydride solution by means of a capillary tube to the bottom of a conical-bottom vessel containing the sample solution. The neutralization reaction and the shape of the vessel produce turbulent agitation, ensuring adequate mixing and a complete reaction. The evolved hydride gases can be measured directly in a heated quartz cell. Alternatively, they can be cold-trapped with liquid nitrogen, or even collected in a balloon for subsequent analysis.

A combination of amalgamation with gold and the cold vapor approach are the basis of the most common highly sensitive method for mercury determination. The gold amalgam is heated rapidly to 500–700°C and the evolved mercury is measured at 253.6 nm. Such an approach has been claimed to detect as little as 0.1 ng of the analyte.

The glow discharge atomic absorption approach requires a flat conductive sample. It opens the atomic absorption technique to the determination of refractory oxide- and carbide- forming elements, like boron and zirconium. Like arc/spark emission and X-ray spectrometry it is subject to surface inhomogeneity effects.

4.4 Detectors and Readout [9,10]

Atomic absorption instruments still utilize photomultipliers even though solid state detectors are beginning to make inroads in emission spectrometry. While one can visualize a certain utility for charge-coupled devices and photodiode arrays, the high sensitivity of the photomultiplier, particularly in the UV, remains unmatched. Instrument manufacturers must select a photomultiplier design with a very broad wavelength range to cover the available resonance lines from cesium (852.1 nm) to arsenic (193.7 nm).

Significant parameters of a photomultiplier are the *dark current* (current flow with no impinging light), the *dark noise* (instability in the dark current), the *shot noise* (the instability in output with impinging light), and the *quantum efficiency* (the number of impinging photons required to release a single electron from the photocathode). The noise parameters are all increased with increasing dynode voltage. Quantum efficiency is an inherent characteristic of the photocathode design. Output from a "quiet" and efficient photomultiplier can be amplified so that radiant fluxes as low as 10^{-11} lumens can be measured.

In all modern instruments the line source is modulated and the modulation is coupled to demodulation in the readout amplifier. The reason, it will be recalled, is to isolate the atomic absorption signal from emission from the atomization source. Modulation may be electronic (an alternating supply current to the line source), or mechanical (a chopper between the line source and the atomizer). The frequency of this modulation must be high enough to avoid low frequency noise components of the atomizer emission. Also, the analyst must be aware that the photomultiplier "sees" all the light. Thus, if the total atomizer emission is very high (say, from a very luminous flame) the photomultiplier may be saturated, yielding an erroneous difference signal.

Today, all instruments are designed to provide a digital output signal. This requires an internal clock to define a highly reproducible integration interval. In measuring a rate, such as in flame work, integration frees the analyst from having to estimate the best average stable reading from a rapidly changing analog display. In discrete measurements, such as in graphite furnace work, the integration sum can compensate for atomization kinetic variation. Thus, within certain limits peak shape effects are eliminated. Now most graphite furnace work is accomplished using integrated peak area rather than peak height.

Noise suppression filtering must be applied *after* the signal has been converted to absorbance by a logarithmic amplifier or software because the signal is then proportional to analyte concentration or analyte mass. Calibration curve nonlinearity may be compensated for by electronic linearization circuitry based on some model of the curvature. Today, calibration curves and other forms of calibration are handled by internal software.

5. CALIBRATION AND MEASUREMENT [9]

There are four basic ways to calibrate for an atomic absorption measurement. Each has a fairly well defined area of applicability. The *calibration curve* technique requires the least effort on the part of the analyst but is usually best when applied to simple alloy solutions or to alloy solutions in which the analyte has been chemically isolated to some degree. For routine use, and when the highest accuracy is not required, the nonlinear portion of the calibration curve can be utilized to extend the working range of the measurement. The more calibration points that are used to define the curvature the greater the resultant confidence when operating in that region. With modern instrumentation, except in special circumstances, calibration curves are rarely drawn manually because calibration is accomplished under software control. Sometimes matrix effects may be compensated for by calibrating with a suite of reference materials that match the sample matrix and represent a graded series of analyte concentrations. Such a suite of standards is rarely available, however. Typically, calibration is with pure solutions of the analyte, sometimes "matrix-matched" with measured amounts of only the major matrix element.

For highly precise work the *bracketting* technique is applied. Here, a blank and two standards are prepared. One standard is about 5% higher than the expected concentration of analyte; the other is about 5% lower. The solutions are diluted to the same final volume. If all solutions yield absorbances in the linear calibration range (commonly, between 0.2 and 0.5 absorbance units) highly precise results can be obtained by solving a simple equation:

$$C_t = \{[(C_h - C_l)(A_t - A_l)]/(A_h - A_l)\} + C_l$$

Where, C_t is the concentration of analyte in the test solution; C_h is the concentration of analyte in the high standard; C_l is the concentration of analyte in the low standard; A_t is the blank-corrected absorbance of the test solution; A_h is the blank-corrected absorbance of the high standard; and A_l is the blank-corrected absorbance of the low standard.

The bracketting technique is also applicable to the nonlinear portion of the calibration curve where it serves to define a short line segment tangent to the true curve. This is less accurate but commonly used when working at higher concentrations.

The *method of additions* makes possible the accurate determination of analyte in highly complex alloys by compensating for matrix effects. In order to utilize it properly, however, it is necessary that the analyst know the exact extent of the linear portion of the calibration relationship to ensure that all spiked solutions remain within it. Also, a true solution blank should be prepared. Ideally, this would be an exact match to the sample matrix but totally analyte-free. In practice one must usually settle for something less.

In the method of additions the analyst spikes several identical aliquots of the sample solution with incremental portions of standard analyte solution and measures these and the blank. The analyst or the software then plots the blank-corrected absorbances versus concentration of added analyte, usually performing a least-squares fit. Solving the resultant linear equation for the condition of absorbance-equals-zero yields a negative number whose absolute value is the concentration of analyte found in the unknown sample solution.

A simplified version of the method of additions is sometimes known as the *spiking technique*. The analyst either weighs two identical test portions of the unknown test material, or volumetrically (or gravimetrically) splits one dissolved portion into two equal volumes (or weights of solution). One portion is spiked with analyte (keeping within the linear calibration range), the other is not. Both are measured along with a blank. The unknown concentration is then calculated from:

$$C_t = A_t S/(A_h - A_t)$$

where, C_t is the concentration of analyte in the test solution; A_t is the blank-corrected absorbance of the test solution; A_h is the blank-corrected absorbance of the spiked test solution; and S is the concentration of the spike.

The spiking technique captures the essential advantage of the method of additions (albeit minus some of its rigor). It has proven to be a "workhorse" approach to the successful determination of many analytes in complex or unfamiliar alloys, compensating for matrix effects.

It should be noted that the presence of a matrix effect can be confirmed by comparing a method of additions line prepared with pure reagents and water with another method of additions line prepared in the same manner with the test solution. If the two curves are not exactly parallel, a matrix effect exists.

6. DEALING WITH INTERFERENCES [9,10]

6.1 Types of Interference

No one any longer claims that atomic absorption spectrophotometry is interference free. However, the analyst should be aware that AA is one of the better analytical methodologies with respect to interference effects. Nevertheless, we can clearly distinguish three broad categories of interference: spectral, chemical, and physical.

6.1.1 Spectral

Most of these effects have been touched upon in the preceding pages. They fall into two groups: atomic line interferences and nonspecific absorbance influences. *Atomic line interferences* include nonanalyte emission lines from the lamp (either from the inert filler gas or from other elements in a multielement hollow cathode). Also, the resonance line can be partially absorbed by overlapping lines from matrix elements in the atom cloud (although the effect is rare). Narrowing the slit width or switching to a different resonance line will solve these problems.

Nonspecific absorbance encompasses molecular bands, smoke, and particle light scattering in the atom cloud. The influence of these spectral interference effects is much more pervasive than that due to atomic lines. These effects are particularly troublesome in electrothermal measurements. They are handled by the use of a background correction system—either continuum-source, Zeeman, or Smith-Hieftje—as described in Section 6.2 below.

6.1.2 Chemical

Chemical interferences can be very generally defined as derived from any chemical reaction that influences the measurement in an uncontrolled manner. One type involves the formation of compounds or complexes that diminish the population of free ground state analyte atoms in the atom cloud. Phosphate influences the measurement of calcium in the air/acetylene flame in this way. Here, raising the temperature by changing to a nitrous oxide/acetylene flame is all that is necessary. An alternative approach is the addition of a *releasing agent*, which forms a more stable compound or complex with the interferent than the analyte does. An example, referred to above, is the use of ammonium fluoride to improve the sensitivity for certain refractory oxide-forming elements in the flame.

In the graphite furnace compound formation can be a problem by two distinct routes: the analyte can form compounds, such as refractory carbides, diminishing analyte atomization, and the analyte can form volatile compounds, such as chlorides, which are lost during the char cycle prior to atomization. In the former case there is usually no recourse, except, perhaps, to substitute a less reactive tube material (e.g., pyrolytic graphite or a tantalum liner), or possibly to coat the inside of the furnace with a refractory oxide. In the latter case a number of *matrix modifier* additives have been used successfully. A great deal of lore has developed on this subject.

Much early work in this area was reported by R.D. Ediger [21]. Chloride can be removed by adding ammonium nitrate to the sample aliquot. The ammonium chloride that forms sublimes at a sufficiently low temperature (340°C) to be easily removed during the char cycle [21,22]. Selenium, arsenic, bismuth, and tellurium analytes can be stabilized with respect to higher char temperatures if nickel is added to the samples. Other important matrix modifiers are phosphate (first studied by Ediger for cadmium determination), magnesium nitrate, and palladium.

Ionization effects may be considered chemical interferences as well. Here, analyte atoms (commonly in a flame) are excited to become free ions, diminishing the number of ground state atoms available for light absorbance. Decreasing the flame temperature is one answer but it may lead to other interference problems, such as incomplete dissociation of analyte molecular species. The addition of an *ionization suppressor* (the same amount to sample, blank, and standard solutions) is usually the best approach. Ionization suppressors are solutions of easily ionized elements. Lanthanum, strontium, cesium, potassium, and sodium solutions are most commonly employed.

6.1.3 Physical

These are interferences due to changes in the transport of analyte to the observation zone of the atom cloud. Differences in the viscosity of sample, blank, and standard solutions are one cause. With aqueous-based work this relates to total dissolved solids differences, although with varying concentrations of organic solvents similar effects occur. Increased viscosity results in decreased aspiration rate and reduced droplet size. Viscosity differences also result in differences in the manner in which aliquots are deposited (and subsequently atomized) in the graphite furnace.

The dissociation of water which takes place in the flame is an endothermic reaction, while most solvents produce an exothermic reaction. Differences in water:solvent ratios can, therefore, affect analyte atom population at the observation height in the flame.

The most practical solution to such problems is to attempt to prepare blanks and standard solutions that match the test solution as closely as possible. Another good choice is the method of additions. If there is sufficient sensitivity, a third course is to dilute the solutions until the effect becomes negligible. If the analyst has access to a dual channel instrument, an internal standard can be added in the same concentration to all solutions and used to eliminate the transport effect. An internal standard will probably be less effective in correcting influences on the optimal observation height.

6.2 Background Correction

The elimination of nonspecific absorbance spectral interferences is a sufficiently large subject to warrant a separate discussion. While these techniques have been applied to a flame they are essential in electrothermal methodologies, where, in most instances, the atomic absorption signal is immersed in smoke, scatter, and molecular bands.

6.2.1 Continuum-source Correction [9]

In this oldest commercial scheme an intense continuum source is optically balanced in transmitted intensity with the line source, both beams traversing the same path through the furnace and the monochromator. Alignment of the two sources is critical since they both must trace precisely the same path. Balance can be achieved by inserting a graded filter in the continuum beam.

The continuum source may be a tungsten iodide lamp for the visible wavelengths, or a hydrogen or deuterium arc lamp for the ultraviolet region. In operation a chopper directs first the line source, then the continuum source through the atom cloud. Smoke, particles, and molecules absorb both beams to an equal degree while only the line source beam is also absorbed by the ground state analyte atoms. The difference between the two signals represents the corrected absorbance of the analyte.

Actually, the portion of the continuum beam passed by the exit slit *includes* the analytical resonance wavelength, and so *some* analyte specific absorbance occurs. But it represents no more than 1.5% of the continuum intensity even in the case where 100% of the line source is absorbed. This is assuming a resonance line with a 0.003 nm line width and an exit slit with a 0.2 nm bandpass. In the concentration range of the graphite furnace technique this error is negligible.

However, the continuum-source solution to the background problem does have some drawbacks. Noise is high in this system, and background levels above about 0.7 absorbance units may not be completely removed. Matrix components in the atom cloud may absorb some (nonresonance) wavelengths in the bandpass of the continuum. Most significantly, the continuum-source correction system will generally over correct in the presence of structured background. This refers to the rotational and vibrational fine structure in the molecular bands. The continuum-source system, in effect, subtracts a mean absorbance background across the (comparatively) broad bandpass, while the resonance line may reside between two closely spaced fine structure lines. The presence of overcompensation due to fine structure may be confirmed by measuring with and without continuum background correction at several different slit widths. If the degree of correction increases with increasing bandpass, structured background is likely to be present.

6.2.2 Zeeman Correction [9,23–27]

In 1897 Pieter Zeeman published his experimental discovery of the long-predicted influence of a magnetic field on a light beam [*Philosophical Magazine* 5, 226 (1897)]. In the so-called *normal Zeeman effect* the energy levels of atoms are split into three components—one remains at the original level, one shifts higher, and one shifts lower. This results in three spectral lines—one at the original resonance wavelength, designated π, one of higher energy (lower wavelength), called $+\sigma$, and one of lower energy (higher wavelength), called $-\sigma$. The σ components are each shifted the same amount, a distance which is proportional to the magnetic field strength. At 8k Gauss (0.8 Tesla) they each shift about 0.01 nm. The π line becomes linearly polarized while the σ components are polarized perpendicular to the magnetic field. With respect to the original resonance line intensity (100%) the split lines show intensities in the ratio $+\sigma$:$-\sigma$:π, 25%:25%:50%.

The normal Zeeman effect is exhibited by barium, beryllium, cadmium, calcium, lead, mercury, magnesium, palladium, silicon, strontium, tin, vanadium, and zinc. Many other elements exhibit an *anomalous Zeeman effect* in which the resonance line is split into various arrays of lines. If the resultant number of lines is odd, the original resonance line position is retained by one of the components (examples are Cr, Fe, Mo, Ni, Se, Te, Ti), but if the resultant number is even the original resonance line position disappears (e.g., Ag, Al, As, Bi, Co, Cu, Mn, Sb). The fine structure due to isotopes also exhibits complex Zeeman splitting.

There are conceptually many distinct ways to utilize the Zeeman effect to correct for nonspecific absorbance in atomic absorption. One of the few that can flawlessly correct for structured background, however, is a design that places an alternating magnetic field perpendicular to the light path through the furnace and a fixed polarizer in the exiting beam. In this design the magnet surrounds the graphite furnace while the line source continually emits the normal resonance line. Such an arrangement is termed *analyte-shifted* because the magnetic field affects the ability of the analyte atoms to absorb the resonance line. When the alternating magnetic field is *off* the sum of the background and atomic absorption are measured from the unsplit wavelength absorbance. When the magnetic field is *on* the resonance wavelength splits into π and σ components and the fixed linear polarizer holds back the π component so that only the background is measured (by the sum of the σ component absorbances). The difference between the two signals is the background-corrected atomic absorption. Since only the ability of the analyte atoms to absorb the line source is effected by the magnetic field the correction is at the true line source wavelength. The σ components in such a system perfectly isolate the analyte resonance from nearby fine structure.

If a longitudinal alternating magnetic field were to be used instead of a transverse field, the polarizer becomes unnecessary since the π component disappears in this case. Design of a longitudinal system, however, requires a magnetic pole at each end of the furnace with an appropriate hole in each to allow passage of the light beam. Such a system would have the advantage that no light throughput would be lost at the polarizer.

Either of these two analyte-shifted designs are highly effective. However, problems can occasionally arise from nearby atomic lines or molecular bands that exhibit Zeeman splitting. There are analyte-shifted, transverse field systems that utilize a constant magnetic field and a rotating polarizer that passes first the π, then the σ components. Such systems require features that allow equalization of the oppositely polarized light components.

Source-shifted systems have the magnetic field at the line source. With a constant field (either transverse or longitudinal) a rotating polarizer is required to pass first the background plus analyte absorbance then the background wavelengths. As mentioned above, one serious problem is that it is difficult to light and maintain stable emission from hollow cathodes in the presence of a strong magnetic field. Only electrodeless discharge lamps run at low power are suitable. Also, since the σ components shift about 0.01 nm from the true resonance line, fine structure within this range is imperfectly corrected. Source-shifted systems, however, appear to be the only background correction approach practical for flame work. Thus, it is fortunate that background is not a serious problem with most flame work.

Calibration curves for Zeeman-corrected absorbances are difficult to treat theoretically, being derived from the difference between two absorption coefficients for analyte atomic absorption and two for the background absorption. *Rollover* is the practical consequence of this situation. This is the phenomenon of calibration curves "topping out" at high concentrations then dipping downward toward the concentration axis with increasing analyte concentration. With electrothermal techniques "high concentration" is still a minute amount, of course. Thus, the analyst must be cautious in data interpretation since one absorbance reading can sometimes represent two distinct concentrations—one high and one low.

6.2.3 Smith-Hieftje Correction

This approach utilizes conventional hollow cathode line sources pulsed with high current to produce momentary bursts of line-broadened, self-absorbed continuum radiation. With the application of high current a cold atom cloud forms immediately in front of the lamp cathode, absorbing some of the intense line radiation from the back of the cathode.

The continuum burst is absorbed by the background while the low current sharp line source resonance is absorbed by both the analyte atoms and by the background. The difference between the two signals is the atomic absorption.

The advantages of this approach are its simplicity and its freedom from the alignment requirements of continuum-source correction. Unfortunately, not every element broadens effectively when its hollow cathode is pulsed at high current. This results in a loss of sensitivity. Like the continuum-corrected systems this approach will over-correct in the presence of structured background.

7. SENSITIVITY, DETECTION LIMITS, AND QUANTIFICATION LIMITS [9,28]

Before beginning a summary of applications it is appropriate to pause briefly and attempt to clarify some points related to the reporting of test results and instrument and method performance data. Much of this short discussion relates to all instrumental methods, but this chapter is an apt setting since some of the literature on atomic absorption has notably muddied these waters.

Sensitivity is a term best defined simply as the slope of the linear portion of the calibration curve. For a nonlinear calibration curve the sensitivity is some higher order function of analyte concentration (or mass). By analytical chemistry convention the instrumental measurement (dependent variable) is the ordinate and so sensitivity should always be of the form, $(\Delta A/\Delta C)$ or $(\Delta A/\Delta m)$, where A is the absorbance, C is the concentration, and m is the mass. To compare instrument response conveniently we should like to invert these terms: $(\Delta C/\Delta A)$ and $(\Delta m/\Delta A)$, but we must call them *reciprocal sensitivities*.

Sometimes (particularly with graphite furnace methods) the term *characteristic mass* is used. This is the mass of analyte that produces a net absorbance of 0.0044 absorbance units (1% absorbance). The corresponding term, *characteristic concentration* has occasionally been applied to flame methods. When peak area is used as the calibration basis the mass or concentration that produces 0.0044 absorbance unit · seconds is the characteristic mass or concentration. These terms have been used to predict method and instrument performance, but they cannot be used alone to measure the detection limit or the quantification limit, which are both functions of instrument noise.

The *detection limit* is defined as the reciprocal sensitivity times the absolute value of the standard deviation of the blank reading times some integer factor (usually 3).

That is, $C_{dl} = (\Delta C/\Delta A)\sigma k$ and $m_{dl} = (\Delta m/\Delta A)\sigma k$

Thus, the detection limit incorporates both reciprocal sensitivity and instrument noise as reflected in the variability in the blank measurement. The detection limit is the analyte concentration or mass that can be distinguished from noise with a given statistical confidence.

The *quantification limit* is a quite different concept, and, perhaps, the most useful of all these terms. The quantification limit defines the lowest concentration or mass that can be reported as a number. It is always higher than the detection limit, and may be significantly higher. It can be best illustrated graphically, as shown in Figure 5.2, which is a hypothetical atomic absorption calibration curve and the associated confidence limits at some statistical uncertainty. As we examine lower and lower absorbance readings we discover that A_{dl} is the first whose uncertainty range includes a concentration of zero. Thus, at A_{dl} (and all lower absorbances) it is impossible to distinguish an actual analyte concentration from zero. Extending a line from A_{dl} to the highest analyte concentration it could represent in the uncertainty envelope defines the *detection limit*. The detection limit concentration then defines the lower limit of a concentration interval spanning the envelope, the mean of which is the *quantification limit*.

In the region below C_{dl} it is appropriate to report "none detected." In the region between C_{ql} and C_{dl} it is appropriate to report "less than" the value of C_{ql}. Above C_{ql} numerical values are reported.

Before closing this discussion it should be noted that the rapid, practical determination of both detection limits and quantification limits has been the subject of much debate. An approach that has been recommended by the Instrument Manufacturers Association is the "mean to sigma" measurement. A series of measurements of the same (low, but easily detectable) level of analyte are made to establish a standard deviation. The detection limit is then defined by the concentration of analyte that produces an average result equal to twice (or three times) the established standard deviation. For no particularly sound statistical reason, ten times the standard deviation of the blank measurement times the reciprocal sensitivity has been widely used as an estimate of the quantification limit.

8. APPLICATIONS TO THE TRACE ANALYSIS OF METALS

Unlike the previous chapter where it was useful to categorize applications by the analyte, here, as with most of the remainder of the subjects in this book, it is more appropriate to classify methods by the alloy matrix. Within an alloy category subheadings will treat different atomic absorption methodologies for various analytes.

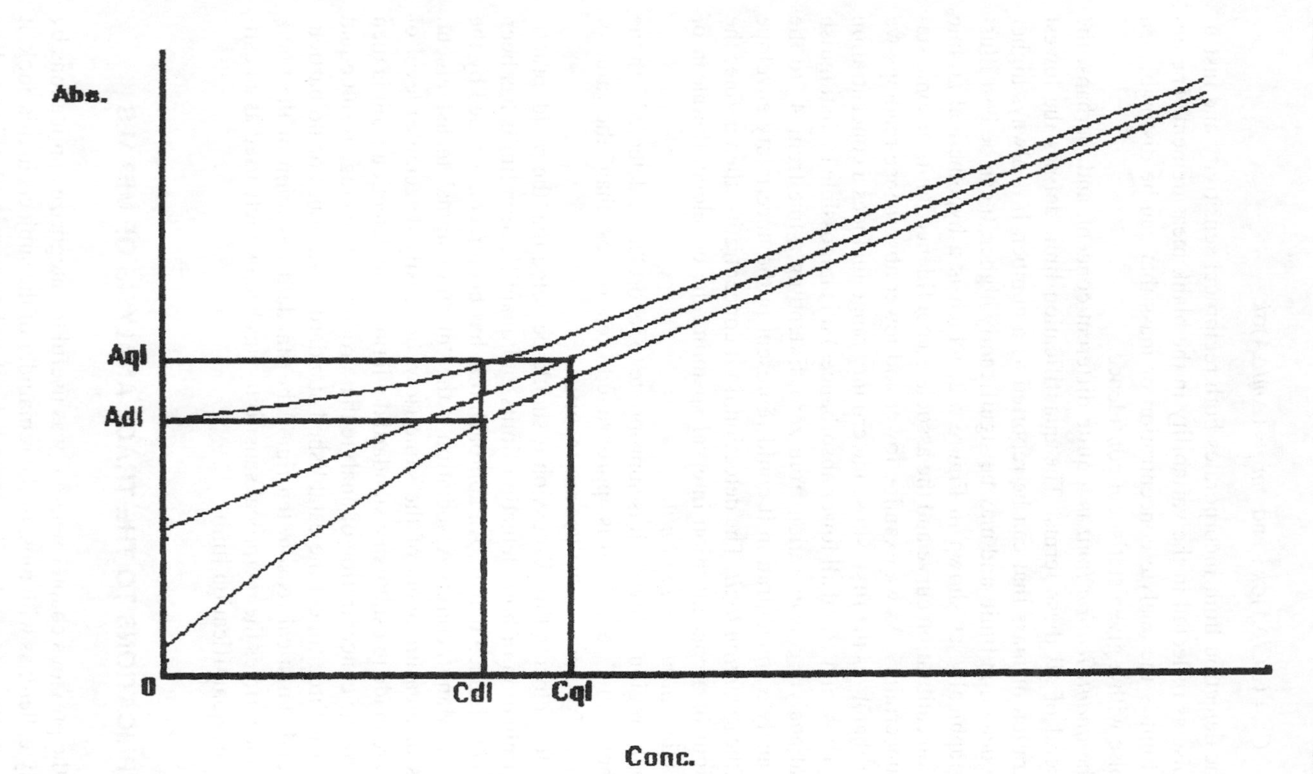

Figure 5.2 Confidence envelope for a calibration curve (after a drawing by Welz–Ref. 9). Used with permission.

8.1 Ferrous Alloys

8.1.1 Flame Methods

The analysis of iron and steel was among the earliest applications of flame atomic absorption spectrophotometry. A review by Scholes summarized the early published work in this area [29]. Early work with air/acetylene flames sometimes made use of certified metal alloy reference materials for calibration as a means of side-stepping matrix interferences [30]. With the advent of hotter nitrous oxide/acetylene flames, while some workers began to explore chemical additives, others found it simpler to make use of chemically analyzed metal standards. Such matrix-matched comparative methods were published for silicon [31], titanium [32], and tellurium (using a deuterium arc background corrector) [33], and for a variety of elements (some at high concentrations) in stainless and tool steels [34,35].

As this suggests early work soon revealed that atomic absorption was hardly interference-free. Molybdenum determination in ferrous alloys with the air/acetylene flame was particularly troublesome. David [36] suggested a luminous, fuel-rich, reducing flame and the addition of aluminum ion. Mostyn and Cunningham [37] found ammonium chloride to be a more effective interference suppressor. Kirkbright, Smith, and West were able to achieve similar results using the nitrous oxide/acetylene flame without an additive [38]. Ramakrishna, West, and Robinson later showed that signal enhancement and depression are, in fact, occurring in the presence of various matrix elements with nitrous oxide/acetylene. They recommended a return to the addition of aluminum [39]. Barnes saw a similar depression of chromium absorption in an air/acetylene flame caused by the iron matrix from low alloy steels. While both aluminum and ammonium chloride were effective additives, the latter was recommended [40].

It became clear that iron enhanced the atomic absorption of both molybdenum and chromium in the nitrous oxide/acetylene flame, while it exhibited a severe depressing effect in the air/acetylene flame. Oxidation states were also important. Thomerson and Price, therefore, recommended a nitrous oxide/acetylene approach that involved preparing calibration standards with matrix-matched iron levels, and fuming all standards and test samples in perchloric acid [41]. In the presence of high tungsten levels a perchloric acid/sulfuric acid/phosphoric acid mixture was utilized. Such an approach can be successfully applied to tool steels, for example [42].

Some investigators dealt with the effect of iron by chemical separation approaches. Chromium was determined successfully with air/acetylene after extraction [as Cr (VI)] with MIBK from 1–3 M hydrochloric acid in the presence of fluoride [to prevent the extraction of iron (III)] [43]. Another approach involved the extraction of chromium (VI) with tribenzylamine/chloroform from 0.7 M sulfuric acid, then back-extraction into an aqueous phase [44]. Molybdenum was isolated by extraction of its α-benzoinoxime complex with MIBK from 1% (v/v) perchloric acid, then measured in the organic phase with nitrous oxide/acetylene [45]. Another approach combined α-benzoinoxime and potassium ethyl xanthate extractions to further isolate molybdenum from tungsten [46].

Direct nitrous oxide/acetylene methods have been developed for titanium in steels, permanent magnet alloys, and cast irons [47], and for vanadium in steels [48]. Less useful today are methods for low sensitivity elements that utilize large samples (Nb) [49], or preconcentration by solvent extraction (W) [50], since these elements can now be determined more efficiently by other techniques.

Nitrous oxide/acetylene made practical the atomic absorption measurement of silicon and aluminum in steels. Price and Roos recommend dissolution with dilute hydrochloric acid and hydrogen peroxide to retain silicon in solution for AA determination [51].

Aluminum in steel is often somewhat more of a sensitivity problem so solvent extraction and other preconcentration schemes are common. Headridge and Sowerbutts utilized an acid digestion bomb with either hydrochloric acid or a mixture of hydrochloric and hydrofluoric acids to completely dissolve refractory aluminum inclusions in a variety of steels. The iron matrix was removed by extraction with MIBK from 6M hydrochloric acid solution. Aluminum was then extracted from pH 4.7 solution with acetylacetone. The organic phase was then measured directly. Good results were obtained down to 0.001% aluminum [52]. Donaldson used open beaker dissolution followed by filtration, ignition, and fusion of the insoluble residue with sodium bisulfate. Iron and other elements were removed by mercury cathode electrolysis, then aluminum was isolated by acetylacetone/chloroform extraction. Aluminum was returned to the aqueous phase and then measured [53].

Another important group of analytes in ferrous alloys are those with their most sensitive resonance lines in the ultraviolet. This list includes antimony (217.6 nm), arsenic (193.7 nm), bismuth (223.1 nm), cadmium (228.8 nm), lead (217.0 and 283.3 nm), selenium (196.0 nm), tellurium (214.3 nm), tin (224.6 nm) and zinc (213.9 nm). This group, with increasingly rare exceptions, is considered objectionable in steel, and so the analyst is generally looking for very low levels. Background interference is an important problem for flame atomic absorption in this wavelength region; and hollow cathode sources for many of these analytes are weak in intensity. Thus, sensitivity was poor until the advent of deuterium arc background correction and electrodeless discharge lamps. Tin is more sensitive in nitrous oxide/acetylene; all the others in the group are more sensitive in air/acetylene. Of this group cadmium and zinc are the most sensitive analytes with the lowest detection limits. About 20 times less sensitive are the subgroup comprised of antimony, bismuth, lead, selenium, and tellurium. Arsenic and tin are about three times less sensitive than these [54].

Various schemes were devised to increase sensitivities and detection limits. Menis and Rains utilized solvent extraction, an electrodeless discharge lamp, and an argon/entrained air/hydrogen flame to determine trace arsenic in cast iron. Iron was removed by extraction with 2-thenoyltrifluoroacetone (TTA)/carbon tetrachloride, first at pH 1.5, then at pH 2.5. Then arsenic was extracted with diethylammonium diethyldithiocarbamate (DDDC)/chloroform from dilute hydrochloric acid/potassium iodide/sodium meta-bisulfite solution. Finally, the arsenic was back-extracted into the aqueous phase by displacement with copper ion [55].

In Chapter 3 we referred to the trioctylphosphine oxide (TOPO)/iodide/MIBK extraction developed by Burke to collect and concentrate tin [56], as well as antimony, lead, bismuth, silver [57], and thallium [58], and extended to tellurium by Bedrossian [59]. The analytes are extracted from a dilute hydrochloric acid/potassium iodide/ascorbic acid solution and measured by flame atomic absorption directly in the organic phase. Dagnall, West, and Young determined trace lead in steels by extracting the iron into isoamyl acetate from hydrochloric acid solution, then evaporating the aqueous layer to dryness. The residue was dissolved in 5% (v/v) hydrochloric acid, potassium iodide was added and the H_2PbI_4 that forms was extracted with MIBK and aspirated directly into an air/acetylene flame. The 283.3 nm resonance line was utilized. A variety of mild steels, high speed tool steels, and stainless steels were successfully analyzed [60]. Vassilaros isolated lead and bismuth from low alloy steels, stainless steels, and high temperature superalloys by coprecipitation with manganese dioxide. The membrane-filtered precipitate was dissolved in 2:1:1 hydrochloric acid:hydrogen peroxide:water and measured by flame atomic absorption [61].

Headridge and Smith extracted antimony as the iodide complex from hydrochloric acid-dissolved mild steel solutions using MIBK [62]. Headridge and Richardson used a similar approach for bismuth [63]. Marcek, Kinson, and Belcher determined trace levels of tellurium by first precipitating the analyte with stannous chloride, then extracting it as the diethyldithiocarbamate complex with amyl acetate [64]. Published flame work for selenium has been largely limited to higher, alloying levels, such as those in free-machining stainless steels [65].

Flame atomic absorption has long been important for the determination of calcium and magnesium in ferrous alloys. Headridge and Richardson determined calcium at ppm levels in stainless steels by removing most of the matrix by means of a acetylacetone/pyridine extraction [66]. Most methods for calcium and magnesium recommend the addition of strontium as an ionization suppressor. A nitrous oxide/acetylene flame is common for calcium, somewhat less so for magnesium [67].

Peterson and Kerber published a valuable review of the atomic absorption analysis of ferrous alloys that largely covers flame work [67]. A number of composite analytical schemes for the analysis of an array of analytes in ferrous materials by flame AA have been published [68–71].

In summary, flame techniques continue to be a valuable asset for the analysis of iron, steel, and ferroalloys. At intermediate to low concentration levels they can serve as a useful complement to ICP-OES, especially where that technique suffers from intractable line interferences. For the great majority of ferrous alloy applications preconcentration techniques are not necessary. But there are many instances where the use of such approaches can extend detection limits and quantification limits near those obtainable by electrothermal AA.

8.1.2 Furnace Methods

Electrothermal methods for ferrous alloys are generally not as straightforward as those for some other alloy systems, such as nickel-base alloys. However, much has been done and continues to be done in this area.

In 1974 Frech published work on the determination of trace levels of antimony in steel using a Massmann-type furnace. Chromium was added to the *aqua regia*-dissolved samples to level out its influence. The results compared favorably with those obtained by the TOPO-MIBK extraction-flame AA approach. The author attributed the lack of interference from chloride to be due to the fact that his furnace was sealed, allowing complete reduction to elemental antimony during the argon-purged, oxygen-free char cycle [72]. Frech published similar work for bismuth in 1975, except in this case the additive was ammonium hydroxide, placed on top of the dried sample to allow a higher char temperature. Values for low alloy steels, stainless steels, and ferromolybdenum again compared well with the TOPO/MIBK flame AA method [73]. The same year Ratcliffe, Byford, and Osman reported on a procedure for low levels of arsenic, antimony, and tin. They concluded that only the exclusive use of dilute nitric acid can be tolerated to dissolve the test portions [74]. Barnett and McLaughlin reached the same conclusion for the determination of antimony, arsenic, bismuth, cadmium, lead, and tin in ferrous and nonferrous alloys. They utilized diluted ultra-purity nitric acid to dissolve the test portions. These authors reported extensive results with iron-base certified reference materials for arsenic, antimony and lead [75]. Dulski and Bixler reported lead, bismuth, and silver results on certified low alloy steel standards using nitric acid dissolution [76].

Frech and others continued to employ hydrochloric acid for lead determinations, observing that drying and charring conditions could be established that showed no lead losses [77]. It was discovered that residual water that remains in pores in the graphite after the drying step forms hydrogen during the char cycle. The hydrogen reacts with iron (II) chloride to form hydrogen chloride, which is evolved away, preventing losses of volatile $PbCl_2$ (and PbCl), which presumably do not form in this case [78,79].

Chloride continued to be a concern, however. Del Monte Tamba and Luperi determined tin in low alloys, stainless steels, and high speed tool steels by removing added hydrochloric acid by evaporation to low volume, then adding additional nitric acid before diluting the sample solutions to final volume [80].

Acid soluble and acid insoluble aluminum in irons and low alloy steels were determined by Shaw and Ottaway. They dissolved a 0.5 g test portion in 40% (v/v) nitric acid, filtered the solution through a hardened, low porosity filter paper (Whatman No.542) and measured the diluted filtrate directly. The residue was ignited, fused with 0.3 g 2:1 sodium carbonate: sodium tetraborate, leached with dilute nitric acid, and measured [81]. Persson, Frech, and Cedergren extended the approach to stainless steels by dissolving the test portion (0.5 g) in 5:2.5:3 hydrochloric acid:nitric acid:water. The solution was filtered on a 0.22 μm membrane filter. Ammonium sulfate was added to the filtrate solution as a matrix modifier to evolve NH_4Cl during the char. The membrane was dissolved in hot nitric acid, evaporated to dryness, fused with 0.5 g of 2:1 sodium carbonate:sodium tetraborate, and leached with dilute nitric acid [82].

Phosphorus was determined in steel by Whiteside and Price. They dissolved a 1.0 g test portion in 2.5 mL of hydrochloric acid, 2.0 mL of nitric acid, and 1.5 mL of perchloric acid. The cooled sample was diluted with 5 mL of water and filtered through Whatman No. 541 filter paper (or equivalent) into a 20-mL volumetric flask. A 20 μL aliquot was measured at 213.6 nm using a phosphorus EDL line source [83]. A more sensitive and practical approach is possible today using a massive pyrolytic graphite L'vov platform in a stabilized temperature porous graphite furnace. Lanthanum is recommended as a matrix modifier. It is claimed that such an approach using Zeeman background correction can accurately determine 0.002% phosphorus in steel [84].

A considerable body of literature has been published on the direct analysis of solid steel samples by graphite furnace atomic absorption. Andrews and Headridge determined ultratrace levels of bismuth in high-purity irons, cast irons, and steels using 2–12 mg test portions in a specially constructed furnace [85]. Andrews, Aziz-Alrahman, and Headridge obtained ultratrace data for lead from reference materials with certified lead levels, including low alloys, cast irons, and stainless steels. Results compared favorably with certificate values. The solids approach proved to be so sensitive that for concentrations between 20 and 100 ppm the less sensitive 261.4 nm lead line was used. For the range 1–20 ppm the normal lead resonance at 283.3 nm was employed [86]. Similar work for the determination of silver by Aziz-Alrahman and Headridge was conducted on 17 steel and iron materials. In five of these, comparison with certified or literature values was possible and the correspondence was good [87].

Lundberg and Frech conducted a trace analyte homogeneity study in order to ascertain if common steels were homogeneous at the milligram test portion sizes required by a solids GFAA approach. They studied lead in stainless steel, mild steel, and ferromolybdenum (as well as nickel-base alloy), and antimony in mild steel by dissolving 2, 5, 10, 25, 50, and 100 mg test portions and measuring them by graphite furnace AA. It was concluded that sample inhomogeneity could account for only a small part of the variability seen in the solids GFAA approach [88].

The observed variability was attributed to the lack of isothermal conditions as the sample is being atomized. These authors inserted solid samples into a furnace maintained at 1800°C to allow atomization under thermal equilibrium conditions. Lead was determined in this way [89], followed by silver, bismuth, cadmium, and zinc (using a modified autosampler to insert the solid into the furnace) [90]. Peak area integration was utilized in all cases. They found that analyte recovery from solids is related to the solubility of the analyte in the hot or molten matrix. Also important are the relative volatilities of analyte and matrix, which should be very different. Thus, only 80% of the lead and bismuth is recovered from solid stainless steel. This means that aqueous-based standards would result in erroneously high results [91]. Backman and Karlsson determined lead, bismuth, zinc, silver, and antimony by a solids approach. They concluded that sample shape (chips, flakes, spheres) affects the results and should be controlled [92].

L'vov and Novichikhin published work on submerged spark slurry formation from solid samples. The slurry is injected into a graphite furnace for analysis. The approach has been applied to steels, as well as other metals [93].

8.1.3 Hydride Methods and Other Methods

Fleming and Ide documented the applicability of hydride generation methods to ferrous alloys with a thorough study of antimony, arsenic, bismuth, lead, selenium, tellurium, and tin in low alloy, stainless, and high speed steels. For arsenic, bismuth, lead, and tin they dissolved a 1.0 g test portion in 20 mL of 1:2:1 nitric acid:perchloric acid:water. For antimony, selenium, and tellurium they used 20 mL of 3:1:1 hydrochloric acid:nitric acid:water. The solution in both cases was diluted to 100 mL and a 2-mL aliquot was transferred to the generator vessel. Depending on the analyte, 10M hydrochloric acid, 9M sulfuric acid, or 250 g/L tartaric acid were added. In addition antimony required potassium iodide and lead required potassium dichromate. Tin also required potassium iodide if the matrix contained high copper. A sodium borohydride pellet was used to generate the gaseous covalent hydride, which was carried by part of the argon stream into an argon/entrained air/hydrogen flame. Results from about 40 reference materials were reported with generally very good correspondence to certified values [94].

Fleming and Ide discovered a number of interferences, some of which were mitigated by the presence of matrix iron. Thus, for example, for arsenic determination nickel, cobalt, and copper interfere without iron, but in the presence of iron only copper interferes by decreasing the arsenic signal. They recommended the method of additions to handle such situations [94]. Welz later published work using an electrically heated quartz tube atomizer that described an interference-free solution matrix for each of the same analytes in low alloy steels [95]. Bettinelli, Spezia, and Bizzarri compared results from GFAA and hydride generation for bismuth [96], and later for arsenic, bismuth, antimony, and selenium [97]. They concluded that while the obtained results were equivalent, the hydride methods had a clear advantage. They were faster, less subject to intractable interferences, and showed one order of magnitude lower detection limits.

Haug and Yiping described studies in which germanium hydride (germane) was evolved from low alloy steels and trapped on zirconium-coated graphite tubes heated at about 600°C. The sequestered and concentrated analyte was then atomized and measured [98]. Despite good published data for a number of important analytes, hydride techniques are not widely used in the ferrous metals industry. Part of this may stem from a bad "early press" the technique received when it was first introduced, when it was believed to be imprecise and cumbersome. Today with well-researched methods and flow injection automation it is neither.

A different technique that also deserves mention here is discrete pulse nebulization flame atomic absorption. This technique is designed to allow measurement of highly concentrated solutions as a discrete aliquot with the flame atomic absorption response measured as a peak. Batistoni, Fuertes, and Smichowski used 5 g test portions dissolved and diluted to 50 mL to measure low levels of aluminum, antimony, arsenic, cobalt, lead, tin, and vanadium in steels [99].

8.2 Nickel- and Cobalt-base Alloys

8.2.1 Flame Methods

One of the earliest papers in this area reported on the determination of magnesium in electronic nickel and nickel alloys. This early work by Andrew and Nichols already suggested the excellent sensitivity for this analyte [100]. Dyck extended this air/acetylene approach to chromium and manganese as well

[101]. In the late sixties several papers appeared in which attempts were made to extend flame atomic absorption to determine major alloying constituents. Mostyn and Cunningham analyzed nickel alloys for titanium using nitrous oxide/acetylene [102]; Arendt measured niobium in Inconel™ 718 alloy [103]; and Welcher and Kriege reported on optimized parameters for ten major analytes in nickel-base high temperature alloys [104] and nine major analytes in cobalt-base alloys [105].

Separation and preconcentration work for tramp levels of antimony, bismuth, cadmium, lead, silver, tin, zinc, and other elements began to appear in the literature around 1970. In Chapter 3 we referred to the manganese dioxide coprecipitation method for traces of antimony, bismuth, lead, and tin in nickel, developed by Burke [106] and elaborated by Blakeley, Manson, and Zatka [107]. Burke recommended the air/acetylene flame for antimony, bismuth, and lead, and the argon/entrained air/hydrogen flame for tin.

In 1971 Newland and Mostyn published a solvent extraction preconcentration scheme for trace and ultratrace levels of bismuth in nickel alloys. A 5 g test portion was dissolved in 35 mL of hydrochloric acid and 6 mL of nitric acid (added in small portions to control the reaction). The solution was transferred to a separatory funnel with 30 mL of hydrochloric acid. Iron and molybdenum were removed by repeated extraction with *n*-butyl acetate. The remaining aqueous phase was reduced in volume, dilute hydrochloric acid, ascorbic acid, and potassium iodide were added, and the solution was repeatedly extracted with MIBK. The combined extracts were reduced in volume and measured; or alternatively, transferred to an aqueous medium and measured. In both cases the final volume was 10 mL [107a]. We have already mentioned the TOPO/iodide/MIBK solvent extraction scheme developed by Burke [(56–58)] and elaborated by Bedrossian [59]. That preconcentration system was well tested for nickel alloys.

Kirk, Perry, and Arritt developed an ion exchange scheme to separate bismuth, cadmium, lead, silver, and zinc from high nickel alloys. Each analyte was measured by flame atomic absorption [108]. White, Harper Jr., Friedman, and Banas co-plated bismuth (from an 8.0 g test portion) with copper on a copper-plated platinum electrode. The copper layers and accompanying trace bismuth were stripped off with dilute nitric acid (8.0 mL total volume) and measured by flame AA without further dilution [109].

8.2.2 Furnace Methods

It was in the field of nickel alloy trace analysis that some of the pioneering work in electrothermal methods was confirmed. In 1974 Welcher, Kriege, and Marks published comparative data for in-house produced high temperature alloys with a range of tramp levels of bismuth, lead, selenium, tellurium, and thallium. The graphite furnace method involved dissolution of a 1 g test portion in 30 mL of 1:1:1 nitric acid:hydrofluoric acid:water and reduction in volume to about 5 mL. The salts were dissolved by warming with 20 mL of water, then the solution was diluted to volume in a 50-mL plastic volumetric flask. The results compared favorably with those generated by preconcentration optical emission and X-ray fluorescence methods [110]. Dulski and Bixler compared the direct technique of Welcher, Kriege, and Marks [110] with several other direct and separation schemes. For bismuth, lead, selenium, silver, tellurium, and thallium in nickel-based alloys it was found that the approach of Welcher, Kriege, and Marks gave accurate results with the least expenditure of time [76].

Work with high purity nickel did not follow until several years later. Studies by Zatka [111] and by Forrester, Lehecka, Johnston, and Ott [112] extended the approach to three additional elements: antimony, arsenic, and cadmium. The absence of major alloying constituents allowed dissolution in diluted nitric acid and simplified the determinations of ultratrace analyte levels. A high purity nickel matrix is almost an ideal sample for graphite furnace work. NIST SRM's 671, 672, and 673 are nickel oxide materials that were certified by a program that included such work.

Kujirai, Kobayashi, Ide, and Sudo isolated traces of tellurium from nickel- and cobalt-base high temperature alloys by coprecipitation of the elemental form with arsenic using hypophosphorous acid. The filtered residue was dissolved in nitric acid and measured [113]. Bosnak, Carnrick, and Slavin studied the determination of bismuth in nickel-base superalloys using a L'vov platform approach. They confirmed that the presence of the nickel matrix is a distinct advantage in that it permits higher char temperatures. They recommended a *wider* (2.0 nm) bandpass when the unresolved doublet at 223.1 nm (222.83 and 223.06 nm) is used with Zeeman background correction. The wider slit results in increased light throughput with an improved signal-to-noise ratio [114]. Trostle, Beals, Kuczenski, and Shaver discovered that the 223.1 nm bismuth resonance line with the 2.0 nm slit yielded erroneous data when iron-bearing alloys were analyzed. The problem involved the fact that baseline shifts were producing negative values for the blank-corrected peak areas. The (twofold) less sensitive 306.8 nm resonance line is free of this effect but requires that peak height be employed to achieve the required sensitivity [115].

Reichardt published work that shows that the 223.1 nm bismuth resonance and 0.2 nm slit with a palladium/magnesium nitrate matrix modifier addition also yields good data with peak height measurement. Reichardt also published L'vov platform measurement conditions for lead, selenium, tellurium, and thallium. Peak area was measured for all analytes except bismuth. Data were reported on ten replicate determinations of NIST SRM's 897 and 898 (nickel-base high temperature alloys). Mean values corresponded well with certified values [116].

Simpson analyzed nickel-base superalloys for tramp levels of gallium. Test portions (0.5 g) were dissolved in 20 mL of 4:1:8 nitric acid:hydrofluoric acid:water and diluted to 50 mL in a plastic volumetric flask. A 5 μL aliquot was diluted with 25 μL of water and 10 μL of 1% (w/v) nickel in 1M nitric acid on the L'vov platform, then dried, charred, and atomized. The excess of nitric acid, the dilution, and the nickel addition are intended to minimize the effect of gallium fluoride loss and improve peak shape. The absorbance was measured at 294.42 nm since iron interference was discovered on the 287.4 nm line [117].

Mile, Rowlands, and Jones found that ammonium dihydrogen phosphate added as a matrix modifier to sample and calibration solutions will allow the accurate determination of trace lead in a wide range of nickel alloys (and steels) from a single calibration curve. All previous work either was matrix-matched to some extent, or utilized some form of the method of additions [118].

The direct determination of tramp elements from solid nickel alloy samples was investigated by Marks, Welcher, and Spellman. These authors determined bismuth, lead, selenium, tellurium, thallium, and tin in a series of in-house prepared alloys that had been inoculated with various levels of these six tramp impurities. A 1 mg sample of chips was weighed on a balance with ± 0.05 mg accuracy and transferred into the graphite tube through the enlarged sample port using a cut-off micropipette tip as a funnel. Dry and char cycles were eliminated as unnecessary. Optimum atomization temperatures ranged from 2200°C (for lead) to 2600°C (for selenium). Due to the absence of salts the signal-to-background ratio was greater than that normally seen with dissolved alloys, especially when lower atomization temperature measurements were compared. The authors found that commercial instrumentation varied in how effectively this approach could be applied [119]. Headridge and Thompson determined bismuth in solid nickel-base alloy samples using a constant temperature induction furnace [120]. Studies such as these have not been extensively pursued probably because the limited array of analytes (those that evolve at low temperatures) limit the interest of commercial instrument designers.

8.2.3 Hydride Methods

Drinkwater published a procedure for bismuth in nickel-base high temperature alloys using EDTA to prevent interference effects from the sample matrix. The atomization chamber was a silica tube mounted over an air/acetylene burner. The bismuth hydride was generated with sodium borohydride [121]. Fleming and Ide [94] had shown that nickel was a serious interferent in the generation of selenium hydride. Welz and Melcher solved this problem for nickel-based materials by incorporating a preliminary precipitation of the nickel with sodium hydroxide. A dissolved sample aliquot was reacted with a slight excess of 6% (w/v) sodium hydroxide solution and diluted to volume when cooled to room temperature. The solution was mixed and filtered through a hardened filter paper. An aliquot of the filtrate was acidified with hydrochloric acid and transferred to the hydride generation flask. The calibration curve must be prepared from pure nickel solutions spiked with selenium and carried through the entire procedure [122].

8.3 Copper Alloys

8.3.1 Flame Methods

One of the very early industrial applications of flame atomic absorption was the determination of zinc in copper-base alloys. Studies by Gidley and Jones applied custom-built equipment to copper alloys (as well as to aluminum- and zirconium-base materials) [123,124]. Sprague, Manning, and Slavin were first to report on the determination of selenium and tellurium in copper [125]. The use of a high intensity line source lamp and an air/hydrogen flame first allowed tin to be determined at low levels. Capacho-Delgado and Manning were able to obtain 1% absorbance for 1 ppm of tin. They analyzed leaded brass and other brass (as well as lead alloys and zirconium) [126]. Sattur analyzed copper-base alloys for a number of analytes. Test portions were dissolved in hydrochloric acid plus a minimum of nitric acid, except when tin and antimony were present. In this case a mixture of fluoboric acid and nitric acid were used. Values for low levels of antimony, cadmium, iron, lead, manganese, nickel, and zinc were reported [127].

In Chapter 3 we mentioned the ion exchange separation of cadmium, tin, and zinc developed by McCracken, Vecchione, and Longo. These elements were measured with an air/acetylene flame [128]. Kallmann and Hobart combined a fire assay procedure with a flame atomic absorption measurement to determine gold, palladium, and silver in copper, electronic scrap, jeweler's sweeps, and other materials. The doré bead that is formed by the fire assay procedure is dissolved in acids, the solution is heated to dryness, and the cold salts are dissolved in 2% (w/v) sodium cyanide solution and measured [129]. Thormahlen and Frank dissolved niobium-bearing copper/nickel alloys and other copper alloys in mixtures of 3N sulfuric acid, hydrogen peroxide, and hydrochloric acid for the determination of silicon [130].

Reichel and Bleakley developed a lanthanum hydroxide coprecipitation procedure for anode and blister copper that allows trace levels of antimony, arsenic, bismuth, iron, lead, selenium, tellurium, and tin to be measured by flame atomic absorption. A 20 g test portion is weighed into an 800-mL beaker and dissolved in dilute nitric acid. NO_x gases are boiled off and the cooled solution is diluted to 300 mL; 20 mL of 5% (w/v) lanthanum nitrate solution is added. The lanthanum is precipitated with 190 mL of ammonium hydroxide, then filtered and washed. The residue is then dissolved with hydrochloric acid and hot dilute nitric acid and washed into a 200-mL volumetric flask. The cooled solution is diluted to the mark and mixed. For work of the best accuracy spikes of lanthanum and copper (which contaminates the precipitate) are added to the blank and calibration solutions [131].

Alumina was determined in dispersion-hardened copper by an evaporation/carbonate fusion procedure published by Celis, Helsen, Hermans, and Roos [132]. Arsenic was determined by Likaits, Farrell, and Mackie at levels above 0.05% in blister copper (and flue dust) by a direct measurement in a nitrous oxide/acetylene flame at 193.7 nm, using an electrodeless discharge lamp as the line source [133]. Donaldson employed a mixed lanthanum/iron ammonium hydroxide coprecipitation to preconcentrate antimony. The precipitate is dissolved in 25% (v/v) hydrochloric acid and antimony is measured in a final solution that is 15% (v/v) hydrochloric acid/0.1% (w/v) tartaric acid [134]. The same author isolated bismuth from copper alloys using only iron as a carrier for the ammonium hydroxide precipitation [135]. In the presence of large amounts of aluminum, antimony, lead, or tin an alternative procedure is recommended. Here, bismuth is extracted with chloroform as the diethyldithiocarbamate complex at pH 11.5–12.0 from a sodium hydroxide/citrate/tartrate/EDTA/cyanide medium. The analyte is returned to the aqueous phase for flame AA measurement at 223.1 nm [135].

Tsukahara and Tanaka determined silver impurity in copper by extracting the analyte from a dilute nitric acid/hydrobromic acid medium with tri-*n*-octylmethylammonium bromide/*n*-butyl acetate and aspirating the organic phase [136]. Persiani and Durkin determined gold, palladium, and silver in high copper electronics scrap by a direct air/acetylene measurement using standards that were matrix matched to the approximate copper level of the test solutions [137]. Phosphorus was indirectly determined in brasses and bronzes by extraction of the phosphoantimonyl molybdate complex with MIBK and measurement of the antimony line at 206.8 nm using an air/acetylene flame [138]. Kharbade and Agarwal utilized a trioctylphosphine oxide/ MIBK extraction for tin in copper art objects (see also Chapter 3) [139]. Kojima and Takayanagi extracted silver from high purity copper test solutions as the 1,2-bis(ethylthio)ethane - picrate ternary complex with chloroform. Discrete nebulization of the 50 μL chloroform extract gave a sensitive signal [140].

8.3.2 Furnace Methods

In 1974 Shaw and Ottaway published a direct graphite furnace procedure for lead in high purity copper and in various brass and bronze alloys. Samples were simply dissolved in dilute nitric acid and diluted. The method is applicable over the range 0.1–0.0001% lead [141]. Mullen determined trace levels of antimony, bismuth, selenium and tellurium in high purity copper by coprecipitation with iron using ammonium hydroxide. The precipitate was filtered on a 21-mm diameter filter paper disk (Whatman No.542) with suction and using special techniques to obtain an even deposit. The disks were dried, then a 2-mm area was punched out of the disk and transferred to a graphite furnace for measurement [142]. Everett determined traces of the platinum group metals (except osmium and ruthenium) in copper (and nickel) following their isolation by ion exchange [143].

Fritzsche, Wegscheider, Knapp, and Ortner showed that impregnation of graphite furnace tubes with tungsten, zirconium, tantalum, or molybdenum produced a surface suitable for the atomization of tin. Such tubes are not reactive with the acids and acid concentrations necessary to prevent tin hydrolysis. They applied the approach to copper (and aluminum) matrices [144].

Haynes analyzed high purity copper for antimony, arsenic, selenium, and tellurium using a direct nitric acid dissolution [145]. However, attempts to apply the procedure to ultratrace levels of selenium and tellurium were unsuccessful. Muir and Anderson then developed a separation scheme based on isolation of these two analytes in elemental form by reduction with hypophosphorous acid using an arsenic carrier [146]. Sentimenti and Mazzetto returned to the direct nitric acid dissolution approach, utilizing a L'vov platform and Zeeman background correction. Their results for NIST SRM's 395 and 398 ("Copper Benchmarks") agreed well for selenium, tellurium, bismuth, antimony, cadmium, arsenic, lead, tin, silver, manganese, nickel, iron, cobalt, and chromium. The latter four elements were atomized off the tube wall. Magnesium nitrate modifier was added for cadmium, cobalt, nickel, and tin, while 1:1 ammonium hydroxide:water was added for manganese [147]. Follow-up work by Milella, Sentimenti, and Mazzetto added zinc as an analyte and included results from NIST SRM 393 (High Purity Copper Benchmark) [148].

Baker and Headridge [149] analyzed 2–30 mg solid samples of copper and low-alloy copper for trace levels of bismuth, lead, and tellurium using an induction furnace design similar to that used by Headridge and Thompson for nickel-base alloys [120], and Andrews, Aziz-Alrahman, and Headridge for steels [85–87].

8.3.3 Hydride Methods

Bedard and Kerbyson determined bismuth at concentrations from 10 to 0.1 ppm by isolation from 10 g copper test portions using a double lanthanum hydroxide coprecipitation. The precipitate was dissolved in dilute hydrochloric acid and diluted to 100 mL. A 20-mL aliquot was transferred to the argon flushed generator vessel and adjusted to 4M in hydrochloric acid. A single sodium borohydride pellet was added and the gases were collected in a balloon reservoir for 20 sec. Then a stopcock was opened to sweep the bismuth hydride with argon gas into an argon/entrained air/hydrogen flame [150]. Castillo, Mir, Vela, and Martinez used a 0.5 g test portion for bismuth and electroplated the copper from dilute nitric acid/sulfuric acid solution. The electrolyte was reduced, then diluted to 20 mL; a 1-mL aliquot was transferred to the sealed generator vessel, then 3 mL of a 4% (w/v) sodium borohydride solution was added. The evolved gases were injected directly into an air/acetylene flame and the results were read as a peak height [151].

Aznarez, Alduan, Castillo Suarez, Bonilla Polo, and Lanaja del Busto determined antimony in copper using a 0.2 g test portion dissolved in 1 mL of 2M nitric acid. NO_x gases were removed by heating, then 1 mL of 2M potassium thiocyanate solution was added to precipitate copper. Stibine (SbH_2) was evolved from the entire sample by injecting 6 mL of 2% (w/v) sodium borohydride solution. An air/acetylene flame was employed for the measurement [152].

Wickstrøm, Lund, and Bye determined arsenic in NIST SRM 398 ("Unalloyed Copper V") by dissolving a 1-g test portion, diluting to 250 mL, and removing a 10-mL aliquot. Sodium hydroxide solution was added to precipitate copper, then sodium borohydride solution was added. The solution was diluted to exactly 50 mL, then filtered. The filtrate was transferred to a generator flask, hydrochloric acid was added, and the evolved arsine (AsH_2) was measured with an air/acetylene-heated 19-cm quartz tube atomizer [153].

8.4 Aluminum Alloys

8.4.1 Flame Methods

Bell reported on the determination of cadmium, chromium, copper, iron, lithium, magnesium, manganese, nickel, and zinc using hydrochloric acid/hydrogen peroxide dissolution [154]. Sattur reported reference material results for copper, iron, and magnesium [127]. Mansell, Emmel, and McLaughlin utilized three different calibration approaches in the determination of copper, magnesium, zinc, and iron in aluminum scrap [155]. Wilson showed that the use of a nitrous oxide/acetylene flame makes matrix matching and the method of additions superfluous, at least for the elements studied: chromium, copper, silver, and zinc. Magnesium determination, even with nitrous oxide/acetylene required the presence of aluminum in calibration standards and blanks. The use of hydrofluoric acid as a releasing agent for zirconium was tried with limited success [156]. Peterson matched the aluminum concentration in preparing standards for beryllium calibration, however, she found no effect from eight other common aluminum alloy matrix components [157].

Price and Whiteside determined copper, iron, lead, magnesium, manganese, nickel, silicon, tin, titanium, and zinc in silicon-aluminum alloys by a microscale dissolution with dilute nitric and hydrofluoric acids in an acid dissolution pressure bomb. Boric acid is added to complex fluoride ion and cesium chloride is added as an ionization suppressor [158].

8.4.2 Furnace Methods

Ashy, Headridge, and Sowerbutts [159] determined zinc in aluminum and aluminum/silicon alloys using 5–6 mg solid samples. They employed the induction-heated graphite well furnace described for work with nickel alloys, steel, and copper alloys [85–87,120,149]. Gries and Norval described two schemes for preparing primary and secondary solid standard reference materials for GFAA work. The example chosen involved pure aluminum that was doped with thallium by an ion-implantation method to produce a primary standard. Secondary standards were produced by spiking molten urea [160]. Langmyhr and Rasmussen determined gallium using solids in a high frequency induction heated graphite furnace. The results compared favorably with results obtained by a conventional solution-based graphite furnace technique [161].

8.5 Lead, Tin, and Zinc Alloys

8.5.1 Flame Methods

Farrar reported on the analysis of lead-base bearing alloys and type metals. A 0.1–0.5 g test portion was dissolved in 15 mL of hydrobromic acid and 0.5 mL of bromine. Heat was applied until the alloy was dissolved and most of the bromine was removed. Data for copper in type metal was presented [162]. Sattur reported flame atomic absorption data for several NIST SRM's, as well as for other samples. Analytes included antimony, bismuth, silver, and tin in lead-base alloys; bismuth and copper in tin-base bearing metal; and copper, magnesium, and manganese in zinc-base alloys [127].

Smith, Blasi, and Feldman employed an internal standard technique with a two-channel atomic absorption spectrophotometer to determine seven elements in zinc die casting alloys. The matrix zinc was selected as the internal standard for certain trace elements: magnesium, iron, and lead. Added chromium proved best as the internal standard for trace cadmium, as well as for the alloying components: aluminum and copper. All analytes were determined with air/acetylene except aluminum, which was determined with nitrous oxide/acetylene [163].

Tin-lead solders were analyzed for low and trace levels of aluminum, antimony, arsenic, bismuth, cadmium, copper, gold, iron, nickel, silver, and zinc, as described in a paper by Hwang and Sandonato. Test portions of 1 g were dissolved in 50 mL of 3:2:5 nitric acid:fluoboric acid:water in a room temperature ultrasonic bath. The samples were diluted to exactly 100 mL with the solvent mixture and measured directly. Nitrous oxide/acetylene was used for aluminum and arsenic; air/acetylene was used for all the other analytes [164].

Quarrell, Powell, and Cluley recommended a mixture of fluoboric acid and hydrogen peroxide for the dissolution of lead alloys used for sheathing electric cables. They used this approach to prepare samples for the flame atomic absorption determinations of tin and antimony using nitrous oxide/acetylene, and air/acetylene, respectively [165].

Llacer and Catala described a stripping method utilizing 6M hydrochloric acid to determine chromium in passivation films on tinplate [166]. Guerra published a method for copper and gold in the lead/tin-based solder sampled from the reservoirs of the high speed wave soldering machines used in the mass production of printed circuit boards. Test portions (0.25 g) were dissolved with 1.0 mL of water, 7.8 mL of hydrochloric acid, and 1.3 mL of nitric acid [167].

8.5.2 Furnace and Hydride Methods

Fox developed both furnace and hydride procedures for selenium and tellurium in lead alloys. For the GFAA method a 2 g test portion is dissolved in 10 mL of perchloric acid and 2 mL of nitric acid, then fumed until reduced to approximately 5 mL. The cooled sample is then boiled, first with 35 mL of water, then again with an addition of 35 mL of hydrochloric acid. The cooled sample is transferred to a 100-mL volumetric flask and diluted to the mark with 1:1 hydrochloric acid:water, and mixed. The lead chloride precipitate is allowed to settle and a 50-mL aliquot of the supernatant solution is transferred to a beaker.

Then 1 mL of arsenic carrier solution (0.25 g high purity arsenic trioxide + 2 g sodium hydroxide, dissolved in 10 mL of water with warming, then diluted to 200 mL) and 10 mL of hypophosphorous acid are added. The solution is stirred and boiled, then filtered while still warm onto a glass fiber filter with suction. The beaker and residue are washed with 1:1 hydrochloric acid:water, then with water. The glass fiber filter is placed in the beaker, treated with nitric acid, and heated to drive off NO_x gases. When cool the filter is removed and rinsed into the solution. The solution is transferred to a 50-mL volumetric flask, diluted to the mark, mixed, and used to measure selenium and tellurium by GFAA.

For the hydride method a 10 g test portion is dissolved in 15 mL of perchloric acid and 5 mL of nitric acid, then fumed, boiled with water and hydrochloric acid, and cooled, as above. The solution is then filtered (Whatman No. 540) into a 100-mL volumetric flask. The beaker and residue are rinsed with 1:1 hydrochloric acid:water. The flask is diluted to the mark with 1:1 hydrochloric acid:water and mixed. A 10-mL aliquot is transferred to a 200-

mL volumetric flask, diluted to the mark with 1:1 hydrochloric acid:water, and mixed. A 20-mL aliquot of this solution is transferred to the hydride generator vessel for reaction and measurement. Or, if an automatic system is used the solution from the final dilution is automatically pumped into the generator system. Calibration standards and blanks must be made up from high purity lead to compensate for the signal depression by the lead matrix [168].

8.6 Refractory Metals and Alloys

8.6.1 Flame Methods

Kirkbright, Peters, and West determined copper in niobium and tantalum metal by extraction of the copper 8-hydroxyquinoline complex from fluoride solution at pH 4.5 using ethyl acetate. They measured the organic phase directly [169]. The same authors determined molybdenum in niobium and tantalum by extraction of the molybdenum (IV) 8-hydroxyquinolate from fluoride/EDTA solution at pH 4.5 using butanol, again measuring the solvent phase [170]. Mantel, Aladjem, and Nothmann dissolved niobium metal electrolytically using a current density of 100 *mA*/cm^2 in a 2:1 ethylene glycol:ethanol solution saturated with ammonium chloride. The sample anode was weighed before and after electrolysis to determine the weight of sample in solution. The electrolyte was diluted with 1:1 ethanol and directly measured for copper and iron concentration by flame AA [171].

Biechler and Long described the determination of low levels of rhenium in the presence of major amounts of molybdenum. The solution was adjusted to pH 12 with sodium hydroxide solution, then extracted with Aliquat 336™/chloroform. The organic phase was evaporated on a steam bath and the residue dissolved in MIBK, which was then directly measured in a nitrous oxide/acetylene flame [172].

Myers determined aluminum, vanadium, and iron in titanium alloys and iron in zirconium alloys. Aluminum and vanadium determination required a 2 g test portion of the titanium alloy; 75–100 mL of water was added, followed by hydrofluoric acid in 1-mL increments until the metal dissolved. Then nitric acid was added dropwise until the titanium was oxidized (colorless). Then 50 mL of 4% (w/v) boric acid solution was added and the solution was transferred to a 500-mL volumetric flask, diluted to the mark, and mixed. This solution was aliquotted and diluted further, if necessary, for higher concentrations of aluminum and vanadium. Iron determination required a 1 g test portion of either titanium or zirconium alloy that was treated in a similar manner, except that a 100-mL volumetric flask was used. If iron concentration exceeded 0.2% a further aliquot/dilution was required [173].

Schlewitz and Shields used a similar dissolution approach for zirconium alloys. They studied the effect of sample matrix, acid concentration, and flame conditions in the determination of chromium, copper, iron, nickel, and tin. They found that the zirconium matrix suppresses the analyte absorbances in the order: Fe>Cr>Ni>Sn>Cu, and that the suppression is strongly influenced by flame conditions [174]. The same authors published a method for low levels of tantalum in zirconium and zirconium alloys. A 5 g test portion was dissolved by adding 30 mL of water, then 10 mL of hydrofluoric acid in small increments. When the reaction ceased 5 drops of 30% hydrogen peroxide was added. The sample was evaporated to dryness, then redissolved with 0.6M hydrochloric acid/1.4M hydrofluoric acid. The solution was transferred to a 125-mL separatory funnel and extracted with MIBK. The organic phase was measured directly. As little as 10 ppm of tantalum could be determined [175].

Tungsten-base heavy alloys (90W/7.5Ni/2.5Cu) were analyzed for iron, copper, and nickel with a procedure published by Pabalkar, Naik, and Sanjana. Test portions were dissolved in nitric acid/hydrofluoric acid. Ammonium citrate and ammonium hydroxide were added, resulting in a pH 5 solution. The authors concluded that of the three analytes only iron was found to be (slightly) affected by the tungsten matrix and (strongly) by the reagent concentrations [176]. Donaldson separated cobalt and zinc from molybdenum and tungsten metals by chloroform extraction of the thiocyanate/diantipyrylmethane ternary complexes from pH 3.5 citric acid/thiourea/fluoride solution. The two analytes were then back-extracted with dilute ammonium chloride/ammonium hydroxide solution. The sample solution was then acidified to 1% (v/v) hydrochloric acid and measured by flame AA [177].

8.6.2 Furnace Methods

Chromium metal, widely used as a ferrous alloy charge material, is commonly listed with ferroalloys; however, it certainly qualifies as a refractory metal. Hamner, Lechak, and Greenberg determined antimony, arsenic, bismuth, selenium, silver, and tellurium in chromium metal by GFAA using a sodium peroxide/sodium carbonate fusion. A 1 g finely divided test portion was weighed into a 50-mL zirconium crucible containing 6 g of sodium peroxide and 2 g of sodium carbonate. It was mixed thoroughly, then covered with 1 g of sodium peroxide. It was cautiously fused, then cooled and leached with water in a stainless steel beaker. The solution was transferred to a glass beaker and cautiously acidified with 20 mL of nitric acid, then boiled, cooled, and transferred to a 200-mL volumetric flask, diluted to volume and mixed. A 20-mL aliquot was transferred to a 25-mL volumetric flask, a standard nickel nitrate solution was added, and the flask was diluted to the mark and mixed.

The aliquot volume injected into the graphite furnace had to be limited to 10 μL to keep the nonspecific absorbance within limits that could be handled by the deuterium background corrector [178].

Docekal and Krivan developed a solid sample GFAA approach for high purity molybdenum metal powder (and molybdenum silicide powder). The maximum sample weight of molybdenum powder that could be processed was 80 mg, which was weighed directly into a cup or platform on a microbalance in a class 100 laminar-flow clean bench. Copper, potassium, magnesium, manganese, sodium, and zinc were determined. Detection limits ranged from 0.06 ppb for zinc to 0.5 ppb for sodium. Results were compared with those from other ultratrace methodologies, such as neutron activation analysis, glow discharge mass spectrometry, isotope dilution mass spectrometry, and others, as well as with GFAA using a sample dissolution/solution-based approach [179].

Frise, Krivan, and Schuierer utilized a modified Grün SMI automated solid sampling atomic absorption spectrophotometer to determine ultratrace levels of copper, iron, manganese, sodium, and zinc in high-purity tantalum powders. Using the maximum practical sample weight of 40 mg, detection limits ranged from 0.1 ppb for zinc to 27 ppb for iron. Results, for the most part, compared favorably with those from a sample dissolution GFAA approach [180].

8.7 Precious Metals

8.7.1 Flame Methods

Kallmann and Hobart studied the effect of various cations and anions, and of flame parameters on the absorbance of rhodium. A fuel-rich (reducing) air/acetylene flame and a nitrous oxide/acetylene flame gave diminished response. Rhodium absorption was significantly increased by a lean, oxidizing air/acetylene flame. These and observations concerning burner height led to the theory that high boiling (>3700°C) rhodium metal was being formed under reducing conditions, resulting in a low population of ground state atoms. The authors observed significant enhancement of rhodium absorbance with the addition of alkali metal, aluminum, and zinc sulfates. They speculated that stable, complex rhodium salts are being formed, then dissociated in the flame, increasing the ground state rhodium atom population [181].

Gupta determined ten elements in osmiridium and native platinum using a dry chlorination procedure. A test portion of 10 mg of osmiridium or 100 mg of native platinum was covered with finely powdered sodium chloride in a quartz boat. The boat was inserted in a tube furnace at 700°C and the system was flushed with chlorine gas for 7 h, passing the effluent through a series of receiving flasks containing water and dilute hydrochloric acid saturated with sulfur dioxide. The system was cooled, flushed with nitrogen, and the sample was dissolved in 1M hydrochloric acid, filtered, and any insolubles rechlorinated. The trap solutions and the main solution were combined in a flask and boiled to fumes of perchloric acid, collecting the distillate in cold hydrobromic acid. Sodium chloride was added to the distillate and it was carefully evaporated to just dryness. For osmiridium the residue was dissolved in 0.1M hydrobromic acid, diluted and aliquotted, a copper/cadmium buffer was added, and it was diluted to volume. For native platinum the residue was dissolved in copper/cadmium buffer and diluted to volume. The solutions were then measured for osmium and ruthenium by flame AA. The distillation flask residual liquid was evaporated to dryness with nitric acid, then twice again with hydrochloric acid. The residue was dissolved in *aqua regia*, evaporated to dryness three times more with hydrochloric acid, then dissolved and diluted to volume with 3M hydrochloric acid. An aliquot was evaporated to dryness, dissolved in copper/cadmium buffer, diluted to volume and measured for silver, gold, palladium, (and platinum in osmiridium). A different aliquot was treated similarly for rhodium and iridium (in platinum). A third aliquot was dried and dissolved in lanthanum or strontium buffer, then diluted and measured for iron, copper, and/or nickel [182].

Adriaenssens and Verbeek published studies that showed that intermetallic matrix interferences of silver, gold, palladium, and platinum for an air/acetylene flame were lower in a 2% (w/v) potassium cyanide solution than in the normal acid media employed [183]. The same authors published a procedure for silver and gold cupellation prills, which were dissolved in acid and converted to chloride salts by repeated evaporations. The dried salts were dissolved in 10% (w/v) potassium cyanide solution; tetrapotassium EDTA was added to clear turbidity due to impurities such as lead, and the solutions were diluted to volume and measured. The final solution volume was 2% (w/v) in potassium cyanide. Silver, gold, palladium, and platinum were determined [184].

Ashy and Headridge determined iridium and ruthenium in rhodium sponge. The test portion was dissolved with hydrochloric acid and sodium chlorate in a sealed glass tube at 250°C. Iridium and ruthenium were extracted from the solution, made 2 molar in hydrochloric acid, with methyltriphenylphosphonium chloride in chloroform. The organic layer was then evaporated to dryness and then dissolved in lithium perchlorate dissolved in acetonitrile. The lithium compound served as an interference suppressant in the subsequent flame AA measurement. If precautions are taken to prevent its loss osmium can also be determined [185].

Bhattacharyya and Das determined iridium in platinum samples dissolved in *aqua regia*. An aliquot was adjusted to pH 2 and reacted with potassium nitrite to mask platinum ion. The solution was adjusted to pH 2.2, phthalate buffer (at pH 2.2) was added and an ethanol solution of Versatic-10™ (liquid chelating exchanger) was added. The mixture was heated under reflux for 15 min, then cooled and extracted with chloroform. The iridium was returned to the aqueous phase by extraction with 20% (v/v) perchloric acid. The aqueous phase was diluted to volume and measured in an air/acetylene flame [186].

8.7.2 Furnace Methods

Kragten and Reynaert described a procedure for the determination of trace iron in gold and silver. Gold is dissolved in hydrochloric and nitric acids, fumed in sulfuric acid; the matrix gold is precipitated as the colloidal element with sulfur dioxide gas, and GFAA aliquots are taken from the supernatant centrifugate. Silver is dissolved in nitric acid, heated to dryness, and the salts dissolved in water. Hydrochloric and sulfuric acids are added; the silver chloride precipitate is centrifuged, and aliquots for GFAA are taken from the supernatant solution [187].

Rowston and Ottaway published the results of fundamental studies on atom formation processes with the noble metals (including silver) in carbon furnaces [188].

8.8 Miscellaneous Alloys

8.8.1 Flame Methods

Humphrey determined magnesium at ppm levels in *uranium metal* using direct acid dissolution with the method of additions [189]. Jursik dissolved uranium metal in nitric acid, then extracted the matrix element with tributyl phosphate/carbon tetrachloride. The aqueous phase was then measured for aluminum, iron, and nickel. Aluminum required nitrous oxide/acetylene, the others air/acetylene [190]. Scarborough reported that molybdenum, ruthenium, palladium, and rhodium could be determined in complex uranium alloys because, fortuitously, uranium eliminates mutual interferences among the four analytes [191].

Mansell, Emmel, and McLaughlin described the analysis of *magnesium alloys* that were dissolved in a minimal amount of hydrochloric acid with some dropwise addition of nitric acid. The solutions were filtered to remove silica, then diluted to create solutions that were 1% of the alloy (i.e., 1g/100mL). Calcium, copper, manganese, and zinc were determined [155].

Scarborough, Bingham, and DeVries determined traces of chromium, cobalt, iron, manganese, and nickel in *sodium metal*. A 1 g test portion is reacted with water vapor-saturated argon in a Vycor crucible, then neutralized with hydrochloric acid. Lanthanum is added as a carrier and precipitated as the hydroxide with either sodium hydroxide or ammonium hydroxide. The supernatant solution is poured off and discarded, and the precipitate is dissolved in 6M hydrochloric acid, diluted to the mark in a 100-mL volumetric flask, and measured with a 3-slot Boling (high solids) burner head [192].

Roth, Bohl, and Sellers determined trace levels of tellurium in high purity *bismuth metal*. Test portions (1 or 10 g, depending on the expected analyte level) were dissolved in 10:1 hydrochloric acid:nitric acid, then additional hydrochloric acid was added until a clear solution was formed. A small volume of water was added and the sample was transferred to a separatory funnel with 6M hydrochloric acid. The solution was extracted with MIBK, and the organic phase was measured by flame AA [193].

Deak developed a procedure for the rapid sorting and identification of alloys using a nitrous oxide/acetylene atomic absorption approach [194]. Hannaford and Lowe determined boron isotopic ratios using neon-filled boron discharge lamps and a nitrous oxide/acetylene flame [195].

8.8.2 Furnace Methods

Yudelevich, Zelentsova, Beisel, Chanysheva, and Vechernish determined aluminum, bismuth, cadmium, calcium, chromium, cobalt, copper, indium, lead, magnesium, manganese, nickel, silver, and zinc in high purity *gallium metal*, and most of the same analytes in *rhenium metal*. Gallium was dissolved in dilute hydrochloric acid/nitric acid. The solution was either diluted and analyzed directly, or the matrix element was removed from 12M hydrochloric acid solution by extraction with β,β′-dichloroethyl ester (Chlorex™). In the latter case the aqueous phase was evaporated to dryness and the residue was dissolved in dilute hydrochloric acid, diluted to volume, and measured by GFAA. Rhenium was dissolved in nitric acid and either analyzed directly, or the matrix element was removed by extraction with tri-*n*-octylamine/toluene. In this latter case the aqueous phase is evaporated to very low volume, then diluted to volume with dilute nitric acid and measured by GFAA [196].

Tin was determined in high purity gallium as described by Sahayam and Gangadharan. The 0.1–0.25 g test portion is dissolved in 3 mL of nitric acid in an acid digestion pressure bomb, concentrated, then diluted to final volume (10 mL). Ammonium molybdate was used as a matrix modifier [197].

Garbett, Goodfellow, and Marshall studied the volatilization of different sodium salts from a graphite rod atomizer with an anticipated goal of predicting char cycle behavior when dissolved *metallic sodium* samples are analyzed [198]. The same authors described a method for iron, nickel, and chromium in sodium chloride that suggests that the sodium chloride matrix is eliminated during the char with acceptable analyte losses [199].

Norval and Gries determined trace thallium in *cadmium metal* by a direct solids approach using primary standards prepared by evaporating standard thallium solutions on high purity cadmium foil. Secondary standards were prepared from urea doped with thallium. Data from the solid sample approach compared favorably with data from an extraction/FAA method [200].

REFERENCES

1. Wollaston, W.H. Philosophical Transactions of the Royal Society, London, Ser. A, 92, 365 (1802).
2. Kirchhoff, G; Bunsen, R. Philosophical Magazine, 20, 89 (1860).
3. Woodson, T.T. Review of Scientific Instruments, 10, 308 (1939).
4. Walsh, A. Spectrochimica Acta, 7, 108 (1955).
5. Alkemade, C.Th.J.; Milatz, J.M.W. Applied Scientific Research, Section B, 4, 289 (1955).
6. Alkemade, C.Th.J.; Milatz, J.M.W. Journal of the Optical Society of America, 45, 583 (1955).
7. Holick, G. "Flame Emission, Atomic Absorption, and Atomic Fluorescence Spectrometry" Instrumental Analysis, 2nd ed. (G.D. Christian and J.E. O'Reilly, eds.), Allyn and Bacon, Boston, 1986, pp. 278–321.
8. Skoog, D.A.; West, D.M. Principles of Instrumental Analysis, 2nd ed., Saunders, Philadelphia, 1980, pp. 302–333.
9. Welz, B. Atomic Absorption Spectrometry (trans. C. Skegg), 2nd ed., VCH, Weinheim, Germany, 1985.
10. Price, W.J. Spectrochemical Analysis by Atomic Absorption, Heyden, London, 1979.
11. Robinson, J.W. Atomic Absorption Spectroscopy, Marcel Dekker, NY, 1966.
12. Fuller, C.W. Electrothermal Atomization for Atomic Absorption Spectrometry, The Chemical Society, London, 1977.
13. Thompson, K.C.; Thomerson, D.R. Analyst, 99, 595 (1974).
14. Gough, D.S. Analytical Chemistry, 48, 1926 (1976).
15. Gillette, R.K. Spectroscopy, 9, 42 (1994).
16. Rains, T.C.; Epstein, M.S.; Menis, O. Analytical Chemistry, 46, 207 (1974).
17. Willis, J.B. Nature, 207, 715 (1965).
18. Bond, A.M. Analytical Chemistry, 42, 932 (1970).
19. Urbain, H. A, Chambosse, A. Atomic Spectroscopy, 3, 143 (1982).
20. Parker, C.R.; Erickson, J.O.; Culver, B.R. American Laboratory, 85 (Nov.) (1974).
21. Ediger, R.D. Atomic Absorption Newsletter, 14, 127 (1975).
22. Manning, D.C.; Slavin, W. Analytical Chemistry, 50, 1234 (1978).
23. Stephans, R.; Ryan, D.E. Talanta, 22, 655 (1975).
24. Koizumi, H.; Yasuda, K. Analytical Chemistry, 47, 1679 (1975).
25. Brown, S.D. Analytical Chemistry, 49, 1269A (1977).
26. Miller, J.D.; Koizumi, H. American Laboratory, Nov., 35 (1979).

27. Carnrick, G.R.; Barnett, W.; Slavin, W. Spectrochimica Acta, 41B, 991 (1986).
28. Rowe, C.J.; Routh, M.W. Research/Development, Nov., 24 (1977).
29. Scholes, P.H. Analyst, 93, 197 (1968).
30. Beyer, M. Atomic Absorption Newsletter, 4, 212 (1965).
31. McAuliffe, J.J. Atomic Absorption Newsletter, 6, 69 (1967).
32. Bowman, J.A.; Willis, J.B. Analytical Chemistry, 39, 1210 (1967).
33. Barnett, W.B.; Kahn, H.L. Atomic Absorption Newsletter, 8, 21 (1969).
34. Panday, V.K.; Gaguly, A.K. Atomic Absorption Newsletter, 7, 50 (1968).
35. Knight, D.M.; Pyzyna, M.K. Atomic Absorption Newsletter, 8, 129 (1969).
36. David, D.J. Analyst, 86, 730 (1961).
37. Mostyn, R.A.; Cunningham, A.F. Analytical Chemistry, 38, 121 (1966).
38. Kirkbright, G.F.; Smith, A.M.; West, T.S. Analyst, 91, 700 (1966).
39. Ramakrishna, T.V.; West, P.W.; Robinson, J.W. Analytica Chimica Acta, 44, 437 (1969).
40. Barnes Jr., L. Analytical Chemistry, 38, 1083 (1966).
41. Thomerson, D.R.; Price, W.J. Analyst, 96, 321 (1971).
42. Husler, J. Atomic Absorption Newsletter, 10, 60 (1971).
43. Fogg, A.G.; Soleymanloo, S.; Burns, D.T. Talanta, 22, 541 (1975).
44. Donaldson, E.M. Talanta, 27, 779 (1980).
45. Castillo, J.R.; Belarra, M.A.; Aznarez, J. Atomic Spectroscopy, 3, 58 (1982).
46. Donaldson, E.M. Talanta, 27, 79 (1980).
47. Headridge, J.B.; Hubbard, D.P. Analytica Chimica Acta, 37, 151 (1967).
48. Capacho-Delgado, L.; Manning, D.C. Atomic Absorption Newsletter, 5, 1 (1966).
49. Schiller, R. Atomic Absorption Newsletter, 9, 111 (1970).
50. Musil, J.; Dolezal, J. Analytica Chimica Acta, 92, 301 (1977).
51. Price, W.J.; Roos, J.T.H. Analyst, 93, 709 (1968).
52. Headridge, J.B.; Sowerbutts, A. Analyst, 98, 57 (1973).
53. Donaldson, E.M. Talanta, 28, 461 (1981).
54. Barnett, W.B.; Kerber, J.D. Atomic Absorption Newsletter, 13, 56 (1974).
55. Menis, O.; Rains, T.C. Analytical Chemistry, 41, 952 (1969).
56. Burke, K.E. Analyst, 97, 19 (1972).
57. Burke, K.E. Talanta, 21, 417 (1974).
58. Burke, K.E. Applied Spectroscopy, 28, 234 (1974).
59. Bedrossian, M. Analytical Chemistry, 50, 1898 (1978).
60. Dagnall, R.M.; West, T.S.; Young, P. Analytical Chemistry, 38, 358 (1966).
61. Vassilaros, G.L. Talanta, 21, 803 (1974).
62. Headridge, J.B.; Smith, D.R. Laboratory Practice 20, 312 (1972).
63. Headridge, J.B.; Richardson, J. Analyst, 95, 930 (1970).
64. Marcec, M.V.; Kinson, K.; Belcher, C.B. Analytica Chimica Acta, 41, 447 (1968).
65. Peterson, E.A. Atomic Absorption Newsletter, 9, 129 (1970).
66. Headridge, J.B.; Richardson, J. Analyst, 94, 968 (1969).
67. Peterson, G.E.; Kerber, J.D. Atomic Absorption Newsletter, 15, 134 (1976).
68. Thomerson, D.R.; Price, W.J. Analyst, 96, 825 (1971).
69. Nall, W. R.; Brumhead, D.; Whitman, R. Analyst, 100, 555 (1975).
70. Damiani, M.; DelMonte Tamba, M.G.; Bianchi, M.G. Analyst, 100, 643 (1975).
71. Klein, A.A. Atomic Spectroscopy, 3, 133 (1982).

72. Frech, W. Talanta, 21, 565 (1974).
73. Frech, W. Zeitschrift für das analitische chemie, 275, 353 (1975).
74. Ratcliffe, D.B.; Byford, C.S.; Osman, P.B. Analytica Chimica Acta, 75, 457 (1975).
75. Barnett, W.B.; McLaughlin, E.A. Analytica Chimica Acta, 80, 285 (1975).
76. Dulski, T.R.; Bixler, R.R. Analytica Chimica Acta, 91, 199 (1977).
77. Frech, W.; Lundgren, G.; Lunner, S.-E. Atomic Absorption Newsletter, 15, 57 (1976).
78. Frech, W.; Cedergren, A. Analytica Chimica Acta, 82, 83 (1976).
79. Frech, W.; Cedergren, A. Analytica Chimica Acta 82, 93 (1976).
80. DelMonte Tamba, M.G.; Luperi, N. Analyst, 102, 489 (1977).
81. Shaw, F.; Ottaway, J.M. Analyst, 100, 217 (1975).
82. Persson, J.-A.; Frech, W.; Cedergren, A. Analytica Chimica Acta, 89, 119 (1977).
83. Whiteside, P.J.; Price, W.J. Analyst, 102, 618 (1977).
84. Welz, B. Atomic Absorption Spectrometry, (trans. by C. Skegg), 2nd ed. VCH, Weinhein, Germany, 1985, pp. 398–399.
85. Andrews, D.G.; Headridge, J.B. Analyst, 102, 436 (1977).
86. Andrews, D.G.; Aziz-Alrahman, A.M.; Headridge, J.B. Analyst, 103, 909 (1978).
87. Aziz-Alrahman, A.M.; Headridge, J.B. Talanta, 25, 413 (1978).
88. Lundberg, E.; Frech, W. Analytica Chimica Acta, 104, 67 (1979).
89. Lundberg, E.; Frech, W. Analytica Chimica Acta, 104, 75 (1979).
90. Lundberg, E.; Frech, W. Analytica Chimica Acta, 108, 75 (1979).
91. Frech, W.; Lundberg, E.; Barbooti, M.M. Analytica Chimica Acta, 131, 45 (1981).
92. Backman, S.; Karlsson, R.W. Analyst, 104, 1017 (1979).
93. L'vov, B.V.; Novichikhin, A.V. Atomic Spectroscopy, 11, 1 (1990).
94. Fleming, H.D.; Ide, R.G. Analytica Chimica Acta, 83, 67 (1976).
95. Welz, B. Atomic Absorption Spectrometry (trans. by C. Skegg), 2nd ed., VCH, Weinheim, Germany, 1985, pp. 397.
96. Bettinelli, M.; Spezia, S.; Bizzarri, G. Atomic Spectroscopy, 13, 75 (1992).
97. Bettinelli, M.; Spezia, S.; Bizzarri, G. Atomic Spectroscopy, 15, 115 (1994).
98. Haug, H.O.; Yiping, L. Journal of Analytical Atomic Spectrometry, 10, 1069 (1995).
99. Batistoni, D.A.; Fuertes, M.I.; Smichowski, P.N. Atomic Spectroscopy, 10, 12 (1989).
100. Andrew, T.R.; Nichols, P.N.R. Analyst, 87, 25 (1962).
101. Dyck, R. Atomic Absorption Newsletter, 4, 170 (1965).
102. Mostyn, R.A.; Cunningham, A.F. Atomic Absorption Newsletter, 6, 86 (1967).
103. Arendt, D.H. Atomic Absorption Newsletter, 11, 63 (1972).
104. Welcher, G.G.; Kriege, O.H. Atomic Absorption Newsletter, 8, 97 (1969).
105. Welcher, G.G.; Kriege, O.H. Atomic Absorption Newsletter, 9, 61 (1970).
106. Burke, K.E. Analytical Chemistry, 42, 1536 (1970).
107. Blakeley, St.J.; Manson, A.; Zatka, V.J. Analytical Chemistry, 45, 1941 (1973).
107a. Newland, B.T.N.; Mostyn, R.A. Atomic Absorption Newsletter, 10, 89 (1971).
108. Kirk, M.; Perry, E.G.; Arritt, J.M. Analytica Chimica Acta, 80, 163 (1975).
109. White Sr., J.A.; Harper Jr., W.L.; Fiedman, A.P.; Banas, V.E. Applied Spectroscopy, 28, 192 (1974).

110. Welcher, G.G.; Kriege, O.H.; Marks, J.Y. Analytical Chemistry, 46, 1227 (1974).
111. Zatka, V. Analytical Chemistry, 50, 538 (1978).
112. Forrester, J.E.; Lehecka, V.; Johnston, J.R.; Ott, W.L. Atomic Absorption Newsletter, 18, 73 (1979).
113. Kujirai, O.; Kabayashi, T.; Ide, K.; Sudo, E.; Talanta, 29, 27 (1982).
114. Bosal, C.P.; Carnrick, G.R.; Slavin, W. Atomic Spectroscopy, 7, 148 (1986).
115. Trostle, D.; Beals, T.; Kuczenski, R.; Shaver, M. Atomic Spectroscopy, 12, 64 (1991).
116. Reichardt, M.S. Atomic Spectroscopy, 13, 178 (1992).
117. Simpson, R.T. Atomic Spectroscopy, 10, 82 (1989).
118. Mile, B.; Rowlands, C.C.; Jones, A.V. Journal of Analytical Atomic Spectrometry, 7, 1069 (1992).
119. Marks, J.Y.; Welcher, G.G.; Spellman, R.J. Applied Spectroscopy, 31, 9 (1977).
120. Headridge, J.B.; Thompson, R. Analytica Chimica Acta, 102, 33 (1978).
121. Drinkwater, J.E. Analyst, 101, 672 (1976).
122. Welz, B.; Melcher, M. Analytica Chimica Acta, 153, 297 (1983).
123. Gidley, J.A.F.; Jones, J.T. Analyst, 85, 249 (1960).
124. Gidley, J.A.F.; Jones, J.T. Analyst, 86, 271 (1961).
125. Sprague, S.; Manning, D.C.; Slavin, W. Atomic Absorption Newsletter, No.20, 1 (1964).
126. Capacho-Delgado, L.; Manning, D.C. Atomic Absorption Newsletter, 4, 317 (1965).
127. Sattur, T.W. Atomic Absorption Newsletter, 5, 37 (1966).
128. McCracken, J.D.; Vecchione, M.C.; Longo, S.L.; Atomic Absorption Newsletter, 8, 102 (1969).
129. Kallmann, S.; Hobart, E.W. Talanta, 17, 845 (1970).
130. Thormahlen, D.J.; Frank, E.H. Atomic Absorption Newsletter, 10, 63 (1971).
131. Reichel, W.; Bleakley, B.G. Analytical Chemistry, 46, 59 (1974).
132. Celis, J.P.; Helsen, J.A.; Herman, P.; Roos, J.R. Analytica Chimica Acta, 92, 413 (1977).
133. Likaits, E.R.; Farrell, R.F.; Mackie, A.J. Atomic Absorption Newsletter, 18, 53 (1979).
134. Donaldson, E.M. Talanta, 26, 999 (1979).
135. Donaldson, E.M. Talanta, 26, 1119 (1979).
136. Tsukahara, I.; Tanaka, M. Talanta, 27, 237 (1980).
137. Persiani, C.; Durkin, F. Atomic Spectroscopy, 3, 194 (1982).
138. Sarkar, A.K.; Parashar, D.C. Atomic Spectroscopy, 12, 19 (1991).
139. Kharbade, B.V.; Agarival, K.C. Atomic Spectroscopy, 14, 13 (1993).
140. Kojima, I.; Takayanagi, A. Journal of Analytical Atomic Spectrometry, 11, 607 (1996).
141. Shaw, F.; Ottaway, J.M. Atomic Absorption Newsletter, 13, 77 (1974).
142. Mullen, J.D. Talanta, 23, 846 (1976).
143. Everett, G.L. Analyst, 101, 348 (1976).
144. Fritzsche, H.; Wegscheider, W.; Knapp, G.; Ortner, H.M. Talanta, 26, 219 (1979).
145. Haynes, B.W. Atomic Absorption Newsletter, 18, 46 (1979).

146. Muir, M.K.; Andersen, T.N. Atomic Spectroscopy, 3, 149 (1982).
147. Sentimenti, E.; Mazzetto, G. Atomic Spectroscopy, 7, 181 (1986).
148. Milella, E.; Sentimenti, E.; Mazzetto, G. Atomic Spectroscopy, 14, 1 (1993).
149. Baker, A.A.; Headridge, J.B. Analytica Chimica Acta, 125, 93 (1981).
150. Bedard, M.; Kerbyson, J.D. Analytical Chemistry, 47, 1441 (1975).
151. Castillo, J.R.; Mir, J.M.; Vela, M.L.; Martinez, C. Atomic Spectroscopy, 7, 85 (1986).
152. Aznarez Alduan, J.; Castillo Suarez, J.R.; Bonilla Pollo, A.; Lanaja del Busto, J. Atomic Spectroscopy, 2, 125 (1981).
153. Wickstrøm, T.; Lund, W.; Bye, R. Analyst, 120, 2695 (1995).
154. Bell, G.F. Atomic Absorption Newsletter, 5, 73 (1966) and 6, 18 (1967).
155. Mansell, R.E.; Emmel, H.W.; McLaughlin, E.L. Applied Spectroscopy, 20, 231 (1966).
156. Wilson, L. Analytica Chimica Acta, 40, 503 (1968).
157. Peterson, E.A. Atomic Absorption Newsletter, 8, 53 (1969).
158. Price, W.J.; Whiteside, P.J. Analyst, 102, 664 (1977).
159. Ashy, M.A.; Headridge, J.B.; Sowerbutts, A. Talanta, 21, 649 (1974).
160. Gries, W.H.; Norval, E. Analytica Chimica Acta, 75, 289 (1975).
161. Langmyhr, F.J.; Rasmussen, S. Analytica Chimica Acta, 72, 79 (1974).
162. Farrar, B. Atomic Absorption Newsletter, 4, 325 (1965).
163. Smith, S.B.; Blasi, J.A.; Feldman, F.J. Analytical Chemistry, 40, 1525 (1968).
164. Hwang, J.Y.; Sandonato, L.M.; Analytical Chemistry, 42, 744 (1970).
165. Quarrell, T.M.; Powell, J.W.; Cluley, H.J. Analyst, 98, 443 (1973).
166. Llacer, J.; Catala, R. Atomic Absorption Newsletter, 15, 113 (1976).
167. Guerra, R. Atomic Spectroscopy, 1, 58 (1980).
168. Fox, G.J. Atomic Spectroscopy, 11, 13 (1990).
169. Kirkbright, G.F.; Peters, M.K.; West, T.S. Analyst, 91, 411 (1966).
170. Kirkbright, G.F.; Peters, M.K.; West, T.S. Analyst, 91, 705 (1966).
171. Mantel, M.; Aladjem, A.; Nothmann, R. Analytical Letters, 9, 671 (1976).
172. Biechler, D.G.; Long, C.H. Atomic Absorption Newsletter, 8, 56 (1969).
173. Myers, D. Atomic Absorption Newsletter, 6, 89 (1967).
174. Schlewitz, J.H.; Shields, M.G. Atomic Absorption Newsletter, 10, 39 (1971).
175. Schlewitz, J.H.; Shields, M.G. Atomic Absorption Newsletter, 10, 43 (1971).
176. Pabalkar, M.A.; Naik, S.V.; Sanjana, N.R. Analyst, 106, 47 (1981).
177. Donaldson, E.M.; Charette, D.J.; Rolko, V.H.E. Talanta, 16, 1305 (1969).
178. Hamner, R.M.; Lechak, D.L.; Greenberg, P. Atomic Absorption Newsletter, 15, 122 (1976).
179. Cocekal, B.; Krivan, V. Spectrochimica Acta, 50B, 517 (1995).
180. Friese, K.-C.; Krivan, V.; Schuierer, O. Spectrochimica Acta, 51B, 1223 (1996).
181. Kallmann, S.; Hobart, E.W. Analytica Chimica Acta, 50B, 517, (1995).
182. Gupta, J.G.S. Analytica Chimica Acta, 58, 23 (1972).
183. Adriaenssens, E.; Verbeek, F. Atomic Absorption Newsletter, 12, 57 (1973).
184. Adriaenssens, E.; Verbeek, F. Atomic Absorption Newsletter, 13, 41 (1974).
185. Ashy, M.A.; Headridge, J.B. Analyst, 99, 285 (1974).
186. Bhattacharyya, S.S.; Das, A.K. Atomic Spectroscopy, 10, 188 (1989).

187. Kragten, J.; Renaert, A.P. Talanta, 21, 618 (1974).
188. Rowston, W.B.; Ottaway, J.M. Analyst, 104, 645 (1979).
189. Humphrey, J.R. Analytical Chemistry, 37, 1604 (1965).
190. Jursik, M.L. Atomic Absorption Newsletter, 6, 21 (1967).
191. Scarborough, J.M. Analytical Chemistry, 41, 250 (1969).
192. Scarborough, J.M. Analytical Chemistry, 39, 1394 (1967).
193. Roth, D.J.; Bohl, D.R.; Sellers, D.E. Atomic Absorption Newsletter, 7, 87 (1968).
194. Deak, C.K. Atomic Absorption Newsletter, 10, 6 (1971).
195. Hannaford, P.; Lowe, R.M. Analytical Chemistry, 49, 1852 (1977).
196. Yudelevich, I.G.; Zelentsova, L.V.; Beisel, N.F.; Chanysheva, T.A. Analytica Chimica Acta, 108, 45 (1979).
197. Sahayam, A.C.; Gangadhara, S. Atomic Spectroscopy, 14, 83 (1993).
198. Garbett, K.; Goodfellow, G.I. Marshall, G.B. Analytica Chimica Acta, 126, 135 (1981).
199. Garbett, K.; Goodfellow, G.I.; Marshall, G.B. Analytica Chimica Acta, 126, 147 (1981).
200. Norval, E.; Gries, W.H. Analytica Chimica Acta, *83*, 393 (1976).

6
Atomic Emission Spectrometry

1. INTRODUCTION [1–4]

Flame emission spectrometry—admittedly, a subset of the current subject—has already been given its due in the previous chapter and will be only lightly reprised here. The reason for this arrangement is that the comparatively low temperature regime of a chemical flame is closely related to atomic absorption in some of the basic methodology, and in much of the equipment used. Moreover, for metals work only a few elements are viable analytes for flame emission today—most notably the alkali metals. In this chapter we will treat the higher temperature phenomena of light emission from electrical discharge sources. These range from the dc arc at 4000 K to the inductively coupled plasma at 10,000 K. Potential analytes in metal matrices nearly run the entire gamut of the periodic chart.

Atomic emission* with an electrical excitation source has several distinct advantages over other techniques that have been applied to the trace analysis of metals. First, it is applicable directly to electrically conducting metal specimens, effectively eliminating chemical preparation (except in some important instances). Second, the test portion size, albeit small compared to that used in wet methods, is still large enough to allow for a valid representation (after thermal homogenization of microscale segregation) in reasonably uniform materials. The technique is sensitive and can be configured for low detection limit operation. Unlike some competitive approaches, once set up routine operation and maintenance does not require highly skilled personnel. Finally, capital equipment costs are now moderate compared to those of many other instrumental approaches.

---

* Elsewhere in this volume the more familiar synonym, "optical emission" and the abbreviation, "OES" have been used freely. We reserve the option of using these terms interchangeably with "atomic emission" with the understanding that they all refer to electromagnetic energy emission in the near infrared/visible/UV range and its measurement.

Today, the vast majority of arc/spark atomic emission spectrometers and a large number of all plasma instruments are utilized for metals analysis. There is a long, rich history here, beginning in 1873 when I.N. Lockyer published the first atomic emission analysis of a metal alloy [5]. In the period from 1900–1920, electrical excitation sources underwent intense development. This culminated in the first practical spark source atomic emission method for metals, published by Meggers, Kiess, and Stimson in 1922 [6]. A few years later Gerlach and Sweitzer described how the proper use of an internal standard could reduce or eliminate the variability in results produced by instrumental instabilities [7].

All of the early work utilized a photographic emulsion as the detection device. "Spectrography" (the term is derived by analogy to "photography"), like classical wet chemistry, contained a large element of art and lore in addition to science. Film or glass plates were developed in a darkroom adjacent to the spectrographic lab, and line densities were measured with a microdensitometer. This methodology has been largely replaced by photoelectric detection systems, which have grown in use steadily since their introduction in the late 1940s.

In the 1960s argon-purged arc/spark chambers were first combined with vacuum optics to allow measurement of lines below 190 nm.* This region includes important lines for carbon, sulfur, phosphorus, arsenic, and other key analytes. In 1965 the first commercial dc plasma/echelle grating spectrometer was introduced, followed about a decade later by the first inductively coupled plasma spectrometers. It was also in the mid-1960s that Grimm published his design for a practical glow discharge lamp [8]. Other developments with excitations followed: a practical, demountable hollow cathode, high energy prespark, and high repetition rate sources, and time-resolved measurement that collects data only from the most sensitive interval of an excitation cycle.

Spark ablation and laser ablation systems have been employed to produce a fine metal aerosol that is transported by an argon stream to an ICP source. Here, the sampling and excitation features of the design are usefully separated. The different characteristics of radial- and axial-viewed plasmas have also been explored in commercial designs. And recent developments in solid state charge transfer device detectors have been incorporated into new instruments.

In addition, there have been and continue to be a number of special subject areas where developments have been driven by somewhat narrower

---

* While the angstrom (Å) persists in the spectroscopist's argot we will apply the SI-derived unit, nanometer (nm) in this chapter.

needs. The laser microprobe, for example, directs light from a precisely aimed laser pulse into a spectrometer. It is applicable to the qualitative identification of inclusions in metal alloys, for example. Alloy sorting is an area that has received a great deal of attention, utilizing atomic emission among other approaches. Molten metal analysis by atomic emission has been attempted with fiber optic probes, aerosol transport probes, and by other means. The glow discharge lamp has been applied to depth profiling, especially of conductive coating layers. Much effort has been expended in accurately modeling the glow discharge sputtering mechanism to allow continuous analyte quantification.

Robotics has ushered in the dawning of the "peopleless" laboratory—often an atomic emission spectrometer figures prominently in the design. Such systems are currently spreading in the basic steel and aluminum industries. The spectrometer system monitors its own performance and calibrates itself, as required. Some "expert system" software even has been designed to sense count rates from totally unknown samples and select the appropriate measurement conditions and calibration curves "on-the-fly."

The day of the master spectrographer, who could read and interpret an alloy spectrum like a page of lucid text, has passed as completely as the day of the classical wet chemical analyst. Their skills are no less needed today, but that fact is much less apparent. The modern laboratory is equipped with analytical engines of amazing power—30–40 channel direct-reading polychromators are not uncommon. And there is reason to believe that even the unique storage medium of the photographic emulsion has been duplicated in the magnetic storage of pixel arrays. The question that remains is, who is left to interpret them?

Instrument manufacturers have assumed more and more of the art, lore, and science of the field in order to design and build "black box" technology for their customers. Industry has responded in kind by allowing the expert skills and knowledge to slip away, partly at least to offset the sizeable pricetag of those black boxes. It is a sad anomaly that in industry today we must look to the laboratories with the most outdated equipment to find people with the skills to interpret spectra.

One goal of this chapter is to answer an obvious question about all of this: why should we care? A major concern that directly affects atomic emission is the eroding infrastructure for producing certified reference materials. Even the instrument manufacturers have begun to show signs of nervousness about the supply and reliability of metal alloy standards. While the master spectrographer was just as dependent on comparative methodology as the keyboard jockey who replaced him, he had some options. A tramp element line was either present or absent. If it was present, semiquantitative estimates could usually be derived by comparison to a moving plate study of a mixed oxide standard. An unexpected element in an alloy could also usually be identified and quantified. Today, 0.1% gold in a piece of stainless steel scrap would probably go unrecognized by a million dollar spectroscopy lab.

With the possible exception of some recent developments with charge transfer detectors, the dc arc with photographic emulsion detection remains the fastest and most complete means of monitoring high purity metals for trace level impurities.

However, provided that appropriate reference materials remain available, clearly there are significant advantages to the more modern spectrometric equipment. Speed is, perhaps, the most important. The melter in the control pulpit in a steel mill melt shop can reasonably expect to see an accurate 25 or 30 analyte report on his cathode ray monitor within three minutes of the time a test casting is sent through the pneumatic tube system to the chemistry lab. A great deal of steel (and money) depend on such timely and correct results.

In this chapter we will attempt to cover the subject of atomic emission as it pertains to the trace analysis of metals. Emphasis will be placed on methodologies as they are currently employed, including some treatment of slower, but more versatile technology. The organization first covers some basic principles that were neglected in the previous two chapters. We will have more to say about instrumentation, including excitation sources, optics, and detectors. Next is a discussion of calibration and measurement, followed by a section on the practical details associated with the methodology. Special topics, such as depth profiling, molten metal analysis, and robotics will follow. An applications section will close this chapter with sections on iron-base alloys, nickel- and cobalt-base alloys, aluminum-base alloys, copper-base alloys, and others.

2. PRINCIPLES

In Chapters 4 and 5 we briefly described some of the ways matter and electromagnetic radiation interact. These included absorption, fluorescence, and emission—all utilized in analytical methodologies of various forms. There remain only a few subjects of special relevance to emission that need to be addressed here.

2.1 Frequency Relationships [2,3]

In Chapter 5 we saw a simplified Grotian diagram that illustrated some of the electron transitions for a sodium atom. If we imagine such a diagram for an atom with three electron energy states: a ground state, E_0, and two excited neutral states, E_1 and E_2, we can describe three events: $E_0 \rightarrow E_2 \rightarrow E_0$; $E_0 \rightarrow E_1 \rightarrow E_0$; and $E_0 \rightarrow E_2 \rightarrow E_1 \rightarrow E_0$. Each drop in energy level from a higher to a lower state results in the emission of light of a characteristic wavelength (or frequency). In 1905 Walter Ritz demonstrated that these frequencies are strictly additive. Thus, the emitted frequency for $E_2 \rightarrow E_0$ is equal to the sum of the emitted frequencies for $E_2 \rightarrow E_1$ and $E_1 \rightarrow E_0$.

Sommerfield and Kossel's Spectroscopic Displacement Law states that the excited neutral atom lines of an element of atomic number Z are similar in character to the singly ionized ion lines of the next higher atomic number (Z + 1) element. Moreover, they will also be similar to the doubly ionized ion lines of the Z + 2 element. Spectral lines of excited neutral atoms are designated with a Roman numeral I, singly ionized lines as II, and doubly ionized lines as III. Such a relationship (e.g., FeI, CoII, NiIII) is termed an isoelectronic sequence. What lies behind it is merely the similar atomic orbital structure of a Z atom, a Z + 1 monovalent ion, and a Z + 2 divalent ion, each having the same number of electrons.

This is sometimes interpreted as similarity between the arc spectrum of element Z and the first spark spectrum of element Z + 1, and the second spark spectrum of element Z + 2. However, while it is true that the dc arc produces primarily transitions to excited neutral states and that sparks produce more singly and doubly ionized ions, no excitation source produces completely uniform transitions for all elements. The terms "arc spectrum" and "spark spectrum," used in this sense, should, therefore, be discouraged. A good illustration of the error of such simplifications is the experimental observation that sodium ion lines are weak when produced by short duration high voltage sparks but dominant in low temperature chemical flames. Also, it is known that in a typical arc discharge, between 0.01% and 0.1% of the species are present as ions since that much ionization is needed for the arc to be self-sustaining.

2.2 Processes in Local Thermal Equilibrium [4,9]

In the last chapter we touched upon the fact that things become a great deal simpler if the analytical measurement is conducted in a thermally stable environment (i.e., an environment that is not heating up or cooling down during the measurement interval). Such a system is said to be in a state of *Local Thermal Equilibrium* (LTE) and is much more easily treated theoretically. In atomic emission, such a state prevails in the observation zone of the dc arc, the dc and ICP plasmas, and in the glow discharge and hollow cathode lamps, but not in the ac spark or in most discrete sampling or direct injection modes of operation. Since subatmospheric processes are difficult to study from first principles it is also expedient to put aside glow discharge and hollow cathode sources until the next section.

What is left can be examined by the Maxwell-Boltzman Distribution Law, the Saha Equation, and other treatments to establish expected concentrations of atoms, ions, and electrons in the plasma, and to predict emission intensities at various wavelengths. The spectrometric processes are modeled exactly as if the analytical species were contained in a constant temperature furnace at atmospheric pressure.

The Maxwell-Boltzman Distribution Law can be understood as a generalization of the Maxwell Distribution Law, which defines the relative number of atoms or molecules of a single type forming a gas that have a given range of kinetic energies under isothermal, nonturbulent (and otherwise ideal) conditions. It can take the form:

$$(\Delta N/N) = (4h^3)(\Delta v)v^2c^{-h^2v^2}/(\pi)^{1/2} \quad (1)$$

where $(\Delta N/N)$ is the fraction of total atoms (or molecules), N, that have velocities confined to the interval Δv, which ranges from $[v - 1/2(\Delta v)]$ to $[v + 1/2(\Delta v)]$, and with v as its midpoint. The constants h (Planck's), c (speed of light) and π have their usual meaning and values ($h = 6.6 \times 10^{-27}$ erg · sec; $c = 3.0 \times 10^{10}$ cm/sec; $\pi = 3.14$).

The Maxwell-Boltzman Distribution Law is expressed as:

$$N(c) = 4\pi N(m/2\pi kT)^{3/2}c^2e^{-mc^2/2kT} \quad (2)$$

where $N(c)$ = the number of atoms (or molecules) with velocities between c and $(c + dc)$; N = the total number of atoms (or molecules); m = the mass of the atom (or molecule); T = the absolute temperature; k = Boltzman's constant (1.38×10^{-16} erg/K).

The Saha equation describes the process of ionization as an equilibrium process in a gas:

$$K_A = (n_e n_{A^+}/n_A) = [2(2\pi mkT)^{3/2}/h^3](Z_{A^+}/Z_A)e^{-E_A/kT} \quad (3)$$

where K_A = the equilibrium constant for the reaction, $A \rightleftarrows A^+ + e^-$; n_e = the electron density (number of electrons/cm^3); n_A = the atom density (number of atoms/cm^3; m = the mass of the electron (9.1×10^{-28} g); k = Boltzman's constant; T = the absolute temperature; h = Planck's constant; Z_{A^+} = the partition function for A^+; Z_A = the partition function for A; E_A = the ionization energy of A.

At 1 atmosphere pressure this expression can be reduced to:

$$K_a = [(4.83 \times 10^{15})T^{3/2}](Z_{A^+}/Z_A)10^{-5040V_A/T} \quad (4)$$

where V_A = the ionization potential of A in electron-volts.

The intensity of light emission associated with a given electron transition is given by:

$$I = (NhcgA/\lambda Z_M)e^{-E_E/kT} \qquad (5)$$

where I = the intensity of light emission (in erg · sec/cm^3 · steradian); N = the upper excited state atom or ion density (number of atoms or ions/cm^3) h = Planck's constant (erg · sec); c = speed of light (cm/sec); λ = wavelength (cm); g = the degeneracy of the upper state (a correction for the equivalence of other transitions); A = the transition probability (sec^{-1}); Z_M = partition function; E_E = the excitation potential (eV); K = Boltzman's constant (eV/K).

Such calculations serve to illustrate the physical basis of the analytical measurement but they have little practical value for the working spectroscopist. Their predictive value is poor with real-world samples because, even with simple, idealized materials, it is impossible to accurately determine the upper state population density.

2.3 Transient and Subatmospheric Processes [4]

Unlike the dc arc, spark sources are discontinuous—in effect, they consist of a series of very short duration dc or ac arcs. The cloud of ions and excited atoms that forms expands outward from the sample surface, producing the measured emission. But unlike the vapor cloud from the dc arc it does not contribute to the current transport. Excitation in the spark is, thus, decoupled from the electrical flow. During the interval of the spark discharge power in the analytical gap and the associated plasma, temperature are both higher than they are in the dc arc. The resulting spectrum is, thus, more complex, even including some ion lines. The analytical measurement is an integrated average of the light emission from a rapidly fluctuating process that is not in Local Thermal Equilibrium and, therefore, impossible to model simply, even under idealized conditions.

Direct insertion probes for the ICP also generate data that cannot be modeled theoretically. Since lengthy preburns are usually not practical for this type of work, LTE has not been achieved when the measurement exposure begins. The same can be said for laser microprobe sources, whether or not they are utilized in conjunction with an auxiliary spark.

Low pressure discharge sources, such as hollow cathode lamps and glow discharge lamps, are difficult to treat theoretically because the equilibria that prevail are far from the thermal atmospheric state where the Maxwell-Boltzman and Saha treatments are applicable. Microwave plasmas operate at very high frequency, again defying any simple theoretical analysis.

The conclusion appears to be that emission spectroscopy, at least for now, must remain an empirical science. For processes that approximate the condition known as Local Thermal Equilibrium we can dimly perceive the quantitative mechanism at work in ideal cases. For other conditions the models fail. Today, optical emission spectrometry must remain dependent upon calibration with reference materials whose compositional values have been established by more fundamental methodologies, most notably, perhaps, classical wet chemical techniques. In the trace realm, where classical techniques are weak, reference materials must be certified using more exotic technologies, or calibration from synthetically prepared solutions must be relied upon.

3. INSTRUMENTATION

3.1 Excitation Sources

3.1.1 Flame

There is no need to repeat here what was said about flame emission sources in the previous chapter. It is sufficient to say that their practical application in trace metal analysis is largely confined to a few analytes, most notably sodium and potassium. Today, nitrous oxide/acetylene premixed flames with a slot burner are most common. A good monochromator is required.

3.1.2 Arc [2,4]

The *dc arc* is by far the most common type of arc excitation source. In most configurations the sample is the anode. The source is a high current/low voltage discharge that can be provided by a simple dc power supply. Voltage ranges from 10 to 50 V and current from 1 to 35 A. The dc arc is hotter than chemical flames but must be considered low temperature among electrical sources. The graphite cathode electrode is heated to about 4000 K. This leads to the thermal emission of additional electrons, which, combined with the normal current flow, can heat the sample anode to about 7000 K. The discharge of electrons from cathode to anode is continuous once initiated by a spark source, or (now rarely) by bringing the electrodes into contact, then separating them. There is also a counter flow of positive ions from the anode to the cathode. The conduction channel attains an equilibrium state somewhere between 4000 and 7000 K. Light emission also occurs from a nonconducting envelope of heated gas that surrounds the conducting channel. Temperatures in this region range from 3000 to 4000 K.

The dc arc is most often employed for quantitative trace or qualitative impurity survey analyses of powder samples, which are loaded into a graphite crater electrode. This consists of a recessed cup-structure at the end of a graphite rod. As the anode heats sample components are sequentially evolved in the order of their boiling points. The introduction of various additives in close admixture* with the sample can be used to tailor the resultant discharge plume to enhance the measurement of the desired analytes. A *carrier*, such as silver chloride, may be added to transport certain trace elements into the vapor state (the *carrier distillation method*). A *spectrochemical buffer*, such as lithium carbonate, may be added to stabilize the discharge by providing copious amounts of ions and electrons. In the absence of a buffer the volatilization of a given boiling range of sample components is likely to affect current flow and the resultant arc temperature. Also, graphite powder is sometimes mixed with the sample and other additives. Its purpose is to serve as a diluent and to promote the "burn" by increasing the anode temperature.

Some practices for powders in crater electrodes specify leveling the sample charge, then inserting a pin into the center to produce a small hole. Often the packed electrodes are dried in a drying oven or under an infrared lamp prior to the measurement. These techniques are intended to prevent the rapid gas evolution that often occurs during arcing from dislodging sample particles.

Since the dc arc powder technique is most frequently employed for the trace determination of an array of analytes representing a wide range of boiling points typical preburns are short (<5 sec) and exposure times are long (1 min or more). Detection limits are good, ranging from fractional ppm levels for the most sensitive element lines (e.g., Ag338.3 nm, Be313.0 nm, Co345.3 nm, Ga294.4 nm with a sodium fluoride buffer) to about 10 ppm for the least sensitive. It is quite possible to adjust the mixture of sample material and additives and the excitation and measurement parameters to maximize the response for certain analytes, and to minimize the response for obtrusive matrix elements. For example, the carrier distillation approach, referred to above, has been utilized to good effect in impurity surveys of uranium metal, where the highly complex base metal spectrum is decreased dramatically in intensity.

When the dc arc is operated in air, intense cyanogen band structures result from the reaction of the graphite electrodes with atmospheric nitrogen. These bands can be troublesome for certain analytes. But they can be eliminated by purging the arc chamber with argon. In another approach (the Stallwood jet) a 75% argon/25% oxygen mixture is utilized in a flowing stream that both eliminates cyanogen and retards wandering of the arc.

Arc wander, however, is but one of the contributions to poor precision with the dc arc. The test portion that is actually measured is large for an atomic emission technique. It is this fact that is most responsible for its high sensitivity. But at the same time it results in a high density plasma whose emission

---

* A small vibrating cup and ball mixer, such as that once used for preparing dental amalgams, can be useful for this purpose.

lines are prone to self-absorption. The fractional distillation of matrix components is a complex process that leaves many analytes only partially volatilized even after a two-minute burn. Moreover, there are synergistic effects due to the presence of easily ionized elements in the sample matrix.

The *globular arc* technique is an approach that has been applied to small metal pieces, especially sections of wire. The test sample is placed in the conical recession of a graphite electrode. The dc arc is struck from a carbon cathode and maintained at a 3-mm analytical gap in air. The sample melts into a globule, then oxidizes, during a 5 to 10 sec preburn. Exposure then begins and lasts about 20 sec. The approach has been effectively applied to the trace analysis of copper.

Direct current arc spectra are characterized by comparatively low background and comparatively small numbers of lines—mostly those of excited neutral atoms. It is a highly sensitive technique for tramps, especially when measures are taken that suppress matrix emission. Perhaps its greatest strength is for the rapid screening of high purity metals and alloys for trace impurities. These procedures can range from visual inspection of a photographic plate or film with a microdensitometer, or even a jeweler's loupe to a computer readout from a 40-channel direct reader.

The dc arc's reputation for poor precision is only partially mitigated in two other forms of arc excitation: the *ac arc* and the *pulsed dc arc*. These both show sensitivities roughly similar to the dc arc. They both operate at somewhat lower temperatures. The ac arc operates at a supply current of about 1 ampere. Because of ions that remain in the arc column as the positively charged voltage drops past zero the arc spontaneously re-ignites as the negative voltage mounts. Pulsed dc arcs are unidirectional discharges of long duration (compared to a spark). They may be electronically triggered or "air-interrupted" continuous discharges. Neither ac arcs nor pulsed dc arcs are much used these days. No commercial power supplies for these two source types are currently being manufactured.

3.1.3 Spark [2–4]

Spark excitation is a high voltage transient pulse applied at high frequency. The light emission produced by the sample is similarly pulsed and must be time-integrated either by photographic exposure or by electronic means. As the current pulse drops the discharge channel collapses and a portion of the plasma cloud atoms are vapor deposited on the counter electrode.

Often a standard mixture of element oxides is measured at the same time at different dilutions on the same plate or film under the same conditions. It is then used for line comparisons and either qualitative or semiquantitative assessments.

Production of a spark discharge is a little more involved than the production of an arc. A charging circuit is used to convert line voltage by means of a transformer to several kilovolts, which is used to charge a capacitor. A high speed trigger switch opens and closes rapidly, allowing the capacitor to discharge across the analytical gap, then quickly recharge for the next cycle. For an ac spark, typically sources utilize a 5–100 nanofarad (nF) capacitor that is charged to somewhere between 1,000 and 30,000 volts. The frequency of the spark discharge ranges from 50 Hertz (Hz) to 2 kilohertz (kHz) or higher, depending on the design and the user-selectable source parameters (these may include the resistance, the capacitance, and/or the inductance).

The spark source for atomic emission has undergone much more evolution than the arc. Today, modern spark sources are dc (or unidirectional), employing a diode rectifier in the circuit. The analytical cycle typically begins with a high energy pre-spark (HEPS), which is designed to rapidly heat, melt, and homogenize a region on the solid sample surface. The effect is to eliminate the influence of the sample's thermal history. HEPS is characterized by a high current flow during the discharge (approximately 130 amp), which, in a very real sense samples a test portion and prepares it for measurement. The duration of the HEPS phase is method- and matrix-specific.

In developing a new procedure one may wish to prepare "burnoff curves"; these are graphs in which the signal for a given analyte is plotted versus time, starting at the initiation of the HEPS. Eventually a "steady state" is reached for each element, at which point no further increase in signal occurs. A compromise HEPS duration must then be selected for use on a direct reader. Durations range from 10 to 60 sec. The use of a so-called high repetition rate source (HRRS)* can reduce these times by more than half, but these sources are expensive to build and buy.

HEPS is an aggressive discharge, penetrating about 50 μm below the surface of a steel sample at the center of the burn crater. The spark gap is flushed with argon for a number of reasons. First, to prevent sample oxidation; second, to allow the use of lines below 200 nm; and third, because HEPS in argon produces high electron field densities at inclusion boundaries, causing preferential attack at inclusions larger than about 1 μm in diameter. This contributes to the homogenization process. In modern instruments the spark chamber is continuously flushed with argon at about 0.5L/min. This is increased to an analytical flow of about 4L/min just before sparking begins.

HEPS is followed immediately by the spark measurement cycle. This is a lower current (approximately 60 amp) discharge. The individual sparks are

---

* A typical HEPS on a conventional source may operate at 300Hz while a HEPS on an HRRS may operate at 400-500Hz.

slightly shorter in duration than those in the HEPS. In a typical modern instrument, software sets the lower energy parameters automatically and simultaneously starts the exposure.

During this measurement the power of each individual spark discharge is greater than that of the dc arc, producing a plasma that is momentarily much hotter than the continuous dc arc plasma. Spark spectra, thus, have more lines, including some ion lines that are absent in arc spectra. The power of the spark determines the average spectrum intensity, but the shortness of the spark duration (i.e., the portion of the individual pulse during which the applied energy remains constant) determines the degree of "spark-like" character in the spectrum. Thus, short, high current discharges produce highly spark-like spectra, while lengthy, low current discharges produce arc-like spectra.

Sometimes both sorts of excitation are used in the same measurement. This proves useful if major, minor, and tramp level analytes are to be measured simultaneously. In this case, after major and minor alloy components are measured by a conventional spark, the software program switches to a 20 ampere, long duration discharge to produce an arc-like spectrum for the tramp element channels.

Examination of the burn spot is always a good practice. Very large burn spots, such as those produced by arcs and arc-like discharges may indicate that complete homogenization of the test portion may not have been accomplished during the preburn. A white ring of surface oxides surrounding the burn crater indicates that a trace (or more) of oxygen has gotten into the spark chamber. In the absence of oxygen a very dark, black ring forms. This is composed of microscopic, vapor-deposited metal particles, and generally indicates a "good burn."

Certain metals of poor thermal conductivity (e.g., sintered metals and cast irons) may get hot enough to crack or break up during sparking. This problem may sometimes be quickly solved by switching to a lower energy per discharge.

The spark is noted for producing precise results, even for major concentration components, like chromium and nickel in stainless steels. It is not particularly sensitive, however, and its detection limits, even under ideal conditions, do not challenge those of the dc arc. Solid samples in the form of rods have been used for *point-to-point* analyses—either using a second piece of sample as its own counter electrode, or using a carbon counter electrode. Much more commonly these days a Petrie stand is utilized to hold a solid sample disk [typically, 1–3 in (2.5–7.6 cm) in diameter] and a thoriated tungsten counter electrode. This is the well-known *point-to-plane* configuration. Occasionally, powders are compressed into pellets for use with a point-to-plane approach, but binders must be carefully chosen so that the resultant solid is electrically conducting.

At one time (before the advent of plasma sources) solution-spark techniques were fairly common. In the *porous cup technique* hollow porous carbon electrodes are soaked in prepared sample solutions, dried, then sparked. The *rotating disk* is an edge-grooved carbon wheel fitted on a motorized spindle in the spark stand. The bottom of the wheel is immersed in a shallow boat of prepared sample solution and the top of the wheel confronts the counter electrode across the analytical gap. The *rotating platform* employs a dried sample residue on a horizontally mounted carbon wheel. The *copper spark* method was developed to determine rare earths in nuclear materials. It is sensitive enough to be employed routinely even today. The prepared sample solution is evaporated drop-by-drop on the flat end of a 1/4 in. (0.625 cm) diameter high purity copper rod that is ultimately sparked as the sample electrode. The copper aids in generating a sensitive, low detection limit measurement. The method is applicable to many other analytes and alloy matrices.

3.1.4 Plasma [2,4,10]

The *dc plasma jet* (DCP) is essentially a uniquely configured dc arc stabilized by flowing argon streams. In the most successful design two carbon anodes and one tungsten cathode form an inverted "Y." Argon is introduced as a vortex flow of 1L/min at each of the anodes, and at 2L/min at the cathode. The argon flows create a thermal pinch of very high current density. The solution sample is introduced as an aerosol in a 4L/min argon flow just under the crook of the inverted "Y," where it is excited and where its emission is observed. This highly localized region of sample emission is below the arc channel so that background emission is low compared to that of a conventional dc arc. Because the excitation region is small, an echelle grating spectrometer is commonly employed to separate and measure the light. Recently, a charge transfer solid state detector has been introduced for such a system.

Calibration of most analytes with the DCP results in three orders of linearity—generally inferior to the ICP, but much better than atomic absorption. Long term drift is endemic to the DCP because the electrodes wear away, eventually requiring replacement. The temperature of the DCP is sufficient to dissociate and excite refractory oxide-forming elements. However, the very short dwell time in a very small observation zone results in detection limits generally higher than those obtained by ICP.

Microwave plasmas include those called *capacitively coupled microwave plasmas* (CMPs), *microwave induced plasmas* (MIPs), and others. They often employ helium instead of argon. Their greatest success (and only commercially available configuration) is as an elemental detector for gas chromatography. Refractory compounds (like Ta_2O_5) do not dissociate well in such plasmas, although a high electron temperature yields satisfactory detection limits for most elements.

The most successful plasma source for liquid samples is the *inductively coupled plasma* (ICP), developed over the decade 1965 to 1975. The ICP torch has assumed various forms, but what is sometimes termed the Fassel Torch can be described as three concentric quartz tubes, the center tube tapered to a fine point. The solution aerosol is transported with about 1 L/min of argon through the tip of the center tube. An auxiliary argon flow of 1–5 L/min is sometimes necessary between the two innermost tubes in order to light and sustain the plasma. Between the middle and outer tubes, a cooling argon flow of about 15L/min is applied tangentially, maintaining a toroidal pattern to efficiently carry away heat. An induction coil, consisting of 3–4 turns of water-cooled copper tubing encircles the torch. This is sometimes referred to as the *load coil*.

Early work by Greenfield, *et al.* [11] utilized a low frequency (7 MHz), high power (5–12 kW) circuit design to sustain an argon/nitrogen plasma. Fassel [12,13] chose to work at high frequency (27–50 MHz), low power (1–2 kW) with pure argon. This proved to be the better design since it results in a high concentration of current flow on the plasma surface.˙ Combined with a high sample gas velocity the center of the plasma is cooled to about 7000 K. This allows the sample aerosol to enter. Since virtually all of the aerosol forms a dense line in the center of the plasma there is no analyte on the plasma surface to cause self-absorption. The analyte requires about 2 milliseconds after entering the plasma to attain the observation height (15–20 mm above the top of the load coil). This is long by spectrometric standards and allows time for complete atomization to occur. The confined shape of what becomes the free atom stream, and its stable temperature profile at the observation height make for a nearly ideal emission source.

The buffering effect of easily ionized elements, which strongly influences chemical flames and electrical arcs, is not a factor with the ICP source. The reason is that the very high electron density of the plasma overpowers any cooling effect from species of low ionization potential. The spectra produced by an ICP source are uniquely rich in ion lines and are most usefully explored with reference to a compendium prepared specifically for ICP work. Several of such references are available [14–16], providing a valuable supplement to the more general *MIT Wavelength Tables* [17].

Modern commercial instruments utilize sources designed to operate at one of two FCC-approved radio frequencies: 27.12 or 40.68 MHz. This radio frequency is applied through an impedance-matching circuit to the water-cooled copper coil that surrounds the torch, producing a strong, rapidly changing magnetic field in the flowing argon. A spark is then applied from a Tesla

---

˙For a history of the development of the ICP source see reference (10).

coil or other source, creating a small, instantaneous channel of ions and electrons. The changing magnetic field induces eddy current flows that rapidly heat the increasingly conductive argon. Very quickly a hot, stable discharge in the shape of an elongated torus is created. The temperatures range from 10,000 K in the center of the load coil to about 7000 K at the normal observation height, where ion lines are abundant.

The ICP source is so hot and stable that it is not unreasonable to expect a linear calibration range extending over five orders of magnitude of analyte concentration. The extreme stability of this source also makes the use of a sequential scanning monochromator practical for multi-element atomic emission work. Detection limits are dependent upon a number of instrumental factors, but most importantly upon the efficiency of nebulization and aerosol transport to the plasma.

Typically, a peristaltic tubing pump draws sample solution and continuously passes it to one of several types of nebulizer. For dissolved metals a conventional *cross-flow nebulizer* in either fixed or adjustable design is usually best. The aerosol passes into a spray chamber that contains baffles or a flow spoiler to remove large droplets. The aerosol that reaches the plasma should have a maximum droplet diameter of 15 μm. The sample solution uptake rate is 1–3mL/min. About 3% of the nebulized sample reaches the plasma. If the sample solution contains high solids a *Babington-type nebulizer* is called for. This design utilizes a "V"-shaped channel instead of an orifice to bring a thin stream of solution in contact with the sample argon flow. While resistant to clogging the Babbington design sacrifices some sensitivity and precision.

There are also HF-resistant nebulizers, utilized in conjunction with a sapphire-tipped torch. These also show degraded precision and some torch designs tend to clog at the tip.

The *concentric glass or Meinhard nebulizer* is intended for very dilute solutions, such as water samples, since its good sensitivity is compensated by a tendency for the fine capillary to become blocked. *Grid and frit systems* rely on the deposit of sample on a large surface area from which it is blown into droplets. *Ultrasonic nebulizers* are highly efficient with about 10% of the sample reaching the plasma. However, they require desolvation by heating, then cooling the aerosol to avoid extinguishing the plasma.

The torch itself takes different forms that are instrument-specific, including a "mini-torch" design that lowers operating costs by utilizing less argon. A recent development is the *axial-viewed plasma* in which the torch is mounted horizontally and the plasma measured by sighting down the sample channel in the plasma torus. This can result in a fivefold detection limit improvement for "clean," low matrix samples, but may produce problems for dissolved metals samples that are easily handled by conventional transverse measurement.

While the measurement precision for transverse and axial viewing is about the same, with axial viewing certain elements show increased self-absorption at high concentrations, in addition to an improved detection limit. The net effect, then, is to shift the working range to lower concentrations. With axial viewing, in general, matrix effects tend to be more severe, with larger enhancements and suppressions. Also, molecular emission due to the OH species in the plasma becomes an annoyance with axial viewing, being threefold more intense than in the transverse arrangement [18,19]. On the whole, there appears to be no distinct advantage to axial viewing for work with complex metals, however, a laboratory that analyzes a variety of samples may want to consider one of the commercial designs that permit both traverse and axial operation.

One manufacturer offers an ICP/OES/MS combined instrument in which the light is sampled transversely. Design considerations for simultaneous MS operation require the OES observation zone to be located at a nonideal, compromise position. When used without the MS, the OES observation zone can be optimized, however.

There are a number of alternative means for introducing sample into the ICP. These may be categorized by sample type as liquid or solid techniques, or by operating mode as continuous or discrete techniques. The direct analysis of solid metal has a distinct advantage for busy labs in terms of the time saved by avoiding the weighing, dissolution, and dilution steps of a solution technique. A disadvantage to most direct solids work is the potential for the test portion (i.e., the actual sample weight that is measured) to be unrepresentative of the bulk material. Discrete measurement techniques tend to be much more sensitive than continuous measurement techniques since the response of a concentrated pulse of signal is integrated. A disadvantage is that discrete techniques seldom show the level of precision shown by continuous techniques (where a rate is being measured). Discrete techniques also usually require very small test portions, which may be unrepresentative (solids) or may be difficult to transfer with high accuracy (liquids).

With these observations in mind we can now briefly consider a few sample introduction schemes that have been employed, and in most cases commercialized, for ICP-OES. The *direct insertion* or *sample elevator technique* is a discrete solid sampling approach in which the sample is placed in a small graphite crucible at the end of a rod, which replaces the innermost injector tube of the torch. The crucible is located below the load coil until the plasma has achieved stable operation. Then the crucible is manually or electromechanically raised into the plasma. *Graphite furnace vaporization* utilizes an apparatus nearly identical to that in a GFAA instrument to vaporize, but not atomize, a 10 mg or smaller solid sample. The central injection hole in the graphite tube must be plugged or absent so that the maximum amount of vaporized sample is swept out with the argon stream into the plasma.

Hydride generation is usually applied to ICP-OES as a continuous flow injection approach in which a flowing sample solution stream and a flowing sodium borohydride solution stream are brought together and the generated gaseous hydrides are introduced into the ICP. It has recently been shown that mixing at a merge point close to the nebulizer orifice minimizes the interference of transition metals (which take longer to react with the reagent than do the hydride-forming analytes) [20]. Finely divided solids have been mixed with liquids and introduced into the plasma by pumping them through a high solids nebulizer in a *slurry sampling* approach.

Perhaps the strongest solids method for metals and alloys is the *spark ablation* approach in which a high voltage spark from a tungsten counter electrode is used to ablate a fine metal aerosol, which is transported by an argon stream into an ICP. Such a system is capable of generating highly accurate and reproducible results with little or no memory effects between samples, provided design criteria for the system are capable of producing aerosol particles of optimal size and shape. Several commercial designs for spark ablation devices (some with associated cyclones and other aerosol-conditioning accessories) are being marketed.

An area receiving a fair degree of study is *laser ablation* for ICP-OES (although the ICP-MS application is even more widespread). A ruby Nd:YAG or excimer laser of several hundred millijoules (mJ) power is typically employed to generate an aerosol from conducting or nonconducting solids. Laser ablation/ICP-OES requires longer "preburn" times than spark ablation/ICP-OES. However, unlike the latter, which is strictly a bulk analysis technique, laser ablation can be used as a probe of small sample areas [21]. The amount of aerosol generated can be measured by light loss in a flow-through cell and used to continuously correct the emission signal for variations in ablation rate [22].

Sample introduction devices, such as the spark ablation and laser ablation accessories described above, operate by separating the sampling and excitation processes that are normally combined in other sources for solids (one notable exception was the AG/CAP* marketed by Applied Research Laboratories [23]). To a degree this separation allows the sampling and excitation processes to be individually optimized.

Optimization of the conventional liquid aspiration mode is also worth noting before we close this section. Aspirating a 1000 ppm solution of sodium or yttrium will readily reveal the interface between the normal observation zone (NOZ) (this ordinarily is 15–20 mm above the top of the load coil) and the initial radiation zone (IRZ) just below it. Adjustment of the sample

---

*"Aerosol Generator/Capillary Arc" was a device that used a dc arc to generate a metal sample aerosol that was then transferred to a dc capillary arc operated in one atmosphere of argon.

(and sometimes the auxiliary) argon flows will move the interface to the correct position. In some cases a mass flow controller is provided to allow very fine adjustment of the sample argon flow. The sodium solution will produce a bullet-shaped visible emission that defines the two zones. The yellow color is due to excited atoms in the IRZ, whereas sodium ions in the NOZ are colorless. The color returns in the cooler tail above the NOZ. Similarly, yttrium is red in the IRZ as excited atoms and blue in the NOZ as ions, and then red again as excited atoms in the tail [24].

3.1.5 Glow Discharge

Glow discharge has been known and studied as an emission source for over 80 years [25], but the form in use today is traceable to the design of Grimm, published in 1968 [26]. Until recently most work has been with dc glow discharge sources, which require a conductive solid pin or disk. Recent ongoing work led by R. Kenneth Marcus at Clemson University has led to the development of a practical radio frequency glow discharge source applicable to nonconducting samples [27].

As with some of the hyphenated techniques discussed in the preceding section, the *glow discharge lamp* (GDL) separates the sampling and excitation processes, but only by a very short distance. The solid test material (externally mounted to a pumped low pressure chamber) is sampled by a sputtering process with low temperature argon ions. The sputtered sample atoms diffuse to a *negative glow* region where high energy electrons excite them into a light emitting plasma. The light passes through a quartz window into the spectrometer.

Since the sputtering process is nonthermal there is no selective volatilization of low boiling elements. And since the high energy negative glow region remains diffuse there is little Doppler line broadening of the spectrum.

One would surmise that the glow discharge source would be widely accepted and used but price and availability have limited its application until recently. Another problem is the fact that the sputtering process is matrix dependent in a complex manner. Quantitative results depend on closely matching standards and/or the normalization of intensities when all sample components are known. The quantification problem increases when it is desired to use this source for depth profile analysis. Bengston and Lundholm have worked out empirical treatments applicable for such work [28].

The glow discharge is an extremely stable source, and as such has been applied to fundamental studies with high resolution Fourier transform atomic emission spectrometers. Detection limits in steel matrices have been reported in the 0.3 to 1 ppm range [29]. Among the particular strengths of this approach in industry is the direct analysis of compacted metals of less than

100% density—a sample type that creates problems for other forms of excitation.

3.1.6 Hollow Cathode [2]

The demountable hollow cathode lamp (HCL) has been commercialized, accommodating 10–30 mg of metal chips. The test sample is weighed into a 4-mm ID hollow graphite electrode, which is mounted in a lamp housing. The chamber is pumped down, then back-filled with helium at 15 mbar (1500 Pa). If 400 volts are applied between the sample cathode and the water-cooled lamp housing a low pressure glow discharge is formed. In the commercial design the cathode itself is not cooled and thus the cavity emission is enhanced by cathodic heating.' It is possible to increase the current gradually so that volatile elements are vaporized sequentially from the molten sample. Nickel-, cobalt-, and iron-base high temperature alloys have been successfully analyzed for tramp elements by this technique. After a 20 sec preburn at 100 mA, measurement is taken over the range 50 mA to 1A for a total exposure of 200 sec. For certain high temperature alloys results are improved by premixing the test sample with 30% by weight of high-purity silicon metal. The most sensitive elements (detection limit: <0.01 ppm) are cadmium, calcium, indium, magnesium, silver, and sodium. Of intermediate sensitivity (0.01–0.1 ppm) are bismuth, copper, gallium, lead, manganese, thallium, and zinc. The least sensitive elements studied (0.1–1 ppm) are antimony, arsenic, barium, germanium, potassium, selenium, tellurium, and tin.

Other hollow cathode designs have been published, some of which resemble the Grimm glow discharge lamp, at least superficially. Use of an internal standard, either added to a compressed metal powder sample or attributed to a major element in the matrix has proven useful [30].

3.1.7 Laser

The direct use of laser light for the routine analysis of metal alloys is still in a development stage. The laser microprobe is a notable exception, but it is primarily designed to study minute areas in a qualitative or semiquantitative mode, as a support for the metallurgical investigation of inclusions and phases. And in some configurations it is not a direct laser technique but utilizes an auxiliary spark to further excite the laser plume.

A pulsed Nd:YAG laser has been used to depth profile zinc/nickel and tin coatings on steel [31]. Various schemes to utilize a laser to obtain

---

Due to inherent design differences the glow from a GDL (with a high gap voltage and long free ion paths) is termed "abnormal" while the glow from an HCL is termed "normal."

quantitative information on the composition of in-process molten metal have been published [32–35]. Laboratory-based work on solid samples [36,37], and the field sorting of mixed alloys [38] are also active development areas. Much of this work involves the use of time resolved measurement in which only the most analytically useful emission from a laser pulse plume is recorded.

3.2 Optics

3.2.1 Definitions [9]

Before describing the optical systems of atomic emission instruments it is appropriate to briefly review some basic terminology.

Astigmatism: An optical aberration in which the image of a point off the optical axis is a pair of lines.

Chromatic aberration: The effect (due to the fact that glass refracts light of different wavelengths by different amounts) in which a lens shows slightly different focal lengths for different colors.

Collimator: A lens arrangement for producing parallel rays of light.

Focal length: The direct line distance from the focal point to the lens (or concave mirror).

Focal plane: The location (e.g., in an atomic emission spectrometer) where separated light is designed to be detected and measured.

Focal point: The point at which emerging light rays from a lens (or reflected light rays from a concave mirror) meet and are in focus.

Focal power: The reciprocal of the focal length; a measure of the converging or diverging effect of a lens or mirror. Commonly expressed in reciprocal meters (or *diopters*).

Optic axis: A reference line through the vertices and foci of lenses and mirrors in the graphical construction of optical systems.

Ray tracing: The geometric construction of the path of a light ray from a point, through an optical system, to an endpoint (in the spectrometer, the detector).

Reciprocal linear dispersion: The number of nanometers (or angstroms) separating two spectral lines that are 1 mm apart on the focal plane.

Relative aperture (f number): The ratio of the focal length to the slit area; a measure of the light throughput, which affects the signal-to-noise ratio, and thus precision.

Resolution: The minimum distance that must separate two spectral lines in order for them to be recognized as separate by a given optical system.

Spherical aberration: The effect in which the outer portions of a spherical lens or mirror produce a focus in a different plane than does the center.

3.2.2 Conventional Plane Mountings [4]

There are many possible spectrometer optical designs, and over the years a number of them have achieved some commercial success. Design configurations tend to acquire the name of their originator or popularizer. Today many of these spectrometer designs have become history: Wadsworth, Eagle, Cornu, Littrow. These have been largely replaced with optical plans that tend to be "folded" and thus more compact, requiring a smaller "footprint" for the optical bench.

There are two conventional plane grating mounts that are currently widely used for sequential optical emission spectrometers. The *Fastie-Ebert* mount uses a large concave mirror to direct incident light (from the sample plume) to a plane grating. The same mirror intersects the dispersed spectrum and directs it to either a photographic plate or to a 45° mirror and photomultiplier. The *Czerny-Turner* mount uses a small collimating mirror to direct the sample light to a plane grating. The dispersed spectrum is directed to a separate focusing mirror, which directs it to a series of exit slits. Alternatively, a photographic plate or film cassette can be inserted in the focal curve. For sequential scanning operation the focusing mirror is replaced by a spherical or parabolic mirror that directs the dispersed light to a single exit slit.

The focal plane image in plane mounting instruments is generally very high quality. Great pains are taken in the optical design to minimize chromatic, spherical, and astigmatic aberrations by careful matching and alignment of the collimating and focusing optical elements.

Sequential OES instruments use software control to slew the grating to the region of the selected line, then initiate a peak-seeking routine that searches for a match to the line profile stored in memory, and finally to a sequence that searches for the maximum within that profile. Despite the fact that computation and slewing rates are designed for speed, when the number of analytes is large, measurement times tend to accumulate.

3.2.3 Echelle Mounting [4]

An echelle grating (the word means "ladder" in French) consists of fine lines that are ruled a much greater distance apart than those of conventional diffraction gratings. About 100 lines per in. is typical. These are very high resolution gratings, but they have a rather narrow wavelength range. Commonly, an echelle grating is crossed with a prism or grating of very much lower resolving power. This serves the function of separating the overlapping orders in steps at a right angle to the individual spectrum dimension. Most echelle mountings use a modified Czerny-Turner arrangement. Such an instrument with a 0.75 m focal length will show the same resolution at 250 nm as a 3.4 m

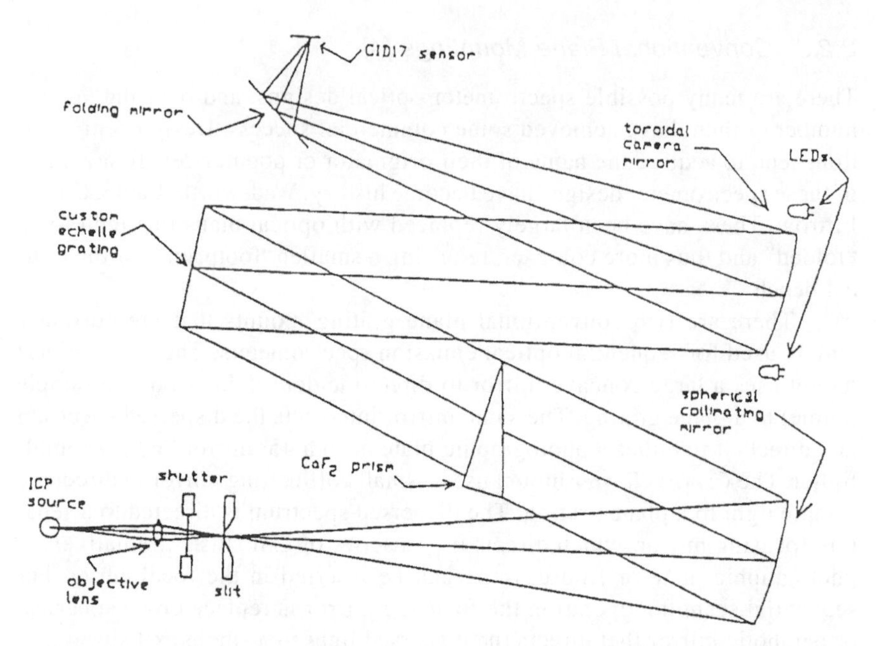

Figure 6.1 Optical diagram of Thermo Jarrell-Ash echelle/CID spectrometer. (Reprinted from J.V. Sweedler, K.L. Ratzlaff, M.B. Denton eds. Charge-Transfer Devices in Spectroscopy, VCH Publishers, New York, 1994. Used with permission.)

Fastie-Ebert conventional plane grating instrument. Wavelength scanning is generally not employed with echelle mountings because the orientation of the crossed dispersing elements must be precisely maintained.

The rectangular array of ordered spectra that is produced at the focal plane can be photographed. Early designs of a commercial instrument utilized a Polaroid camera and a series of prepared masks for rapid qualitative survey work. In the 1970s a random-access digital television camera was utilized to quantitatively process information from the focal plane [40]. Today, the new charge transfer solid state detectors are ideally suited for this application (see Figure 6.1).

3.2.4 The Rowland Circle [3,4,41]

The greatest number of commercial atomic emission instruments sold today employ a concave spherical grating in a Rowland Circle mounting. This type of grating is an arc of a sphere and produces a spectrum with a series of foci that describe a circle in space. The entrance slit sits on the Rowland Circle, as

Henry A. Rowland (1848–1901) was an American physicist who developed this optical plan.

does the concave grating. The dispersed light is measured along an arc of the Rowland Circle as well, either through a series of fixed exit slits and photomultipliers, or with a fixed or moveable photographic cassette. With the *Abney* mounting the camera is fixed, but there are two alternative entrance slits used to cover two complementary wavelength ranges. With the *Paschen-Runge* mounting the camera is moveable. Commercial direct reading instruments generally utilize a modified Paschen-Runge design with fixed exit slits, each with a photomultiplier mounted behind it. The entrance slit position may be optically folded from a different position using one or more plane mirrors. See Figure 6.2.

There is a minor price to pay for the simplicity of the Rowland Circle design, however. The entrance slit image at the exit slits is astigmatically stretched. This requires making the exit slits longer and using photomultipliers with appropriately shaped photocathodes in order to capture the majority of the light.

Fine adjustment of the observation zone in the source plume in relation to the entrance slit height combined with using an aperture at the exit slit can sometimes improve detection limits. This strategy becomes difficult to employ with a Rowland Circle design due to the astigmatism.

Direct reading polychromators generally utilize somewhere between 20 and 40 exit slits on the Rowland Circle. These are preselected before the instrument is assembled at the factory but can be changed after installation by a field service engineer. Frequently, two or more slit/photomultiplier channels are devoted to the same analyte in order to provide flexibility for both concentration range and sample matrix. While there are clever means to arrange components using mirror assembles, sometimes space constraints limit the line selection and compromise lines must be used.

There is always some means of *profiling* a direct reader. This involves displacing the entire spectrum a short distance across the slit array to center line profiles in the slit openings. This may involve rotation of the grating or of a refractor plate, or movement of the entrance slit.

The optical bench of a direct reader is mounted on a heavy metal frame of great dimensional stability. The entire assembly is likely to be in an evacuated chamber to allow access to low wavelength lines. In some designs the chamber is thermostated slightly above ambient temperature to isolate it from the effects of minor room temperature changes.

3.3 Detectors

3.3.1 Photographic Emulsions [1,42]

A photographic emulsion, looked at objectively as a detection device for spectroscopy, has many advantages. Its sensitivity is readily extended by increas-

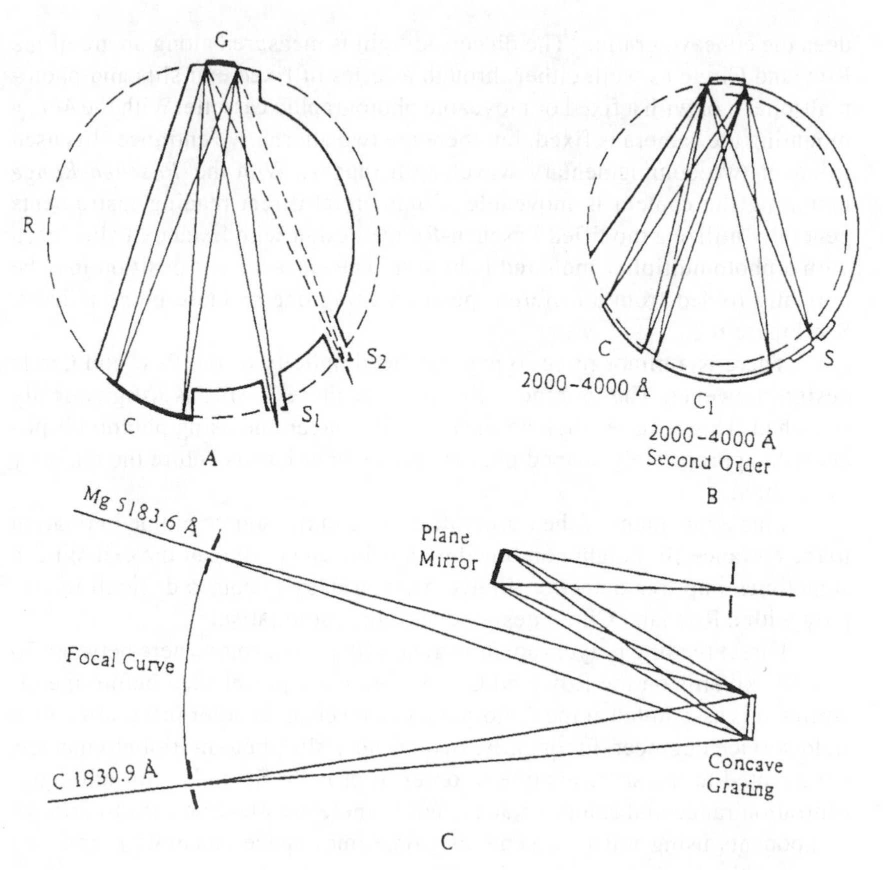

Figure 6.2 Rowland Circle optical diagrams. A: modified Abney mounting (R: Rowland Circle; G: concave grating; C: fixed camera; S_1 and S_2: alternative entrance slits to cover different wavelength ranges). B: Paschen-Runge mounting (C: moveable camera; S: entrance slit). C: modified Paschen-Runge used in direct readers (exit slits with photomultipliers mounted behind them located along the Rowland Circle; folded optics allows the relocation of entrance slit).
(Adapted from E.L. Grove, ed. Analytical Emission Spectroscopy, Vol. 1, Part I, Marcel Dekker, New York, 1971. Used with permission.)

ing the exposure time, and the potential resolution is staggering compared to any other detector.* However, the response to light is nonlinear and may be affected by temperature and other factors (these effects are known, collectively, as *reciprocity failure*). Some newer detectors have a broader dynamic

* With the pace of miniaturization in solid state devices one learns to "never say never," but pixel elements have a long way to go to rival the minute size of silver grains in a fine grain emulsion.

response range than emulsions. And then, of course, film and plates have to be developed before they can be made to release their information.

In the mid-1960s when atomic absorption was gaining wide acceptance and the dc plasma was a brand new emission source, and the ICP had yet to be commercialized, most laboratories maintained a darkroom for the development of photographic spectra plates. It is true that there were direct readers—room-size devices with impressive banks of clock dials or counter tubes—but these were most likely to be found in process control laboratories. In most other laboratories optical emission meant photographic detection.

Besides the darkroom facilities with their solutions and timers, a microdensitometer was generally required. This device was used to identify the analyte line by comparison to a "master plate," then scan across it to measure the maximum blackening of the line profile. A calculating table was also necessary to establish and apply an emulsion calibration for each batch of photographic plates or film. Intensity ratio curves were plotted by hand. And finished plates were stored in files for future reference.

Today there are very few laboratories remaining that have retained these capabilities. In most cases both the equipment and the expertise to use it have been lost. The reason is that photoelectric detection combined with cheap computer power have made it easy to perform most of the required analyses with a minimum of time and special skills. Direct reading spectrometers have come down in price and size, and very stable sources, like the ICP, have made even less expensive sequential units practical.

But there were advantages to the use of a photographic emulsion. An experienced spectrographer[*] could, in only a few minutes time visually scan a familiar matrix spectrum for foreign impurities. With only a little more time and effort a total unknown could be qualitatively surveyed. The "semi-quantitative survey" was another comparatively rapid process. Here rough estimates of the concentration levels of the major and minor components were assigned, usually based on a visual comparison of the line blackness between the unknown and a graded series of standards. Quantitative work took a great deal more time, but it was reliable. And the storage of plate or film spectra (always painstakingly identified) was a valuable archive for future rechecks, the analysis of additional analytes, and even, occasionally, as legal evidence.

Another point that is rarely made any longer is that the optical emission spectrographer understood and controlled the optics of his instrument to a greater extent than is common today. There were procedures for measuring and calculating half-width resolution, tests for proper slit illumination, and

---

[*] Technically, this term applies *only* to someone who photographs spectra. However, everyone who works in spectroscopy is properly termed a "spectroscopist."

even tests for determining the degree of astigmatism [43]. Today's instruments are high performance analytical engines that must be fine-tuned and serviced by experts. However, there may still remain someone who misses the days when they could raise the hood and tinker.

3.3.2 Photomultipliers [42,44]

Although today their position is being nudged by new solid state technology, photomultipliers remain the dominant light measuring transducer in atomic emission spectroscopy. This class of detector was briefly discussed in Chapter 4. Only a short review and a few additional details need to be added here.

A photomultiplier is unchallenged as a single channel detector. It is highly sensitive with good detection limits, showing low dark current and no read noise. Typical photomultipliers show a *quantum efficiency* of 5–20%, but some are as high as 30%. In the near infrared, however, this figure will drop below 1%.

A photomultiplier consists of an evacuated glass envelope with a window for the entrance of light and electrical connections for input power and output signal. The internal structure consists of a photocathode, a series of 4–16 electron-multiplier dynodes, and an anode where the multiplied current is read. The light window may be glass, quartz, lithium fluoride, or other material, chosen for its transmittancy. The window may be positioned on the envelope to receive light end-on or side-on. The photocathode generally consists of some cesium compound, although other substances are sometimes used. A photomultiplier's *S-number* refers to a spectral response associated with a photocathode surface material. Thus, an S-1 response designation refers to an AgOCs photocathode, which shows maxima near 380 and 810 nm and a minimum near 500 nm.

The dynodes are commonly copper-beryllium or nickel substrates coated with CsSb or BeO. Each successive dynode stage is at a higher applied potential than the previous one. The stages are geometrically arranged so that impinging electrons strike a dynode surface in an area where the electric field directs ejected secondary electrons toward the next dynode stage. Dynode stage designs are known by descriptive terms, such as "venetian blind" or "circular cage."

Due to design considerations the gain per stage is intentionally limited. The total gain, G, for a photomultiplier is simply $G = g^n$, where g is the mean

---

Quantum efficiency is a measure of the efficiency with which incident photons are converted to measurable signal.

Generally, spectral response is measured in milliamperes/lumen or milliampere/watt.

gain per dynode and n is the number of dynodes. Typically, G is in the range 10^6–10^8. The gain can be adjusted to some degree by varying the voltage in the voltage divider circuit that supplies the dynodes. The usefulness of a given photomultiplier design, however, depends most importantly on the quantum efficiency of the photocathode material and the design of the dynode array.

Instrument manufacturers select photomultipliers based on their spectral response curve and on a signal detection threshold defined as *equivalent noise input*. This latter term is the light flux that produces a signal equal to the root mean square current from all noise sources.

While component selection is seldom a consideration for the working analyst there are three features of the photomultiplier that he needs to remain aware of. The first is *saturation*—this is the point at which the radiant power on the photomultiplier is so high that the anode current will no longer increase with increasing light intensity. Photomultipliers can be damaged even before this point is reached if they are operated in the region of nonlinear response. The second is the change in the gain response known as *fatigue* (long-term drift) and *hysteresis* (short-term drift or "sinusoidal" drift). For critical measurements the photomultiplier should be pretreated by exposing it to light of approximately the same wavelength and intensity before an actual measurement is made. Many instruments have a built in "fatigue light." Finally, there is the need to be aware that photomultipliers have a limited useful service life due to leaking of the vacuum seal. Traces of atmospheric gases will increase the background noise, signalling the need for replacement.

3.3.3 Photodiode Arrays [42]

For many years spectroscopists and instrument designers have been looking for a means of recording and measuring electronically an entire spectrum at one time, mimicking in a more convenient form the best features of a photographic emulsion.

The vidicon tube, which used an electron beam to read out an array of light sensitive diodes, was an early attempt at producing such a device. While this technology enjoyed a short period of interest, it suffered from limited dynamic range and a number of other technical restrictions (image blooming, image burning, magnetic field interference, and time lag). It also required an expensive high voltage power supply.

The *photodiode array* (PDA) was a significant improvement over the vidicon. The PDA is a solid state assembly with addressable, hard-wired connections. A typical device for spectrometric applications consists of a linear

---

The effective value of the current in an alternating-current circuit.

array of 1,024 pixels,[*] each 25 μm wide and 2.5 mm high. The high length to width aspect ratio is suited to capturing the slit image from a conventional spectrometer.

In operation photons generate a current in a pixel, which, in turn, deletes the charge on a minute capacitor. After a suitable integration period the current required to recharge the capacitor is measured. The entire array can be recharged in about 10 milliseconds. The quantum efficiency of the PDA is 50–70%—better than the best photomultipliers. However, the recharging process is very noisy, resulting in read noise that amounts to 1000–1200 electrons. The dark noise can also be high because of leakage from the capacitors.

PDAs have four times the dynamic range of a vidicon tube. The also do not suffer from most of the problems encountered with vidicon technology; however, they are slower. The read noise problem with PDAs is especially troublesome with low light levels. As a result, in some applications, an image intensifier is used in front of the PDA. This usually consists of a photocathode to convert impinging photons to an electron signal, then a *microchannel plate* (MCP), which amplifies the electrons, then a phosphor layer to convert the electrons back into photons. These then strike the PDA. In such a device (sometimes known as an IPDA—intensified photodiode array) the impinging radiant flux is amplified 3,000 times.

3.3.4 Charge Transfer Devices [42,45]

There are two different types of charge transfer devices (CTD's), both invented in the early 1970s. Neither was originally conceived as an atomic emission detector. The *charge coupled device* (CCD) grew out of inspired musings about the Picturephone™ and bubble memories at Bell Labs. The *charge injection device* (CID) was originally developed at General Electric as a single-element device to monitor incident light levels during the testing of experimental memory chips. Today both technologies seem to be poised to revolutionize atomic emission spectrometry.

Before describing their differences it is important to highlight what these two new technologies bring in common to the field. Combined with an echelle optical system CTD's have transferred to electronic media most of the advantages of the photographic emulsion. Both CCDs and CIDs are available in rectangular arrays, which are ideal to receive the stacked orders of spectra that a predispersing prism and an echelle grating produce. See Figure 6.1 for one such arrangement. Broad spectral regions can be recorded and stored on tape or disk for future reference. Thus, new analytes can be quantified days or years later without re-analyzing the sample. Large numbers of emission lines

[*] Some may have forgotten that this term is a contraction of "picture element."

for a given element can be used simultaneously to calibrate and quantify, then compare, results for the analyte. In this way an assortment of wavelengths can be rapidly screened for interelement interferences. And when several different wavelengths yield the same result a high level of confidence accrues by inference to the value.

There are currently a small number of manufacturers marketing this technology in commercial atomic emission spectrometers, often in combination with other innovations, like an axial-viewed ICP torch or an ICP/OE/MS interface. Regardless of the merit of these combinations, the value of this detector technology seems certain to ensure its growth.

3.3.4.1 Charge Coupled Devices. The CCD has to be regarded as the more mature of the two CTD innovations since its widespread applications in consumer camcorders and in professional astronomy have led to numerous refinements. For example, of special relevance to atomic emission spectrometry is the development of the slow-scan camera with 50–500 kHz pixel read-rates. There are many manufacturers of CCDs worldwide.

The CCD is available in a wide range of geometries, including linear, square, and rectangular arrays of a variety of sizes, some with millions of pixels. Pixel size normally ranges between 10 and 30 μm square for scientific applications.

A CCD consists of a *p*-type silicon substrate topped with a *p*-type silicon epitaxy layer. On top of this is an insulating layer of SiO_2, and then another of SiN_2. Sitting on these refractory layers is the gate, composed of polycrystalline silicon, heavily doped with boron. If the gate is made positive and the epitaxy layer is made negative electrons can be stored in a potential well in the epitaxy. An overlapping gate from the neighboring cell, also positive, but biased slightly, allows the well of electrons to move to its well. In readout the charge packet of electrons moves from cell to cell down the rows on the chip, then picks up an end column "conveyer" that reads the row end cells and transports the charge to a MOS-FET transistor, which reads each charge packet and dumps it to ground.

The CCD pixel read is very accurate but not recoverable. Electrons, not holes, are the charge carrier, allowing faster read rates. But in the CCD architecture charges have to travel long distances. And usually much useless information must be read before the needed charge packets can be accessed.

Dark current (or, more appropriately, dark charge) in the CCD arises mostly from defects and discontinuities in the silicon. These create energy states intermediate between the valence and conduction bands, allowing heat to promote noise. Readout noise is dominated by what is termed KTC noise.

Equal to $(KTC)^{1/2}$, where K = Boltzmann's constant; T = the absolute temperature; C = the capacitance of the node being reset.

It can be eliminated by the slow-scan camera systems by a technique known as *correlated double sample* readout. Here the output node potential is read before and after a charge packet is shifted from the serial resister to the output node. The difference is the noise corrected signal.

A major problem for emission spectrometry is that CCDs are subject to blooming—the leakage of charge between cells in areas of high contrast (such as often occurs when bright and weak spectral lines occupy the same spectral region). This has been addressed with a number of anti-blooming designs, involving either changes to the standard chip (inclusion of electron overflow drains), or the timing of polarity changes during photon integration (clocked recombination, in which excess charge is neutralized by recombination with holes at the surface of the epitaxy layer).

The quantum efficiency of CCDs ranges from 40–60% at around 650 nm, dropping rapidly below about 500 nm. This shortcoming is corrected by the application of organic phosphor coatings that absorb lower wavelength photons and re-emit them at wavelengths where the CCD shows good quantum efficiency. CCDs are thus useable well into the vacuum UV.

The strength of the CCD detector is its sensitivity to low light levels. One ICP instrument manufacturer has eliminated the need to readout a full array at a series of different integration times by using a number of subarrays that cover wavelength regions where prominent ICP lines occur.

3.3.4.2 Charge Injection Devices. The CID presents the appearance of a newer technology with under-utilized potential. In fact it is about the same age as the CCD but has been closely held by patents to one manufacturing source. The available CID array sizes and formats have been very limited compared to CCDs, and slow-scan cameras with CIDs have not been marketed. M. Bonner Denton and his group at the University of Arizona have pioneered analytical work with this technology.

A CID consists of a *n*-type silicon epitaxy layer deposited on a *p*-type silicon substrate. On top of the epitaxy are deposited the same two insulators (SiO_2 and SiN_2) as for a CCD. Two gate electrodes are deposited on the insulators. These have been called by various names—the least confusing approach is to call them the collection gate and the sensing gate. During photon absorption both are negatively charged, but the collection gate has twice the negative potential of the sensing gate. As with the CCD, charges accumulate in a potential well below the gate. To read the charge on the pixel the sensing gate is disconnected from its power source. The sensing gate potential is then measured (by select registers at the edges of the array). The negative charge on the collection gate is then raised to zero, causing the charge to move under it. This results in a change in the sensing gate potential, which is measured again. The difference between the two measured potentials is a *correlated double sample* noise corrected signal.

The charge can be destroyed or retained at this point. To retain the charge the collection gate is returned to its original negative potential. To destroy the charge both gates are biased to zero potential, which allows the charge to drain off into the substrate.'

In the above description we speak of "charge" since (for historical, not technical reasons) holes, not electrons are the charge carriers in a CID. Holes have a lower mobility than electrons, but unlike the CCD, charge mobility is not an issue with the CID since only a single transfer is required for each read. The select registers at the edges of the array effectively isolate a single pair of collection and sensing gates for readout.

Since charge mobility and transfer speed are not concerns a CID can be operated at liquid nitrogen temperatures. This reduces the dark current (dark charge) to very near zero. CIDs show an average output signal two orders of magnitude less than that of CCDs. Their readout noise is also about two orders of magnitude greater than that of CCDs. However, a new design CID with a MOS-FET buffer transistor between the sensing gate and the select register promises to reduce readout noise to about the level of a CCD. Unlike the CCD, the CID is a surface channel device. This means that charge is stored higher in the epitaxy layer, closer to the gate; this allows the CID to hold three times the charge in each pixel that a CCD can hold.

When charge is dumped, however, the empty CID pixel must be primed before it resumes linear charge accumulation. This is done automatically, usually by flashing a light emitting diode at it, under software control. The CID is inherently free of blooming problems. The quantum efficiency of a typical CID is about 47% at 500 nm. Like CCDs, CIDs are coated with organic phosphors to improve their vacuum UV performance.

3.3.4.3 CID Versus CCD. Presently, CCDs have a clear advantage at low light levels, but there seems to be good reason to believe that the CID is the superior technology for atomic emission. Since the CID pixels can be read rapidly and either destructively or nondestructively, random access variable integration of different spectral lines is possible. A sensitive line can be integrated for ten sec six times and averaged while an insensitive line is being integrated for 60 sec once. Any pixel can be integrated until a desired signal-to-noise criterion is met. The greater charge storage per CID pixel combined with the readout flexibility should lead to superior precision and wider dynamic range. The use of segmented CCD arrays, in fact, is an attempt to mimic some of the advantages of the CID.

The name "charge injection device" derives from an early design in which the charge *injected* into the substrate was measured.

4. CALIBRATION AND MEASUREMENT

4.1 Qualitative and Semiquantitative Work

There was a time when photographic emulsion optical emission work was the principal means that a metals analysis laboratory utilized when asked to identify a totally unknown alloy, to sort a collection of mixed grades, or to confirm the true identity of mislabeled scrap. An experienced spectrographer could "shoot a plate and read it" in less than half a work day and be able to report with a fair degree of confidence what was in the alloy. Sometimes a series of suspected matching alloys would be included on the plate for comparison; sometimes just mixed metal oxides were included to bring up the elemental spectra. Qualitative results generally took the form, "yes, its type 316 stainless steel" or, perhaps, something like: "major amounts of copper and molybdenum, minor amounts of silicon, manganese, and zirconium in a cobalt-based matrix." Semiquantitative analysis would assign a crude set of values: "5% chromium, 1% nickel, <0.2% manganese, balance iron."

The key to good, rapid qualitative and semiquantitative work was the spectrographer's familiarity with the spectra of the elements—the characteristic doublets and triplets and single lines that were like a recognized signature. At least three lines were commonly needed to positively identify an element. But much of this work was as much art as science.

When laboratories closed their darkrooms they had to rely on X-ray fluorescence or possibly sequential ICP-OES with dissolved samples for this sort of information. The direct reader was always a perilous option here because it is possible that no channels exist for some of the unknown's important components.

In the previous sections we have seen how this picture may be changing again. The new solid state detectors promise to provide an electronic snapshot of the complete spectrum. Future expert software systems may do all or most of the spectral interpretation for us. It is possible then that once again emission spectroscopy may become the prime screening source for mixed, mislabeled, or totally unknown metals.

4.2 Quantitative Work

4.2.1 Internal Standard [2–4]

Until the recent advent of very stable sources like the ICP the greatest cause of instrumental variability in atomic emission has been the source. It could be argued, however, that the ICP has, to some degree, replaced instabilities in the discharge with mass transfer variabilities associated with sample solution viscosity. Other sources of error are endemic to the field: long-term drift due to aging in the electronics or fogging of the optics; short-term or sinusoidal drift associated with temperature or humidity changes in the room, and others. In spectrographic work differences in film emulsion batches and in developing the latent image introduce errors.

A solution to all of these problems is the use of the internal standard. In solids work the internal standard is typically the base metal of the alloy (iron for steels, aluminum for aluminum alloys, etc.).* In solution work the internal standard is sometimes an added inoculant, chosen, in part, because it does *not* occur in the as-received sample.' Whether it is native or foreign to the sample an internal standard and its associated *internal standard line* are selected to meet a number of criteria, most of which are fairly obvious:

1. An added internal standard must be available in high purity form and must be chemically compatible with the sample solution matrix (e.g., silver is a bad choice if the samples are in hydrochloric acid).
2. An added internal standard must vaporize in a similar manner to the analyte element (however, in ICP work all elements vaporize well).
3. An added internal standard must not have emission lines that interfere with principal lines of the analytes. An element with a comparatively simple spectrum is best.
4. The selected internal standard line must be free of interferences from the sample matrix, and from any lines or bands from the source.
5. The selected internal standard line must vary with changing spectrometric conditions in the same way that the analyte varies.

This last criteria means that the internal standard line and the analyte line represent a *homologous line pair*, as described by Walther Gerlach and Eugen Schweitzer in 1929 [7]. The ratio of the intensity of the analyte line to the intensity of the internal standard line should, thus, remain constant, or change very little over widely varying excitation conditions. The intensity ratios of known homologous line pairs vary less than 10% between arc and spark excitation. Homologous line pairs always represent similar energy transitions; both are excited, neutral atomic lines, or both are ion lines.

---

* It should be noted that the degree to which the base metal concentration is known varies considerably. Sometimes it is analyzed by other means, but more typically it is taken by difference. So long as the line is not self absorbed its absolute concentration is not important. However, the base metal concentration may also vary somewhat among a group of samples. This is a source of potential error but is generally negligable due to the great difference in concentration between the analyte and the internal standard.

' Occasionally, in trace analysis a portion of the spectrum that is line-free is used as the internal standard. This approach in situations where the backround is a significant part of the total analyte signal can sometimes compensate for spectral variability better than an internal standard line.

The intensity ratio of the analyte line to the internal standard line is also immune to optical and electronic drift and temperature effects. Moreover, when the same amount of internal standard is added to calibrants and to test solutions viscosity effects on the ICP disappear. It is thus possible (though not recommended as routine practice) to prepare calibrants in one acid mixture, and samples in a totally different acid mixture.

Unfortunately, there is no known magic (or scientific) formula for the selection of homologous line pairs. However, one can rely upon guidance from the technical literature or perform a simple test by measuring the proposed line pair under two widely different excitation conditions. Arc and spark excitation is one common choice. Another that has been recommended is to measure the spark spectra of the line pairs with and without argon flow. One can confirm the choice of a homologous line pair by comparing the relative standard deviation of the absolute intensity of the analyte line to the relative standard deviation of the intensity ratio of the analyte line to the internal standard line. With an increasing number of replicates, the relative standard deviation of the intensity ratio measurements should drop significantly below the relative standard deviation of the absolute analyte intensity measurements.

The absolute intensity of the internal standard line under the chosen analytical conditions is also somewhat of concern. A very large denominator in the analyte/internal standard ratio will introduce inaccuracies. However, so long as the internal standard line is not self-absorbed, it is perfectly acceptable to introduce a factor into the software to reduce the number of counts for the internal standard.

In the most widely accepted mode of calibration the intensity ratio of the analyte/internal standard is plotted versus the concentration ratio of the analyte/internal standard. Homologous line pairs typically plot very well. Nonhomologous line pairs are more likely to plot poorly, or require separate plots for alloy subtypes. This relates to the fact that homologous line pairs are less subject to interelement interferences.

There is clearly a range of usefulness for different homologous line pairs. Some track changing spectral conditions in lock-step, others may pair up for some instrumental variables but fall out of step with respect to others. The ICP is reasonably tolerant in regard to homologous line pairs due to its high temperature and extreme stability. However, the complexity of the ICP spectrum and the large number of ion lines require some judgement to avoid spectral overlaps. Scandium, yttrium [46], and indium are commonly added as internal standards in ICP solution work.

Various studies have appeared in the literature suggesting the use of multiple internal standards in a multi-element ICP assay. Some of these are known by acronyms, such as GIRM (generalized internal reference method) and PRISM (parameter-related internal standard method); they attempt to account for all aspects of method variability by combining responses from internal standards that respond principally to one source of variation each [47,48]. Others have stressed the fact that any form the internal standard correction takes is more effective under robust operating conditions: high plasma power and long analyte residence times in the plasma (i.e., low sample carrier flow) [49].

Before closing this section it should be noted that there is a dichotomous antipode to the homologous line pair: the *fixation line pair*. These lines are paired *because* they vary considerably differently in the event of changing spectrometric conditions. A fixation pair is sometimes used to monitor the stability of the excitation source in arc/spark work since any variation in this ratio is a sign that the energy in the analytical gap has varied. The fixation pair nearly always consists of an ion line and an excited atom line of the same element. For example, the ratio, iron (II) 273.0 nm(ionic)/iron (I) 360.8 nm (atomic) increases by about 0.1 for every 5 amp increase in peak spark source current.

It should also be noted before moving on that a few laboratories still operate using an older convention of measuring a *typical intensity* for the internal standard alloy matrix element. For this purpose they take great pains to obtain and meter out for use a bar of high-purity iron, or aluminum, or copper, or whatever their sample matrix happens to be. The intensity obtained for the internal standard line for the pure material becomes the typical intensity. The obtained value is ordinarily rounded to the nearest 100 digits. The intensity ratios of the analyte/internal standard are then multiplied by this value so that they become whole numbers. The concentration ratios of the analyte/internal standard are then multiplied by 100.

4.2.2 Background [3,50]

Background is the baseline noise, excursions from which represent signal. Fundamentally, there are only two important sources of background: thermal continuum radiation and the "wings" of adjacent spectral lines or bands. The argon thermal continuum in ICP-OES is an example of the baseline elevation produced by the source itself. Forward power and viewing height are adjusted to maximize signal-to-background for this source of background noise. What results is generally accepted as a fact of life without much possibility for further improvement. The effects from lines and bands are quite another matter, however.

Light scatter as a source of background is not a large factor in modern commercial instruments.

Table 6.1 Some Examples of ICP-OES Background Influences. [51]

Influence	Example of Affected Line	Remedy
Carbon impurities in the argon	Arsenic 193.70 nm	Use liquid argon boil-off gas.
Silicon lines due to HF use	Molybdenum 202.0 nm	Use HF-resistant torch & spray chamber
NO bands from air entrainment	Phosphorus 213.62 nm	Quartz torch extension; increased argon flow
OH bands	Vanadium 309.31 nm	Lower observation height
O_2 absorption	Tin 189.99 nm	Use vacuum spectrometer

It should be noted that an intense broad line, such as a matrix element line or a molecular band structure can seriously elevate the background on *both* sides of a nearby analyte line. And some matrix species (e.g., magnesium) can influence background as much as 10 nm away from their strongest lines.

Examples of emission lines and bands that strongly influence background structure are numerous, but many are not obvious. Moreover, in trace level analysis where background constitutes a large majority of the total analytical signal it is imperative that the analyst understand what is happening in the immediate vicinity of the analyte line. As an illustration of easily overlooked background influences consider the brief list in Table 6.1.

In addition there are N_2^+ bands from plasma-entrained air between 350 and 400 nm and several strong argon lines that can affect background in their immediate vicinity [51]. Arc and spark emission in air or nitrogen will also have to face (or, preferably, avoid) the background influences of the cyanogen band structure. However, most arc/spark work is now performed in argon.

Modern instruments allow the analyst to inspect a "window" on the region surrounding the emission line of interest. The analyst sees the line shape and the background on both sides of the line. He can then use the software to select baseline reference locations on one or both sides of the line for the purpose of baseline correction.* The selection of these reference points involves some critical judgement. For example, it is prudent to avoid specifying a baseline reference in a region of rapidly changing background, such as on the steep downslope of an adjacent peak. The reason is that is likely that location will not show very reproducible counts from sample to sample.

---

* The software uses these selected locations to perform some simple geometry, then integrate the designated area.

Elements whose concentration varies 500- to 1000-fold during the course of analyzing a set of samples may produce background interference by altering the background in the vicinity of the analytical line in a manner inconsistent with the selected locations for baseline correction. It is best to avoid such situations by selecting a different analyte line, if possible, or by subdividing the sample set, using different background correction locations.

The concentration of analyte that must be present to produce the same response as the background at the location of the analyte line is termed the *background equivalent concentration* or *BEC*. This quantity can be determined by simply measuring the background signal with an analyte-free matrix solution and using an analyte calibration to determine how much analyte that signal represents. BEC is related to the detection limit as follows.

$$C_{DL} = k\sigma_B(\mathrm{BEC})/C_o \quad (6)$$

where C_{DL} is the detection limit; σ_B is the standard deviation associated with the measurement of some very low concentration, C_o; and k is a statistical factor (usually taken as 3, though more rigorously it is $2(2)^{1/2}$). Another expression of the same concept is:

$$C_{DL} = 3 \times (\mathrm{RSD})_B \times (\mathrm{BEC})/100 \quad (7)$$

where $(\mathrm{RSD})_B$ is the relative standard deviation of the background signal.

Boumans published an interesting discussion of the concept of the *detection limit* as it relates to background and spectral interferences [52]. He pointed out that there are two different but equally correct approaches to the calculation of detection limits from spectral observations. The approach based on *signal-to-noise ratio* (SNR), widely used in North America, is essentially that which was outlined in the previous chapter:

$$C_{DL} = (\Delta C/\Delta x)\sigma_B k \quad (8)$$

where $(\Delta C/\Delta x)$ is the slope of the calibration curve (the change in concentration, C, with the change in the analyte signal, x). Again, $k = 3$ is recommended. Once again, a reasonable assumption is made here that the detection limit is solely limited by the background.

A second, more involved, approach has been favored by European workers. It is called the SBR/RSDB or *signal-to-background ratio/relative standard deviation of the background* method. If we rewrite equation (8) for concentration C_o yielding signal x_A:

$$C_{DL} = [\sigma_B k/(x_A/C_o)] \quad (9)$$

then normalize σ_B and x_A to the background signal x_B by dividing numerator and denominator by x_B:

$$C_{DL} = (\sigma_B k/x_B)(C_o x_B/x_A) \quad (10)$$

or $$C_{DL} = 0.01k(100\sigma_B/x_B)[C_o/(x_A/x_B)] \quad (11)$$

This, then is:

$$C_{DL} = 0.01k(\text{RSD})_B[C_o/(\text{SBR})] \quad (12)$$

and $$C_{DL} = 0.01k(\text{RSD})_B(\text{BEC}) \quad (13)$$

If we now insert 3 for the value of k equation (13) becomes identical to equation (7).

The advantage to this approach is that it allows the detection limit to be calculated from relative quantities that can be easily and accurately quantified, and which can be independently related to other variables. For example, (SBR) can be expressed as a function of bandwidth, and $(\text{RSD})_B$ can be expressed as a function of absolute background signal. Instruments can be compared by this means.

In Ref. 52 Boumans also addresses measurement precision in terms of the relative standard deviation of the net line signal, which, in turn, depends upon a combination of background fluctuations and fluctuations in the emitted line signal.

4.2.3 *Line Interferences [3]*

We have just seen how structured background and variability in the background influence how we must measure an analyte and how small an amount of analyte we are capable of measuring. But we know that lines from the sample matrix can have an even stronger and more direct influence on the analytical outcome.

There are literally millions of emission lines, and so it is not unexpected that in an alloy of any complexity overlaps with the chosen analyte lines should occur. Overlaps can be direct line overlaps, in which the interferent line is directly superimposed on the analyte line. They can also be "wing" overlaps, in which the interferent line only partially obscures the analyte line. In both cases the intrusion into the analyte's measurement window results in an erroneously high signal. If one were to plot calibration data from standards prepared with and without a line interference the result would be two linear curves shifted parallel to one another. In other words the effect is strictly additive.

Confronted with a line interference the analyst has several options. But he must know the problem exists in order to take action on it. The literature sources listing emission lines are an invaluable resource [14–17], but it is also essential to take the time at some point (preferably during setup of a new instrument) to investigate interferences on the analyte lines that are expected to be used. This can be a large enough task for a 30 or 40 channel ICP direct reader. For a sequential ICP it may be an ongoing project for the entire service life of the instrument. With a direct reader, the analyst makes up standard working solutions for each channel's analyte (500 or 1000 ppm may be suitable) and analyzes them all for all elements in the array. A clever worker may even be able to coax the instrument's software to present the output in a convenient form (say, a 30 x 30 array of analytes versus interferences). Values above some preselected level would be flagged for special attention. Or the high values could be dumped into a special table listing only the likely troublemakers. Table 6.2 is an illustration of such a concentrated table of line overlaps.

Such a table, once it is prepared and verified can be an extremely useful guide that should remain valid throughout the operating life of the instrument. However, using an arbitrary cutoff (such as 1.5 ppm in Table 6.2) can underestimate the influence of the base metal contribution. For example, a low alloy steel company would likely want to list iron as an interferent using a much lower criterion, say 0.3 ppm of apparent analyte. In that event the same array of wavelengths listed in Table 6.2 would include iron as an interferent for antimony, arsenic, bismuth, boron, cadmium, calcium, lead, magnesium, nickel, silicon, tin, and tungsten.

By the same token, many of the interferents listed in Table 6.2 are unlikely to occur in sufficient concentrations in commercial alloys so as to require correction for their presence. There is always the unusual case, however, and new alloys are continually being developed. A table of line overlaps should serve as a reference, but the actual overlap corrections are most effective when they are custom tailored to a specific group or groups of alloys. Applying corrections for elements not likely to occur at significant levels is hazardous because baseline perturbations caused by unrelated effects may result in false corrections being applied to the analyte signal.

Line overlap corrections are easily formulated. First, the test standard containing the analyte in the presence of the line overlap interference is determined using the analyte calibration. This verifies the existence of the interference. Then a series of standard samples containing various ratios of analyte to interferent [ideally, ranging from (0%)/(100%) to (100%)/(0%)] are analyzed on the same calibration. A curve is then prepared of [(% analyte found)–(% analyte taken)] versus (% interferent taken). The slope of this curve is [(% apparent analyte)/(% interferent taken)]. For unknown sample analysis the software is programmed to extract a value for (% interferent) from *its* calibration, multiply it by the above slope and subtract this value from the (% analyte found) from the analyte calibration. This correction procedure is applied to the test standard to confirm that it yields the expected value for the analyte.

The above procedure is the most rigorous approach to line overlap correction. It applies equally well to arc/spark solids work and to ICP-OES

Table 6.2 Example of an Empirically Derived Table of Line Overlaps.

Analyte/λ(nm)	Interferents
Ag/328.0	Zr
Al/396.1	Ca, Ce, *Mo*, Si, W, Zr
As/193.6	*Al*, Co, Cr, Mo, Ta, *V*, W
B/249.6	Co, Hf, Nb, W
Ba/493.4	Si, W
Be/313.0	V
Bi/223.0	Cu, Ni, Ta, Ti, W
Ca/317.9	Cr, Mo, Nb, Ta, Ti, V, W, Zr
Cd/226.5	—
Ce/418.6	*Zr*
Co/228.6	Cd, Fe, Hf, La, Nb, *Ni*, Ta, W
Cr/267.7	V, W
Cu/324.7	Nb
Fe/259.9	—
Hf/273.8	Fe, Mn, Mo, Ta, V, Zr
In /451.1	Ce, Ta, Ti
La/398.8	Ca
Mg/279.0	Ce, Cr, Hf, Mn, Mo, Nb, Ta, V, Zr
Mn/257.6	—
Mo/202.0	Si, Ta
Nb/309.4	Ta, *V*
Ni/231.6	Co, Ta
Pb/220.3	Al, Ce, Co, Cu, Mo, *Nb*, Ni, Ta, W
Sb/206.8	Al, Cr, Mo, W, Zn
Se/196.0	Al, Co, Fe, Hf, Mg, Mn, Ta, W, Zn
Si/251.6	Hf, Mo, *Nb*, *Se*, *Ta*, Ti, W
Sn/189.9	Ti, W, Zr
Ta/240.0	Co, Fe, Hf, Mg
Ti/ 334.9	—
V/292.4	Hf, Mo, Zr
W/207.9	Al, Hf, Mo, Ta, *Zn*
Zn/213.8	Ni
Zr/339.1	—

Note: Apparent concentration of analyte > 1.5 ppm per 1000 ppm of interferent; italics: apparent concentration of analyte > 10 ppm per 1000 ppm of interferent.

solution work. It allows for a curvilinear relationship in the interference (which is not often encountered); however, in that case the correction factor is best read back from the curve for each test sample. Because most ICP calibrations are linear over several orders of magnitude, practitioners of that art are likely to utilize a single solution of pure interferent at some concentration slightly in excess of that found in a typical test solution. They measure its generated signal on the analyte channel and formulate a factor for (% apparent analyte/ % interferent taken) from it directly.

There can be several spectral line overlap corrections on an analyte line, each calculated and subtracted from the net analyte found. Thus, for a large array of analytes on a direct reader the correction scheme can be quite involved, but it is generally handled seamlessly by the software.

It has been the author's experience that spectroscopists tend to become sanguine over line correction factors. And yet there is little evidence to suggest that these corrections, once formulated, will never change. It is prudent to keep a logbook of all line overlap correction factors, and to check and update them on some regular basis. For critical work, and especially for work at interferent and analyte concentrations far removed from those at which the extant correction was formulated, it is important to reformulate the correction just before the test sample analysis.

Most books on the subject distinguish between the type of additive line overlap interferences we have been discussing and multiplicative "interelement effects." These are sample or source effects that result in calibration curve rotation (i.e., slope change). The origin of such effects is only dimly understood. They tend to be much more important in arc/spark emission work than in ICP work—a fact, which, itself, gives a clue to their origin.

Slickers [2] lists some of the probable causes for interelement effects. The thermal history of solid samples can result in differing inclusion forms or differing solidification forms (e.g., dendrite formation). An inadequate preburn on such samples is likely to cause interelement effects. The discharge can be affected by oxygen or hydrogen in the argon, forming compounds in the sample crater. Slight variations in sample composition can affect both the alloy melting point and the energy of vaporization for certain analytes. Such variations can also affect the sample's thermal conductivity; this leads to higher or lower crater temperatures and increased or decreased analyte vaporization. The presence of easily ionized elements in the sample can lower the discharge temperature, leading to decreased analyte vaporization. The presence of certain elements can alter the optimum viewing height for the analyte. The presence of certain elements can also alter the background. Interelement effects frequently exert their influence over many lines of the analyte element.

The correction for interelement effects is accomplished in a manner analogous to spectral overlap corrections. A series of standard samples exhibiting varying degrees of the multiplicative interelement effect are analyzed on the same calibration. This time a curve is prepared of: [(% analyte found) – (% analyte taken)/(% analyte taken)] versus (% interferent). The slope of the curve can then be inserted in:

$$(\%\ \text{analyte})_c = (\%\ \text{analyte found})_x\ [1 \pm (\text{slope})(\%\ \text{interferent})_x] \quad (14)$$

where the subscript c designates the corrected test sample value and the subscript s designates the values associated with the test sample. The "+" and "–" are used to handle the direction of the curve rotation. As with spectral overlaps, there may be several interelement corrections per analyte line.

4.2.4 Calibration and Standardization [3]

There is a great deal of confusion among the uninitiated about the peculiar use to which spectroscopists put these two terms. Calibration traditionally involves a plot of intensity ratios versus concentration ratios.* As previously mentioned, for many alloy systems, and particularly for low alloys of iron, aluminum, and copper metal, the base metal is used as the internal standard. Curves are fit to the data points using various software-generated models: linear, curvilinear, weighted curvilinear, and others. Modern software allows the analyst to view a proposed fit graphically, to add and subtract various interferent corrections, and to "read-back" the calibrant concentrations from a proposed model before trying another or finally accepting it. The analyst should handle this computer power wisely. For example, a third or fourth order polynomial fit that gives good data for two or three reference materials that are a poor match for the sample will probably generate sample data that is wildly wrong.

Busy operations, like a steel mill control laboratory, do not calibrate their direct readers daily, or even weekly. What they do is *standardize* every time they operate it. At the time of calibration two standardization reference materials were analyzed—one with a low concentration of analyte and one with a high concentration of analyte. The recorded results were, say, 395 and 57,600 intensity ratio counts. Today, the same two reference materials were analyzed yielding, say, 423 and 58,225 intensity ratio counts. The software merely solves two linear simultaneous equations:

$$y = mx + b \tag{15}$$

$$395 = m(423) + b$$

$$57{,}600 = m(58{,}225) + b$$

$$m = 0.975 \text{ and } b = +807.6$$

This new curve is then used to convert test sample intensity ratios to concentration ratios.

Today's direct reading arc/spark instruments sometimes utilize two integration periods—one for an initial spark-like exposure, followed by an arc-

* And, in fact the interference corrections described in the previous section are traditionally performed on intensity ratios and concentration ratios.

like exposure for tramps. The software is programmed to decide which exposure to apply to the final report for a given analyte based on the calculated analyte concentration. Often two or more channels are devoted to the same element because of sensitivity considerations, or to avoid interferences. The software can, likewise, be programmed to assign a channel to a final result based on calculated concentration, and on the presence of critical levels of signal on interferent channels.

In the past arc/spark spectrometry utilized many *curve sets* to handle the calibration of a given analyte in different subclasses of test sample. There might be one curve for manganese in carbon steel, another for manganese in low alloy steel, a third for hadfield or rail steel, and, perhaps, three or four curves for manganese in stainless steels. The entire curve set for all analytes could be enormous. Today, the trend is toward the concept of *global calibration*, which drastically reduces the number of needed calibration curves for spark instrumentation. K.A. Slickers has shown that with appropriate instrumentation, in particular with the use of the high-energy prespark, a single curve can serve for several orders of magnitude in several different categories of alloys. Chromium, for example, can be plotted successfully for low alloy steels, tool steels, and stainless steels on a single curve.

Conversion of test sample results from concentration ratios obtained from conventional calibration curves to weight percent in the alloy is a fairly straightforward process. The standard approach for a direct reader makes the assumption that all sample components have been determined:

$$C_b + \Sigma C_i = 100 \tag{16}$$

where C_b is the concentration of the base metal and C_i represents an analyte concentration. Dividing both sides of the equation by C_b:

$$1 + \Sigma(C_i/C_b) = 100/C_b \tag{17}$$

$$C_b[(1 + \Sigma(C_i/C_b)] = 100 \tag{18}$$

$$C_b = 100/[(1 + \Sigma(C_i/C_b)] \tag{19}$$

Note that (C_i/C_b) is the concentration ratio of all analytes. Once C_b is calculated by this means (or known by a wet chemical assay), the individual analyte concentrations are easily calculated:

$$C_i = (C_i/C_b)C_b \tag{20}$$

Concentration ratios are often calculated as $100(C_i/C_b)$ and, if so, then here: $C_i = (100(C_i/C_b))C_b/100$.

5. PRACTICAL DETAILS [2]

There are a large number of quality control and system maintenance measures that need to be managed in atomic emission spectrometry. We can only touch upon a few basics here.

With direct readers it is necessary to check the system *profile* frequently to ensure that all lines are centered in their bandpass window. This process is instrument specific. It frequently involves a mercury line source that is used to manually dial in a maximum reading on the mercury channel. Most manufacturers recommend profiling before beginning a calibration or a standardization and sample run, and repeating the profiling after every four hours of continuous operation. More frequent profiling is indicated if there are temperature or humidity variations in the instrument's environment.

Argon purity is an important concern since all atomic emission instruments are sensitive to carbon or oxygen impurities. A characteristic white ring burn mark is a sign of oxygen in arc/spark work. The highest purity argon is generally obtained from the gaseous "boil-off" from liquid.

In arc/spark work magnetized samples will deflect the discharge in an uncontrolled manner. Care should be taken to avoid inadvertent magnetization of susceptible alloys by avoiding magnetic holders for sample polishing. Sample preparation, in general, is a critical step in solids work, and considerable thought should be given to grinding and polishing operations. Soft inclusions, such as lead in aluminum, will tend to smear across the sample surface. Hard inclusions, such as silicon in hypereutectic aluminum/silicon alloys, are likely to tear the belt or disk during polishing. Corundum or other forms of alumina are the standard abrasives for polishing most alloy solids for most analytes. Silicon carbide is standard for aluminum determination. Safety issues are a major concern in solids preparation. Protocols to prevent eye and hand injuries, and dust inhalation, as well as addressing other hazards, must be in place.

Maintenance is the key to optimal performance with atomic emission spectrometers. An ICP-OES instrument requires regular cleaning of the torch, spray chamber, and nebulizer. The peristaltic pump tubing should be changed before each use to prevent memory effects. Boron is a particularly difficult analyte in this regard and always requires a new pump tube.

6. SPECIAL TOPICS

There are a few ancillary topics that require brief mention before we proceed to a discussion of specific applications to trace metal analysis. A few of these are very active areas at this time and are likely to be areas of rapid developments.

The use of lasers in atomic emission has been a trendy topic for decades, but practical results have been sparse. Part of the problems seem to be related to the fact that focused laser light produces a molten puddle with mirror-like properties. Defocused beams, pulsed beams, and raster scanning have shown mixed results. Laser work, however, has been in the forefront of time-resolved emission (sometimes called time-gated emission), in which high speed, computer-controlled circuitry ignores all but the most analytically useful portion of the emission from the plume that results from a laser pulse.

Depth profiling, particularly using a glow discharge source, is developing into a viable, practical tool, although interest seems to be concentrated outside of the U.S. Various inorganic coatings applied to metal surfaces ("hot dipped," electrolytically- or vapor-deposited) can be studied by this means. Other surface effects, such as carburizing, nitriding, deboronization, and chromium depletion can also be investigated. The technique has been beset with a fundamental problem, however: the rate of ion milling is composition-related in a complex way. Now, however, fundamental parameters to describe what is occurring have been formulated.

Molten metal analysis is a major trend and atomic emission approaches are in the forefront of this work. Fiber optics to transport light and plastic tubes to transport ablated metal aerosols are recent approaches to interface with a nearby spectrometer. Furnace-side atomic emission spectroscopy is not yet routine, but much development work continues in this area.

Robotic laboratories, on the other hand, are already an important trend in certain metals industries. "Container labs" are basically trailers that are wheeled onto the furnace floor to accept production test pieces, analyze them, and report results—all without human intervention. These units contain highly automated sample preparation equipment, an array of standards, and, usually, an atomic emission spectrometer. The software does the rest. Investment in this type of facility is becoming common in the basic steel and aluminum industries in the U.S., and particularly in Europe where the first such installations were located.

7. APPLICATIONS

7.1 Iron-base Alloys

Atomic emission spectrometry has been associated with the iron and steel industry for a large part of the twentieth century. The big steel companies were also among the first industries to adopt photoelectric direct readers when they entered the marketplace in the early 1950s. Today it is difficult to imagine a ferrous alloy producer operating without this methodology in some form.

7.1.1 Arc/Spark

Still the mainstay for most of the iron and steel industry, arc and spark excitation engenders few new papers in the literature any more since it is a mature technology. The American Society for Testing and Materials' *Annual Book of ASTM Standards*, Volumes 03.05 and 03.06 are good sources for test methods. There the reader will find detailed procedures for "point-to-point" [53] and "point-to-plane" [54] analyses of carbon and low alloy steel. The former is a photographic plate technique while the latter is a direct photoelectric readout method. There are also two "point-to-plane" methods for stainless steels. One utilizes either an intermittent arc or an air-interrupted spark for 18Cr/8Ni alloys [55]. The other is a vacuum spectrometer method utilizing a self-initiating or a triggered high voltage spark [56]. These books also include arc/spark methods for blast furnace iron [57] and for high-purity iron [58], and a few analyte specific methods [59].

There is also a publication by ASTM (*Suggested Methods of Analysis of Metals, Ores, and Related Materials* [60]), which contains a number of arc/spark procedures for ferrous based materials. These lack the consensus and supporting data of standard methods, representing as they do, for the most part the experience of one laboratory. They may be a useful starting point for in-house development work, however. There are about a dozen methods included that involve the arc/spark analysis of iron-base alloys.

We have already referred to the aerosol generator/capillary arc technique, which was an early approach to the concept of separating sampling (by a dc arc) and excitation of the generated metal aerosol (by an argon-flushed dc arc confined in a narrow channel). The work, as applied to low alloy, stainless, and tool steels, was published by Jones, Dahlquist, and Hoyt in 1971 [23].

7.1.2 Direct Current Plasma (DCP)

The DCP enjoyed an immediate popularity in the steel industry when the technology was first commercialized with an echelle optical system in 1965. Today its position has partly been displaced by the ICP, although there are still many installations in use. The technique, still capitalizing on its price advantage, is expected to enjoy somewhat of a renaissance with the introduction of CTD-based readout instruments.

Acid digestion bomb dissolution [61] and microwave dissolution [62] have been applied to the DCP-OES analysis of low alloy and silicon steels. A variety of analytes were determined, including aluminum, titanium, silicon, phosphorus, tin, manganese, chromium, nickel, and molybdenum. In both dissolution schemes a mixture of hydrochloric, nitric, and hydrofluoric acids were used. It should be noted that this work by Fernando, Heavner, and Gabrielli was an early application of microwave dissolution to metals. Potter and Vergosen used DCP emission to determine neodymium and boron in iron-neodymium-boron permanent magnet alloys [63].

Hydride generation using a flow injection approach has been combined with DCP measurement for the determination of antimony in iron- (and copper-) based samples. In this approach L-cysteine is used to reduce antimony (V) to antimony (III) prior to "on-line" generation of stibene with sodium borohydride [64].

7.1.3 Inductively Coupled Plasma (ICP)

The ICP has become a dominant feature of many iron and steel analysis laboratories. If there is time to dissolve and dilute a sample there is reason to consider this technology. While many interferences can be readily handled by mathematical corrections, a few ferrous labs continue to regard chemical separations as an option. There are also solids based systems that utilize a variety of technologies.

Wallace evaluated six different lines for *boron* in regard to interference from matrix iron [65]. All the usable lines in boron's simple spectrum show some problems with iron. The conventional solution is distillation of the analyte as the methyl borate ester. This can be done as an "off-line" separation, as described by Hosoya, Tozawa, and Takada [66], or by a direct injection, discontinuous flow approach, as published by Lopez, Molinero, Ferrer, and Castillo [67]. Another approach is the removal of iron and other transition metals on the mercury cathode. Ion exchange and solvent extraction separations are also likely to be directly adaptable from spectrophotometric methods for boron.

Above 200 nm *phosphorus* exhibits a number of lines with severe interferences. In particular the sensitive line at 253.57 nm shows a serious iron line overlap. The sensitive lines at 213.62 nm and 214.91 nm are no help since they show a massive effect from copper, which is often present in iron-base alloys at levels high enough to influence phosphorus results. Nakahara relates these problems and shows how he solved them by using the vacuum UV line at 178.29 nm, which is free of iron and copper interferences [68]. Of course, there usually are chemical solutions to spectroscopic problems if one has the time, the knowledge, and the will to pursue them. Wittman and Schuster adapted an indirect atomic absorption method for phosphorus to ICP-OES. A 1.5 g steel sample is dissolved in a mixture of nitric acid, perchloric acid, and water, heated to fumes, and cooled. The solution is oxidized with permanganate and then reduced with nitrite. After boiling out NO_x gases, and cooling, the solution is reacted with an ammonium meta-vanadate/ammonium molybdate mixture, then extracted with MIBK in the presence of citric acid. The organic phase is measured by ICP for molybdenum at 281.62 nm. One atom of phosphorus is present in the phosphovanadomolybdate complex for eleven atoms of molybdenum [69].

Del Monte Tamba, Falciani, Dorado Lopez, and Gomez Coedo used a similar microwave dissolution procedure to that described in reference (62) to determine *aluminum* in steels [70]. *Lanthanum, cerium, praseodymium, and neodymium* were determined in mild steel without prior separation by Grossman, Ciba, Jurczyk, and Spiewok [71]. Vaamonde, Alonso, Garcia, and Izaga reported on ICP techniques for ferroalloy analysis. Dissolution procedures and line selection were described for the determination of major and minor elements in ferromanganese, ferrochromium, and ferrosilicon [72].

Umemoto and Kubota used a graphite cup (3.5 mm deep by 3mm wide) as a direct insertion device to determine *arsenic and antimony* in iron and steels. The cup containing the acid-dissolved sample aliquot was inserted into the torch, the torch was lit and stabilized, then the cup was raised into the plasma. The method of additions was used to quantify results [73]. *Vanadium and titanium* in low alloy steel were determined by electrothermal vaporization/ICP-OES. Okamoto, Murata, Yamamoto, and Kumamaru used a tungsten boat furnace with an aliquot of acid-dissolved low alloy steel. The vaporized sample was carried into the ICP by the argon stream mixed with a low flow of hydrogen [74].

Duffy and Thomas used a dual-view axial/radial ICP emission spectrometer to determine *boron, phosphorus, and sulfur* in low alloy steels. The instrument utilized an echelle grating and CCD segmented subarrays for wavelength coverage. The transfer optics and the polychromator were purged with inert gas. It was concluded that the best approach is to utilize radial viewing unless insufficient sensitivity is observed. Axial viewing shows 5–10 times lower detection limits but tends to exhibit more matrix effects and a shorter useful analyte concentration range. Data were generated using the NIST low alloy series, SRM's 361–365 [75].

Hydride methods for the ICP followed earlier work that utilized atomic absorption measurement. Work by Walton employed a sequential ICP to determine *arsenic, antimony, and bismuth* in low alloy steels. The hydrides were generated by mixing a sample stream containing 0.1 g sample/100 mL in the presence of hydrochloric acid and potassium iodide with a stream of sodium borohydride solution and passing the product stream into an ICP spray chamber. Interferences from chromium, copper, manganese, molybdenum, nickel, and vanadium were not significant at this dilution [76]. Ozaki and Oliviera determined the same elements in steel but utilized a Teflon™ membrane gas/liquid phase separator so that only the gaseous hydrides and none of the solution were directed to the ICP torch (a nickel high temperature alloy was also analyzed) [76A].

Wickstrøm, Lund, and Bye noted a serious depression of *selenium* signal caused by tin and arsenic when low alloy steels were analyzed by hydride generation-atomic absorption. The effect was greatly diminished when the same samples were analyzed by hydride generation-ICP. Detection limits for selenium were also found to be lower when the ICP measurement was used [77]. Chen, Chen, Feng, and Tian determined *arsenic* in low alloy steel and cast iron using a continuous flow system that incorporates a reagent/sample reaction chamber and a gas/liquid separator maintained in an icebath [78].

An in-line alumina column has been utilized in ICP methods for trace levels of *phosphorus and sulfur*. McLeod, Cook, Worsfold, Davies, and Queay determined phosphorus in steel without interference from iron, using the very sensitive 213.62 nm phosphorus line. The nitric acid-dissolved sample was injected into a flowing stream of distilled water that carried it onto a microcolumn of activated alumina. The phosphate ion was adsorbed and the iron matrix passed through. Then 0.5M potassium hydroxide was injected into the flowing stream. The phosphorus was eluted and passed into the ICP torch [79]. Yamada, McLeod, Kujirai, and Okochi employed the same type of column to adsorb trace sulfur from acid-dissolved high purity irons. The sulfate ion was released from the column by injecting dilute ammonium hydroxide. The sulfur was measured at 180.73 nm in the vacuum UV [80].

Flow injection has been utilized in combination with ion chromatography and a sequential ICP. Giglio, Mike, and Mincey reported that they could utilize the most sensitive line for each analyte after cation separation. There was adequate time between peaks to slew the grating to the next wavelength [81]. Souza, *et al.* used electrolytic on-line dissolution of stainless steels in combination with ICP-OES for rapid compositional analysis [82].

Perhaps the most mature and commercially practical solids technique with the ICP at this point is spark ablation. Here, a separate sampling unit utilizes a high repetition rate spark to produce and sustain a stream of metal aerosol particles, which is carried to the ICP torch through a plastic tube by an argon gas flow. Gomez Coedo, Dorado Lopez, Jimenez Seco, and Guitierrez Cobo evaluated this technology for the determination of *aluminum, titanium, niobium, and vanadium* in low alloy steels. They also compared the technique's performance to the ICP as a solution technique, and to solid spark atomic emission. In regard to limit of detection, spark ablation/ICP-OES did nearly as well as the solution-based ICP approach, and significantly better than direct solid spark atomic emission [83]. The same laboratory successfully applied the technique to remelted and iron-diluted ferrovanadium [84], and to *boron* in steels [85]. Gagean and Mermet compared spark ablation and laser ablation as sampling sources for an axially viewed ICP-OES. Spark ablation required much less "preburn" time, was cheaper, and easier to use. The laser required less surface preparation. Ablation with an excimer UV laser (not a Nd:YAG laser) can be more sensitive and show lower detection limits than spark ablation [21].

7.1.4 Microwave Plasmas

This technology, as applied to metals analysis, has not reached a commercial stage. However, the work is being pursued by researchers, especially with steels. Masamba, Smith, and Winefordner used a 94.5% helium/4.5% hydrogen capacitively coupled microwave plasma and directly inserted machined solid steel reference materials pressure-fitted into the end of a hollow graphite electrode. The source was operated between 350 and 500 W. Linear calibration curves were obtained for *manganese, chromium, lead, and tin* [86]. Jin Zhang, Liang, Yang, and Jin determined *boron* in steels using a flow injection system that utilized two ion exchange columns in series. Measurement was with an argon microwave plasma torch of unique design [87].

7.1.5 Glow Discharge

Publications on the use of glow discharge for metals have been sporadic since the introduction of the lamp design by Grimm [8]. Radmacher and Swardt compared results for steel and pig iron by GD-OES with the use of a high repetition rate spark source [88]. Since the introduction of the commercial (Leco Corp.) instrument an increasing number of papers have appeared on this subject. Weiss describes a morphology-based matrix effect with graphitized cast irons. Initially, there is a decreasing metal/increasing graphite surface layer exposed to the excitation as the metal species are preferentially vaporized. This is followed by an increasing carbon signal. After a lengthy preburn, however, a steady state condition is achieved [89].

7.1.6 Lasers

Lasers still represent a largely unfulfilled promise for metals analysis. A continuous beam melts a puddle with reflecting properties. And in high power, Q-switched, pulsed mode operation a high density plasma forms that emits a strong continuum background. Also the plasma is encased in a sheath of cool atoms so that a great deal of self absorption of analyte lines occurs. Separating sampling by laser and excitation by spark discharge is only a partial answer since precision and sensitivity problems persist with this approach as well. Kagawa, Hattori, Ishikane, Ueda, and Kurniawan overcame these problems with a laser-induced shockwave plasma approach. They employed a TEA CO_2 laser with a short pulse duration for a low alloy sample target in a 1 torr (133 Pa) vacuum chamber. A small primary plasma near the sample surface emits continuum radiation, but an expanding hemispherical secondary plasma emits sharp lines against a low background [90]. Campos and coworkers have applied a Nd:YAG laser to the determination of carbon [91,92] and sulfur [93] in steel.

7.2 Nickel- and Cobalt-base Alloys

7.2.1 Arc/Spark

A carrier distillation method for trace levels of *lead, bismuth, and tin* in high temperature nickel-base alloys was first published by Atwell and Golden. A 0.2 g sample was dissolved in nitric and hydrofluoric acids. Indium (40 μg) was added as an internal standard and the solution was transferred to a platinum crucible, heated to dryness, then ignited. To 9 parts of the oxides thus formed the authors added 1 part of a blended carrier consisting of 1 part lithium fluoride and 11 parts silver chloride. After grinding and mixing 50 mg were weighed into an anode cup electrode. The charge was vented by inserting a venting tool and then excited by dc arc. Synthetic solutions for calibration were prepared from pure solutions in an analogous manner. A bias was noted for tin. Results for lead and bismuth were good [94]. The detection limit was about 0.5 ppm.

Later work by the same authors improved the speed by using chips directly, along with lithium carbonate flux. The detection limits were improved as well: 0.2 ppm for bismuth and 0.4 ppm for lead. Chips were crushed and screened and the 10–40 mesh cut was utilized. 0.1 g of the screened chips and 7.5 mg of lithium carbonate were mixed and then transferred to a crater electrode. Once again the dc arc was used. Calibration curves were based on a range of weights of a single matrix-matched standard [95].

Marks, Cone, and Leao published a dc arc procedure for trace levels of *tellurium* in nickel-base high temperature alloys. Tellurium was precipitated as the metal with stannous chloride. The tellurium was collected on a 25-mm diameter 0.45 μm membrane filter. The filter was then carefully folded and placed in a crater electrode. The electrode was then placed in a 185°C oven to char the membrane, then in a muffle furnace at 400°C until ashing was completed. Arsenic was found to be the most suitable internal standard [96].

7.2.2 Inductively Coupled Plasma (ICP)

Kujirai, Yamada, Kohri, and Okochi analyzed numerous nickel- and cobalt-base high temperature alloys by a direct ICP approach. Test portions of 0.5 g were dissolved in 1:1:1 hydrofluoric acid:nitric acid:water, spiked with 1 mg of magnesium, and diluted to 100 mL in polypropylene volumetric flasks. Precision and accuracy were improved by the use of an HF-resistant Teflon™

spray chamber that had been treated with metallic sodium to improve its wettability [97]. Whitten utilized microwave dissolution of 0.5 g samples of superalloy scrap. First, 5mL of a 7:3 mixture of hydrochloric acid:hydrofluoric acid, then 2 mL of nitric acid were added before the vessel was sealed and heated at 50% power (325 W) for 30 sec. Any remaining insolubles were ignited and fused with 0.5 g of sodium peroxide and recombined with the filtrate (plus an additional 3 mL of hydrochloric acid). Copper was used as the internal standard [98]. Gomez Coedo, Dorado Lopez, and Vindel Maeso dissolved cobalt-base superalloys by reacting a 0.5 g test portion with 50 mL of *aqua regia* and 2 mL of hydrofluoric acid. When dissolution is complete the solution is fumed with 50 mL of 1:1 sulfuric acid:phosphoric acid. Then the cooled solution was diluted to about 40 mL with 10% (v/v) hydrochloric acid. Undissolved residue was filtered, ignited, reacted with 2 mL of sulfuric acid, 5 mL of hydrofluoric acid, and, dropwise, about 1 mL of nitric acid. After evaporation to dryness on a hotplate, and ignition, the residue was fused with 0.5 g of sodium carbonate. The cooled melt was leached in the filtrate and the solution was diluted to 500 mL with 10% (v/v) hydrochloric acid. A 25-mL aliquot of this solution was spiked with 5 µg of yttrium as an internal standard and diluted to 100 mL with 10% (v/v) hydrochloric acid [99].

Vozzella and Condit determined *yttrium* as a constituent of complex nickel alloys. Hydrochloric, hydrofluoric, and nitric acids were used to dissolve 0.5–1.0 g samples in a microwave oven. By employing the minimum volume of hydrofluoric acid needed to effect dissolution and prevent "earth acid" precipitation, as little as 10 ppm of yttrium could be determined. The line at 360.07 nm was used [100].

7.2.3 Hollow Cathode

Thornton published a paper in 1969 on the application of this methodology to the atomic emission of tramp elements in high temperature alloys (and steels). Test portions of 10–100 mg of millings were weighed into a small graphite hollow cathode electrode that fit on the end of a tungsten rod. The chamber is pumped down, flushed with helium, and then re-evacuated. A current of 200 mA was used initially. The spectrum was photographed, the power was increased, and the spectrum was rephotographed. This process was continued, racking the photographic plate between each exposure until it was reasonable to assume that all the analytes had been vaporized. The most easily distilled out are *silver, cadmium, mercury, thallium, and zinc* (complete at 500 mA). The next group is *bismuth, indium, potassium, sodium, lead, tin, and tellurium* (complete at 750 mA). Then *arsenic, copper, gallium, magnesium, manganese, and antimony* (complete at 1A). The addition of 30 weight percent silicon metal to the test portion was found to mask the interference effects of matrix elements [101].

7.3 Copper-base Alloys

7.3.1 Arc/Spark

The globular arc technique described in Section 3.1.2 has been successfully utilized for high purity copper alloys. Publicover reported on a method for oxygen-free electrolytically pure copper that applies the approach to trace and ultratrace levels of *33 elements, including cadmium, mercury, phosphorus, and zinc at* <*0.5ppm*. Two grams of millings, clippings, or chips are briquetted into a 3/8 in. (0.95 cm) diameter, 3/16 in. (0.48 cm) thick solid that is placed on a platform-type graphite electrode. The sample is dc arced with a graphite counter electrode for 1.5 min at 15 amp [102]. The ASTM Suggested Methods book [60] contains a modified version of this procedure, as well as three spark methods (for brasses, cathode copper, and copper alloys in general).

7.3.2 Plasma

Epstein, Koch, Epler, and O'Haver developed a dc plasma emission method for *phosphorus* in copper-base alloys that employs on-line ion exchange to remove copper and iron overlap interference on the highly sensitive phosphorus lines at 213.62 and 214.91 nm. Two schemes were utilized: a cation exchange system that retained copper and iron, and an activated alumina column that retained phosphorus. The latter system was an effective preconcentration step (phosphorus was eluted with 1.5M sodium hydroxide) that allowed 20 ppm to be accurately determined [103]. Carpenter and Till published an ICP method for brasses that provided data for pattern recognition analysis as applied in forensic studies [104]. King and Wallace used ICP-OES to determine *aluminum* (396.15 nm), *iron* (273.96 nm), *lead* (283.31 nm), *nickel* (231.60), *phosphorus* (178.28 nm), *silicon* (288.16 nm), *tin* (189.99 nm), and *zinc* (334.50 nm) in eight samples of copper powders, copper oxides, and brasses. Samples were simply dissolved in hydrochloric and nitric acids with the addition of hydrofluoric acid (without heating if silicon was to be determined)[105].

7.3.3 Other Methods

Glow discharge was found to be a rapid and accurate means of determining the concentrations of major and minor elements in a wide range of copper-base alloys. Kruger, Butler, Liebenberg, and Bohmer reported on results from alloys containing 56–99% copper, including rolled, chill-cast, continuous cast, and extruded samples. *Lead, iron , nickel, zinc, and tin*, as well as copper, were determined. Internal standard calculations were based on concentration ratios from a reference material containing all analytes. An assumption was made that all analytes sum to 100% [106].

Sabsabi and Cielo determined small amounts of *iron, nickel, and silver* in copper alloys using a focused Nd:YAG *laser*. Time-gating was necessary to avoid the intense continuum between 50 nanoseconds and 1 microsecond after the firing of the laser pulse. After 1 microsecond a neutral excited atom spectrum predominates. Typically, the authors chose to record data between 5 microseconds and 25 microseconds [107].

7.4 Aluminum-base Alloys

7.4.1 Arc/Spark

There are three point-to-plane spark discharge methods for aluminum and aluminum alloys in the *Annual Book of ASTM Standards* [108]. The most recent of these is Standard Method E1251, which specifies an argon atmosphere and a self-initiating capacitor discharge. The Suggested Methods book contains a powder/dc arc technique that can be calibrated by synthetic solution standards. The sample (with appropriate additions of pure aluminum) is dissolved in 1:4 sulfuric acid:water, evaporated to dryness, ignited at 600°C, and ground in an agate mortar. Then a 400 mg portion is ground with 800 mg of graphite powder in a second agate mortar. The sample is then packed into three identical graphite crater electrodes. One drop of 1:3 phosphoric acid:water is added to each, then the electrodes are dried in a drying oven at 150°C. A water-cooled electrode holder is utilized for a 60 sec exposure with no preburn.

7.4.2 Plasma

A "U"-shaped dc plasma, which can be used both as an emission source and as an atom reservoir for atomic absorption measurements, was applied to the determination of *lithium* in aluminum alloys. Pavlovic, Pavlovic, and Marinkovic demonstrated the applicability of this design for acid-dissolved samples introduced as a liquid aerosol in an argon stream into the center of the arc cavity where a gas vortex stabilizes the arc. Emission was measured at 610.36 nm and 670.78 nm [109]. Batistoni, Farias de Funes, and Smichowski studied the determination of *magnesium and manganese* in aluminum and aluminum alloys. They noted detection limits of 0.003% for magnesium and 0.005% for manganese. They also observed that sample size (for a concentric Meinhard nebulizer) was limited to 0.1 g/100 mL and that acid concentration (HCl) must be kept to a minimum [110].

Tao and Kumamaru determined sub-ppm levels of *chromium* in aluminum alloys by forming the 8-hydroxyquinolate in the acid-dissolved sample, vaporizing it in a tungsten boat furnace, and directing the vapor into an ICP torch [111]. Yamamoto, Obata, Nitta, Nakata, and Kumamaru determined trace levels of *beryllium* in magnesium-aluminum alloys (and in copper metal) using a microporous Teflon™ tube separator for the continuous solvent extraction of the acid-dissolved sample. The extraction system used was acetylacetone/carbon tetrachoride [112]. A flow injection approach that utilizes the on-line electrolytic dissolution of solid samples of aluminum and direct ICP-OES determination of seven elements has been described. Yuan, Wang, Yang, and Huang used 1.0M nitric acid as the electrolyte with a current density of 1150 mA/cm^2. *Chromium, copper, iron, magnesium, manganese, silicon, and zinc* were measured with silicon and iron showing the greatest variability [113].

Steffan and Vujicic compared commercial aluminum reference materials from five different producers using both laser ablation-ICP-OES and conventional spark-OES. In particular they examined aluminum-silicon and aluminum-silicon-magnesium alloys for their content of *calcium, chromium, cobalt, copper, iron, lead, magnesium, manganese, nickel, silicon, sodium, tin, titanium, and zinc*. The two methods utilized were compared and found to show similar detection limits. Reproducibilities ranged from 0.75–4.0% R.S.D. No significant deviations from the certified values of the reference materials were found [114].

7.4.3 Glow Discharge

Dogan, Laqua, and Massmann described the use of the glow discharge lamp for the analysis of alloying elements in aluminum-base metals [115]. Naganuma, Kubota, and Kashima studied a large number of parameters that affect the performance of glow discharge for the analysis of aluminum alloys. Pure aluminum, pure copper, an 8% copper-aluminum alloy and a 4.9%Cu/7.2%Si—aluminum alloy were used in these studies. Surface roughness, grain size, and sample thickness were all found to be important. The surface is better with a rough finish, the grain size should be as fine as possible (remelting, if necessary) and samples and standards should be of the same thickness (to compensate for heating effects) [116].

7.4.4 Laser

Ishizuka and Uwamino determined eight elements (*Cr, Cu, Fe, Mn, Ni, Pb, Ti, and Zn*) in aluminum alloys by Q-switched ruby laser ablation-microwave-induced plasma. Precision ranged from 2.4 to 10.4% R.S.D. and the detection limit ranged as low as 0.9 ppm (for zinc) [117]. Sabsabi and Cielo used a Q-switched Nd:YAG laser to produce a plasma from the surface of aluminum alloys whose light was directly measured.* Time resolution measurement was

---

* Nomenclature is not consistently applied in this new area: however, "laser breakdown spectroscopy" is becoming common for the direct measurement of the light from a laser-produced plume.

applied to avoid problems related to self-absorption, line broadening, and matrix effects. Precision and detection limits were found to be about equivalent to those of conventional atomic emission techniques [118]. Andre, Geertsen, Lacour, Machien, and Sjostrom used a XeCl excimer UV laser emitting at 308 nm to analyze four different aluminum alloys (aluminum-magnesium, aluminum-zinc, aluminum-copper, and aluminum-copper-silicon). Approximately 50 laser pulses are delivered to the sample surface during the 1-sec integration period. Each pulse is preceded by a 1 microsecond delay and is gated to collect data for 1 microsecond. Data collection is preceded by a 20 sec "preburn" (consisting of 1000 laser pulses). The authors observed good reproducibility and attribute it largely to the use of a laser operating in the UV [119]. Geerstsen, Lacour, Mauchin, and Pierrard conclude that the basic technique has great potential for elemental microanalysis. It is quantitative and insensitive to the alloy matrix [120].

7.5 Miscellaneous Alloys

7.5.1 Arc/Spark

The *Annual Book of ASTM Standards* contains a method for the analysis of *silver* by a powder/dc arc technique [121]. There are also numerous arc/spark procedures described in the Suggested Methods volume [60], including point-to-plane spark methods for: *antimony metal, antimonial lead alloys, type metal, tin-lead solder, magnesium and its alloys, titanium alloys, zinc and zinc alloys.* Also included are dc arc methods for: *beryllium metal, bismuth/cadmium eutectic alloy, indium metal, lead-base alloys, molybdenum metal, palladium, platinum, rhenium, 99.99 grade fine silver, tantalum, tin-base alloys, tungsten metal, zinc-base alloys, and zirconium metal.* The book also contains numerous other arc/spark procedures for metals. For example, *gallium/lead and antimony/lead alloys* by a solution porous cup spark procedure, *plutonium* by an ion exchange/graphite spark technique, and *gold, silver, and platinum* by a copper fluoride carrier distillation technique.

7.5.2 Plasma

The *Annual Book of ASTM Standards* contains a method for the analysis of *refined gold* by DCP [122]; a DCP method for hafnium in *zirconium and zirconium alloys* [123]; and a method for the analysis of *zinc-5%mischmetal alloys* by ICP [124]. The Suggested Methods volume [60] contains a method for unalloyed titanium by DCP.

Harrington, Jones, and Bramstedt used ICP-OES (and flame AA) to establish values for chromium and zinc in *corrosion resistant coating* reference materials for X-ray fluorescence analysis [125]. Kujirai, Yamada, and

Hasegawa determined trace levels of arsenic, cobalt, copper, iron, manganese, nickel, titanium, vanadium, zinc, and zirconium in *chromium metal* by ICP. The 1 g test portion was dissolved in 3 mL of hydrochloric acid and 10 mL of perchloric acid, then heated to strong fumes to oxidize chromium. Then 60 mg of lanthanum were added along with a small amount of filter pulp. The solution was then adjusted to pH 11.0 with sodium hydroxide solution and filtered. The lanthanum hydroxide precipitate was dissolved with dilute hydrochloric acid. The coprecipitation is then repeated. The final solution is evaporated nearly to dryness, then diluted to 20 mL with 10% (v/v) hydrochloric acid and measured by ICP-OES [126].

Steffan and Vujicic determined 18 elements in *zirconium alloys* by a direct ICP approach. The samples were dissolved with hydrofluoric and nitric acids, the solutions were evaporated to dryness, then the residue was dissolved and diluted to 100 mL with 1.2M hydrochloric acid. Of the lines selected for use (Al308.22 nm, B182.9 nm, Cr205.55 nm, Nb309.42, Ni 221.65 nm, and Sn181.11 nm) all required mathematical correction because of zirconium line overlaps. Results obtained from three NIST SRM's were reported [127].

Silve Kallmann was an early advocate for combining classical approaches and plasma-OES techniques for *precious metal* analysis. In a 1984 paper that reviewed the field he compared and contrasted classical wet chemistry, fire assay, atomic absorption, and plasma emission for *gold, silver, and the platinum group metals*, illustrating how these techniques can be made to work together [128]. Kato published a review paper with 165 references on the use of ICP-OES for the analysis of radioactive material, including *uranium and plutonium metal* [129].

Lajunen, Kokkonen, Karijoki, Porkka, and Saari published a method for the determination of aluminum, arsenic, and phosphorus in *elemental silicon*. The test portion for aluminum and phosphorus by DCP was dissolved in nitric acid/hydrofluoric acid/hydrogen peroxide, evaporated to near dryness (total dryness for phosphorus determination), then diluted to 10 mL with dilute hydrochloric and nitric acids. For arsenic by hydride-DCP-OES the sample was dissolved in nitric acid/hydrofluoric acid, then perchloric and sulfuric acid were added. The sample was carefully fumed to dryness, then diluted to 100 mL with dilute hydrochloric acid [130]. Yudelevich, Zaksas, Shaburova, and Cherevko developed ICP and DCP procedures for several different *high temperature superconducting materials*. Solutions, prepared by dissolving the test portion in a small quantity of hydrochloric and nitric acids, were analyzed by ICP using molybdenum as an internal standard for the $YBa_2Cu_3O_x$ composition, and yttrium as the internal standard for the *Bi-Ca-Cu-Sr-O* and *Tl-Ca-Pb-Sr-Cu-O* materials. A two-jet argon DCP was utilized to directly analyze powdered samples that had been blended with graphite powder containing germanium and palladium as internal standards, and dispersed into the plasma by means of a high frequency discharge [131].

7.5.3 Glow Discharge

Jager reported on the analysis of *gold* using the Grimm-design glow discharge lamp for atomic emission. Poor reproducibility was corrected by relating all measurements to the sputtering rate. This procedure also reduced the effects of metallurgical history. More than two decades later Harville and Marcus reported that radio frequency glow discharge atomic emission shows distinct advantages over conventional dc glow discharge that extend beyond the ability to analyze nonconducting samples. In particular, they described work on the trace analysis of *high-purity gold, platinum, and silver and sterling silver alloys*. They employed the background in the vicinity of the analyte line as the internal standard. The emission stabilizes rapidly and shows good long-term and short-term precision. The detection limits were shown to be superior to those of dc glow discharge, as well as to those of conventional arc/spark atomic emission [133].

7.5.4 Laser

Goodall, Johnson, and Wood studied *uranium/zirconium alloys* by laser ablation-ICP-OES and discovered anomalous concentration readings at different laser power densities. Chemical analysis of the aerosol showed it to be identical to the bulk composition. The effect was attributed to a variation in aerosol particle size affecting the atomization efficiency in the ICP torch [134]. Nemet and Kozma applied the direct time-resolved measurement of emission from laser excited plumes to the analysis of *gold jewelry alloys*. A low energy (10 mJ) Q-switched Nd:YAG laser was used, leaving the sample undamaged. Spectra were detected by means of a photodiode array with a microchannel plate image intensifier. For each analyte/sample type an optimized time delay and measurement time-gate was established that produced the best line signal-to-background ratio [135].

REFERENCES

1. Ohls, K.D. Spectrochimica Acta, 51B, 245 (1996).
2. Slickers, K. Automatic Emission Spectroscopy, Bruhlsche Universitatsdruckerei, Geissen, Germany, 1980.
3. Thomsen, V.B.E. Modern Spectrochemical Analysis of Metals, ASM International, Materials Park, OH, 1996.
4. Barnes, R.M. "Emission Spectroscopy" Instrumental Analysis, 2nd ed. (G.D. Christian and J.E. O'Reilly, eds.) Allyn and Bacon, Boston, 1986, pp. 322–355.

5. Lockyer, I.N. Philosophical Transactions, 163, 253, 639 (1873).
6. Meggers, W.F.; Kiess, C.C.; Stimson, F.S. National Bureau of Standards (NIST) Scientific Paper No.18, Washington, D.C., 235 (1922).
7. Gerlach, W.; Schweitzer, E. Foundations and Methods of Chemical Analysis by the Emission Spectrum, Vol. 1, Voss, Adam Hilger, London, 1929.
8. Grimm, W. Spectrochimica Acta, 23B, 443 (1968).
9. Van Nostrand's Scientific Encyclopedia, 4th ed., D. Van Nostrand, Princeton, NJ, 1968.
10. Fassel, V.A. Analytical Chemistry, 51, 1290A (1979).
11. Greenfield, S.J.I.L.; Berry, C.T. Analyst, 89713 (1964).
12. Wendt, R.H.; Fassel, V.A. Analytical Chemistry, 37, 920 (1965).
13. Dickinson, G.W.; Fassel, V.A. Analytical Chemistry, 41, 1021 (1969).
14. Boumans, P.W.J.M. Line Coincidence Tables for Inductively Coupled Plasma Emission Spectrometry, Vols. I and II, Pergamon, New York, 1980.
15. Parsons, M.L.; Forster, A.; Anderson, D. An Atlas of Spectral Interferences in ICP Spectrosdcopy, Plenum, New York, 1980.
16. Winge, R.K.; Fassel, V.A.; Peterson, V.J.; Floyd, A. Inductively Coupled Plasma Emission Spectroscopy, Elsevier, New York, 1985.
17. Massachusetts Institute of Technology Wavelength Tables, The MIT Press, Cambridge, MA, 1969.
18. Ivaldi, J.C.; Tyson, J.F. Spectrochimica Acta, 50B, 1207 (1995).
19. Alavosus, T.J.; Murphy, R. Schatzlein, D. American Laboratory, July, 1995, 31.
20. Ding, W.W.; Sturgeon, R.E. Analytical Chemistry, 69, 527 (1997).
21. Gagean, M.; Mermet, J.M. Journal of Analytical Atomic Spectrometry, 12, 189 (1997).
22. Richner, P.; Borer, M.W.; Brushwyler, K.R.; Hieftje, G.M. Applied Spectroscopy, 44, 1290 (1990).
23. Jones, J.L.; Dahlquist, R.L.; Hoyt, R.E. Applied Spectroscopy, 25, 628 (1971).
24. Wallace, G.F. Atomic Spectroscopy, 4, 188 (1983).
25. Paschen, F. Annals of Physics, 50, 901 (1916).
26. Grimm, W. Spectrochimica Acta, 23B, 443 (1968).
27. Marcus, R.K.; Harville, T.R.; Mei, Y.; Shick Jr., C.R. Analytical Chemistry, 66, 902A (1994).
28. Bengston, A.; Lundholm, M. Journal of Analytical Atomic Spectrometry, 3, 879 (1988).
29. Broekaert, J.A.C. Applied Spectroscopy, 49, 12A (1995).
30. De Marco, R.; Kew, D. Spectrochimica Acta, 41B, 591 (1986).
31. Anderson, D.R.; McLeod, C.W.; English, T.; Smith, A.T. Applied Spectroscopy, 49691 (1995).
32. Kim, Y.W. Proceedings of the Electrochemical Society, 90–18, 175 (1990).
33. Carlhoff, C.; Kirchhoff, S. Progress of Analytical Chemistry in the Iron and Steel Industry, Commission of European Communities, EUR14113, 1992, pp. 150–3.
34. Tappe, W.H. Progress of Analytical Chemistry in the Iron and Steel Industry, Commission of European Communities, EUR14113, 1992, pp. 142–9.
35. Aragon, C.; Aguilera, J.A.; Campos, J. Applied Spectroscopy, 47, 606 (1993).
36. Leis, F.; Sdorra, W.; Ko, J.B.; Niemax, K. Mikrochimica Acta, 2, 185 (1989).

37. Uebbing, J.; Brust, J.; Sdorra, W.; Leis, F.; Niemax, K. Applied Spectroscopy, 45, 1419 (1991).
38. Lee III, Y.; Sneddon, J. Spectroscopy Letters, 25, 881 (1992).
39. Sdorra, W.; Brust, J.; Niemax, K. Mikrochimica Acta, 108, 1 (1992).
40. Wood, D.L.; Dargis, A.B.; Nash, D.L. Applied Spectroscopy, 29, 310 (1975).
41. Annual Book of ASTM Standards, Vol.03.06, American Society for Testing and Materials, West Conshohocken, PA, 1997, designation E1507.
42. Charge Transfer Devices in Spectroscopy (J.V. Sweedler, K.L. Ratzlaff, M.B. Denton, eds.) VCH, New York, 1994.
43. Annual Book of ASTM Standards, Vol. 03.06, American Society for Testing and Materials, West Conshohocken, PA, 1997, designation E356.
44. Annual Book of ASTM Standards, Vol. 03.06, American Society for Testing and Materials, West Conshohocken, PA, 1997, designation E520.
45. Hanley, Q.S.; Earle, C.W.; Pennebaker, F.M.; Madden, S.P.; Denton, M.B. Analytical Chemistry, 68, 661A (1996).
46. Wallace, G.F. Atomic Spectroscopy, 5, 5 (1984).
47. Lorber, A.; Goldbart, Z.; Harel, A.; Sharvit, E.; Eldan, M. Spectrochimica Acta, 41B, 105 (1986).
48. Ramsey, M.H.; Thompson, M. Journal of Analytical Atomic Spectrometry, 1, 185 (1986).
49. Romero, X.; Poussel, E.; Mermet, J.M. Spectrochimica Acta, 52B, 487 (1997).
50. Annual Book of ASTM Standards, Vol. 03.05, American Society for Testing and Materials, West Conshohocken, PA, 1997, designations E158, E305.
51. Ediger, R.D.; Hoult, D.W. Atomic Spectroscopy, 141 (1980).
52. Boumans, P.W.J.M. Analytical Chemistry, 66, 459A (1994).
53. Annual Book of ASTM Standards, Vol. 03.05, American Societry for Testing and Materials, West Conshohocken, PA, 1997, designation E212.
54. Annual Book of ASTM Standards, Vol. 03.06, American Society for Testing and Materials, West Conshohocken, PA, 1997, designation E415.
55. Annual Book of ASTM Standards, Vol. 03.05, American Society for Testing and Materials, West Conshohocken, PA, 1997, designation E327.
56. Annual Book of ASTM Standards, Vol. 03.06, American Society for Testing and Materials, West Conshohocken, PA, 1997, designation E1086.
57. Annual Book of ASTM Standards, Vol. 03.06, American Society for Testing and Materials, West Conshohocken, PA, 1997, designation E485.
58. Annual Book of ASTM Standards, Vol. 03.06, American Society for Testing and Materials, West Conshohocken, PA, 1997, designation E421.
59. Annual Book of ASTM Standards, Vols. 03.05 and 03.06, American Society for Testing and Materials, West Conshohocken, PA, 1997, designations E293, E404, E1009.
60. Suggested Methods for Analysis of Metals, Ores, and Related Materials, 9th ed., American Society for Testing and Materials, West Conshohocken, PA, 1992.
61. Fernando, L.A. Analytical Chemistry, 56, 1970 (1984).
62. Fernando, L.A.; Heavner, W.D.; Gabrielli, C.C. Analytical Chemistry, 58, 511 (1986).

63. Potter. N. M.: Vergosen III. H.E. Talanta. 32. 545 (1985).
64. Chen. H.: Brindle. I.D. Zheng. S. Analyst. 117. 1603 (1992).
65. Wallace. G.F. Atomic Spectroscopy. 2. 61 (1981).
66. Hosoya. M.: Tozawa. K.: Takada. K. Talanta. 33. 691 (1986).
67. Lopez Molinero. A.: Ferrer. A.: Castillo. J.R. Talanta. 40. 1397 (1993).
68. Nakahara. T. Spectrochimica Acta. 40B. 293 (1985).
69. Wittmann. A.A.: Schuster. L.G. Spectrochimica Acta. 42B. 413 (1987).
70. DelMonte Tamba. M.G.: Folciani. R.: Dorado Lopez. T.: Gomez Coedo. A. Analyst. 119. 2081 (1994).
71. Grossman. A.M.: Ciba. J.: Jurczyk. J.: Spiewok. W. Talanta. 37. 815 (1990).
72. Vaamonde. M.:Alonso. R.M.: Garcia. J.: Izaga. J. Journal of Analytical Atomic Spectrometry. 3. 1101 (1988).
73. Umemoto. M.: Kuboto. M. Spectrochimica Acta. 42B. 491 (1987).
74. Okamoto. Y.: Murata. H.: Yamamoto. M.: Kumamaru. T. Analytica Chimica Acta. 239. 139 (1990).
75. Duffy. M.: Thomas. R. Atomic Spectroscopy. 17. 128 (1996).
76. Walton.S.J. Analyst. 111. 225 (1986).
76A. Ozaki. E.A.: Oliveira. E. Journal of Analytical Atomic Spectrometry. 8. 367 (1993).
77. Wickstrøm. T.: Lund. W.: Bye. R. Journal of Analytical Atomic Spectrometry. 10. 809 (1995).
78. Chen. H.Y: Chen. H-W: Feng. Y-L: Tian. L-C Atomic Spectroscopy. 18. 29 (1997).
79. McLeod. C.W. Cook. I.G.: Worsfold. P.J.: Davies. J.E.: Queay. J. Spectrochimica Acta. 40B. 57 (1985).
80. Yamada. K.: McLeod. C.W.: Kujirai. O.: Okochi. H. Journal of Analytical Atomic Spectrometry. 7. 661 (1992).
81. Giglio. J.J.: Mike. J.H.: Mincey. D.W. Analytica Chimica Acta. 254. 109 (1991).
82. Souza. I.G.: Bergamin. F.H: Krug. J.A.: Nobrega. J.A.: Oliveira. P.V.: Reis. B.F.: Gine. M.F. Analytica Chimica Acta. 245. 211 (1991).
83. Gomez Coedo. A.: Dorado Lopez. M.T.: Jiminez Seco. J.L.: Guitierrez Cobo. I. Journal of Analytical Atomic Spectrometry. 7. 11 (1992).
84. Gomez Coedo. A.: Dorado Lopez. T.: Gutierrez Cobo. I.: Escudero Baquero. E. Journal of Analytical Atomic Spectrometry. 7. 247 (1992).
85. G. Coedo. A.: Dorado. T.: Escudero. E.: G. Cobo. I. Journal of Analytical Atomic Spectrometry. 8. 827 (1993).
86. Masamba. W.R.L.: Smith. B.W.: Winefordner. J.D. Applied Spectroscopy. 46. 1741 (1992).
87. Jin. Q.: Zhang. H.: Liang. F.: Yang. W.: Jin. Q. Journal of Analytical Atomic Spectrometry. 11. 331 (1996).
88. Radmacher. H.W.: deSwardt. M.C. Spectrochimica Acta. 30B. 353 (1975).
89. Weiss. Z. Spectrochimica Acta. 51B. 863 (1996).
90. Kagawa. K.: Hattori. H.: Ishikane. M.: Ueda. M.: Kurniawan. H. Analytica Chimica Acta. 299. 393 (1995).
91. Aguilera. J.A.: Aragon. C.: Campos. J. Applied Spectroscopy. 46. 1382 (1992).
92. Aragon. C.: Aguilera. J.A.: Campos. J. Applied Spectroscopy. 47. 606 (1993).

93. Gonzalez, A.; Ortiz, M.; Campos, J. Applied Spectroscopy, 49, 1632 (1995).
94. Atwell, M.G.; Golden, G.S. Applied Spectroscopy, 24, 362 (1970).
95. Atwell, M.G.; Golden, G.S. Applied Spectroscopy, 27, 464 (1973).
96. Marks, J.; Cone, R.; Leao, E. Applied Spectroscopy, 25, 493 (1971).
97. Kujirai, O.; Yamada, K.; Kohri, M.; Okochi, H. Applied Spectroscopy, 40, 962 (1986).
98. Whitten, C.W. Atomic Spectroscopy, 8, 81 (1987).
99. Gomez Coedo, A.; Dorado Lopez, M.T.; Vindel Maeso, A. Journal of Analytical Atomic Spectrometry, 2, 629 (1987).
100. Vozzella, P.A.; Condit, D.A. Analytical Chemistry, 60, 2497 (1988).
101. Thornton, K. Analyst, 94, 958 (1969).
102. Publicover, W.E. Analytical Chemistry, 37, 1680 (1965).
103. Epstein, M.S.; Koch, W.F.; Epler, K.S.; O'Haver, T.C. Analytical Chemistry, 59, 2872 (1987).
104. Carpenter, R.C.; Till, C. Analyst, 109, 881 (1984).
105. King, A.D.; Wallace, G.F. Atomic Spectroscopy, 6, 4 (1985).
106. Kruger, R.A.; Butler, L.R.P.; Liebenberg, C.J.; Böhmer, R.G. Analyst, 102, 949 (1977).
107. Sabsabi, M.; Cielo, P. Journal of Analytical Atomic Spectrometry, 10, 643 (1995).
108. Annual Book of ASTM Standards, Vols.03.05 and 03.06, American Society for Testing and Materials, West Conshohocken, PA, 1997, designations E227, E607, E1251.
109. Pavlovic, M.S.; Pavlovic, N.Z.; Marinkovic, M. Journal of Analytical Atomic Spectrometry, 4, 587 (1989).
110. Batistoni, D.A.; Farias de Funes, S.S.; Smichowski, P.N. Atomic Spectroscopy, 11, 85 (1990).
111. Tao, S.; Kumamaru, T. Journal of Analytical Atomic Spectrometry, 11, 111 (1996).
112. Yamamoto, M.; Obata, Y.; Nitta, Y.; Nakata, F.; Kumamaru, T. Journal of Analytical Atomic Spectrometry, 3, 441 (1988).
113. Yuan, D.; Wang, X.; Yang, P.; Huang, B. Analytica Chimica Acta, 251, 187 (1991).
114. Steffan, I.; Vujicic, G. Journal of Analytical Atomic Spectrometry, 9, 1117 (1994).
115. Dogan, M.; Laqua, K.; Massmann, H. Spectrochimica Acta, 27B, 65 (1972).
116. Naganuma, K.; Kubota, M.; Kashima, J. Analytica Chimica Acta, 98, 77 (1978).
117. Ishizuka, T.; Uwamino, Y. Analytical Chemistry, 52, 125 (1980).
118. Sabsabi, M.; Cielo, P. Applied Spectroscopy, 49, 499 (1995).
119. Andre, N.; Geertsen, C.; Lacour, J-L; Mauchien, P.; Sjostroms, S. Spectrochimica Acta, 49B, 1363 (1994).
120. Geertsen, C.; Lacour, J-L; Mauchien, P.; Pierrard, L. Spectrochimica Acta, 51B, 1403 (1996).
121. Annual Book of ASTM Standards, Vol.03.06, American Society for Testing and Materials, West Conshohocken, PA, 1997, designation E378.

122. Annual Book of ASTM Standards. Vol.03.06. American Society for Testing and Materials. West Conshohocken. PA. 1997. designation E1446.
123. Annual Book of ASTM Standards. Vol.03.06. American Society for Testing and Materials. West Conshohocken. PA. 1997. designation E1552.
124. Annual Book of ASTM Standards. Vol.03.06. American Society for Testing and Materials. West Conshohocken. PA. 1997. designation E1277.
125. Harrington. D.E.: Jones. J.S.: Bramstedt. W.R. Atomic Spectroscopy. 4. 171 (1983).
126. Kujirai. O.: Yamada. K.: Hasegawa. R. Journal of Analytical Atomic Spectrometry. 8. 481 (1993).
127. Steffan. I.: Vujicic. G. Journal of Analytical Atomic Spectrometry. 9. 785 (1994).
128. Kallmann. S. Analytical Chemistry. 56. 1020A (1984).
129. Kato. K. Atomic Spectroscopy. 7. 129 (1986).
130. Lajunen. L.H.J.: Kokkonen. P.: Karijoki. J.: Porkka. V.: Saari. E. Atomic Spectroscopy. 11. 193 (1990).
131. Yudelevich. I.G.: Zaksas. B.I.: Shaburova. V.P.: Cherevko. A.S. Atomic Spectroscopy. 13. 108 (1992).
132. Jager. H. Analytica Chimica Acta. 58. 57 (1972).
133. Harville. T.R.: Marcus. R.K. Analytical Chemistry. 67. 1271 (1995).
134. Goodall. P.: Johnson. S.G.: Wood. E. Spectrochimica Acta. 50B. 1823 (1995).
135. Nemet. B.: Kozma. L. Journal of Analytical Atomic Spectrometry. 10. 631 (1995).

7
Mass Spectrometry

1. INTRODUCTION [1–4]

Producers and users of metals and alloys are, for the most part, not yet completely comfortable in the domain of mass spectrometry. Thermal ionization "high-res" instruments have always been a necessary investment for the nuclear industry; and a few specialty metal producers have always opted for spark source instruments to screen ultratrace level impurities. But for the large majority of companies the considerable expense of this technology could not be justified.

The decision seemed logical. Commercial instruments were not only costly, but complicated, and procedures were lengthy and subject to numerous errors. Under the best of conditions the precision was still no match for more familiar methodologies. And the instruments had a reputation for excessive down-time.

In the last decade or so this picture has changed dramatically. The inductively coupled plasma ion source, the quadrupole mass analyzer, and to some extent, developments in glow discharge and spark- and laser-ablation sampling have made mass spectrometry an attractive option for the trace and ultratrace analysis of metals. The obtainable precision has improved nearly to the point where at times it is sample limited, with variation primarily due to heterogeneity of the analyte in the alloy. Perhaps most significantly for laboratories with constrained budgets, it is now possible to get a working system for about the cost of a simultaneous X-ray fluorescence spectrometer. This is not pocket change, but for many labs it has entered the realm of the possible.

It is appropriate to begin here with a brief outline of the historical developments in inorganic mass spectrometry that led up to the present renaissance of this technology for the analysis of metals. Like the history of so many other instrumental approaches, it is a story of stops and starts and strange turnings.

Mass spectrometry had its origins in the inorganic realm nearly a century ago. Joseph John Thomson (J.J. Thomson), associated principally with the discovery of the properties of the free electron, also built the first mass spectrometer in 1910. At the Cavendish Laboratory at Cambridge University, Thomson measured the mass to charge ratio of positive ions by the "parabola method." He employed a magnetic field and a parallel electrical field (the x coordinate), both perpendicular to the direction of an ion beam (z coordinate). The ions were deflected along a parabola in a perpendicular plane (y coordinate). Thomson found evidence with his new technique that not only the (then newly discovered) radioactive elements but also many stable elements occurred as mixtures of isotopes. The term "isotopic"' was first used in 1913 by Frederick Soddy (a chemist). Also in 1913 Thomson published *Rays of Positive Electricity and Their Application to Chemical Analysis* (Longmans, Green, London, 1913) in which he predicted the analytical power of the technique.

One of Thomson's many distinguished pupils was F.W. Aston, who built a series of mass spectrometers of increasing refinement and determined most of the natural abundance isotope ratios for the elements. A.J. Dempster also contributed to this list, which was complete by 1933. In 1935 Dempster built the first double-focusing mass spectrometer . In 1936 the so-called Mattauch-Herzog design was first realized. This instrument allowed the entire mass spectrum to be recorded simultaneously with a photoplate on its unique focal plane. In 1940 A.O. Nier designed and built a double-focusing mass spectrometer that replaced the bulky 180° deflection ion path with a 60° deflection, using a wedge sector magnet.

Most of the early work by Thomson and Aston utilized gas discharge ion sources, but the continuing efforts to identify natural abundances of the elements required innovative source designs. Dempster utilized a vacuum spark source for the last stable elements studied: platinum, palladium, gold, and iridium. In 1954 N.B. Hannay built a practical, general purpose spark source instrument based on the Mattauch-Herzog design. A commercial version first appeared in 1958.

In the meantime developments in organic analysis had been advancing rapidly. Early workers had been beset by problems due to extraneous peaks that were traced to the oils and greases that were used in the vacuum systems of the mass analyzer. Perceptive observers realized that these "noise signals" were the signatures of organic structure. By this route the enormous field of organic mass spectrometry was born.

From the Greek: *isos* (same) + *topos* (place)—meaning occupying the same place in the periodic chart.

The first 180° deflection commercial instrument was installed in the petroleum industry in the U.S. in 1940, but Nier's 60° sector design, as well as his new electron impact ion source, soon became widespread. In 1948 A.E. Cameron and D.F. Eggers built one of the first nonmagnetic mass spectrometers—a time-of-flight instrument. From 1953 to 1955 W. Paul and coworkers published papers on the first work with quadrupole designs. Later innovations include the ion trap and ion cyclotron resonance spectrometers (the latter work derived from work at the National Bureau of Standards—now the National Institute of Standards and Technology— in 1948).

In 1970 J.W. Coburn first applied a glow discharge ion source, and in 1973 W.W. Harrison and E. H. Daughtrey first applied a hollow cathode ion source to mass spectrometry. A.L. Gray first coupled a dc (capillary arc) plasma (1974), then an inductively coupled plasma (1978) to a mass spectrometer. The first commercial ICP-MS instrument was offered for sale in 1983. At about the same time the first commercial GD-MS was also marketed.

We could end this brief history here since the glow discharge MS and the inductively coupled plasma MS designs now dominate in the trace analysis of bulk metals and alloys. However, it is probably worth our time to briefly trace related developments.

Secondary ions, produced by bombarding a sample surface with a primary ion beam had been investigated by J.J. Thomson and others, but R.E. Honig in the mid-1950s first recognized their potential for the very sensitive analysis of surfaces and for depth profiling. In the early part of the decade of the 1970s commercial SIMS instrumentation became available. The ion microscope was described by R. Castaing and G. Slodzian in 1962. The first commercial ion microprobe for the analysis of minute areas on a sample was designed by H.J. Liebl in 1967. Laser microprobe mass spectroscopy was investigated by R.E. Honig and co-workers in the early 1960s, and commercial instrumentation followed 20 years later. These devices provide information about both the organic and inorganic composition of microscopic features.

In organic analysis, the development of so-called soft ionization sources (field desorption, plasma desorption, laser desorption, fast atom bombardment, and others) in the 1970s reduced the degree of fragmentation experienced by large or meta-stable molecules, yielding more tractable structural information. Ion cyclotron resonance led to Fourier transform mass spectrometry—a technique with the potential for an extremely high degree of mass resolution. Mixture analysis was addressed by an array of hyphenated techniques: GC-MS, LC-MS, and MS-MS.

The state-of-the-art in surface analysis is also a growing list of acronyms: SNMS (secondary neutral mass spectrometry), RIMS (resonance ionization mass spectrometry), and MPRIMS (multiphoton resonance ionization mass spectrometry), among others.

The majority of these developments will not concern us here since our purpose is quantitative elemental trace analysis of bulk material, and our focus is on routine, rapid procedures. In this chapter we will briefly discuss some of the theory associated with the formation, separation, and detection of ions. It will be necessary to define a number of terms that are part of the jargon of this discipline. Next, we will consider the instrumentation—ion sources, ion introduction systems, mass analyzers, detectors, and data systems. We will devote some time to quantification issues, including the use of internal standards and the powerful technique known as isotope dilution. We will look at interferences of various sorts and highlight their implications. As in previous chapters we will conclude with the metal and alloy applications.

2. PRINCIPLES OF THE INSTRUMENTATION [5,6]

As we saw in the Introduction above there are many diverse technologies encompassed by the field of mass spectrometry (see Figure 7.1). However, three fundamental processes characterize them all: ionization, separation, and detection. The sample ions that are separated and detected comprise a highly rarified gas traveling through a high vacuum chamber where the pressure must not exceed about 10^{-5} torr (about 10^{-3} Pa). A mass spectrum (computer readout or developed photoplate) is a plot of ionic abundance versus mass-to-charge ratio. In inorganic mass spectrometry some high energy source is employed to form ions from the analyte atoms. These are largely mono-atomic ions, although dimers and other polyatomic ions often form. And they are mostly singly charged. Usually only positive ions are measured, although it is possible (and sometimes necessary) to separate and measure negative ions. In inorganic mass spectrometry we simply count detected analyte ions and relate them to analyte concentration in the sample.

In organic mass spectrometry a weaker energy source is applied to the sample creating a number of molecular fragment ions. In the argot of the discipline the ionized whole molecule is called the *molecular ion* and the ionized molecule fragments are called *fragment ions.* The organic mass spectrometry analyst looks to the fragmentation pattern of the spectrum to provide clues to the molecular structure of an unknown compound. Since the mass-to-charge (m/z) ratio scale is precisely calibrated this becomes a powerful quantitative tool. Thus, if the unknown compound is acetone (CH_3-CO-CH_3), the molecular ion (CH_3-CO-$CH_3^{\cdot}$) occurs at $m/z = 58$. The molecular ion is often not the most abundant ion. Prominent peaks at $m/z = 43$ ($CH_3CO^{\cdot}$) and $m/z = 15$ ($CH_3^{\cdot}$) document the loss of one methyl group. By such means the puzzle is pieced together. In organic compound identification, usually mass spectra are not the only qualitative methodology applied to a problem. A complex molecule is also likely to be scrutinized by Fourier transform infrared

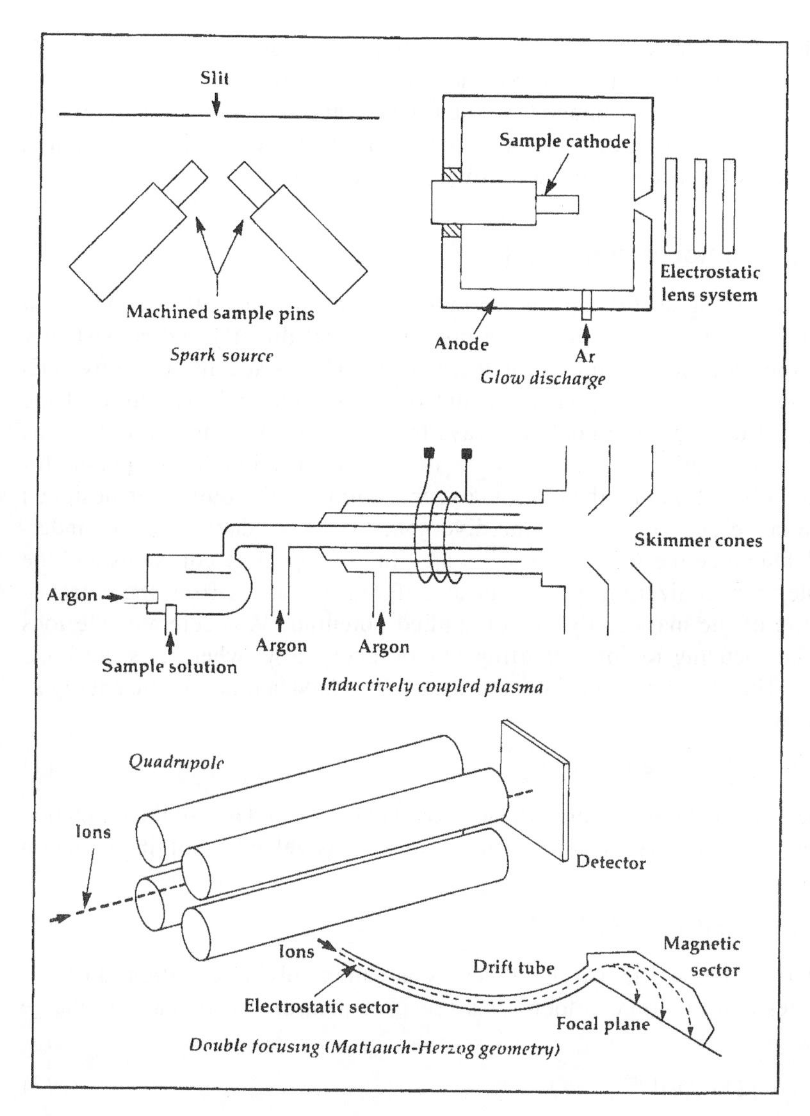

Figure 7.1 Three different ion sources (top) and two types of mass separator (bottom). (From T. R. Dulski, *Advanced Materials & Processes*, 141, 20 (1992); used with publisher's permission.)

absorption spectrometry, UV-visible molecular absorption spectrophotometry, and possibly others.

Mass spectrometry is also an extremely sensitive quantitative tool for organic analysis provided that an interference-free peak can be found that uniquely characterizes the analyte concentration. The hyphenated techniques—

gas chromatography MS, liquid chromatography MS, and tandem-MS (or MS-MS) allow complex mixtures to be accurately quantified.

In order to better understand the fundamentals common to both organic and inorganic mass spectrometry we will examine the ionization, separation, and detection processes in the following subsections.

2.1 Formation of Ions [7,8]

Before anything useful can happen in a mass spectrometer, the analyte must be converted to gaseous ions in a proportion that directly and consistently relates to its concentration in the test portion. This is accomplished by various means depending upon the nature of the sample and the nature of the analytical task, but the goal is always to create ions that are as uniform as possible in kinetic energy. Otherwise, the ion beam cannot be adequately focused for injection into the mass analyzer section. While sources are designed to minimize energy spread some divergence is unavoidable. This is understandable since the ionizing process may involve particle collisions and the sample atom ionization may occur at different distances from the entrance aperture of the mass analyzer. An applied potential, V, accelerates the ions into the focusing region imparting kinetic energy, zV, where z is the ionic charge. The ions emerging from the source thus have a total kinetic energy E_k given by:

$$E_k = zV + E_o = 1/2mv^2 \quad (1)$$

where E_0 is the kinetic energy the ion may have acquired due to the ionization process, m is the mass of the particle, and v is its velocity. Solving equation (1) for v,

$$v = [2(zV + E_0)/m]^{1/2} \quad (2)$$

We can readily see that if E_0 is very small then solving equation (1) for v indicates that the ionic velocity is a sensitive measure of its mass-to-charge ratio:

$$v = [2V(z/m)]^{1/2} \quad (3)$$

In the dawn of this discipline the ion sources utilized by Thomson and Aston were gas discharge tubes. These were suitable only for relatively volatile elements. However, in 1918 Dempster developed the *thermal ionization (TI) source*, a methodology that is still used for the most accurate inorganic work today. This source is generally used for high purity analytes that have been isolated chemically from their sample matrix. It is most often applied these days to isotope ratio or isotope dilution work (to be discussed later). The best results are obtained for analyte elements with a first ionization potential that is no higher than 6.5 eV (this group includes the alkali metals, the alkaline earths, and the rare earths). A microdrop of sample solution is dried onto a metal filament. Under high vacuum conditions the filament is resistance heated to an exact temperature at which vaporization and ionization are known to occur for the analyte compound involved. If the vaporization and ionization processes are known to occur at very different temperatures a two-filament source is utilized. The sample vaporizes from the filament onto which it was coated to create an atom cloud that surrounds the second filament, which is held at the higher ionization temperature.

In this lengthy but elegant technique there is much lore related to the compound form of the analyte, the filament material used, and the addition of "activator" substances that facilitate the ionization process. Both the *work function* (ω) of the filament material (the energy needed to release an electron from its surface) and its melting point should be as high as possible. Rhenium ($\omega = 5.0eV$; m.p. = 3453 K), tantalum ($\omega = 4.8eV$; m.p. = 3269 K), and tungsten ($\omega = 4.3eV$; m.p. = 3683 K) are common choices for filaments. The mechanism of the "activators" is obscure for the most part.

In the previous two chapters we have seen several expressions of the Saha equation, which is used to describe the degree of ionization in gases. Since mass spectrometry can be based on the production, separation, and detection of *either* positive or negative ions two expressions of the Saha-Langmuir equation apply. The production of positive ions (which is most common) is given by:

$$N_+ = (N_0)^{z(\omega-\varphi)/kT} \tag{4}$$

where N_+ = the number of positive ions; N_0 = the number of atoms vaporized per unit time; z = the charge on the ion in coulombs; ω = the work function of the filament; φ = the ionization potential of the analyte in electron volts; k = the Boltzman constant; T = the temperature of the hot filament in °K. With negative ions the expression is:

$$N_- = (N_0)^{z(\varepsilon-\omega)/kT} \tag{5}$$

where N_- = the number of negative ions; and ε = the electron affinity of the analyte in electron volts.

The lower price and multi-element flexibility of the *ICP-MS* make it the most common inorganic mass spectrometer in use today. In the last chapter we discussed the ICP as an excitation source. The comparatively long residence time of the sample species in the central plasma channel and the high,

---

Platinum has a work function of 5.7eV but is limited by its melting point of 2045 K.

stable temperature regions make the ICP a good source for both atomic emission and for mass spectrometry. Few molecular fragments can survive the sample dissociation process.

As an ionization source the ICP is highly efficient in producing analyte ions (50–100% conversion to singly charged ions for every element in the periodic table). But it is highly inefficient in the way the test portion represents the test sample. The nebulizer/spray chamber takes a variety of forms, all designed to produce aerosol droplets smaller than about 8 μm. These devices typically reject 99% of the test sample to waste. Since there is a substantial flow of argon in the plasma the sample atoms commonly account for less than 1 ppm of the total atom population of the plasma. The process of sampling the atmospheric pressure plasma by differential pumping for separation and analysis at a pressure 10 orders of magnitude lower is also very inefficient. The sampler cone touches the hot atmospheric plasma on one side and is pumped to about 1 torr (133 Pa) on the other. Its orifice diameter ranges from about 1.0 to 1.2 mm, depending on the design. The skimmer cone is usually more sharply angled toward the direction of the plasma than the sampler cone. It experiences the 1 torr region on one side and about 10^{-5} torr (1.3×10^{-3} Pa) on the other. Its orifice size ranges from 0.04 to 1.0 mm but, most typically, is just slightly smaller than the sampler orifice.

The plasma forms a cooler boundary layer near the outer surface of the sampler cone where oxide formation and other reactions are believed to occur. An expanding supersonic jet of argon and sample ions forms in the region between the cones. Only the central portion of this jet enters the mass analyzer region through the skimmer orifice. It has been estimated that ions traverse the two cones in about 3 microseconds. The sampler cone samples only a small portion of the analyte ion population in the plasma and more analyte is lost to ion-electron recombination reactions. Moreover, only about 1% of the ions passing through the sampler orifice also pass through the skimmer orifice. It is a tribute to the extreme sensitivity of mass spectrometry that so much of the sample can be discarded while part-per-trillion detection limits in the test solution are routinely attained.

Another bulk analysis inorganic mass spectrometric approach that has found recent wide acceptance is that based on the *glow discharge* source. Since this solids-based source is also utilized for atomic emission it was described in some detail in the previous chapter. The glow discharge source operates in a low pressure gas medium [usually 0.1–10 torr (13–1330 Pa) argon] and, like the ICP, must be differentially pumped using sampler and skimmer cones, albeit with more moderate pumping. The conventional dc glow discharge source requires a conductive sample. For this reason GDMS has been almost exclusively utilized for metals. More recent work with a radio frequency glow discharge (also described in Chapter Six) is directly applicable to the mass spectrometric analysis of both conducting and nonconducting solids.

The key to the usefulness of both types of glow discharge lies in their short-distance decoupling of the sampling and excitation processes. The *cathode fall* (also called the *cathode dark space*) is a nonluminous region immediately above the sample surface with a high discharge potential across it. Positive ions of argon are formed at the interface with the negative glow region above it and are accelerated by a large potential drop toward the sample surface. They recombine with Auger electrons near the surface and strike the surface of the sample. The kinetic energy of the particles impacting the sample surface is significantly smaller than that predicted by the cathode fall potential because energy is lost in particle collisions. It is sufficient, however, to cause the lattice to recoil from these impacts, ejecting neutral atoms, positive and negative ions, and electrons from the sample surface. The released electrons and negative ions are drawn toward the anode. The positive sample ions are ballistically redeposited on the sample surface by the cathode fall potential. Their impact similarly disrupts the lattice in a process termed "self-sputtering." The ejected neutral atoms leave the surface of the sample with an average energy in the range of 5–15 eV. The effectiveness of this entire process—the so-called *sputter yield*—is a function of the mass and energy of the incoming argon ions, the lattice energy of the sample, and the electronic orbital configuration of the sputtered atoms.

Some of these ejected neutral sample atoms make their way out of the cathode dark space by diffusion. These atoms encounter the *negative glow region*, where they are ionized along with argon gas atoms. The negative glow region is where atomic emission occurs. It is nearly free of electrical and magnetic fields. Ionization occurs here by two mechanisms: by electron impact ($M^{\circ} + e^{-} \rightarrow M^{+} + 2e^{-}$) and, principally, by Penning ionization ($M^{\circ} + Ar^{*} \rightarrow M^{+} + Ar^{\circ} + e^{-}$, where Ar^{*} designates an excited state of an argon atom). The extent of ionization is modest—1% or less of the neutral atoms become ionized—but it is a sufficient yield for both atomic emission and mass spectrometry. Nearly all the resultant sample atoms are singly charged because the energy states of the excited metastable argon atoms (11.5 and 11.7 *eV*) are unable to remove two electrons from most elements. The GDMS spectrum is thus free of *most* multiply-charged ion interferences.

The glow discharge source offers other advantages as well. Unlike other solids techniques (e.g., spark source and secondary ion mass spectrometries) the process of ionization in GDMS is unrelated to the extraction of the test portion, which occurs by a nonthermal sputtering process. This produces a much more uniform concentration response for all elements. The sample ions produced show a low and narrow range of energies and are sampled conveniently from the negative glow region. However, molecular interferences from argon, residual atmospheric gases, and sample matrix ion combinations, as well as interferences from a few doubly-charged ions, can present resolution problems.

An older methodology, *spark-source* mass spectrometry, is still utilized in some laboratories. It employs a vacuum spark discharge at several tens of kilovolts and a repetition rate of approximately 1 MHz on average. However, excitation conditions are varied widely to accommodate variation in sample types. Frequently the sample comprises both anode and cathode, although a point-to-plane arrangement with a tungsten electrode is sometimes used. Automated dynamic control of the gap distance is frequently employed.

The physics of ionization by a spark discharge in a vacuum have been described in detail by Ramendik, Verlinden, and Gijbels [9]. Ionization here is a thermal process and many multiply-charged ions are produced. As the plasma expands recombination processes reduce the number of multiply-charged ions by reaction with free electrons. Polyatomic cluster ions also form during the expansion of the plasma.

The vacuum spark produces a beam of sample ions with a significant energy spread (2–3 kV), requiring velocity focusing by an electrostatic analyzer, such as that incorporated in the Mattauch-Herzog and Nier-Johnson double-focusing mass spectrometer designs.

2.2 Separation of Ions [5,9,10]

Before the beam of ions that emerges from the source can enter the mass analyzer region of the instrument, it must be processed in several ways. The extent of "beam conditioning" required is associated with the energy spread of the ions as they leave the source and certain specific requirements of the type of mass analyzer used.

All instrument types have some sort of ion velocity limiting device that also prevents photons from reaching the detector. One common type used with quadrupole instruments is the *Bessel box*, which receives the source ion beam through an aperture in-line with a central plate on which a retarding potential is applied. The plate absorbs photons and fast neutral atoms and repels ions below the retarding potential. Ions above the retarding potential will only be slightly deflected and will impact the box wall. Ions of the exact energy of the retarding potential are deflected just enough to pass out through an exit aperture. Such a device may be regarded as a kinetic energy filter. It tends to defocus the ion beam, however, and generally requires a preliminary mass filter ("pre-quad") to recollimate the beam.

Another approach is to use a continuous off-axis series of electrostatic filters. Sometimes called an "omega lens" for its similarity in shape to the Greek letter (Ω). The curved array provides an effective block for photons and neutral atoms (which travel in a straight line and strike the wall) while a continuously focused beam of kinetic energy-selected ions traverses the sinuous path. The manufacturers who use this technology (there were two in 1998) claim that it eliminates "fringe effects" at the entrance to the quadrupole and that it limits the formation of polyatomic ions. Its detractors claim that it leads to severe space charge effects (see below) due to ion crowding.

The ultimate kinetic energy analyzer is the *electrostatic sector* of a *double-focusing mass spectrometer*, of course. This is designed with two electrodes that are sections of a toroid, a cylinder, or, most typically, a sphere. These are maintained at equal but opposite voltage. As the source ion beam passes between these electrodes, high velocity positive ions impact the positive electrode. Their kinetic energy is so great that they are only slightly deflected by the potential field. Low velocity ions are strongly deflected by the field and impact the negative electrode. Ions will successfully traverse a path between the two geometrically parallel surfaces if the centripetal and electrical forces are equal:

$$mv^2/r = zE \qquad (6)$$

where m = mass of the ion; v = velocity of the ion; r = radius of the ion's path; z = ionic charge; E = electric field strength.

If we neglect the excess kinetic energy, E_0, from Eq. (1) we can combine it with Eq. (6) as follows:

$$1/2(mv^2) = zV \qquad (1)$$

$$mv^2 = 2zV$$

Substituting in (6): $2zV/r = zE$

$$2V/r = E$$

$$Er = 2V$$

$$r = 2V/E \qquad (7)$$

Thus, for a given electrostatic field strength, E, only ions with kinetic energy equal to the accelerating potential, V, can successfully traverse a circular path of radius, r, between the electrodes.

There are generally slits before and after the electrostatic sector. The entrance slit is called the alpha (α) slit, the exit slit is called the beta (β) slit. This serves to further limit the kinetic energy spread of the ion beam. It should be noted that in the reverse Nier-Johnson design the electrostatic sector follows the magnetic sector's separation of ions by their momentum.

In the *magnetic sector* the ions are directionally focused in accordance with their mass-to-charge ratios. Such a device is, in effect, a momentum analyzer. The magnetic sector of a double-focusing mass spectrometer (whether it is the 180° sector of a Mattauch-Herzog design or the 60° sector of a Nier-Johnson design) exposes the sample ions to a centripetal force, Hzv, where H is the magnetic field strength. This product is equal to the centrifugal force, mv^2/r_m, where r_m is the radius of curvature of the magnetic field.

$$mv^2/r_m = Hzv \qquad (8)$$

The momentum, mv, of the ions is derived by rearranging this equation:

$$mv = Hzr_m \qquad (9)$$

and their velocity is:

$$v = Hzr_m/m \qquad (10)$$

Once again substituting in equation (1) (without the E_0 term):

$$1/2(mv^2) = zV \qquad (1)$$

$$[m(Hzr_m/m)^2]/2 = zV$$

and simplifying: $(z/m)[(H^2r_m^{\ 2})/2] = V$

and $(m/z) = (H^2r_m^{\ 2})/2V$ (11)

Therefore: $r_m = \{[2V(m/z)]^{1/2}\}/H$ (12)

which means that for a given applied accelerating potential, V, and a given magnetic field strength, H, ions of a given mass-to-charge ratio, (m/z), will follow a unique path of radius, r_m. It also tells us that the mass scale on the focal plane of a Mattauch-Herzog instrument varies with the square root of the ionic mass-to-charge ratio.

With instruments that utilize electrical detection V is held constant while H is scanned, or vice versa. This brings ions of successively higher (m/z) ratios to the detector.

Except for 180° sector instruments in which the entire ion path and both the source and the exit slits are imbedded in the magnetic field all designs experience some degree of *edge effect*. This is a resolution limitation due to the fact that the magnetic field boundaries are not sharp.

A *quadrupole mass analyzer* accomplishes the separation of ions without the use of heavy magnets. Four exactly parallel, precisely machined metal rods' are wired so that diagonally opposite pairs experience the same dc and radiofrequency signals. The ion beam passes down through the center of these rods. The mutual effect of the dc and rf fields influences the ions' lateral

---

Molybdenum metal is a useful material for this application. Gold metallized ceramic has also been used.

position so that ions of only one mass-to-charge ratio successfully negotiate the journey. All others collide with each other or impact the rods.

The ideal cross-sectional shape of the rods is hyperbolic. In that case the potential, V_q, at any point (x,y) in the gap between the rods at time t (in sec) is given by:

$$V_q = [(V_{dc} + V_{ac} \cos \omega t)(x^2 - y^2)]/r^2 \quad (13)$$

where V_{dc} = the applied dc voltage; V_{ac} = the amplitude of the applied rf (ac) signal; ω = the frequency of the rf (ac) signal in radians/sec; and r = the radius of the gap between the rods. Hyperbolic rods are becoming more common; however, the above relationship is a reasonable estimate even for the more easily manufactured cylindrical rods. The forces acting on an ion are calculated by differentiating equation (13) with respect to x and with respect to y and multiplying each result by the ionic charge:

$$F_x = -z(\delta V_q/\delta x) = -z[(V_{dc} + V_{ac} \cos \omega t)2x/r^2] \quad (14)$$

and $$F_y = -z(\delta V_q/\delta y) = +z[(V_{dc} + V_{ac} \cos \omega t)2y/r^2] \quad (15)$$

It can be shown that the ion trajectories are not simple curves; rather, a sinusoidal component is superimposed on their flight path. The optimal resolution is obtained when V_{dc}/V_{ac} is 0.168. If this limiting value is exceeded, however, no ion of any mass-to-charge ratio can traverse the rods. The frequency of the ac signal (ω) is held fixed while the amplitude of both the ac and dc potential is varied while maintaining a constant ratio between them. The higher the frequency used the better the abundance sensitivity (see below). Commercial quadrupole instruments for trace inorganic work may use frequencies which range from 1.2 to 3.0 MHz.

Edge effects (or "fringe fields") can prevent low energy ions from entering the quadrupole. This can be eliminated by the use of a short "prequad." This is a short quadrupole array carrying only the ac signal. This allows the dc potential (dc pole bias) to be slightly lowered below the average ion energy of the entering ions. The ions thus travel at a lower velocity and spend more time traversing the rods and getting sorted. These design features are responsible for extending the useful range up to 300 daltons.

2.3 Detection of Ions [3,5,9,11]

Currently, detectors for inorganic mass spectrometry fall into three general categories: photographic emulsions, the Faraday cup, and electron

---

Imagine an $x - y$ coordinate system in a cross-sectional slice through the gap with the origin at the exact center of the gap.

multipliers.' Photoplate work with mass spectrometry shows little advantage over electronic detection except for the convenient archival information storage that it provides. A line in order to be measured on a microdensitometer requires at least 10^5 singly charged ions to impact the ion-sensitive emulsion. However, only about 3000 ions are required to form a visually detectable line. Sometimes, when a limited size sample is all that is available, it is possible to calculate a useful estimate of the amount of material that needs to be ionized to obtain a measurable analyte signal:

$$v = (M10^{20}) / (A\rho\Delta cI) \quad (16)$$

where v = the volume of sample consumed during exposure (in μm^3); M = the average atomic weight of the matrix (in g); A = Avogadro's number (6.0221×10^{23}); ρ = the density of the matrix (in g/cm^3); Δ = the number of ions that must be formed for one ion to reach the emulsion (generally: 10^7–10^8) ; c = the estimated concentration of the analyte in parts-per-million (atomic); and I = the % abundance of the measured isotope line.

The blackening of a line is inversely related to the size of the analyte species. Thus, compact, small-radius ions produce darker lines than large-radius ions of the same charge, mass, and abundance. Also, monatomic ions produce darker lines than the same number of polyatomic ions of the same charge and mass.

Ion sensitive emulsions are nonlinear in their response and have a short dynamic range. They are also notoriously nonuniform. Plate fogging is a problem. And there is little agreement about the best mathematical approach to quantification. This accounts for at least some of the variability in results from Mattauch-Herzog spark source instruments.

For quantitative work with photoplates it is necessary to take a graded series of exposures (as many as 15) of a known or a pure material. After the developed plate is read on a microdensitometer a plot is prepared of the logarithm of exposure time versus the obtained transmission. One of several different linearization methods is employed. Corrections are applied for line width, the emulsion calibration, and other factors. The corrected ion intensities are then ratioed to known concentrations of analyte to obtain a *relative sensitivity factor.*

Sometimes an excessive amount of confidence is placed in these numbers since it has been repeatedly demonstrated that relative sensitivity factors are strongly influenced by the sample matrix.

Solid state electro-optical arrays consisting of two microchannel plates, a phosphor screen, fiber optics, and a photodiode or CCD array have been tried on the focal plane of a Mattauch-Herzog instrument. No great improvement in detection limits over photoplate work has been realized. However, there may be further developments in this field.

The *Faraday cup* consists of a flat, obliquely inclined metal electrode in a cup-like confinement known as a Faraday cage. The separated positive ions enter the cage and strike the electrode. They are neutralized by electrons drawn from ground after passage through a resistor. The electrode is inclined to prevent reflected ions or released secondary electrons from leaving the cage. The voltage drop across the resistor is then amplified and measured as the ion current, which is, in turn, a measure of the analyte ion's abundance and charge. It is *not* related to the kinetic energy, the mass, or the electronic configuration of the analyte ions.

The Faraday cup is not a sensitive detector, and its principal use is for the measurement of relatively high analyte concentrations. The required means of amplification makes its response comparatively sluggish, and so it is unsuitable for the measurement of transient signals, such as those from a flow injection or laser ablation accessory. When used to measure analytes at the "high trace" to "minor" concentration levels the Faraday cup detector is a robust approach, exhibiting low noise and consistent sensitivity. In general purpose instruments the Faraday cup is generally used in combination with some type of electron multiplier detector, which then handles the measurement of trace and ultratrace level analytes.

There are currently three types of *electron multipliers* applied in commercial inorganic mass spectrometers. The *channeltron* is essentially a continuous dynode of semiconductor-coated glass in the shape of a cornucopia. The large open end of the cone is biased at a high negative potential, while the small end is grounded. Separated ions enter the mouth of the cornucopia and strike the curved wall, releasing secondary electrons. These travel in a line along the gradient of decreasing negative charge and strike the wall further on. The cascade continues, producing a pulse of current at the far end, which is processed by a fast preamplifier and counted. This type of detector is also sometimes called a channel electron multiplier (CEM) or a secondary electron multiplier (SEM). This type of ion measurement (like most) must be done in the dark, since even a single errant photon from the source will produce a cascade of electrons.

The channeltron is a highly sensitive and responsive, but somewhat fragile detector. Saturation occurs at an output current of about 0.01 microamperes. Moreover, the gain varies with usage, dropping to a consistent 10^5 only after several months of use. There is some degree of mass discrimination, as well. That is, low mass-to-charge ratio ions generate more output response than an equal number of high mass-to-charge ratio ions.

The *discrete dynode detector (DDD)* is essentially similar to a photomultiplier tube without the glass envelope, for which there is no need since it is housed in the high vacuum of the spectrometer. Separated ions impact the first (conversion) dynode, which emits electrons directed toward the second dynode where they generate more electrons. The cascade continues, each step at a successively higher potential until a measurable pulse is collected at the anode. Typically, there are between 10 and 20 dynodes in the chain. The DDD is generally 1.5 to 2 times more sensitive than a channeltron, but the main advantage is its service life, which is triple that of the other detector.

The latest development in this area is the *dual state discrete dynode detector*. This device measures the ion current as an analog signal at a mid-point dynode. If a limiting current is exceeded the measurement is handled as an analog signal, otherwise the signal cascades through the rest of the dynodes and is measured as a pulse count digital signal. The clear advantage is that a broad range of concentrations can be measured in a single scan, saving valuable time for transient event measurement, such as that needed for graphite furnace, laser ablation, or flow injection accessories.

2.4 Definition of Terms [3,5,12–16]

The preceding sub-sections were intended as a broad outline of some of the principles behind some of the major features of commercial inorganic mass spectrometers. A great deal has been omitted, some of which will be incorporated in Section 3.0 (Instrumentation). However, there remain a number of subjects that are properly included here. For our purposes these can be appropriately handled by a glossary format. Some of these terms have already been utilized in the previous text, others will be used in Section 3.0. Most will only occur here. Readers who proceed to other, more advanced texts on this topic are likely to encounter most of these terms.

Abundance Sensitivity: A parameter that describes the extent of tailing on both sides of a mass-to-charge peak. While poor abundance sensitivity can adversely affect the ability to accurately quantify adjacent peaks it is a distinct concept from *resolution* (see below). A higher frequency quadrupole rf signal tends to improve this parameter. Good abundance sensitivity means that if A (an ultratrace analyte peak) can be resolved from B (a matrix peak) at 10 ppm B, it is also likely to be resolved from B at 1000 ppm B.

amu: This is an abbreviation for "atomic mass unit"—a term that is *not* approved by IUPAC.* Before 1961 the physics community defined an amu as 1/16 the mass of ^{16}O and chemists defined it as 1/16 the average mass of natural abundance oxygen. Sometimes today the symbol u is employed. It corresponds to 1/12 the mass of ^{12}C, which has been assigned the value of 12.000000 by IUPAC consensus.

Base Peak: The base peak is the largest peak in the mass spectrum; it is sometimes used to normalize the intensities of all the other peaks. In or-

* The International Union of Pure and Applied Chemistry.

ganic mass spectrometry it is important to distinguish the base peak from the molecular ion peak, since they are often not the same peak.

Cool Plasma: In ICP-MS this is a reduced temperature inductively coupled discharge designed to prevent capacitive coupling between the plasma and the load coil and the Mach disk (see below) that forms between the sampler and skimmer cones. A cool plasma retards the formation of ArO^+ and other polyatomic ions, which interfere with the analysis of trace levels of alkali metals, iron, calcium, and magnesium. Cool plasmas are achieved by a number of different strategies. They typically operate between 600 and 700 watts. They are meta-stable plasmas, being easily extinguished by a single change of matrix. Cool plasmas are used extensively to analyze semiconductor materials.

Dalton (abbrev.: Da): A dalton is the mass of one nucleon (proton or neutron). Since much of inorganic mass spectrometry deals with singly charged ions, short-hand usage sometimes refers to a peak's mass in daltons rather than its mass-to-charge ratio.

Desorption Ionization: This is a generic term that refers to a host of methodologies by which sample atoms are simultaneously evolved from a surface and ionized. Secondary ion mass spectrometry (SIMS) utilizes a process that belongs to this category. SIMS utilizes a beam of ions to achieve desorption ionization. Other techniques, more frequently used in organic applications, utilize neutral atoms (fast atom bombardment or FAB), photons (laser desorption), electric fields (field desorption), and other processes. In some cases the goal is to ionize large or fragile organic molecules without severely fragmenting them.

Dynamic Lens Optimization: Automatic computer control of the ion focusing process prior to the magnetic sector, which is incorporated in some commercial instruments. This includes rapid defocusing to protect the detector from high signals.

Hull Function: This is a commonly used mathematical treatment to linearize the calibration function for ion spectra recorded on photographic plates. Specifically, it is:

$$E = (1 - T) / (T - T_s)^{1/R} \tag{17}$$

where E = the exposure time; T = transmission; T_s = transmission at saturation; R = a linearization parameter.

Isobaric Interference: Peak overlap interference due to the presence of a monatomic ion that has the same mass-to-charge ratio as the analyte peak being measured. An example is the interference of ^{58}Fe on ^{58}Ni. As long as the analyte is not monoisotopic an alternative analyte isotope peak can be selected. In the case of a monoisotopic analyte a line overlap correction can often be made using known amounts of the interferent (which must also be quantified in the test sample) in a manner analogous to that used in atomic emission (see Chapter Six).

Mach Disk: This is the shock front that occurs between the sampler and the skimmer cones of an ICP-MS. In this region ions are traveling at supersonic velocity. Commonly, the radiofrequency field of the ICP load coil induces a voltage on the rapidly expanding plasma that creates a discharge between the cones. Doubly charged ions are formed here which are the source of potential interferences. Design features of certain instruments eliminate or minimize this effect.

Mass-To-Charge Ratio (m/z): This is the ratio of the total number of nucleon masses to the total number of electrostatic charge units for a given ion. Older texts inappropriately used the symbol (*m/e*) for this quantity, but e should be reserved for the charge of an electron in coulombs (1.6×10^{-19}). It should be noted that, except in very high resolution work, the quantity, *m/z*, is referred to as a whole (integer) number.

Molecular Ion Peak: In organic mass spectrometry the molecular ion peak is the peak produced by the unfragmented, entire analyte molecule. To calculate the mass-to-charge ratio the molecular weight of the analyte is determined by the sum of the lightest isotopes, not the average of the natural abundances. It must be noted that the molecular ion peak is not always the most abundant peak in the spectrum and it should never be referred to as a parent ion peak or a base peak.

Oscillator, Crystal Controlled: This is a radiofrequency generator impedance matching circuit that automatically controls the rf frequency to match the impedance of an ICP plasma load coil under varying conditions (ignition, aqueous or organic sample introduction, etc.) The two Federal Communication Commission (FCC)-approved frequencies are 27.12 MHz and 40.68 MHz. At high signal levels the 27.12 MHz plasma produces less interferences than the 40.68 MHz plasma.

Oscillator, Free Running: This is a radiofrequency source design that allows the frequency to vary (within FCC-approved limits) by balancing it against a network of capacitors and inductors. Free running oscillators produce an ICP plasma that is difficult to extinguish. They are more typical at 40.68 MHz since the frequency limits at 27.12 MHz (near channel 2) are strict.

Percentage Total Ionization: In organic mass spectrometry this is the abundance of a given (*m/z*) peak in proportion to the sum of all (*m/z*) abundances in a stated mass range. The low end of the range is specified (as, for example, $\%\Sigma_{159}$), the high end of the range is implicitly the (*m/z*) of the molecular ion.

Polyatomic Interference: This is peak overlap interference due to a molecular ion with the same (*m/z*) ratio as the analyte peak being measured.

These polyatomic ions are formed in the plasma, on the sampler cone outer surface, or between the sampler and skimmer cones from alloy matrix components, solution reagent species, and support gases. Some design parameters will minimize but usually not eliminate such effects. An alternate analyte line or a blank correction are the appropriate remedies.

Relative Intensity: The intensity of a mass spectrum peak relative to its base peak.

Resolution: In the past there has been some confusion about the application of this term. Modern texts define resolution in mass spectrometry as the ratio, $m/\Delta m$, where Δm is the difference in mass number between two equal height peaks, m and $(m + \Delta m)$. The "valley" produced by peak overlap must be specified. A 10% valley (i.e., peaks that are 90% resolved) is common, but 15%, and even 50% valleys are sometimes used. A more rigorous approach also uses the $(m/\Delta m)$ ratio but defines Δm as the width of the peak at m at half of its height. This is referred to as the full width at half maximum (FWHM) approach. Resolution is regarded as acceptable as long as Δm is 1 or less.

Resolution generally deteriorates with increasing peak masses (with quadrupole instruments it is feasible to find parameters that result in uniform resolution across the usable mass peak range). Thus, it is useful to describe an operating range for an instrument by referring to its upper limit of *unit resolution*. If Δm is less than 1 up to mass 301 the instrument is said to show a unit resolution of 300. High resolution instruments can meet this criterion up to tens of thousands of mass units (far above their actual useful mass range) and show Δm values in the inorganic, low mass region as low as 0.010, allowing the separation of nominally identical masses.* (Cf.:abundance sensitivity, above).

Space Charge Effects: Unlike the charge on a conductor, which is treated mathematically as an infinitely thin two-dimensional surface layer, ions in vacuum occupy volume where their electrostatic charge is not balanced by opposite charges. Such a region is said to have a space charge. In ICP-MS bias in sensitivity toward higher mass analytes is produced by a tendency for ions to move away from the center of the extracted ion beam in *inverse* proportion to their mass. This inverse, mass-related, coulombic repulsion is described collectively as "space charge effects." Interference on the measurement of light mass analytes by the presence of heavy mass matrix elements is a result of this phenomenon. It has been shown that instruments

---

*Fourier transform ion cyclotron resonance mass spectrometers have demonstrated the ability to separate $^{35}Cl^{+}$ and $^{35}Cl^{-}$—the mass and charge of two electrons, and a resolution in excess of 1,000,000! (Ref.3).

that employ a high accelerating potential minimize these effects. Carefully prepared matrix-matched blanks and synthetic standards, or the use of the method of additions are two answers to the problem.

3. INSTRUMENTATION

3.1 Ion Sources [5,6,9,10,17]

Electron ionization (or electron bombardment) is the most commonly used source for organic mass spectrometry. The sample is introduced as a gas or vapor into a small box under high vacuum. A beam of 70 eV electrons passes across the sample stream, ionizing and fragmenting the organic molecules. The source of the electron beam is a heated tungsten or rhenium filament that produces electrons by thermionic emission. After crossing the sample stream the electrons are collected on an electron trap anode. Since most systems are configured for positive ion spectra the electron trap collects negative sample ion fragments as well. Neutral fragments of the sample are cleared by the vacuum system. The remaining positive ions are accelerated through a slit or aperture by the joint action of a positive ion repeller electrode in the source box and a negative accelerator electrode (or draw-out plate) outside the source. The positive ion beam then passes through focusing, collimating, and beam centering electrostatic lenses, prior to entering the mass analyzer region.

Electron ionization is a very energetic process that often leaves only a small amount of the intact molecular ion. This greatly complicates the elucidation of the structure and molecular weight of an unknown organic substance. "Softer" sources were developed to address this problem. *Chemical ionization* uses ionized methane or a similar reagent gas to ionize the organic sample by the comparatively "soft" process of *proton transfer*. *Desorption ionization* techniques, such as *fast atom bombardment* (FAB), *field desorption*, and *electrospray* achieve a similar result by a variety of techniques.

In inorganic mass spectrometry one of the most important ion sources is still *thermal ionization*. Today, it is used principally for the most accurate single-analyte isotope ratio and isotope dilution studies with high resolution double-focusing instruments. Typically, a small volume of the sample solution is evaporated onto a tantalum, tungsten, or rhenium filament. In a single-filament design the filament is resistance heated in vacuum and the analyte is vaporized and ionized in one combined process.

It is also used to determine traces of inorganic gases, especially the inert gases.

In a few procedures the analyte is electrodeposited onto the filament using microscale techniques.

In a double-filament design the analyte atoms are vaporized from one (cooler) filament and adsorbed onto another (hotter) filament where they are ionized. In a triple-filament design two different sample solutions are deposited on each of two evaporation filaments so that they can be compared under nearly identical ionization conditions on a third filament. Separating the vaporization and ionization processes in this way reduces ion fractionation in the vaporization process. It also facilitates the determination of low boiling analytes that are otherwise difficult to analyze.

Thermal ionization shows the most abundant ion yields for analyte elements with a first ionization potential of less than about 7 eV (the alkali metals, alkaline earths, lanthanides, and actinides). Analytes of higher ionization potential generally require specialized approaches, such as adsorption on an ion exchange resin bead. The bead is then fixed to a filament. Carbon from the resin plays some role in retaining and analyzing the analyte.

The *inductively coupled plasma* has become the most important ion source for inorganic mass spectrometry. Commercial torch designs are in most respects similar to those for atomic emission, except that the torch and load coil are mounted horizontally. Typically, the sampler cone orifice is positioned 1-2 mm before the edge of the initial radiation zone (IRZ) and fully immersed in the normal observation zone (NOZ). As we saw in Chapter six, molecules and neutral atoms prevail in the IRZ, while ions are produced in the NOZ. The exact relative position of the IRZ and the NOZ is influenced by argon flow rate, sample solution flow rate, and power to the load coil. Most ions are singly charged and the small number of doubly charged ions can be further reduced by varying operating parameters. The ICP can efficiently ionize all analyte atoms whose ionization potential is below 10.5 eV.

At a typical nebulization rate of 1 mL/min a 1 part-per-billion solution would result in 10^7 ions/cm^3 in the plasma for a fully ionized analyte. If it were also monoisotopic this would lead to a peak count rate of 100–1000 counts/sec. The great abundance of argon, water, and acids in the plasma when dissolved metal samples are nebulized leads to the formation of many condensation species in the extracted ion beam, which are the source of interference problems. Many of these interfering species are formed at the sampler cone. The instrument should be designed to prevent a voltage bias at the grounded sampler cone. This is usually accomplished by center-tapping the load coil of the torch to a ground, although there are other design solutions to the problem.

This does *not* include carbon (11.26eV), bromine (11.85eV), chlorine (13.02eV), oxygen (13.62eV), and nitrogen (14.55eV); however, some of these can be detected.

The ICP source is unique in accommodating numerous accessories to increase its versatility. A graphite furnace, direct injection probe, flow injection system, or laser ablation attachment will greatly enhance its lower limit sensitivity and its detection limits. The laser ablation device allows it to be used as a probe of microsized areas. A spark ablation device will yield good precision directly from solid samples. Manufacturers vary in their strengths and degree of sophistication in regard to sample introduction accessories. Some of these devices are also supplied by specialized firms for interfacing to many commercial ICP-MS instruments.

The *glow discharge source* requires a lengthy "pre-burn" to prepare the sample surface for analysis. Both pin and disk samples are commonly accommodated. GD-MS shows a unique advantage in its ability to determine gases in metals. However, it is ordinarily constrained by the availability of closely matching standards, and the powerful approach, isotope dilution, is ordinarily closed to it. The hollow cathode source may be regarded as a glow discharge source that operates in the *normal mode* (voltage does not increase with increasing current). The conventional glow discharge source requires an increase in voltage with an increase in applied current since the glow completely covers the cathode.

The *vacuum spark* utilized in spark source mass spectrometry is often a discharge between two machined pieces of the same metal sample. Sometimes, however, a point-to-plane approach is utilized with a refractory metal counter electrode. While numerous installations remain, the spark source is rapidly losing ground to the ICP and the GD sources. The spark source has a well-deserved reputation for poor precision. The related dc arc source was never much used.

3.2 Ion Introduction Systems

Both ICP and GD ion sources utilize a differentially pumped interface with sampler and skimmer cones, although the pumping requirements are much more stringent for the ICP. The cones are generally nickel or platinum with a shape and aperture size unique to the interface design. A vacuum gate is provided so that the cones can be conveniently removed for cleaning without losing vacuum in the mass analyzer. In the case of the ICP the interface region is the location of lower temperature condensation reactions where many potentially interfering polyatomic ions form.

The plasma enters the sampler cone aperture at about 6000 K, but it cools rapidly as it expands past the orifice. The sampler cone temperature has been measured at about 500°C. The so-called first stage of the interface between the sampler and skimmer cones is maintained between 1 and 3 torr (133–400 Pa) by a mechanical pump of about 300L/min capacity. The second stage, which contains the ion lenses, the mass analyzer, and the detector, is maintained at 1×10^{-5} to 5×10^{-9} torr (1.3×10^{-3} to 6.7×10^{-7} Pa) by turbo-molecular pumping. The aerodynamic shapes of the sampler and skimmer cones are designed to reduce reaction time in the interface region. The discharge that forms between the sampler and skimmer cones is an area of particular concern in regard to interferent formation and many instruments are designed to reduce or eliminate it.

In all spectrometers an accelerating voltage is applied to propel the ions through a slit or aperture. There is always some means of preventing photons from reaching the detector (Bessel box, omega lens, or curved electrostatic sector) and some magnetic lensing to maintain or restore focus to the ion beam. Ion lenses consist of meshes or cylinders maintained at appropriate bias voltages to produce a circular cross-section ion beam at the entrance of the mass analyzer region (quadrupole). In the case of a magnetic sector instrument the focus of a slit image is involved. In a quadrupole system low velocity ions are preferred since they experience a longer dwell-time between the rods.

3.3 Mass Analyzers [3,5,9]

Organic mass spectroscopists currently have more instrument types to use in their work than do inorganic mass spectroscopists. This can and probably will change as the methodologies are extended. In addition to the magnetic sector and quadrupole instruments, the ion trap, the time-of-flight, and the Fourier transform ion cyclotron resonance spectrometers are all viable technologies. We will briefly consider each in turn.

3.3.1 Magnetic Sector

These are the large, traditional instruments that have supported the field of mass spectrometry since its inception. Single-focusing designs can be made to scan their mass range by continuously varying either the magnetic field strength or the accelerating potential. Scan rate, however, is limited by the *reluctance* of the electromagnet (its resistance to a change in field strength), and by hysteresis effects, which necessitate collecting data only when scanning in one direction and discharging and "resting" the magnet between scans.

Double-focusing instruments utilize an electrostatic sector to select ions of only a very narrow range of kinetic energies. The two most prominent designs are the Mattauch-Herzog and the Nier-Johnson (sometimes, also the *reverse* Nier-Johnson, in which the relative positions of the magnetic and electrostatic sectors are switched). The Mattauch-Herzog is principally a photographic instrument since the separated ion beams are focused on a plane.

Eventually, solid state array detection will likely appear in commercial instruments. The Nier-Johnson design is strictly for electrical detection since the final focus is a point at the detector. The reverse Nier-Johnson design has been used in a major commercial high resolution instrument (the VG 9000™ GD-MS).

3.3.2 Quadrupole

The quadrupole mass filter utilizes dc and radiofrequency fields between four (ideally) hyperbolic cross-section rods to achieve mass selection. Opposite rods are electrically connected. Ions are accelerated from the source to the quadrupole by a 5-15 V potential. The rf amplitude and the dc voltage are ramped together in a fixed ratio to scan the masses. The rf frequency is held constant; however, it is a sensitive function of the abundance sensitivity of the instrument.

The quadrupole has the advantage of being able to scan its mass range rapidly. It has less stringent vacuum requirements than many other designs. Ion optics focus is also less of a consideration. There is some tendency to discriminate masses (high mass peaks show less intensity than low mass peaks of the same abundance). However, the space charge effect described in Section 2.4 produces the opposite trend. Design parameters have been developed to minimize both effects. Finally, the quadrupole is one of the more cost effective designs and its use in ICP-MS instruments has contributed greatly to the growth of that methodology.

3.3.3 Ion Trap

Technically, the ion trap is a three-dimensional quadrupole in the sense that an oscillating radiofrequency field is utilized to produce a stable circular ion orbit in a manner analogous to the stable linear path that a given (m/z) ion traverses between the rods of a quadrupole mass filter. The ion trap uses a ring electrode with an annulus of hyperbolic cross-section. Two hyperbolic end-cap electrodes—one above and one below the ring—are electrically connected and together with the ring form the enclosed space in which the ions orbit. A radiofrequency signal of sinusoidal amplitude and fixed frequency is applied to the ring while a dc or ac potential is applied to the end-caps.

The top end-cap has an aperture for the introduction of pulses of accelerated ions, and the bottom end-cap has an aperture for the exit of selected ions with a detector immediately below it. By varying the rf and end-cap electrical parameters sample ions of a given mass-to-charge ratio can be collected in a stable orbit until a sufficient number are accumulated, then the parameters are changed to direct them into the detector. In such a mode of operation the ion trap exhibits extremely low detection limits. However, the space charge (mutual repulsion of like charges) places a limit on the number of ions that can be accumulated. The space charge effect can lead to poor resolution, mass scale calibration errors, and a loss of calibration linearity. Closed-loop automatic control of the ion source is necessary to ensure that too many ions do not accumulate in the trap.

The ion trap has been used mostly in organic work, however, it appears to have great potential in trace and ultratrace inorganic analysis. Limiting the space charge effect may be the key to its expanded application.

3.3.4 Time-of-Flight

Time-of-flight mass spectrometers were first developed in the early 1950s and went through a period of rapid growth, followed by one of declined interest. They are currently experiencing a limited resurgence in organic analysis and in specialized fields, like ion microprobe work. One manufacturer has introduced a glow discharge TOF instrument for inorganic work.

In these devices the sample ions receive a large potential accelerating pulse and then are allowed to drift through a field-free region of defined length (L) to a detector. We have seen that the ion's velocity must be inversely related to (m/z):

$$v = [2V(z/m)]^{1/2} \tag{3}$$

Therefore, their time-of-flight must be equal to:

$$(L/v) = L[m/(z2V)]^{1/2} \tag{18}$$

and thus, is directly proportional to (m/z). Such instruments are calibrated with ions of known mass. Resolution is a key problem with this design since the imparted kinetic energies of ions of the same (m/z) cannot be made perfectly uniform.

3.3.5 Ion Cyclotron Resonance

ICR-MS is a unique methodology that has yet to find widespread use even in the organic realm. It requires conditions of extreme vacuum [10^{-9} torr (1.3×10^{-7} Pa)] and is typically operated in a Fourier transform mode. The sample ions are made to precess in a magnetic field. Energy is applied as a "chirp" of frequencies. Ions of an (m/z) that absorb one of these frequencies will move to an orbit of larger radius, impacting, or at least affecting the confining cell walls and imparting a detectable signal related to that (m/z) abundance. Fourier transform analysis of the resulting time-domain signal converts it to the frequency realm, resulting in a recognizable (m/z) spectrum. The technique shows extremely high resolution, but appears to be some time away from commercialization.

3.4 Detectors [3,5,12]

3.4.1 Photographic Plates

Specialized photographic emulsions have been developed for the detection of ions: Ilford Q2, Kodak SWR, Ionomet IM, and others. These have been used in conjunction with Mattauch-Herzog double-focusing instruments. An endemic problem has been plate fogging by photons either leaking through from the source or (more commonly) created by ionic collisions within the mass analyzer itself. Like photoplate spectrography the application and interpretation of photographic plate mass spectra is rapidly becoming a lost art.

3.4.2 Faraday Cup

This rugged and reliable technology is still employed in some commercial instruments. The Faraday cup still provides a ready means of quantifying high abundance ions with no concerns about overload damage or maintenance. The slow response, however, requires operation in a peak hopping mode since rapidly scanned spectra will be recorded inaccurately.

3.4.3 Channeltron

This cornucopia-shaped detector is gradually being replaced by discrete dynode technology. The channeltron becomes saturated when the output current exceeds 10^{-8} amperes. It is also affected by magnetic fields and must be placed at some considerable distance from sector magnets.

3.4.4 Discrete Dynode Detector

Generally considered more durable and longer-lived than the channeltron, the discrete dynode detector still must be protected from a massive flux of ions. This detector can be up to twice as sensitive as the channeltron and requires replacement less frequently. The dual stage version, which treats ion signals as either analog or digital data, is the current state of the art, eliminating the need for two detectors.

3.5 Data Systems

Modern instruments are sold with a formidable software package. Particularly valuable are tables of isobaric and polyatomic interferences on every conceivable mass peak. They are large, but not exhaustive. Advanced systems, such as dynamic lens optimization, remove much of the art in obtaining a high quality usable response. The learning curve for much of this can be quite formidable, however, and the uninitiated should be aware that the technology is still very far from automatic.

4. CALIBRATION AND MEASUREMENT [5,9,10,17]

4.1 Internal Standards

As in atomic emission, ratioing to an internal standard has proven to be a valuable means of dealing with instrumental variability in mass spectrometry, including both long- and short-term drift. In the case of ICP-MS, mass transfer variation due to viscosity-related changes in the nebulization rate would produce an even greater effect on results than that shown in atomic emission. Fortunately, use of an internal standard efficiently corrects for it in both methodologies.

Generally, 3 to 5 internal standards are added if wide spectral coverage is needed. These elements are selected because they are represented by a "clean" line in distinct regions of the spectrum. The entire group of internal standards collectively provide coverage for the entire (*m/z*) range of the instrument.

The internal standard is an element that is ordinarily not a component or a contaminant of the sample alloy. The atomic emission practice of using the base metal or a major component of the alloy as the internal standard is seldom used in ICP-MS. As we saw in the previous chapter, the internal standard also must not react chemically with the components of the sample (or, in this case, with the other added internal standards), and it must behave in a manner similar to the behavior of the analyte or analytes to which it is applied. In general that means that the internal standard must be similar to the analyte in mass and ionization potential. Due to the extreme sensitivity of ICP-MS it is critical that only very high purity internal standard solutions are utilized.

In the case of ICP-MS the nebulization of the sample solution is an important variable that is easily controlled by this means. Variations in background and mass-dependent sensitivity shifts can be compensated for in this way. One frequently used internal standard element is indium (others sometimes used include rhenium, rhodium, terbium, and tantalum). Monoisotopic internal standards are popular because they introduce no additional, potentially interfering, lines.

4.2 General Calibration

The level of sophistication in approaches to calibration varies widely among the inorganic mass spectrometric methodologies. Solids-based work, such as that from spark source instruments, has tended to rely heavily upon relative sensitivity factors (RSF's) for the various analytes. These are often established by the use of reference materials far removed in composition from the sample matrix. Many studies have clearly shown that results improve when more appropriate reference materials are utilized.

Some authors have advocated the use of the method of additions in ICP-MS work. When accompanied by an appropriate blank, the multipoint method of additions is a nearly ideal way to compensate for matrix suppression effects, which can be problematic, especially for light analytes. A further elaboration, the Generalized Standard Addition Method (GSAM), has been proposed as a means of automatically correcting for isobaric and chemical interferences by the use of "training data sets" [18].

The mode of data collection also affects the quality of the results. It has been found that integrating the area under the entire mass peak is not the optimal approach. The best results are generated by collecting data from only the central portion of the peak [19].

4.3 Isotopic Ratio

The most accurate isotopic work is performed with a thermal ionization source and a double-focusing magnetic sector instrument. Nearly equivalent work can be achieved with an ICP coupled to such a "high res" device. However, a reasonably reliable result can be achieved with an ICP-quadrupole instrument.

If interferences are absent or controlled it is a relatively simple matter to obtain the areas of two isotopic peaks of the same element and calculate their relative abundance. As with the use of internal standards, isotopic ratio work usually compensates automatically for instrumental drift, variations in the uptake rate of solution into the ICP plasma, and (uniquely) even sloppy solution preparation work. Sometimes only the isotopic ratio is needed, but, more commonly, the result is combined with a total analyte determination by another technique so that the atomic and/or weight percent of each isotope can be calculated. To the uninitiated the process of calculating such a result is nonintuitive. Figure 7.2 illustrates the calculation for a metal alloy analyzed for boron enriched in the ^{10}B isotope.

4.4 Isotope Dilution

Isotope dilution mass spectrometry (IDMS) is a definitive technique, that is, a procedure that can achieve an accurate result without the use of reference materials. A stable enriched isotope of the analyte element is added to the sample and the mixture is equilibrated. Then the ratio of the abundance of the added isotope to another analyte isotope is measured by mass spectrometry. This ratio (R) is given by:

$$R = [(A_t C_t W_t) + (A_s C_s W_s)]/[(B_t C_t W_t) + (B_s C_s W_s)]$$

where, A_t = atom fraction of isotope A in the test material

B_t = atom fraction of isotope B in the test material
A_s = atom fraction of isotope A in the spike
B_s = atom fraction of isotope B in the spike
C_t = concentration of total analyte in the test material
C_s = concentration of total analyte in the spike
W_t = weight of the test material test portion
W_s = weight of the spike

Therefore:

$$C_t = [(C_s W_s)/W_t][(A_s - RB_s)/(RB_t - A_t)]$$

Enriched isotopes are commercially available for about 60 elements. There are 272 stable isotopes plus a sizeable number of long-lived radioactive

---

An alloy has been produced with a boron content that has been enriched in the ^{10}B isotope. Atomic emission work, based on calibration with *natural abundance* boron reference materials yielded a result of 1.350 % (w/w) total boron (call this variable: $\%B_r$). Mass spectrometry yielded the following results:

	^{10}B	^{11}B
Wt%	20.56	79.44
Atomic %:	22.15	77.85

The following are literature values:

Atomic Weight of Boron (natural abundance): 10.811g/g-atom [Call this: $(AW)_b$].
Atomic Weight of ^{10}B: 10.01294 g/g-atom [Call this: $(AW)_{b10}$].
Atomic Weight of ^{11}B: 11.00931 g/g-atom [Call this: $(AW)_{b11}$].

Calculate the atomic weight of boron in the alloy, $(AW)_a$:

$$\begin{aligned}(AW)_a &= [(AW)_{b10}(\text{Atomic\% } ^{10}B)/100] + [(AW)_{b11}(\text{Atomic\% } ^{11}B)/100] \\ &= [(10.01294)(22.15)/(100)] + [(11.00931)(77.85)/(100)] \\ &= 10.7886 \text{ g/g-atom.}\end{aligned}$$

Calculate the true value of % (w/w) total boron ($\%B_t$):

$$\%B_t = [(\%B_r)(AW_a]/(AW)_b = (1.350)(10.7886)/(10.811) = 1.347\%$$

Calculate the true value of % (w/w) ^{10}B ($\%^{10}B_t$):

$$\%^{10}B_t = (\%B_t)(\text{Wt\% } ^{10}B)/100 = (1.347)(20.56)/(100) = 0.277\%$$

Calculate the true value of % (w/w) ^{11}B ($\%^{11}B_t$):

$$\%^{11}B_t = (\%B_t)(\text{Wt\% } ^{11}B)/100 = (1.347)(79.44)/100 = 1.070\%$$

Figure 7.2 Illustration of isotopic concentration calculations.

isotopes. For monoisotopic analytes it is possible to employ an artificial (man-made) isotope as a spike. Isotopes for IDMS are expensive, but since only a minute amount is used for each analysis, the purchase of a few milligrams will likely amount to a "lifetime supply." Isotopes are available from: Cambridge Isotope Laboratories (Woburn, MA), Mound Laboratory (Miamisburg, OH), Oak Ridge National Laboratory (Oak Ridge, TN), A. Hempel GmbH (Dusseldorf, Germany), and Oris Stable Isotopes (Gif-sur-Yvette, France).

The equilibration step is the key to this technology. In some cases the analyte element's oxidation state must be changed to ensure its completion. But once the solution is equilibrated any further needed chemical separations or other manipulative steps need not be quantitative since only the *ratio* of the two isotopes is significant.

The technique is also highly dependent upon well-characterized, stable natural abundances. Elements vary in the extent to which their isotopic abundances are known. Elements with variable natural abundances include: hydrogen, boron, carbon, nitrogen, oxygen, sulfur, argon, calcium, strontium, and lead. The accuracy of the isotope dilution approach for these analytes is, thus, inherently reduced.

Sometimes there is a slight separation of isotopes in chemical separations. This comparatively rare phenomenon can effect results. It is particularly a concern when dealing with light elements with large mass differences between the two isotopes. The use of a nonrepresentative blank is another potential error source. Also, fractionation of the isotopes in the ion source has been identified as a potential source of error in the most accurate work.

Thermal ionization, which can yield the most accurate isotopic ratios, has been the traditional ion source for most IDMS work. But, interestingly, the isotope ratio measurement is rarely the limiting factor in the IDMS approach. Spark source, electron impact, and field desorption sources have all been tried, but the ICP holds the most promise in this arena. By spiking with several isotopes multi-element IDMS is possible with the ICP. For applications that require the most accurate results, for the certification of reference materials, and for settling disputed results between methods, labs, or analysts, isotope dilution ICP-MS will likely become an increasingly valuable tool.

WARNING: While isotope dilution can be accomplished with the addition of long-lived radioactive isotopes these require special knowledge and precautions concerning safe handling, personal protection and instrument ventilation requirements. In the absence of specialized training the analyst should never use them.

5. INTERFERENCES [21–23]

5.1 Spectroscopic Interferences

There are two general categories of ion line overlap interferences in mass spectrometry: *isobaric (or isotopic) interferences* and *polyatomic (or molecular) interferences*. The isobaric interferences are caused by monatomic isotope ions that overlap the analyte ion peak. They are a limited and widely published set of effects that are incorporated into the software of most modern instruments. They originate from the sample matrix or the support gas of a plasma source. These interferences can usually be handled easily by measuring the analyte on a different isotope peak. Some examples of isobaric interferences are: $^{70}Ge^+$ on $^{70}Zn^+$, $^{58}Ni^+$ on $^{58}Fe^+$, $^{54}Fe^+$ on $^{54}Cr^+$, $^{40}Ar^+$ on $^{40}Ca^+$, and $^{39}Ar^+$ on $^{39}K^+$.

Polyatomic interferences are quite another, more complicated problem, however. These occur in all forms of mass spectrometry, although their abundance, diversity, and species types vary with instrumental conditions. There are dimers like $^{40}Ar^{36}Ar^+$ (which overlaps $^{76}Se^+$), $^{40}Ar^{38}Ar^+$ (which overlaps $^{78}Se^+$), $^{40}Ar_2^+$ (which overlaps $^{80}Se^+$), $^{16}O_2^+$ (which overlaps $^{32}S^+$), $^{24}Mg_2^+$ (which overlaps $^{48}Ti^+$), $^{37}Cl_2^+$ (which overlaps $^{74}Se^+$), $^{14}N_2^+$ (which overlaps $^{28}Si^+$), and $^{33}S^{34}S^+$ (which overlaps $^{67}Zn^+$). But much more common are oxide, nitride, and hydroxide species (from air, acids, or water), carbon species (from organic materials), and halogen species (from acids). If argon is part of the polyatomic ion it is referred to as an "argide" ion. Table 7.1 lists some common argide ion interferences observed in ICP-MS.

Table 7.1 Selected Argide Ion Interferences in ICP-MS

Interferent	Ion Peak Affected
$^{36}Ar^{1}H^+$	$^{37}Cl^+$
$^{40}Ar^{12}C^+$	$^{52}Cr^+$
$^{40}Ar^{16}O^+$	$^{56}Fe^+$
$^{40}Ar^{23}Na^+$	$^{63}Cu^+$
$^{40}Ar^{35}Cl^+$	$^{75}As^+$
$^{40}Ar^{37}Cl^+$	$^{77}Se^+$
$^{40}Ar^{52}Cr^+$	$^{92}Mo^+$
$^{40}Ar^{56}Fe^+$	$^{96}Mo^+$
$^{40}Ar^{58}Ni^+$	$^{98}Mo^+$
$^{40}Ar^{61}Ni^+$	$^{101}Ru^+$
$^{40}Ar^{63}Cu^+$	$^{103}Rh^+$
$^{40}Ar^{65}Cu^+$	$^{105}Pd^+$

Table 7.2 Miscellaneous Selected Polyatomic Interferences in Mass Spectrometry

Interferent	Ion Peak Affected
$^{12}C^{16}O^{+}$	$^{28}Si^{+}$
$^{14}N^{16}O^{1}H^{+}$	$^{31}P^{+}$
$^{12}C^{16}O_2^{+}$	$^{44}Ca^{+}$
$^{28}Si^{16}O^{1}H^{+}$	$^{45}Sc^{+}$
$^{30}Si^{16}O^{1}H^{+}$	$^{47}Ti^{+}$
$^{32}S^{16}O^{+}$	$^{48}Ti^{+}$
$^{35}Cl^{16}O^{+}$	$^{51}V^{+}$
$^{35}Cl^{16}O^{1}H^{+}$	$^{52}Cr^{+}$
$^{39}K^{16}O^{+}$	$^{55}Mn^{+}$
$^{44}Ca^{16}O^{+}$	$^{58}Ni^{+}$
$^{24}Mg^{35}Cl^{+}$	$^{59}Co^{+}$
$^{46}Ti^{16}O^{+}$	$^{62}Ni^{+}$
$^{31}P^{16}O_2^{+}$	$^{63}Cu^{+}$
$^{48}Ti^{16}O^{+}$; $^{32}S^{16}O_2^{+}$; $^{32}S_2^{+}$	$^{64}Zn^{+}$
$^{35}Cl^{16}O^{+}$	$^{67}Zn^{+}$
$^{57}Fe^{19}F^{+}$	$^{76}Se^{+}$
$^{62}Ni^{16}O^{+}$	$^{78}Se^{+}$
$^{12}C^{35}Cl_2^{+}$	$^{82}Se^{+}$
$^{56}Fe^{35}Cl^{+}$; $^{54}Fe^{37}Cl^{+}$	$^{91}Zr^{+}$
$^{56}Fe^{37}Cl^{+}$	$^{93}Nb^{+}$

Table 7.2 lists some other commonly encountered polyatomic interferences.

Although conditions are usually adjusted to minimize their production, doubly charged ions also occur and are potential interferents.

Literature sources vary in identifying the place where polyatomic ions form. Clearly, in ICP-MS only a few refractory oxide ions are likely to survive in the plasma itself. Condensation reactions in the expansion region between the cones has been widely suggested; however, some authors believe that few reactions occur there. This leaves collisional reactions around the outside of the sampler cone to account for what results.

The best way to avoid polyatomic interferences is to switch to an alternative analyte line. Unfortunately, this is not always possible. Twenty-two elements are monoisotopic in the sense that they have only one stable, naturally occurring isotope (^{4}He, ^{9}Be, ^{19}F, ^{23}Na, ^{27}Al, ^{31}P, ^{55}Mn, ^{59}Co, ^{75}As, ^{89}Y, ^{93}Nb, ^{103}Rh, ^{115}Sc, ^{127}I, ^{133}Cs, ^{141}Pr, ^{159}Tb, ^{165}Ho, ^{169}Tm, ^{197}Au, ^{209}Bi, and ^{232}Th). And, even when other natural abundance isotopes exist an *interference-free* alternative line may not be available. And if it is the isotope may constitute such a small percentage of the total natural abundance composition of the analyte that instrumental sensitivity becomes a factor. In such a situation one drastic measure is to do some chemistry.

Sometimes merely switching to a different dissolution scheme is all that is required. For example, to measure arsenic, selenium, and vanadium by ICP-MS the sample solution must be free of chlorine. In such cases hydrochloric and perchloric acids must be avoided, or, if possible, removed without analyte loss. Sulfuric and phosphoric acids tend to form many polyatomic ions and are usually not used. Hydrofluoric acid and/or nitric acid are commonly used and create few problems.

Occasionally, the full complement of separation science must be brought to bear in order to successfully analyze a material by ICP-MS. This might entail ion exchange, solvent extraction, distillation, or precipitation in order to sufficiently isolate the analyte from interferents. The analytical scheme used should be driven by necessity and involve as few manipulative steps as possible to achieve the desired separation. "Separate as much as needed, but no more," is the admonition for all trace level work, but with mass spectrometry it carries extra weight. The extreme sensitivity of the technique leaves it prone to improperly compensated blank values due to impurity contamination from reagents, labware, etc.

Cooling the spray chamber in ICP-MS has proven to be remarkably effective in reducing oxide and hydroxide polyatomic ions. This strategy simply condenses out most of the water or other solvent before it reaches the plasma. Some instrument designs utilize refrigerant chilled recirculating water to achieve this result while others employ Peltier effect (thermoelectric) coolers.

Some workers have had success in reducing argide interferences by switching to alternatives to pure argon as the plasma support gas. Montaser and coworkers have shown that helium can be a viable alternative plasma gas provided that certain modifications are made to the instrument.' Not only are argon interferences avoided, but also the abundances of analytes with a high first ionization potential are improved. However, the helium discharge is not as robust and stable as the argon discharge [22]. Other workers have employed mixtures of argon with nitrogen, helium, xenon, or hydrogen, reporting reduced problems with both argides and chlorine-containing polyatomic species.

Of course, the use of a double-focusing magnetic sector mass analyzer will provide the resolution to "see around" some potential line overlaps, but by no means all of them. Mathematical correction remains a viable alternative, but to be effective it requires comprehensive knowledge of the

---

Increased first stage pumping and a reduced sampler orifice.

interferences that must be dealt with. If a polyatomic ion has two forms, e.g., $^{40}Ar^{35}Cl^{\cdot}$ and $^{40}Ar^{37}Cl^{\cdot}$, one overlapping the analyte peak and one by itself in the spectrum, the analyst can measure the "clean" polyatomic ion line and use the result to calculate an expected effect from the other polyatomic ion at the analyte ion peak. The calculated effect is then subtracted from the measured analyte signal.

5.2 Transport Effects

By this we mean the influence of the sample matrix on the introduction of ions into the mass analyzer. In glow discharge and spark source instruments this involves the high energy processes intended to produce an ion beam that is representative of the composition of the solid specimen. In ICP-MS this concerns the nebulization, aerosol transport, and plasma processes that achieve the same result for a liquid sample. A great deal of information about these subjects is directly transferrable from corresponding work in atomic emission, which has been much more intensively studied.

The vacuum spark achieves local thermal equilibrium after some interval, but produces a complex array of sample ion species. Glow discharge is basically a nonthermal process, but the etch rate is mass-dependent. With closely matching certified reference materials, both solid state spectrometries can transcend their shortcomings. With ICP-MS, a thoughtfully selected internal standard will overcome any difficulties resulting from changes in solution viscosity.

The clogging of the sampler cone orifice due to salt buildup is an endemic problem in ICP-MS. It has been suggested that a partially clogged orifice is much more of a steady state condition than a clean one. The suggestion then seems to be that *very* frequent maintenance cleaning of the sampler orifice is counter-productive.

In ICP-MS the aqueous sample solution matrix can cool the plasma, causing a drop in the ionization efficiency, and thus in the abundance of the analyte ion. Organic solvents can have a profound effect in this way, as well.

5.3 Ion Beam Effects

This topic refers to the suppression of analyte ion signal by the sample matrix ions. High mass, easily ionized matrix elements have the greatest effect in this arena. Low mass analyte elements that are difficult to ionize show the greatest suppression. The ratio of high mass matrix to analyte is not as important as the absolute amount of matrix. Thus, dilution is a solution to the difficulty.

There is, however, little agreement as to why this suppression effect occurs. Some authors have proposed mechanisms in the central channel of the inductively coupled plasma. But the consensus appears to be that space charge effects in the expansion region between the sampler and the skimmer cones is responsible.

Matrix-induced suppression can be corrected by the use of an appropriate internal standard (i.e., an added precise amount of an element very similar in mass and ionization potential to the analyte), by the use of an isotope dilution approach, by the use of the method of additions, by chemically separating the analyte from the matrix, or by altering ion lens parameters to minimize it.

6. PRACTICAL DETAILS

Inorganic mass spectrometry is still an evolving discipline. There are many instrumental features and accessories among which the potential user can select to customize an installation for his application. We will briefly review some of these options and then close this section with a short literature survey of some of the major topics we have so far covered.

6.1 Nebulizers [12,25]

ICP-MS offers generally the same array of nebulizer styles that are available for ICP-atomic emission. The *Meinhard* or *concentric glass nebulizer* is commonly aspirating (rather than pumped). It consists of a fine capillary inside a hollow glass tube through which argon enters. The opening of the outer envelope is drawn down until it is just larger than the capillary orifice, which is just a short distance away. The Venturi effect draws liquid and nebulizes it. This device offers good sensitivity but is subject to clogging at even moderate solids loading.

The *fixed cross-flow nebulizer* may be either pumped by a peristaltic tubing pump or aspirated. It consists of two (usually platinum) small tubes at right angles to one another. One draws argon, the other draws liquid. It is somewhat more tolerant of dissolved solids than the Meinhard, but lacks its sensitivity. The fixed-cross flow is a versatile and widely used nebulizer.

The *Babington* or *high solids nebulizer* utilizes a groove cut in a Teflon cap to create a falling film of the sample solution over the argon inlet. The Babington design resists clogging by high solids samples. While earlier versions of the Babington style required a considerable sacrifice in analytical performance the modern versions show about the same precision and sensitivity as the fixed cross-flow nebulizer. The *grid nebulizer* also performs well with high solids solutions. It consists of two fine-mesh platinum screens mounted vertically. The sample solution coats the first screen. High velocity argon is blown against the wetted screen producing an aerosol. The second screen reduces the number of large droplets, producing a more uniform mist. There are other, patented designs which effectively handle high solids solutions.

In a class by itself is the very high efficiency *ultrasonic nebulizer*. It consists of an arrangement in which the sample solution is dropped on the transducer of an ultrasonic generator. Samples containing free fluoride ion and high solids samples are not suitable for use with this device. The high density aerosol must be heated, then cooled, to condense out most of the water while the dry aerosol particles pass on to the plasma.

For most systems used for ICP-MS some mechanism for removing large liquid droplets immediately after nebulization is required. This is usually accomplished by means of a spray chamber containing baffles, a flow spoiler, or some sort of tortuous path. Oxide and hydroxide interferent species can be dramatically reduced if the spray chamber is cooled to about 3°C.

6.2 Spark Ablation

Spark ablation ICP-MS is a conductive solids sampling approach that allows rapid, low interference work to be done. This accessory has taken some diverse forms, differing mostly in the way (if any) that the aerosol is pretreated before introduction into the plasma. Generally, a high voltage, low amperage spark is utilized, producing a dry argon-metal aerosol with a particle size <1 μm. The shape and size of the particles is critical, leading some manufacturers to incorporate a cyclone sizing device between the spark and the plasma.

A valuable comparative study with glow discharge MS showed that spark ablation ICP-MS was faster and less subject to argide interferences while being only slightly less sensitive [26,27]. In contrast to some other accessory approaches to ICP-MS, spark ablation is modestly priced and simple to operate. It has been shown to produce accurate and precise data [28]. Earlier work utilized an arc as the aerosol generation source [29,30]. Developments in this area are likely to be critical to the future of mass spectrometry in the metals industry.

6.3 Laser Ablation

Laser ablation ICP-MS may be regarded as a microprobe technique. In commercial instruments the laser is typically a Nd:YAG* device, capable of operating at 266, 354, and 1064 nm. It can operate in *free-running mode* (150

---

* Neodymium-doped Yttrium Aluminum Garnet.

millisecond pulses) or *Q-switched mode* (in which large high energy pulses of very short duration are emitted). In Q-switched mode the energy density is sufficient to ionize the argon carrier stream as it passes across the impact region. This process can ablate optically transparent samples.

The argon stream carries the ablated material directly from the sealed sample chamber into the central channel of the ICP. When a sample is changed a valve switches the argon flow to vent, to prevent an extinguished plasma. A television-based microscope allows the laser to be aimed at an inclusion, corrosion product, or surface blemish by means of a precision stage translator that allows movement in three directions. The laser spot size should be able to be focused down to a 10 μm diameter spot (or smaller).

While commercial instruments comply with Class I laser standards and have interlocked laser safety features, it is prudent to be cautious, especially since the power of these devices may be a significant fraction of a joule [12].

Laser ablation ICP-MS must be clearly distinguished from laser mass spectrometry, which has evolved into laser microprobe mass spectrometry [31]. In laser ablation ICP-MS the laser is strictly a sampling device, while in the latter techniques the laser serves as both a sampling device and an ionization source.

Response factors in laser ablation ICP-MS have been shown to be more uniform across the periodic chart when the laser is operated in a Q-switched mode [32]. The technique is also known for high speed and relatively poor precision. In this regard it has been compared favorably with dc arc atomic emission for rapid semi-quantitative survey work [33]. One author claimed that for metals the Q-switched mode should be used exclusively [34].

Rubbing an unknown metal with a diamond lapping stick and then applying laser ablation ICP-MS to the residue on the abrasive has been shown to be a reliable means of identification and analysis [35].

Much of the variability in the results obtained by laser ablation has been attributed to segregation in the solid and the molten sample. Low laser power and multiple laser pulses in the same area seem to increase the problem [36].

The relative standard deviations of results for eight impurities in uranium metal using an XeCl excimer laser was 1–3% for most analytes [37].

6.4 Other Accessories

Flow injection is probably the most important development outside of those covered above. It offers a means of injecting high solids solutions far beyond the concentration range that can be accommodated by the conventional means of sample solution introduction. The sample solution is introduced as a discrete aliquot in a flowing stream of blank solution or dilute acid, which results in a measured pulse of analyte response. The rapid scanning of the quadrupole can easily handle the concentration gradients that result. And because the total solution volumes are minute no orifice clogging results.

The normal continuous nebulization mode in ICP-MS cannot tolerate much more than 0.1% total solids (1g/L), but the flow injection approach can accommodate much higher solution concentrations. Of course, the software must be designed to process discrete signals, and the appropriate solution pumps must be available and properly interfaced.

Electrothermal vaporization is another option. The hardware here is nearly identical to that used in graphite furnace AA work, except that the furnace is utilized strictly to generate an atom cloud, which is swept into the plasma plume where ionization takes place. *Hydride generation* is also possible using accessory hardware similar to that used in AA and OES. There are *direct insertion* options that allow the introduction of solid specimens directly into the plasma using a graphite rod. *Slurry nebulization* is quite feasible and has gained favor in some quarters.

There are complications with all of these options, but in some cases the analyst needs to wade through the associated difficulties in order to obtain the needed results.

6.5 Survey of the General Literature

Before beginning a review of the practical applications of mass spectrometry as applied specifically to the trace analysis of metals we will pause here to highlight some of the more general papers that contain a great deal of useful detail on the subjects we have covered so far in this chapter.

6.5.1 Plasma Source Mass Spectrometry

Niu and Houk summarized the current state of knowledge about the process of ion extraction from the inductively coupled plasma [38]. ICP-MS has been utilized for the ultratrace analysis of high purity acids [39] and for the certification of geochemical reference materials [40] by workers at NIST. Both papers represent useful illustrations of the methodology. A double focusing magnetic sector ICP was compared to a quadrupole ICP in another study. In high resolution mode the double focusing instrument demonstrated the ability to resolve many (but not all) polyatomic ion line interferences. In low resolution mode ($m/\Delta m = 300$) the double focusing instrument yielded much lower detection limits that the quadrupole [41].

Two papers compared the two FCC-approved* radio frequencies for the generation of an inductively coupled plasma: 27.12 MHz and 40.68 MHz.

* Federal Communication Commission

Both concluded that there were significant advantages in detection limits for the lower frequency. Also, at 40.68 MHz the discharge between the cones (and, thus, the potential for some types of interferences) tends to be greater [42,43]. Besides those texts already referenced, two volumes on the specific subject of plasma source mass spectrometry have appeared [44,45].

The glow discharge as a source for both atomic emission and mass spectrometry has been reviewed [46,47]. Argide interferences and dimer interferences are the primary concern in GD-MS. A paper by Shao and Horlick [48] describes some of the interferences encountered in analyzing a variety of metals. A planar magnetron, such as that utilized in high frequency circuits, has been used to produce glow discharge for mass spectrometry [49]. Relative sensitivity factors have been published for 22 elements determined by a dc glow discharge double focusing magnetic sector instrument [50]. Marcus and co-workers at Clemson University have shown that both metals and nonmetals can be analyzed using a radiofrequency glow discharge source [51,52].

6.5.2 Spark Source Mass Spectrometry

For those interested in learning more about this older technology an abundant literature remains to be plumbed, with applications ranging from steel [53] to the Apollo 11 lunar basalts [54]. There is a great deal of discussion pertaining to relative sensitivity factors [55-57]. A comprehensive review of the field appeared in 1984 [58].

6.5.3 Isotope Dilution

In 1989 Fassett and Paulsen summarized 20 years of experience with this methodology at the National Institute of Standards and Technology [20]. Most of the work up to that period utilized a thermal ionization source, but the potential of ICP-MS for this application was beginning to be recognized. In the same year Longerich cautioned that because ICP-MS shows greater deviations in measured isotope ratios than thermal ionization, special attention to optimizing the spike volume must be given [59]. Thermal ionization has also been compared to secondary ion mass spectrometry for isotope dilution work. The thermal ionization approach showed six times better precision [60].

The careful use of ICP-MS was shown to be capable of a precision of 0.15% relative standard deviation in isotope dilution work [61]. Ultratrace levels of neodymium in high-purity lanthanum compounds were determined using ion exchange preconcentration and isotope dilution [62]. Workers at the Laboratory of the Government Chemist in the U.K. developed methods for magnesium, cadmium, and lead that showed relative standard deviations around 0.5% [63]. Ultratrace amounts of ruthenium, palladium, iridium, and platinum in geological samples were determined simultaneously by ICP-MS after ion exchange isolation [64]. Other workers have developed an apparatus for the on-line addition of an isotope spike in an attempt to make isotope dilution practical and routine [65].

7. APPLICATIONS

As has been our practice in the previous chapters we will conclude here with a summary of practical applications for the trace and ultratrace analysis of metal commodities. While mass spectrometry is certainly not a new technology, it is only relatively recently that industrial producers and consumers of alloys have had access to this methodology. So, recent references will far outnumber older ones. One can only surmise that the list will continue to grow rapidly. We are here, in all likelihood, looking at the early stages of a body of literature that will peak at several decades into the next century.

7.1 Ferrous Alloys

7.1.1 Spark Source

Van Hoye, Adams, and Gijbels reported relative sensitivity factors for 16 elements (Ti, V, Cr, Mn, Co, Ni Cu, As, Zr, Nb, Mo, Sn, Sb, La, W, and Ta) in steel using NIST SRM's 661–665 [66]. A pure sample analyzed by neutron activation was used to confirm accuracy [67]. Yanagihara, Sato, Oda, and Kamada studied the effects of many instrumental parameters on the resultant relative sensitivity factors for steels. NIST SRM 462 (low alloy steel) and SRM 443 (stainless steel) were used in this investigation. Spark gap size, the distance from spark gap to the slit, and sample electrode size were found to be the most important [68]. Saito developed a procedure for the determination of trace levels of hydrogen in stainless steel. Reduction of the blank required baking out the sample and the ion source chamber at 323–343° K for 7 hours. The sample was machined into a two-millimeter diameter cylinder. The counter electrode was a "dehydrogenated" high purity gold wire [69].

7.1.2 Plasma

High resolution glow discharge mass spectrometry was evaluated for low alloy steel applications using the NIST 1760 series of SRMs. Results were reported for 30 analytes, including carbon, oxygen, and nitrogen, using the VG 9000 reverse Nier-Johnson instrument. It was calibrated with solid steel reference materials [70].

Vaughan and Horlick performed a similar evaluation of an ICP-MS quadrupole system (Perkin Elmer Elan 250) utilizing three NIST low alloy SRMs

(361–363) and three NIST stainless steel SRMs (121d, 123c, and 160b). In this case 1 gram test portions of chips were dissolved in a mixture of nitric, hydrochloric, and hydrofluoric acids—first at atmospheric pressure, then at elevated pressure in an acid digestion bomb. Sample solutions were diluted to various steel concentration levels ranging from 0.01 to 1.0% (w/v) based on the concentration range of the analyte element. Rhodium was added as an internal standard. Both synthetic solution calibration and the method of additions were tried. Interferences were extensively evaluated. Under the conditions used it proved difficult to determine certain analytes due to direct line overlaps: silicon [$^{14}N_2^+$ on $^{28}Si^+$ (92.21% natural abundance)], phosphorus [$^{14}N^{16}O^1H^+$ on $^{31}P^+$ (100% natural abundance)], sulfur [$^{16}O_2^+$ on $^{32}S^+$ (95.02%)], vanadium [$^{35}Cl^{16}O^+$ on $^{51}V^+$ (99.76%)], arsenic [$^{40}Ar^{35}Cl^+$ on $^{75}As^+$ (100%)], and germanium [$FeO^+/FeOH^+$ on $^{70}Ge^+$, $^{73}Ge^+$, and $^{74}Ge^+$; $^{36}Ar^{40}Ar^+$ on $^{76}Ge^+$][71].

Kuss, Bossmann, and Müller were able to successfully determine silicon in a series of nitric acid/hydrogen peroxide-dissolved low alloy steel CRMs utilizing the $^{28}Si^+$ line, despite the interference from $^{14}N_2^+$. The blank response was found to be much more stable at mass 28 than at mass 29 (where $^{29}Si^+$ shows a moderate effect from $^{14}N_2{}^1H^+$). The silicon response is nearly 20 times greater at mass 28 due to its high natural abundance (92.21%). The mass 29 peak has a natural abundance of 4.67%. Use of the $^{30}Si^+$ was not considered because of its low natural abundance (3.10%) and the massive interference from the $^{14}N^{16}O^+$ ion [72].

Coedo and Dorado determined 15 elements in unalloyed steels using a flow injection ICP-MS approach. Test portions of 0.25 g were dissolved in 7 mL of nitric acid and 0.2 mL of hydrofluoric acid in a microwave oven using sealed vessels. The solutions were spiked with three internal standards: beryllium (the mass 9 peak was used for ^{75}As, ^{90}Zr, ^{93}Nb, ^{98}Mo, and ^{184}W), rhodium (the mass 103 peak was used for ^{48}Ti, ^{51}V, ^{52}Cr, ^{59}Co, and ^{208}Pb), and thallium (the mass 205 peak was used for ^{27}Al, ^{55}Mn, ^{60}Ni, ^{63}Cu, and ^{209}Bi). The solutions were diluted to a final volume of 50 mL. The flow injection accessory was set to inject a volume of 250 μL into a flowing stream of blank solution that was being continuously nebulized into the plasma. Detection limits were improved six-fold over conventional continuous flow nebulization of the sample at the recommended total solids loading [73].

The same authors optimized a similar procedure for trace levels of niobium, vanadium, and titanium in microalloyed high strength low alloy steels. Test portions of 0.20 g were dissolved in 5 mL of nitric acid, 5 mL of water, and 0.1 mL of hydrofluoric acid in a 30 min. sealed-vessel microwave heating program. Yttrium was added as the internal standard. The solution was diluted to 50 mL and a 500 μL aliquot was injected into the carrier solution stream (flowing at 2.8 mL/min). Close matches to the certified values were reported for NIST SRMs 2165, 2166, and 2167 [74].

Coedo, Lopez, and Alguacil determined hafnium, niobium, tantalum, tungsten, and zirconium in high purity iron using an on-line anion exchange column. Limits of quantification ranged from 5 ppm (Nb) to 14 ppm (W). The test portion (0.25 g) was dissolved in a high pressure microwave vessel using 6.0 mL of nitric acid, 2.0 mL of hydrochloric acid, 1.0 mL of hydrofluoric acid, and 2.0 mL of water. The solution was then evaporated to dryness and the cooled salts dissolved in 4.5 mL of 2M hydrofluoric acid and 0.5 mL of hydrogen peroxide. A 200–μL aliquot was injected into the carrier stream, which passed through a mini-column of Teflon PTFE containing approximately 2.0 mL total volume of strongly basic anion exchange resin (Dowex 1X8, 50-100 mesh), previously washed with 5 mL of 2M hydrofluoric acid. Iron passes through and the analytes all remain on the column. They are then eluted into the nebulizer with 200 μL of a 3:1 mixture of 0.7 M nitric acid:0.5 M hydrochloric acid plus 0.05 moles/liter hydrogen peroxide [75].

Coedo, Dorado, Padilla, and Alguacil determined calcium in steels using a combination of chemical separations and a flow injection approach. Since the ion of the isotope with the highest natural abundance ($^{40}Ca^{+}$ = 96.97%) is directly overlapped with $^{40}Ar^{+}$, measurements were made at the $^{44}Ca^{+}$ peak (2.06% natural abundance). Because $^{44}Ca^{+}$ shows interferences from $^{28}Si^{16}O^{+}$ and $^{12}C^{16}O_2^{+}$ it was necessary to treat the microwave-dissolved samples with sulfuric and hydrofluoric acids. The low natural abundance of the mass 44 peak also made mercury cathode electrolysis necessary so that higher concentration analyte solutions could be prepared while controlling total solids loading. Finally, a flow injection approach using a 200–μL aliquot loop was used with a stream of air as the carrier. Results were reported from steel certified reference materials in the 12–28 ppm calcium range [76].

Coedo, Dorado, Fernandez, and Alguacil determined trace and ultratrace levels of boron in iron and steel using an isotope dilution approach in combination with chemical separation and flow injection ICP-MS measurement. Test portions of 0.25 g were dissolved in closed vessels in a microwave oven using a mixture of 1.0 mL hydrochloric acid, 0.5 mL nitric acid, 0.35 mL sulfuric acid, and 5.0 mL water. The solution was then spiked with 250 ng of ^{10}B and evaporated to salts. The cooled salts were dissolved in 10 mL of water, the solution was adjusted to pH 1.4 and extracted for 3 min. with 20 mL of a 1:1 acetylacetone:chloroform mixture. The organic layer was discarded and the aqueous layer was extracted with 10 mL of 1:1 acetylacetone:chloroform. The organic layer was again discarded and the aqueous layer was heated in a Teflon beaker with 1.0 mL of nitric acid. The solution was reduced to 2 mL, cooled, and diluted to exactly 5 mL, and then measured by flow injection ICP-MS using a 250–μL aliquot.

An instrument/method-specific mass discrimination factor must be determined by measuring a standard of known ^{10}B:^{11}B ratio. NIST SRM 951 was utilized for this purpose [77].

Naka and Gregoire determined trace levels of sulfur in steel using graphite furnace atomization with matrix modification, isotope dilution, and solvent extraction for iron removal. For samples estimated to contain >5 ppm a 0.1 g test portion was dissolved in 4 mL of 3:1 hydrochloric acid:nitric acid. A spike of 0.1 μg of ^{34}S was added and the solution was diluted to 10 mL. For samples containing <5 ppm sulfur a 1.0 g test portion was dissolved in 10 mL of 3:1 hydrochloric acid:nitric acid, a 0.25 μg spike of ^{34}S was added, and the solution was diluted to 25 mL. In both cases a 3-mL aliquot was extracted with 6 mL of acetylacetone (the extraction was repeated in the case of the higher sample concentration solution) and the organic layer was discarded. A 10–μL aliquot of the aqueous phase was deposited in a graphite furnace and dried. Then potassium hydroxide solution was added to convert sulfur in the sample to potassium sulfate to prevent loss during the char and to enhance sensitivity during the volatilization. The furnace was then ramped to 300°C for 20 sec, then to 2500°C for 5 sec. The ($^{33}S/^{34}S$) ratio was measured by ICP-MS and used to calculate the weight percent sulfur in the sample. Good results were obtained on steel standards over the range 1.9–60.0 ppm [78].

Yasuhara, Okano, and Matsumura optimized a laser ablation–ICP-MS approach to analyze a series of carbon steel certified reference materials for ten elements (Al, B, Co, Mn, Nb, P, Sb, Si, V, and Zr). In general results agreed well with certified values. The use of a cyclone to remove large aerosol particles and thus create a narrower distribution of particle sizes proved essential. This device diminishes the potential for solids being deposited on the sampler cone surface. A 0.28% natural abundance isotope of the matrix element ($^{58}Fe^{+}$) was used as the internal standard. Detection limits are comparable to those obtained by the continuous nebulization of optimal concentration dissolved alloy solutions [80].

7.2 Precious Metals

This commodity area has shown a great deal of interest in mass spectrometry, due to the need to provide testing to confirm purity levels. This represents a long-standing situation, as evidenced by a 1972 paper on the analysis of trace contaminants in electroplated *gold* by spark source mass spectrometry [81].

Wayne, Yoshida, and Vance reported on the determination of trace impurities in *palladium* metal powders by glow discharge high resolution mass spectrometry. Pressed sample powders were "baked out" under vacuum and presputtered extensively but still yielded a significant H_2O^{+} peak [82].

Longerich, Fryer, and Strong determined trace level elements in archeological native *silver* artifacts by ICP-MS. Test portions of 4–16 mg were dissolved in 5 mL of 8M nitric acid, then diluted with dilute nitric acid to yield solutions that contained 50 ppm total dissolved solids. Memory effects due to high levels of mercury (and silver) and a suppression effect by the silver matrix were observed [83]. Precious metal assays of geological samples were among the earliest applications of the commercial ICP-MS quadrupole instruments. Denoyer, Ediger, and Hager recounted this work, which included ICP-MS as an analytical finish for fire assay procedures for *platinum*, *palladium*, and *gold* [84].

Another area of great interest is the ICP-MS analysis of spent automobile exhaust catalysts. The results of such studies are utilized both to increase their design efficiency and to assay their value for recycling. Brown, Kunz, and Belitz determined palladium, rhodium, cerium, nickel, iron, and barium at low and trace levels in *platinum*-based automotive catalysts coated on a ceramic substrate. Two slightly different microwave acid dissolution procedures were recommended. Each involves a small quantity of hydrofluoric and hydrochloric acid and a small quantity of finely powdered test sample. One method uses a small quantity of nitric acid, as well. After several heating and cooling cycles the solutions may be fumed on a hot-plate with perchloric acid, or reacted with boric acid solution and heated again in the microwave. Perchloric acid must never be used in a sealed pressure digestion vessel due to the likelihood of a severe explosion [85].

Beary and Paulsen used a hot-plate dissolution IDMS approach for lead in which 0.1 g test portions of used auto catalyst were spiked with ^{206}Pb and then dissolved in nitric and hydrofluoric acids, plus either hydrochloric or perchloric acid. Platinum and palladium could also be quantified by isotope dilution, but rhodium, being monoisotopic, could not. Test portions for precious metal assay were spiked with ^{198}Pt, ^{104}Pd, and In, then sealed in a glass Carius tube with hydrochloric and nitric acids. Following the Carius procedure the solution was reacted with hydrofluoric acid in a Teflon beaker. Mass discrimination factors for platinum and palladium were determined by measuring isotopic ratios in test portions of the high purity metals and relating the results to the most current IUPAC* tabulation of natural abundances. The mass discrimination factor for lead was determined from a measurement of NIST SRMs 981 and 982 (lead isotopic standards). Interference from $^{208}Pb^{2+}$ on $^{104}Pd^{+}$ (both m/z = 104) and from $^{206}Pb^{2+}$ on $^{103}Rh^{+}$ (both m/z = 103) produced a significant error. Chemical separation of the lead was feasible to correct the effect, but was not pursued due to concerns about rhodium recovery (IDMS for palladium would not require quantitative palladium recovery). Therefore, mathematical corrections for the effects were made [86].

Borisov, Coleman, Oudsema, and Carter also determined platinum, palladium, rhodium, (and titanium) in auto catalysts by ICP-MS. Their best procedure utilized a 0.05–0.25 g test portion, 1 mL of water, 4 mL of nitric acid, and 5 mL of 7:3 hydrochloric acid:hydrofluoric acid in a sealed microwave

---

* International Union of Pure and Applied Chemistry

vessel. The vessel was heated at full power (2.5 min.), then 25% power (60 min.), then 65% power (20 min.). The vessel was cooled and 50 mL of 0.35 M boric acid was added. The vessel was then resealed and heated at full power (10 min.), then 35% power (50 min.) [87].

Borisov, Coleman, and Carter, in addition, have published work with spark ablation ICP-MS for the determination of vanadium, rhodium, and platinum in automobile catalysts using pressed sample/graphite pellets. Polyatomic ion interferences were drastically reduced and sample preparation times were significantly shortened. However, their observed detection limits were inferior to those from optimized liquid nebulization samples [88].

Van Hoven, Nam, Montaser, Doughten, and Dorrzapf also utilized spark ablation ICP-MS, but for *gold* and *silver* fire assay beads. They observed the following detection limits: platinum—0.6 ppm; palladium—1.2 ppm; rhodium—0.2 ppm; and iridium—1.1 ppm [89]. Kogan, Hinds, and Ramendik employed laser ablation ICP-MS to determine bismuth, copper, iron, lead, nickel, palladium, platinum, and zinc in *gold* and *silver* and their *alloys*. The authors concluded that the relative sensitivity factors in gold and silver are close enough so that silver reference materials can be used for gold and vice versa [90]. In the light of the expected precision of the technique their conclusion can probably be supported, however, better quantitative work is expected of solution-based ICP-MS.

Graham and Robert developed a procedure for high purity noble metals (and their salts) by solution-based continuous nebulization ICP-MS. Scandium, indium, and rhenium were added as internal standards to the acid-dissolved test portion. Dissolution schemes are given for *rhodium, platinum, palladium*, and *gold* metal. The introduction of 5%(v/v) nitrogen gas into the argon stream reduces the amount of $ArCl^+$ (NCl^+ forms more readily), which removes the line overlaps with the $^{75}As^+$ and $^{77}Se^+$ peaks [91].

Becotte-Haigh, Tyson, Denoyer, and Hinds determined arsenic at trace levels in *gold* by hydride generation ICP-MS. A flow injection system was used to generate arsine from the diluted sample solution after the interfering gold matrix was removed by reduction to the metal with L-ascorbic acid, potassium iodide, and hydrochloric acid [92]. Sun, Mierzwa, Lin, Yeh, and Yang determined aluminum, cadmium, cobalt, copper, gold, iron, lead, magnesium, manganese, nickel, and tin in high purity *silver* by first removing the matrix by precipitation with dilute hydrochloric acid. The solution was membrane-filtered, then the filtrate was measured by ICP-MS [93].

7.3 Aluminum

There have not been a great deal of published articles on the analysis of aluminum by ICP-MS. This may be related to the fact that aluminum proves to be a troublesome element for the technique. Hutton and Eaton describe the orifice-clogging properties of aluminum and other refractory oxide formers (zirconium, uranium, and thorium). Volatile elements (like sodium, potassium, and lithium) tend to "burn off" the sampler cone while aluminum and the other high-boilers tend to accumulate around the orifice, restricting flow. The solution to this problem appears to be flow injection, which allows high total solids solutions to be analyzed as a discrete pulse injected into a carrier stream of dilute acid. Aluminum solutions as high as 2% (w/v) totals solids can be handled readily with such a system [94].

Makishima, Inamoto, and Chiba determined trace gallium in pure aluminum by an isotope dilution/solvent extraction ICP-MS approach. A 0.26 g test portion of high purity aluminum was dissolved in 20 mL of 5M nitric acid. An appropriate spike of enriched ^{71}Ga was added and the solution was evaporated to dryness. The cooled salts were dissolved in 20 mL of 7M hydrochloric acid; then 20 mL of isopropyl ether was added and the solution was extracted for 5 min. The aqueous phase (containing the aluminum) was discarded. The organic phase was extracted for 2 min. with 10 mL of water to return the gallium to the aqueous phase. The extraction was repeated, the final aqueous portions were combined and evaporated to dryness. The residue was dissolved in 20 mL of 1M nitric acid and measured by ICP-MS [95].

Takeda, Yamaguchi, Akiyama, and Masuda determined ultratrace levels of thorium and uranium in high purity aluminum by a solvent extraction ICP-MS method. A 10 g test portion was dissolved in 200 mL of 10M hydrochloric acid and 0.2 mL of 3% hydrogen peroxide with 0.1 mL of a 1mg/mL copper solution was added as a catalyst. The solution was extracted for 3 min. with 50 mL of a 10%(v/v) solution of tributyl phosphate in cyclohexane. The aqueous phase was re-extracted with another 50 mL of 10%(v/v) TBP in cyclohexane and the organic phases were combined. These were then extracted with 50 mL of 0.1M hydrochloric acid for 30 sec. to return thorium and uranium to the aqueous phase. The extraction was repeated and the aqueous portions were combined and 0.5 mL of a 100 ppb bismuth solution was added as an internal standard. Then 20 µL of sulfuric acid was added and the solution was evaporated to light fumes of SO_3. To the cooled sample solution 3 mL of 0.27M nitric acid was added and the solution was again evaporated. Then another 3 mL of 0.27M nitric acid was added and the solution was diluted to 5 mL with 0.27M nitric acid. The solution was then measured by ICP-MS [96].

Van de Weijer, Vullings, Baeten, and deLaat also determined thorium and uranium in aluminum with ICP-MS. They utilized two approaches. The first was a flow injection procedure using test portions dissolved in a 1:3:5 nitric acid:hydrochloric acid:water mixture. The injection volume was 250 µL. While the sampler orifice did not clog in this discrete analysis mode a series of "spikes" in the spectra were attributed to Al_2O_3 particles breaking loose from the orifice edges. The second approach utilized laser ablation ICP-MS of solid samples etched with the dissolution mixture described above, then "preburned" for 2 min. before measurement. Spikes were less frequent with this approach. The authors rejected "spiked" measurement data and were able to achieve a detection limit of 0.2 ppb for both thorium and uranium [97].

7.4 Copper

Swenters, Verlinden, and Gijbels determined the diffusion coefficient of beryllium in copper at different temperatures using spark source and secondary ion mass spectrometries [98].

Chiba, Inamoto, and Saeki determined ultratrace levels of antimony and silver in pure copper using a preconcentration/isotope dilution technique. For silver a 1 g test portion was dissolved in 10 mL of 1:1 nitric acid:water and diluted to 150 mL. A suitable spike of enriched ^{107}Ag was added, then 10 mL of a 1% (w/v) tellurium (IV) chloride solution was added (as a coprecipitant). The solution was evaporated to dryness, the salts dissolved in hydrochloric acid, and the solution diluted to 100 mL. Then 20 mL of 20% (w/v) stannous chloride in 6M hydrochloric acid was added. The precipitate was filtered, dissolved in 10 mL of 1:1 nitric acid:water, diluted to 50 mL, and measured by ICP-MS. For antimony a 1 g test portion was dissolved in 10 mL of 1:1 nitric acid:water and diluted to 50 mL (insolubles were filtered, dissolved in sulfuric acid/nitric acid and recombined). A suitable spike of enriched ^{123}Sb was added, then 10 mL of 0.2% (w/v) lanthanum chloride was added (as a coprecipitant). The pH was adjusted to 9 and the precipitate which formed was filtered, dissolved in 10 mL of 1:1 nitric acid:water, and diluted to 50 mL, and measured by ICP-MS. It was possible to detect 20 ppb of silver and 5 ppb of antimony [99].

Park, Park, Yang, Han, and Lee determined ten elements (Ag, As, Bi, Cd, Cr, Fe, Ni, Pb, Sb, and Zn) in pure copper by an electrodeposition/isotope dilution approach. A test portion of 5 g was dissolved in 35 mL of nitric acid and 3 mL of hydrochloric acid (HCl is not used if Ag is the analyte), and diluted to 250 mL. A 50 mL aliquot was removed and spiked with appropriate amounts of isotopes. The copper matrix is removed by electrolysis at an applied potential of −0.25 V versus a saturated calomel electrode (SCE). Silver is plated at −0.1 V versus the SCE in a separate sample (a small deposit of copper accompanies it). The silver is dissolved off the platinum electrode with 1.5M nitric acid. Results for NIST SRMs 394 and 395 (unalloyed copper) were reported [100].

7.5 Nickel-base Alloys

The complexity of most nickel-base superalloy compositions has been a daunting challenge to this field. Publications concerning these alloy types have been sparse, although mass spectrometers are in place in many of the industries that produce and use these alloys. In 1986 McLeod, Date, and Cheung warned of the potential problems with metal oxide interferences. Their examples include the effect of $^{59}Co^{16}O^+$ on (monoisotopic) $^{75}As^+$; the effect of $^{62}Ni^{16}O^+$ on $^{78}Se^+$; the effect of $^{53}Cr^{16}O^+$ on $^{69}Ga^+$; and the interference of a molybdenum oxide ion on each of the six isotopic ions of cadmium. They also have tabulated the resolution required of a high resolution mass analyzer to overcome the effect in each case [101].

P. Richner studied a flow injection ICP-MS approach to high purity nickel analysis. Nickel chips were rinsed in dilute nitric acid, then in deionized water, to remove surface contaminants. A weighed test portion was dissolved in 30% (w/w) nitric acid. The diluted solutions were injected into a flowing stream of 1% (w/v) nitric acid, using either a 40-, or 50-, or 200-μL sample loop. The method of standard additions was required to compensate for concentration-dependent signal depression effects due to the presence of large amounts of nickel in the plasma [102].

Hinds, Gregoire, and Ozaki utilized a solid sampling (filings or chips) graphite furnace vaporization ICP-MS approach to the determination of lead and bismuth in nickel-base alloys. Bismuth could be successfully analyzed using either synthetic solutions or solid reference materials to calibrate. Lead required solid reference materials for a sound calibration. Tellurium was attempted but could not be analyzed by either approach. The authors speculated that tellurium was not being completely released from the matrix [103].

7.6 Refractory and Reactive Metals

Luo and Chang analyzed *zirconium alloys* for trace levels of chromium, copper, hafnium, iron, manganese, nickel, titanium, and uranium by ICP-MS. Rhenium was utilized as an internal standard, however, the authors contend that for the most accurate work calibration solutions should be matrix-matched to the samples. Chips were washed with acetone, dilute nitric acid, then water. Test portions of 0.25 g were dissolved in 5 mL of 1:1 nitric acid:water with hydrofluoric acid added dropwise. The solution was diluted to 250 mL and measured for all elements except uranium. Hafnium was determined by the method of additions. The others were determined by a direct calibration. For uranium 3, 1 g test portions were weighed into 8-oz. polyethylene bottles. Each was spiked with 500 ng of NIST SRM U-500. Then 5 mL of 1:9 nitric acid:water and 2 mL of hydrofluoric acid (dropwise) were added to dissolve the alloy. Then 25 mL of aluminum nitrate solution was added and the solution was mixed. Then 15.0 mL of methylisobutyl ketone was added and the capped bottle was extracted for 3 min. A 10.0 mL aliquot of the organic phase was transferred to a platinum crucible. The solution was dried under an IR lamp, then heated in a muffle furnace at 600°C. The residue was dissolved in

5% nitric acid and diluted to 10.0 mL for uranium isotope dilution measurement [104].

Panday, Becker, and Dietze used a solvent extraction approach to determine trace and ultratrace levels of 34 analytes in *zircalloy* samples. All the analytes were determined using external calibration; 21 of them were also determined by an isotope dilution procedure. The samples were freed of surface contamination by rinsing with acetone, then 1M nitric acid, then deionized water. The dried chips were dissolved in 5 mL of 1:1 nitric acid:water with the dropwise addition of hydrofluoric acid. Rhodium was added as an internal standard. An aliquot of the diluted sample was adjusted to pH 1.5–2.0 with ammonium hydroxide, then extracted with 120 mL of 50% bis-(2-ethylhexyl)-orthophosphoric acid. The aqueous phase was re-extracted with another 10 mL of the reagent. The organic extracts were combined and extracted twice with a 5-mL volume of water. All the aqueous extracts were combined, evaporated, and diluted to 25 mL. The aqueous solution contained the following: Ag, Ba, B, Ca, Cd, Co, Cs, Cu, K, Li, Mg, Mn, Na, Ni, Pb, Pt, Rb, Sb, Sn, Sr, Tl, and Zn. To the combined organic extract 5 mL of 6M nitric acid and 2 mL of dioctyl alcohol was added and the mixture was extracted for 3 min. The organic phase was re-extracted and the combined aqueous phases were evaporated, then diluted to 10 mL. This aqueous solution contained the following: Al, Bi, Ga, In, Pb, rare earths, and Y. The two aqueous solutions were separately measured by ICP-MS [105].

Vanhaecke, Vanhoe, Vandecasteele, and Dams determined boron in *titanium* by ICP-MS. Solid 80 mg specimens were dissolved in 5 mL of water and 700 μL of hydrofluoric acid, added incrementally. Beryllium was added as an internal standard and the solution was diluted to 50 mL with 0.14M nitric acid and measured [106].

Held, *et al.* determined scandium in high-purity *titanium* utilizing both ICP-MS and GDMS methods. For ICP-MS 0.1 g pieces were cleaned with rinses of water, ethanol, boiling hydrochloric acid, then water, and finally ethanol. The dried and weighed specimen was dissolved in 15 mL of 8M hydrochloric acid by heating in Teflon under reflux for 25 hours. The solution was transferred to a column of Bio-Rad AG 50W-X12 (200-400 mesh) cation exchange resin in a plastic column (prepared by eluting 4M hydrochloric acid/0.1M hydrofluoric acid, then 8M hydrochloric acid). Titanium was eluted with 40 mL of 8M hydrochloric acid and discarded. Scandium was eluted with 4M hydrochloric acid/0.1M hydrofluoric acid. This solution was evaporated to dryness, dissolved in 0.14M nitric acid in an ultrasonic bath. Cobalt was added as an internal standard and the solution was diluted to 10 g (note: weight basis) and measured. Glow discharge work on the $^{45}Sc^{+}$ peak was hindered by the interference of $^{50}Ti^{40}Ar^{2+}$ and by the high background from $^{46}Sc^{+}$, but confirmed the ICP-MS results [107].

7.7 Miscellaneous Alloys and Related Materials

Clegg, Gale, and Millett utilized spark source mass spectrometry to determine ultratrace levels of oxygen, nitrogen, and carbon in semiconductor materials (*silicon, germanium, and indium phosphide*). Since blanks were found to be related to the ionization of residual gas in the sparking chamber it was necessary to increase the pumping efficiency and employ a high speed liquid helium cryogenic pump. Surface contaminants on the sample, including oxides, were removed during presparking [108].

Haney and Gallagher evaluated three spark source MS approaches to the trace analysis of bullet *lead*: photographic plate analysis, electrical detection, and isotope dilution. Photographic detection of 26 analytes showed an average precision of 9% (RSD). Electrical detection used with peak switching yielded a precision of 5.7%. Isotope dilution (for Cu, Ag, and Sb) involved dissolution of 10 mg test portions in a few drops of nitric acid and sufficient water to dissolve the precipitate that formed. Dissolved isotopic spikes and the sample solution were combined with 10–15 mg of graphite and the resultant slurry was dried and compacted into electrodes [109].

Liu, Verlinden, Adams, and Adriaenssens determined ultratrace impurities in *arsenic metal* by spark source MS using photographic detection. An intense background shift to the high mass side of the primary arsenic line was mainly the result of secondary ions and electrons sputtered from the plate emulsion by the primary ion beam. The sensitivity limitation that this effect imposed was corrected by cutting the photoplate to eliminate the $m/z = 73–76$ mass range prior to the analysis [110].

Hecq, Hecq, and Fontignies used glow discharge quadrupole mass spectrometry to analyze a sequence of six different thin *metallic films (aluminum, cobalt, gold, molybdenum, silver, and tantalum)* on a copper substrate. They found that an oxygen plasma yielded better sensitivity than an argon plasma [111].

Vijayalakshmi, Prabhu, Mahalingam, and Mathews employed ICP-MS to determine trace and ultratrace impurities in *uranium nitrate, uranium oxide, and sodium metal* in connection with a fast breeder reactor program. In the presence of the uranium matrix part-per-million level detection limits were achieved. When the matrix was removed by solvent extraction with 60% tributyl phosphate in carbon tetrachloride part-per-billion detection limits could be achieved. Sodium metal samples were vacuum distilled to remove the matrix; the residue was dissolved in *aqua regia* and analyzed for 13 nonvolatile elements (Al, Ba, Bi, Cr, Co, Cu, Fe, In, Li, Mg, Mn, Mo, and Sn). For the determination of zinc and cadmium the sodium metal test portion was dissolved carefully in water in a quartz vessel, neutralized with nitric acid, and diluted to give a 0.5% (w/v) sodium concentration. This solution was analyzed by flow injection ICP-MS [112].

Denoyer, Jacques, Debrah, and Tanner described the development of dynamic lens optimization to control space charge effects for the determination of light mass elements in a *uranium* matrix. Very low detection limits for lithium (0.003 ppb) in the presence of 1000 ppm of uranium were reported [113].

Taddia, Bosi, and Poluzzi compared graphite furnace atomic absorption spectrophotometry (GFAAS) with ICP-MS (without an internal standard) for the determination of tin in *indium phosphide*. For the conditions of this study ICP-MS showed the better detection limit, but GFAAS was significantly better in calculated precision (RSDs: GFAAS—2.0–3.4%; ICP-MS—4.3–12.9%) [114].

Broekaert, Brandt, Leis, Pilger, Pollmann, Tschöpel, and Tölg analyzed *aluminum oxide and silicon carbide* for trace contaminants by ICP-MS. A 1 g test portion of the alumina powder was dissolved in hydrochloric and sulfuric acids, cooled, and appropriately diluted, rhenium having been added as an internal standard. A 0.250 g test portion of silicon carbide was reacted with hydrofluoric and nitric acids, and the solution was diluted to 500 mL [115].

Pollmann, Leis, Tölg, Tschöpel, and Broekaert utilized trace preconcentration by means of reversed phase liquid chromatography and measurement by ICP-MS to characterize the trace impurities in *alumina ceramic powders*. A 1 g test portion was dissolved over 6 hours by heating at 240°C with 10 mL of hydrochloric acid and 5 mL of water. To the clear solution 500 µL of a 4% hexamethylenedithiocarbamate in methanol solution was added. The pH was adjusted to 2–3 and the solution was injected into the liquid chromatograph. The impurity complexes were eluted with a methanol:water mixture and introduced into the ICP-MS [116].

REFERENCES

1. Cornides, I. in Inorganic Mass Spectrometry (F. Adams, R. Gijbels, R. Van Grieken, eds.) John Wiley, New York, pp. 1-15.
2. Segre, E. From X-rays to Quarks, W.H. Freeman, San Francisco, 1980.
3. Watson, J.T. Introduction to Mass Spectrometry, 3rd ed., Lippincott-Raven, Philadelphia, 1977.
4. Gijbels, R.; Adams, F. in Inorganic Mass Spectrometry (F. Adams, R. Gijbels, R. Van Grieken, eds.)John Wiley, New York, 1988, pp. 377-393.
5. Ewing, G.W. Instrumental Methods of Chemical Analysis, 4th ed., McGraw-Hill, New York, 1975, pp. 412–437.
6. What Is Mass Spectrometry?, American Society for Mass Spectrometry, East Lansing, MI, 1989.
7. Colodner, D.; Salters, V.; Duckworth, D.C. Analytical Chemistry, 66, 1079A (1994).

8. Harrison. W.W. in Inorganic Mass Spectrometry (F. Adams. R. Gijbels. R. Van Grieken. eds.) John Wiley. New York. 1988. pp. 85–123.
9. Ramedik. G.: Verlinden. J.: Gijbels. R. in Inorganic Mass Spectrometry (F. Adams. R. Gijbels. R. Van Grieken. eds.) John Wiley. New York. 1988. pp. 17–84.
10. Gray. A.L. in Inorganic Mass Spectrometry (F. Adams. R. Gijbels. R. Van Grieken. eds.) John Wiley. New York. 1988. pp. 257–300.
11. Bacon. J.R.: Ure. A.M. Analyst. 109. 1229 (1984).
12. Annual Book of ASTM Standards Vol.03.06. American Society for Testing and Materials. West Conshohocken. PA. 1999. Practice for Describing and Specifying Inductively-Coupled Plasma-Mass Spectrometers.
13. Denoyer. R.C. Atomic Spectroscopy. 12. 215 (1991).
14. Burgoyne. T.W.: Hieftje. G.M.: Hites. R.A. Analytical Chemistry. 69. 485 (1997).
15. Tanner. S.D.: Paul. M.: Beres. S.A.: Denoyer. E.R. Atomic Spectroscopy. 16. 16 (1995).
16. Van Nostrand's Scientific Encyclopedia. 4th ed.. D. Van Nostrand. Princeton. NJ. 1968.
17. Heumann. K.B. in Inorganic Mass Spectrometry (F. Adams. R. Gijbels. R. Van Grieken. eds.) John Wiley. New York. 1988. pp. 301–376.
18. Selby. M. Atomic Spectroscopy. 15. 2 (1994).
19. Denoyer. E.R. Atomic Spectroscopy. 13. 93 (1992).
20. Fassett. J.D.: Paulsen. P.J. Analytical Chemistry. 61. 643A (1989).
21. Evans. E.H.: Giglio. J.J. Journal of Analytical Atomic Spectrometry. 8. 1 (1993).
22. Zhang. H.: Nam. S-H. Cai. M.: Montaser. A. Applied Spectroscopy. 50. 427 (1996).
23. Gustavsson. I.: Larsson. H. in Progress of Analytical Chemistry in the Iron and Steel Industry (R. Nauch. ed.) Commission of the European Communities (EUR 14113 EN). 1992. pp. 193–200.
24. Kuss. H-M: Müller. M.: Bossmann. D.: Petin. J.: Jiminez Seco. J.L. in Progress of Analytical Chemistry in the Iron and Steel Industry (R. Nauch. ed.) Commission of the European Communities (EUR 14113 EN). 1992. pp. 201–209.
25. Analytical Methods Committee. The Royal Society of Chemistry Analyst. 122. 393 (1997).
26. Jakubowski. N.: Feldmann. I.: Sack. B.. Stuewer. D. Journal of Analytical Atomic Spectrometry. 7. 121 (1992).
27. Jakubowski. N.: Feldmann. I.: Stuewer. D. Spectrochimica Acta. 50B. 639 (1995).
28. Ivanovic. K.A.: Coleman. D.M.: Kunz. F.W.: Schuetzle. D. Applied Spectroscopy. 46. 894 (1992).
29. Jiang. S-J: Houk. R.S. Analytical Chemistry. 58. 1739 (1986).
30. Jiang. S-J: Houk. R.S. Spectrochimica Acta. 42B. 93 (1987).
31. van Doveren. H. Spectrochimica Acta. 39B. 1513 (1984).
32. Denoyer. E.R.: Fredeen. K.J.: Hager. J.W. Analytical Chemistry. 63. 445A (1991).
33. van de Weijer. P.: Baeten. W.L.M.: Bekkers. M.H.J.: Vullings. P.J.M.G. Journal of Analytical Atomic Spectrometry. 7. 599 (1992).
34. Paul. M. Atomic Spectroscopy. 15. 21 (1994).
35. Raith. A.: Hutton. R.C.: Abell. I.D. Journal of Analytical Atomic Spectrometry. 10. 591 (1995).

36. Cromwell, E.F.; Arrowsmith, P. Applied Spectroscopy, 49, 1652 (1995).
37. Leloup, C.; Marty, P.; Dall'ava; Perdereau, M. Journal of Analytical Atomic Spectrometry, 12, 945 (1997).
38. Niu, H.; Houk, R.S. Spectrochimica Acta, 51B, 779 (1996).
39. Paulsen, P.J.; Beary, E.S.; Bushee, D.S.; Moody, J.R. Analytical Chemistry, 60, 971 (1988).
40. Kane, J.S.; Beary, E.S.; Murphy, K.E.; Paulsen, P.J. Analyst, 120, 1505 (1995).
41. Moens, L.; Vanhaecke, F.; Riondato, J.; Dams, R. Journal of Analytical Atomic Spectrometry, 10, 569 (1995).
42. Vickers, G.H.; Wilson, D.A.; Hieftje, G.M. Journal of Analytical Atomic Spectrometry, 4, 749 (1989).
43. Uchida, H.; Ito, T. Journal of Analytical Atomic Spectrometry, 9, 1001 (1994).
44. Evans, E.H.; Giglio, J.J.; Castillano, T.M.; Caruso, J.A. Inductively Coupled and Microwave Induced Plasma Sources for Mass Spectrometry The Royal Society of Chemistry, Cambridge, U.K., 1995.
45. Holland, G.; Eaton, A.N., eds. Applications of Plasma Source Mass Spectrometry The Royal Society of Chemistry, Cambridge, U.K., 1991.
46. Harrison, W.W.; Barshick, C.M.; Klingler, J.A.; Ratliff, P.H.; Mei, Y. Analytical Chemistry, 62, 943A (1990).
47. Marcus, R.K. Spectroscopy, 7, 12 (1992).
48. Shao, Y.; Horlick, G. Spectrochimica Acta, 46B, 165 (1991).
49. Shi, Z.; Brewer, S.; Sacks, R. Applied Spectroscopy, 49, 1232 (1995).
50. Saito, M. Spectrochimica Acta, 50B, 171 (1995).
51. Marcus, R.K. Journal of Analytical Atomic Spectrometry, 9, 1029 (1994).
52. Shick Jr., C.R.; Marcus, R.K. Applied Spectroscopy, 50, 454 (1996).
53. Ito, M.; Sato, S.; Yanagihara, K. Analytica Chimica Acta, 120, 217 (1980).
54. Morrison, G.H.; Kashuba, A.T. Analytical Chemistry, 41, 1842 (1969).
55. Jaworski, J.F.; Morrison, G.H. Analytical Chemistry, 46, 2080 (1974).
56. Van Hoye, E.; Adams, F.; Gijbels, R. Talanta, 26, 285 (1979).
57. Verlinden, J.A.A.; Swenters, K.M.E.; Gijbels, R. Analytical Chemistry, 57, 131 (1985).
58. Bacon, J.R.; Ure, A.M. Analyst, 109, 1229 (1984).
59. Longerich, H.P. Atomic Spectroscopy, 10, 112 (1989).
60. Adriaens, A.G.; Fassett, J.D.; Kelley, W.R.; Simons, D.S.; Adams, F.C. Analytical Chemistry, 64, 2945 (1992).
61. Beary, E.S.; Paulsen, P.J.; Fassett, J.D. Journal of Analytical Atomic Spectrometry, 9, 1363 (1994).
62. Beary, E.S.; Paulsen, P.J. Analytical Chemistry, 66, 431 (1994).
63. Catterick, T.; Handley, H.; Merson, S. Atomic Spectroscopy, 16, 229 (1995).
64. Yi, Y.V.; Masuda, A. Analytical Chemistry, 68, 1444 (1996).
65. Klinkenberg, H.; Van Borm, W.; Souren, F. Spectrochimica Acta, 51B, 139 (1996).
66. Van Hoye, E.; Gijbels, R.; Adams, F. Talanta, 23, 369 (1976).
67. Van Hoye, E.; Adams, F.; Gijbels, R. Talanta, 23, 789 (1976).
68. Yanagihara, K.; Sato, S.; Oda, S.; Kamada, H. Analytica Chimica Acta, 98, 307 (1978).
69. Saito, M. Analytica Chimica Acta, 236, 351 (1990).

70. Clark, J.; Walsh, A.; Wheeler, D.; Seeley, C.R. Advanced Materials & Processes, 11/90, 65 (1990).
71. Vaughan, M-A; Horlick, G. Journal of Analytical Atomic Spectrometry, 4, 45 (1989).
72. Kuss, H-M; Bossmann, D.; Müller, M. Atomic Spectroscopy, 15, 148 (1994).
73. Coedo, A.G.; Dorado, T. Journal of Analytical Atomic Spectrometry, 10, 449 (1995).
74. Coedo, A.G.; Dorado, M.T. Applied Spectroscopy, 49, 115 (1995).
75. Coedo, A.G.; Lopez, T.D.; Alguacil, F. Analytica Chimica Acta, 315, 331 (1995).
76. Coedo, A.G.; Dorado, M.T.; Padilla, I.; Alguacil, F.J. Journal of Analytical Atomic Spectrometry, 11, 1037 (1996).
77. Coedo, A.G.; Dorado, T.; Fernandez, B.J.; Alguacil, F. Analytical Chemistry, 68, 991 (1996).
78. Naka, H.; Gregoire, D.C. Journal of Analytical Atomic Spectrometry, 11, 359 (1996).
79. Yasuhara, H.; Okano, T.; Matsumura, Y. Analyst, 117, 395 (1992).
80. Coedo, A.G.; Dorado, M.T.; Fernandez, B. Journal of Analytical Atomic Spectrometry, 10, 859 (1995).
81. Vasile, M.J.; Malm, D.L. Analytical Chemistry, 44, 650 (1972).
82. Wayne, D.M.; Yoshida, T.M.; Vance, D.E. Journal of Analytical Atomic Spectrometry, 11, 861 (1996).
83. Longerich, H.P.; Fryer, B.J.; Strong, D.F. Spectrochimica Acta, 42B, 101 (1987).
84. Denoyer, E., Ediger, R.; Hager, J. Atomic Spectroscopy, 10, 97 (1989).
85. Brown Jr., J.A.; Kunz, F.W.; Belitz, R.K. Journal of Analytical Atomic Spectrometry, 6, 393 (1991).
86. Beary, E.S.; Paulsen, P.J. Analytical Chemistry, 67, 3193 (1995).
87. Borisov, O.V.; Coleman, D.M.; Oudsema, K.A.; Carter III, R.O. Journal of Analytical Atomic Spectrometry, 12, 239 (1997).
88. Borisov, O.V.; Coleman, D.M.; Carter III, R.O. Journal of Analytical Atomic Spectrometry, 12, 231 (1997).
89. Van Hoven, R.L.; Nam, S-H; Montaser, A.; Doughten, M.W.; Dorrzapf Jr., A.F. Spectrochimica Acta, 50B, 549 (1995).
90. Kogan, V.V.; Hinds, M.W.; Ramendik, G.I. Spectrochimica Acta, 49B, 333 (1994).
91. Graham, S.M.; Robert, R.V.D. Talanta, 41, 1369 (1994).
92. Becotte-Haigh, P.; Tyson, J.F.; Denoyer, E.; Hinds, M.W. Spectrochimica Acta, 51B, 1823 (1996).
93. Sun, Y-C; Mierzwa, J.; Lin, C-F; Yeh, T.I.; Yang, M-H Analyst, 122, 437 (1997).
94. Hutton, R.C.; Eaton, A.N. Journal of Analytical Atomic Spectrometry, 3, 547 (1988).
95. Makishima, A.; Inamoto, I.; Chiba, K. Applied Spectroscopy, 44, 91 (1990).
96. Takeda, K.; Yamaguchi, T.; Akiyama, H.; Masuda, T. Analyst, 116, 501 (1991).
97. Van de Weijer, P.; Vullings, J.M.G.; Baeten, W.L.M.; deLaat, W.J.M. Journal of Analytical Atomic Spectrometry, 6, 609 (1991).
98. Swenters, K.; Verlinden, J.; Gijbels, R. Spectrochimica Acta, 39B, 1577 (1984).
99. Chiba, K.; Inamoto, I.; Saeki, M. Journal of Analytical Atomic Spectrometry, 7, 115 (1992).

100. Park, C.J.; Park, S.R.; Yang, S.R.; Han, M.S.; Lee, K.W. Journal of Analytical Atomic Spectrometry, 7, 641 (1992).
101. McLeod, C.W.; Date, A.R.; Cheung, Y.Y. Spectrochimica Acta, 41B, 169 (1986).
102. Richner, P. Journal of Analytical Atomic Spectrometry, 8, 927 (1993).
103. Hinds, M.W.; Gregoire, D.C.; Ozaki, E.A. Journal of Analytical Atomic Spectrometry, 12, 131 (1997).
104. Luo, S.K.; Chang, F.C.; Spectrochimica Acta, 45B, 527 (1990).
105. Panday, V.K.; Becker, J.S.; Dietze, H-J Atomic Spectroscopy, 16, 97 (1995).
106. Vanhaecke, F.; Vanhoe, H.; Vandecasteele, C.; Dams, R. Analytica Chimica Acta, 244, 115 (1991).
107. Held, A.; Taylor, P.; Ingelbrecht, C.; DeBievre, P.; Broekaert, J.A.C.; Van Straaten, M.; Gijbels, R. Journal of Analytical Atomic Spectrometry, 10, 849 (1995).
108. Clegg, J.B.; Gale, I.G.; Millett, E.J. Analyst, 98, 69 (1973).
109. Haney, M.A.; Gallagher, J.F. Analytical Chemistry, 47, 62 (1975).
110. Liu, X.D.; Verlinden, J.; Adams, F.; Adriaenssens, E. Analytica Chimica Acta, 180, 341 (1986).
111. Hecq, M.; Hecq, A.; Fontignies, M. Analytica Chimica Acta, 155, 191 (1983).
112. Vijayalakshmi, S.; Prabhu, R.K.; Mahalingam, T.R.; Mathews, C.K. Atomic Spectroscopy, 13, 61 (1992).
113. Denoyer, E.R.; Jacques, D.; Debrah, E.; Tanner, S.D. Atomic Spectroscopy, 16, 1 (1995).
114. Taddia, M.; Bosi, M.; Poluzzi, V. Journal of Analytical Atomic Spectrometry, 8, 755 (1993).
115. Broekaert, J.A.C.; Brandt, R.; Leis, F.; Pilger, C.; Pollman, D.; Tschöpel; Tölg, G. Journal of Analytical Atomic Spectroscopy, 9, 1063 (1994).
116. Pollmann, D.; Leis, F.; Tölg, G; Tschöpel; Broekaert, J.A.C. Spectrochimica Acta, 49B, 1251 (1994).

8

Miscellaneous Measurement Techniques

1. INTRODUCTION

This chapter has been reserved to cover those trace procedures that find important and practical niches in metal analysis, but yet for one reason or another do not warrant the level of coverage of the preceding chapters. In some cases, such as for the thermal evolution methods, the range of applicable analytes is narrow. In other cases, the potential applications are broad, indeed, but the methodology is impractically slow, or too expensive for widespread use. Throughout this book it has been our intention to keep our focus on the practical, and so here, as well, we will keep our attention on procedures and techniques that the working analyst is likely to encounter. For the arcane or the cutting edge state-of-the-art the reader will have to look elsewhere, although occasionally we may point out where information on such rarified regions may lie.

Because the subjects covered in this chapter are so diverse there is a danger that a great deal of time and space will be wasted on a topic of little interest to the majority of readers. For that reason the extent of the coverage will be to some degree proportional to a perceived level of utility to a laboratory engaged in the routine analysis of metals. For example, few metals labs utilize polarography any longer, and few have access to a neutron activation facility, and so these potentially very large subjects will receive light coverage.

In an endeavor of this kind it is inevitable that some topic of importance to a given reader will be overlooked. For such omissions the author accepts the blame and offers sincere apologies.

2. THERMAL EVOLUTION METHODS

This category represents those methods that have been more loosely (and incorrectly) called "combustion" methods in the jargon of some laboratories.

Methodologically, it has never been a completely clearly delineated category; however, most analysts in the field of metals work understand that it represents a narrow range of technologies applicable to the determination of carbon, sulfur, nitrogen, oxygen, and hydrogen, and rarely, inert gases, such as argon. The sample is never dissolved, of course, but is utilized in the form of chips, powder, or a small solid. The analyte—in some form—is evolved with heat into a vacuum or a flowing inert or reactive gas stream. A number of different technologies may then be used to measure it. Its pressure-volume product might be measured in a vacuum system, or mass spectrometry might be applied. Much more commonly these days a thermal conductivity or infrared absorption approach is utilized with a flowing gas carrier stream.

For our discussion it is appropriate to distinguish between bulk or gross compositional methods, whose purpose is to determine the total amount of the analyte element present in the sample, and speciation methods, which distinguish between various forms of the analyte present in the alloy. The former will be treated here; details of the latter will be reserved for the next chapter (on inclusions and phases).

2.1 Carbon and Sulfur [1–7]

These two analytes are frequently determined together by the same instrument from the same test portion. An induction or resistance furnace is used to rapidly heat the test portion in a flowing stream of purified oxygen. Most often an accelerator additive is included on top of the sample charge to aid in the inductive coupling and to facilitate the formation and evolution of the analyte oxide species. These may include carbon monoxide (CO), carbon dioxide (CO_2), and sulfur dioxide (SO_2). In most instrumental configurations, chemical converters are used to complete the reaction of CO to form CO_2, and to trap out moisture and any sulfur trioxide (SO_3). A dust trap removes particulate matter from the gas stream. Infrared absorption cells are most commonly used to measure both components, although thermal conductivity detectors are a useful alternative. Programmed heating has been used to distinguish surface and matrix carbon, although other methodology is also applicable. The separation and measurement of graphite, carbides, and sulfides will be discussed in Chapter 9.

2.1.1 Instrumentation

The most common modern instruments are induction furnace/infrared absorption units. The sample may be a cast or trepanned pin or other monolithic solid, usually cut to approximately 1 g. It may also be millings, drillings, or powder. Whatever its final form it is essential that the test portion not be exposed to excessive heating during preparation, which can result in analyte

Table 8.1 Change in Gibb's Free Energy of Formation for Common Accelerator Reactions

Accelerator	Combustion Product	ΔG_f (25°C)
Tungsten	WO_3	−182.62 kcal/mole
Tin	SnO_2	−124.2 kcal/mole
Iron	Fe_3O_4	−242.7 kcal/mole
Copper	CuO	−31.0 kcal/mole

loss. For an accurate matrix value for carbon and sulfur it is also critical that the sample surface be thoroughly degreased. Cutting oils, which may be high in sulfur and are certainly high in carbon, should not be used in the milling and drilling operations. Tool speed should be low to prevent overheating.

The test portion is generally weighed on a balance that is integral to the instrument directly into a tared zirconia crucible of special design. An accelerator is added by means of a measured scoop. It may be high purity tungsten powder or a mixture of tin and tungsten. Sometimes, especially with low-iron alloys, a volume or weight of high-purity iron chips or low-carbon granular copper is added.

The accelerator serves several purposes. First, it aids in achieving the inductive coupling necessary to rapidly heat the test portion. In this regard solid samples are more difficult and may require a larger charge or a different mix of accelerators than fine chips or powder. Second, the oxidation of the accelerator contributes heat to the system. The change in the Gibb's free energy of formation for the oxidation of common accelerator substances is listed in Table 8.1.

This is not the whole story, however, since the accelerator and its reaction product also contribute to the mechanism of evolution of the analyte substances—CO, CO_2, and SO_2. Tungsten is particularly useful when low concentrations of carbon and sulfur are to be determined. Tin produces a high initial temperature, overcoming what would otherwise be a sluggish response to the induced field.

The crucible with its charge of sample and accelerator is raised into the center of the load coil on a ceramic pedestal within a quartz combustion tube. A jet of purified oxygen* impinges from above as the rf field induces melting. Gaseous reaction products are swept out below. From this point there are several design options. Generally, first the gas stream encounters a dust trap

* Generally standard grade oxygen is passed through a trap of sodium hydroxide-impregnated clay (Ascarite II™) to remove CO_2, and one of magnesium perchlorate to remove moisture.

to remove SnO_2 and/or WO_3 from the accelerator reaction and particulate oxides from the sample reaction. This may be nothing more than cellulose fibers, although other dust removal systems can be utilized. When a combination of tin and tungsten are used as an accelerator (for difficult to melt samples) a particularly large amount of dust is generated.

The furnace out-gas stream next encounters a series of traps and converters for the gaseous components. A magnesium perchlorate trap removes moisture, which forms from any hydrogen in the sample. Since the gas stream contains a mixture of carbon monoxide and carbon dioxide, a heated catalyst (at approx. 400°C) is generally used to convert CO to CO_2. Both a rare earth/copper oxide mixture and silica impregnated with platinum have been utilized as the catalyst material. These catalysts may produce an undesirable side reaction, however, by oxidizing sulfur dioxide to sulfur trioxide (SO_3), the anhydride of sulfuric acid. Sulfur trioxide must be trapped out on cellulose or it will contaminate the infrared absorption cell that is used to measure CO_2. For this reason, in some dual (carbon and sulfur) determinators sulfur is measured first, then the gas stream encounters the catalyst and SO_3 trap. There are at least three system designs for the IR measurement of carbon and sulfur. In the simplest arrangement the furnace out-gas stream passes once through a single IR cell that serves as both the reference and measurement device. An emitter source at one end of the cell is mechanically chopped. A band-pass filter and solid state detector at the opposite end produce a pulsed output signal that is diminished by the absorbance of the analyte molecule in the gas stream. A reference baseline is established with pure oxygen before combustion begins.

In closed-loop systems the oxygen stream recirculates through the system until a maximum absorbance, steady state condition is attained. Here, too, the reference condition of the absorbance cell is established before the combustion begins. Closed-loop systems sometimes measure both CO and CO_2 in separate IR cells, thus carbon is measured from the sum of the outputs.

In the third basic design purified oxygen passes through one side of a dual chamber IR cell to serve as a reference; the sample gas passes through the opposite side. Both sides are connected to a diaphragm that is actually half of a sensitive parallel plate capacitor microphone. Distortion of the diaphragm caused by the imbalance in transmitted infrared energy in the two halves of the IR cell is measured as the analyte concentration.

Thermal conductivity detection is still utilized in some modern carbon determinators, especially in those designed to measure very low analyte concentrations. In these devices the excellent sensitivity of microthermistors is utilized. However, the selectivity of IR detection is completely lacking, and so the system must be designed to completely isolate the analyte. In a typical design, purified oxygen is used to combust the sample and then the furnace out-gas flow passes through: 1) a dust trap; 2) a manganese dioxide trap (to remove SO_2: $MnO_2 + SO_2 \rightarrow MnSO_4$); 3) a catalyst to convert CO to CO_2; 4)magnesium perchlorate (to remove H_2O); and 5) a molecular sieve (5A) trap (to collect all of the CO_2). This last is a zeolite material of a very narrow range of pore sizes, which retains carbon dioxide at room temperature, but releases it at elevated temperature.

When all the CO_2 has been collected oxygen flow from the furnace is diverted to vent and purified helium gas is passed through the molecular sieve trap as it is rapidly resistance heated. The carbon dioxide is released and swept by the helium flow through a magnesium perchlorate trap (to remove traces of water), then into a silica gel column, which passes CO_2 while retarding traces of oxygen and other gases. The carbon dioxide is then swept by the helium flow into the measure side of a thermal conductivity detector (purified helium passes continuously through the reference side). The detector is contained in a metal block maintained at a constant temperature. The presence of CO_2 causes an imbalance in a Wheatstone bridge circuit and thus carbon is measured by the small current required to restore the balance.

These days, trace levels of sulfur may also be measured by a similar preconcentration on a special trap material, but when it is released by heating the SO_2 is typically measured by IR absorption.

Resistance-heated instruments are still sold. Their primary use has been in programmed heating applications in which it is important to distinguish between the matrix carbon and surface carbon in the form of contamination, lubricants, or organic surface coatings. Some collaborative work on an international scale by ISO[*] has suggested that calibration for ultra-low carbon by means of water-soluble organic compounds, such as sucrose, requires a modern resistance-heated instrument. Evidently, heating is too rapid in induction furnace instruments to prevent the loss of incompletely oxidized carbon compounds when such a calibration approach is utilized. Ordinarily, induction furnace instruments are calibrated with certified metal reference materials, or (rarely) by dosing the detector with a precise aliquot of CO_2 gas.

There are other, older, instrument designs for the thermal evolution measurement of carbon and sulfur in alloys. Carbon contents as low as 50 ppm were once routinely measured by means of a coulometric titration instrument. The carbon dioxide from an induction or a resistance furnace, scrubbed free of SO_2 by some means (such as a trap of urea impregnated with hydrogen peroxide) was carried by the oxygen stream to the measuring unit. This consisted of a two-chamber titration vessel containing solutions of barium

---

[*] The International Organization for Standardization (Geneva, Switzerland).

hydroxide and barium perchlorate. The barium hydroxide reacted with the CO_2, forming insoluble barium carbonate and releasing hydroxyl ions. A sensitive pH measurement circuit allowed for a controlled electrolysis to restore the original pH. The measured current flow to achieve this was proportional to the carbon content.

Older methods for trace levels of sulfur included the use of a sensitive colorimetric reagent for sulfur dioxide. An induction or resistance furnace is used to combust the sample and the out-gas is bubbled through an aqueous solution of p-rosaniline (fuchsine), containing formaldehyde, ethanol, and sulfuric acid. After dilution to volume the color absorbance was measured at 580 nm on a spectrophotometer, as described in Chapter 4. [8, 9] The method was capable of measuring sulfur levels in the 5–10 ppm range. Most other sensitive chemical methods for sulfur in metals are solution-based and involve the evolution and collection of hydrogen sulfide [10–14].

2.1.2 Practical Details

As mentioned above, carbon and sulfur determinations by thermal evolution techniques are almost always comparative in that certified metal alloy reference materials are used to calibrate them. It is a simple fact that these instrumental methods usually function best in that way, both in the obtained accuracy and the ease of use. Like all comparative methods they, thus, become completely dependent upon the reliability of the certified values of the calibration standards. Therefore, it behooves the analyst to pass critical judgement on the reference materials employed. Despite the fact that optical emission techniques for these two analytes have improved dramatically, there remain few other readily accessible alternative technologies upon which to base a judgement. And so the best course is to use many and varied CRMs as validation standards. In this sense there is some safety in numbers, particularly for these techniques since the "combustion" methods are not particularly matrix-sensitive.

The low trace and ultratrace realm is somewhat problematical here because there are few available CRMs. A reliable blank (usually subtracted by the software) is critical, and considerable thought should be given to the material used to obtain it. The best blank is usually obtained from the replicate average result of a certified trace-level material, rather than from something asserted to be carbon- or sulfur-free.

The best course for the analyst is to utilize an iterative procedure, beginning with a zero blank, then calibrating at some level slightly above the expected range needed, then measuring replicates of a certified low level material. The certified value is (usually automatically) subtracted from the replicate average and the difference (if any) is entered into the instrument software as the first approximation of the true blank. Various CRMs are then analyzed and the results examined. If a bias is observed the instrument is recalibrated using the existing blank and the blanking material is re-analyzed as a sample. If this results in a non-zero bias the material is analyzed again and the difference between the replicate average and the certified value is entered as the new blank. The low level CRMs are re-analyzed and the results are again examined for bias. If any exists the entire process is repeated.

The high temperature combustion process in an induction furnace instrument results in a variable, but always significant, amount of spattered slag and metal that become fused to the wall of the quartz combustion tube. In a short time of continuous operation this coated surface can have somewhat of a gettering-type effect, preventing the quantitative measurement of the evolved analyte gases. Modern determinators either automatically brush out the tube with a wire brush or signal the operator that it is time to brush out the tube manually. This significantly prolongs the useful life of a quartz combustion tube and prevents deterioration in the instrument performance.

Trap and converter maintenance are also critical to optimum instrument performance. Expendables need to be replaced or regenerated on an exacting schedule. The prudent laboratory manager will establish an instrument log for scheduled maintenance and check it regularly.

The ultra-low carbon and sulfur realm (this generally means levels below 10 ppm) has some special precautions associated with it. In Chapter 1 we discussed molten metal sampling of ultra-low carbon (ULC) steels and mentioned the special, noncontaminating samplers required. The samples, as either solids or chips, must be cleaned by some regimen, sometimes involving both dilute acid surface etching and solvent rinsing. Some analysts advocate heating the test portion to drive off surface contamination prior to the higher temperature analysis of low carbon levels in the base metal. Resistance-heated surface/matrix carbon determinators can be useful in this approach.

For ultra-low levels of carbon and sulfur induction furnace crucibles or resistance furnace boats must be pre-ignited at 1000°–1350° C in air or oxygen, and cooled and stored in a desiccator prior to use. Of course, the prepared sample, accelerator, and crucible must not be touched by hand. Most analysts have experienced the best results with tin-tungsten accelerator,* however some claim that ultra-low sulfur determination works best using plain tungsten. Some analysts employ an ultra-high purity grade of oxygen for the combustion of very low carbon and sulfur samples.

The fact that in certain alloys carbon and sulfur levels below 10 ppm have been shown to be metallurgically critical will, no doubt, continue to drive work in this area. In ULC steels the formability of sheet low alloys

---

* A highly-purity version of this product is available for this application.

depends on low interstitials (e.g., carbon, sulfur, and nitrogen). In nickel-base high temperature alloys good hot working characteristics are strongly dependent on very low sulfur levels. There are numerous other examples.

When the sample contains a halogen, such as the chlorine found in certain titanium alloys, or the fluorine in slags, it is important to include a halogen trap in the system to prevent corrosion in the determinator system. Antimony metal—with its ability to collect chlorine and fluorine—has been used in this application.

2.2 Nitrogen and Oxygen [2,15–20]

Inert gas fusion has become the method of choice for these two analytes, which are frequently determined together from the same test portion. This methodology has largely supplanted vacuum fusion (which also permitted hydrogen determination) because of speed, safety, and ease of use. Unlike carbon and sulfur, for which optical emission is the only viable, fundamentally distinct, alternative technique, it is possible to cross-check results for nitrogen and oxygen by a number of methodologies. These include vacuum fusion (using either pressure-volume product measurement or mass spectrometric measurement), kjeldahl chemical determination for nitrogen, and neutron activation analysis for oxygen. In addition, nitrogen can now be determined accurately in most metal matrices by optical emission. The problem here is largely accessibility since few laboratories maintain vacuum fusion or kjeldahl equipment any longer, and neutron activation is far outside the budget constraints of most labs. An answer lies in contract commercial facilities that provide such services for a fee, allowing periodic checks on reliability and the resolution of vendor/customer result disagreements.

2.2.1 Instrumentation

Inert gas fusion utilizes the as-received metal—chips or solid for nitrogen, but almost always a solid for oxygen.[1] Chips are solvent-degreased and dried, solids are prepared by various techniques depending on the matrix. Steels are best hand-filed, especially for oxygen determination. High temperature alloys may be polished on a silicon carbide belt or disk sander. Many contend that titanium alloys require chemical etching with hydrofluoric acid/hydrogen peroxide or hydrofluoric acid/nitric acid. The water-rinsed samples are

---

Oxygen by optical emission is also a distinct possibility, especially at levels above 50 ppm. Powder metals are an exception and a special problem. In this case, however, surface oxygen is generally, the critical parameter.

always washed in a low residue solvent, such as HPLC-grade acetone or methylene chloride. Modern practice utilizes the graphite resistance-heated impulse furnace.

The weighed test portion (0.5–1.0 g) is placed in an out-gassed high purity graphite crucible. The preparation procedure takes two forms, both involving an out-gassing step in the apparatus. The empty crucible is inserted in the instrument between two water-cooled copper electrodes. With a continuous flow of high purity helium a high current (650 A or more) is passed through the electrodes, extracting and sweeping away residual atmospheric gases. The crucible is cooled in the apparatus, under helium flow. At this point the test portion may be introduced into the crucible from above through a helium-swept rotating air lock, or the furnace may be opened, the crucible removed with forceps, the test portion introduced, using a small glass funnel, and the crucible immediately returned to the furnace.

The analyze cycle begins with a purified helium purge of increased flow to sweep away any introduced atmospheric gases and moisture. This is followed by an impulse current of 600–1300 A, which is sufficient to heat the crucible and its contents to 3000°C. Nitrogen in the alloy, either as an interstitial element or an inclusion compound is evolved as the diatomic molecule, N_2. Oxygen in the alloy (primarily as an oxide compound) is converted to CO and CO_2 by reaction with the carbon in the crucible.

The evolved gases pass through a dust filter, then into a heated rare earth/copper oxide converter, which oxidizes CO to CO_2. The gases then pass into an infrared detector cell tuned to the carbon dioxide absorption resonance. An algorithm converts the integrated CO_2 peak to concentration of oxygen in the sample. The gas next passes through sodium hydroxide-impregnated clay to remove the CO_2, and through magnesium perchlorate to remove the moisture. The gas stream then enters one side of a sensitive thermal conductivity detector. The other side receives a continuous stream of purified helium. The difference in signal between the two sides is integrated and calculated as the nitrogen content of the sample. There are other designs utilized, as well. For example, a silica gel column can separate CO_2 and N_2 so that both can be detected with a thermal conductivity detector.

Earlier designs utilized an induction furnace with a large graphite crucible centered inside the load coil. Argon was usually used as the carrier gas. With this design lower temperatures (approx. 2500° C) were typical and extraction times were longer.

The complete extraction of nitrogen and oxygen from the alloy sample requires the dissociation of any nitrides, carbonitrides, and oxides that may be present. Some of these may be extremely refractory in nature. The extreme temperature of the graphite crucible/impulse furnace is usually all that is required. But occasionally techniques employed in the lower temperature induction furnace systems must be resorted to. These include the addition of solid platinum or nickel, or graphite powder. The first two lower the solubility of the gases in the molten pool; the last contributes additional carbon for reduction processes. Platinum and nickel exhibit the high temperature characteristics that Martin and Melnick [15] described as desirable for molten bath additives in vacuum fusion: miscible with most sample alloys, nonreactive with carbon and oxygen (sic) (and nitrogen), low vapor pressure, a poor getter of the analyte gases,[*] and a poor solvent for carbon. Low residual gas platinum and nickel are obtainable for this application, most usefully as rod or foil. Nickel baskets formed from wire have been used to contain large chips for transport through the sample loading air lock. Nickel and platinum foil can be formed into capsules to contain powders.

Induction furnace/gas chromatography devices for nitrogen and oxygen in metals were built by Dallman and Fassel at Iowa State University and by Hanin and Villeneuve at IRSID,[†] both in the mid-1960s [16,17]. Shortly afterwards, Leco Corporation (St. Joseph, MI) and Strohlein & Co. (Dusseldorf, Germany) began marketing commercial inert gas fusion equipment for the determination of gases in metals. In the late 1960s Leco introduced an instrument that utilized the impulse graphite furnace principle. This led to a series of improvements in a product line that is still in use today. Leco also markets an instrument that allows the controlled temperature programming of the sample crucible. This can yield information about the nitrides and oxides present in the alloy (see Chapter 9). It is also possible to obtain information about the residual argon in inert gas-atomized metal powder.

Earlier thermal evolution techniques for nitrogen and oxygen centered around vacuum fusion, which was applicable to both elements, as well as hydrogen. It was slow and cumbersome, but could potentially be *definitive*, as when measuring the pressure-volume product of the analytes using a McLeod gauge. These can be used to read pressures over the range 10^{-5}–10 torr (.001–1333 Pa), but in vacuum fusion equipment usually a range of only 10^{-2}–1.5 torr (1.3–200 Pa) was required. It was also fairly common to couple a mass spectrometer to a vacuum fusion apparatus. Some laboratories devised a means of coupling a gas chromatograph to the equipment. This involved freezing out the vacuum-extracted gases in an isolatable cold trap, sealing off the vacuum, then heating the trap and sweeping it with carrier gas into the GC.

---

[*] Contact getterning, or the absorption of gaseous atoms by a metallic film, is primarily a vacuum system phenomenon, and, as such, is not regarded as a significant factor in inert gas fusion.
[†] *Institute de Recherches de la Siderugie* (at St. Germain-en-Laye, France).

The vacuum fusion thermal extraction of gases from metals was first reported and studied in the early decades of the twentieth century. Commercial equipment was at one time available from the National Research Corporation (Newton Highlands, MA), Leco Corporation (St. Joseph, MI), Baltzers Akiengesellschaft für Hochvakuumtechnick und Dünne Schichten (Liechtenstein), and W.C. Heraeus GMBH (Hanau, Germany). It was generally a quite complicated arrangement of glass tubing, glass stopcocks, and specialized glass vessels, all interconnected, and under extremely high vacuum. There was a great deal of mercury to control, move, and measure the gases. And there was a high frequency induction furnace with a graphite crucible of special design centered in the load coil. One or more "sample arms" were sealed into the vacuum system above the furnace.

Solid specimens could be held in "dimples" in the arm and pushed and manipulated with a magnetic "pusher," which was moved by external magnets outside the glass. Solid test pieces were, thus, dropped into the furnace, where they were melted in a large molten bath, usually of iron, nickel, and/or platinum. Vacuum was maintained by several large mechanical forepumps and one or more mercury diffusion pumps.

In pressure-volume measurement the total pressure due to hydrogen, carbon monoxide, carbon dioxide, and nitrogen was measured. The sequence that followed varied, but one general approach was to oxidize the hydrogen to water and the carbon monoxide to carbon dioxide with a heated, rare earth-doped copper oxide catalyst. The H_2O and CO_2 gases were frozen out in a liquid nitrogen cold trap, which was then heated, and their combined pressure was measured. Then the water vapor was trapped out (with magnesium perchlorate, for example) and the CO_2 was measured alone. From the ideal gas laws:

$$PV = nRT$$

where, P = the pressure in mm of mercury; V = the volume in milliliters; n= the number of moles of gaseous analyte; R = the gas constant (82.06 mL · atm/mole · deg); and T = the temperature in K,

or: $$PV = (g/M)RT$$

where, g = the weight of the analyte gas in grams; and M = the number of grams/mole of the analyte.

Thus, $$g = MPV/760RT$$

and, $$\%\ \text{analyte gas} = 100MPV/760RTW = \text{MPV}/7.6RTW$$

where, W is the sample weight in grams.

2.2.2 Practical Details

Vacuum fusion was a tricky and potentially dangerous procedure since the entire apparatus could (and did) implode from time to time. The operation was usually shielded to protect the operator from glass shards flying past one another, but the health hazard from mercury vapor (as droplets got into the heated parts of the equipment) was an incalculable concern. Today, the inert gas fusion approach has completely displaced it in all but a few isolated niches. We will confine the remainder of our discussion to that technique.

Inert gas fusion is fundamentally a comparative technique, despite some efforts to achieve independence by the application of gas dosing or pure salts (e.g., potassium nitrate) in the calibration. Such approaches always leave some question that the dynamics of gas evolution and measurement for samples do not reflect the calibration dynamics, leading to erroneous results. A more secure approach is the use of metal reference materials, provided that accurately certified, homogeneous reference materials of the appropriate analyte concentration are available.

As with all thermal evolution techniques an accurate blank becomes more and more critical as the work approaches and enters the trace and ultratrace realm. One common source of an oxygen blank involves the relative humidity in the laboratory. If the crucible out-gassing approach is the one that involves opening the furnace to load the test portion, there is a possibility that the water-cooled copper electrodes will condense moisture out of the air. The water will be trapped in the instrument when the graphite crucible is resealed into place. Some of it may be decomposed to O_2 and H_2 by the heat and some of the oxygen may be converted to CO_2 by the graphite. The solution is to control the laboratory humidity, or at least to ensure that it is low.

Trap and catalytic converter maintenance are just as critical here as for carbon and sulfur determinators. In addition, the upper furnace section needs to be brushed out with a metal brush between each measurement. Since a considerable amount of graphite dust is generated it is prudent to periodically clean up the upper and lower electrode sections and the environs of the instrument using a shop vacuum cleaner. It is not wise to keep carbon determinators in the same room, but if they must operate nearby it is particularly critical to control the graphite contamination from the inert gas fusion equipment. Some analysts choose to wear disposable vinyl gloves when operating these instruments since the fine graphite dust adheres tenaciously.

While nitrogen is routinely and accurately measured using a test portion of machined chips, conventional wisdom decrees that oxygen shall only be measured using a solid specimen. This decree is ignored by many labs that are expected to report results on gas-atomized or water-atomized metal powders. In this case, surface oxygen on the powder particles is the needed analyte value since it directly affects powder compaction properties. Oxygen on metal powders can range from 10 ppm to 1000 ppm and higher. With the exception of some titanium reference materials metal standard calibrants for levels above 200 ppm are rare, and so the analyst usually needs to resort to the use of lower sample weights, gas dosing, or perhaps even evaporated potassium nitrate solutions to generate the needed results.

Both sampling and sample preparation are critical in the inert gas fusion determination of oxygen. As we discussed in Chapter 1, pin samples taken from molten metal must be examined for entrained slag and "blowholes" and rejected if any are found. Solid specimens must be cut from finished metal by a procedure that avoids excessive heating. If it is necessary to sample low-oxygen metals that are readily subject to air oxidation the best course is to use a blanket of inert gas. The safest approach for this is a glove box operation. However, most metals and alloys can be correctly sampled for oxygen by simply using a sufficiently slow tool speed. Cooling liquids are not generally recommended.

The preparation of the solid test portion takes various forms. Sometimes it is matrix-specific. Acid pickling, solvent degreasing, hand filing, and polishing on a silicon carbide belt or disk sander have all been suggested for various matrices. For example, Takahari, Abiko, Okochi, and Furuya reported on a sample preparation procedure used by twelve laboratories for the inert gas fusion determination of oxygen in high purity iron.* The sample specimen was cut with a degreased saw, then rinsed in low residue solvents, then pickled in dilute hydrofluoric acid/hydrogen peroxide. The pickling was halted with 1:1 hydrogen peroxide (30%): water. Then the specimen was rinsed in water, then methanol, then acetone, and finally dried in a current of hot air [19].

Much more typically, iron-, nickel-, and cobalt-base alloys are hand filed in a small vise with a clean file or held in pliers and polished on a silicon carbide belt or disk. Titanium alloys may be either abraded or pickled (with either HF/H_2O_2 or HF/HNO_3). In all cases the solid test portions are solvent rinsed with a low residue solvent as a final step. They can be conveniently stored in the numbered wells of a porcelain spot plate in an argon-purged desiccator.

Finally, it should be mentioned that commercial equipment is available that allows the ramping of the heating in the graphite impulse furnace by programming a gradual increase of current through the crucible. Such a device, at least in principle, allows the monitoring of the evolution of the various forms of nitrogen and oxygen present in the test portion: surface

---

* The interlaboratory study was conducted for the Chemical Characterization Subcommittee of the 19th Committee (steelmaking) of the Japan Society for the Promotion of Sciences.

contamination, interstitial nitrogen,* and all the various nitride, carbonitride, and oxide inclusion compounds that may be present. We will return to this subject in the next chapter. One study applied such an instrument to the correction of surface oxygen contamination in the bulk determination of oxygen in iron-boron-silicon chip samples [20].

2.3 Hydrogen [15,21–24A]

At one time hydrogen was commonly determined by thermal vacuum techniques. At least three different manufacturers marketed such equipment. Today, hydrogen is usually determined by inert gas fusion instruments. Sampling and sample handling are critical for many common matrices, such as iron, because the analyte moves freely though the atomic lattice of the base metal at room temperature. This picture is valid for the *ferritic* (alpha) form of iron, as it occurs in low alloys and simple iron/chromium systems, but not for the *austenitic* (gamma) form, as it occurs in chromium/nickel/iron stainless steels. Hydrogen is much less mobile in austenite. Iron alloys that undergo the ferrite → austenite transition at elevated temperatures are, therefore, best analyzed below that temperature. If it is possible to completely extract the hydrogen at a lower temperature, that is usually the prudent course since a high vapor pressure of the metal matrix is likely to produce thin metallic film coatings on cooler regions of the apparatus that may getter the analyte.

Hydrogen is present in solid solution in the transition metals, such as iron, nickel, cobalt, and copper (also: Cr, Mn, Mo, W, Rh, Ag, Pt, and Au) with a relatively low concentration present at saturation. The so-called *pseudohydride-forming metals*, however, (Ti, V, Zr, Nb, Hf, Ta, Pd, the rare earths, and the actinides) can dissolve a great deal more of the element (up to 10 atomic percent, and more). And some metals (e.g., Zn, Ga, Cd, In, Hg) do not absorb hydrogen to any significant extent (except by occlusion when they are deposited by electroplating processes).

In alloy systems the situation becomes quite complicated. As an example, nickel alloying additions greatly increase the solubility of hydrogen in copper metal, but they have no effect in iron metal. Aluminum additions, on the other hand, decrease the solubility of hydrogen in both metals. Moreover, it has been shown that hydrogen may have different solubilities in the different metallic phases that may be present in an alloy.

---

* Hydrogen hot extraction (for interstitial nitrogen, among other applications) is technically a thermal evolution method, but it more properly belongs to Chapter 9, where it will be discussed.

2.3.1 Instrumentation

The inert gas fusion instrument for hydrogen in common use is a graphite crucible/impulse furnace system. The crucible, containing a tin pellet, is outgassed in a carrier stream of purified argon. Then the (typical) 2-g test portion is placed in the crucible and the analytical process is initiated. A large current passes through the crucible, rapidly heating it to about 2500°C. The effluent gas passes through a converter packed with Schütz reagent (I_2O_5 and H_2SO_4 on silica gel) to convert carbon monoxide to carbon dioxide.* The stream then passes through an Ascarite II™ trap (NaOH on clay mineral) where the carbon dioxide is absorbed. Nitrogen, which accompanies the analyte, is slowed down by a short molecular sieve column, while the hydrogen goes through to be measured by a microthermistor thermal conductivity cell. A version of this device has been sold that includes a much larger crucible that is able to accommodate up to a 6-g test portion.

The tin flux serves several purposes, particularly in the analysis of iron-base and titanium-base samples. It facilitates the melting of a large solid test portion, it aids in the dissociation of hydride-type species, and it expedites the evolution of hydrogen from the molten bath.

Sometimes it becomes important to distinguish between *mobile hydrogen*, which, as we mentioned, is free to move through the metal matrix, and *trapped* or *fixed hydrogen*, which is locked in place by lattice dislocations, grain boundaries, or inclusion compounds. Leco Corporation designed an instrument that accepts a special molten metal sampler. The sampler is used to obtain a portion of the metal bath, and it remains sealed for transport to the laboratory. The ceramic tube is placed in the instrument, where it is crushed by a special device. A low temperature heating program produces an output for the mobile hydrogen. When the signal returns to baseline the test piece is heated more vigorously, generating output proportional to the trapped hydrogen.

Currently, the main competitive technique for hydrogen determination is a process monitoring system, primarily used by the steel and aluminum industries, that measures the analyte in the molten metal. One popular design utilizes a bell-shaped vessel that covers a portion of the molten metal and recirculates nitrogen gas in a closed system over the gas/metal interface, passing through the measure side of a thermal conductivity detector, while pure nitrogen passes through the reference side. Hydrogen values obtained by such a

---

* Unlike other conversion systems, I_2O_5 does not oxidize H_2 to H_2O. The reagent darkens with use, due to the formation of I_2: $5CO + I_2O_5 \rightarrow 5CO_2 + I_2$. It was originally developed as a carbon monoxide indicator for mine atmospheres.

system are generally slightly higher than those obtained by laboratory measurement of properly preserved solids analyzed by inert gas fusion. The difference may reflect hydrogen loss during cooling and solidification of the test specimen.

Before the commercial development of inert gas fusion instruments hydrogen was determined by vacuum fusion or vacuum hot extraction techniques. For vacuum fusion a tin bath was commonly used to lower the temperature at which a sample alloy matrix, such as an iron-base alloy, would become molten. With a large excess of tin steels could be melted at 1100–1500° C in an induction heated graphite crucible. The lower temperature minimized the extraction of nitrogen and oxygen, while all the hydrogen was being extracted. In some cases the carbon monoxide that did form could be oxidized to carbon dioxide by I_2O_5, then absorbed on sodium hydroxide-impregnated clay (Ascarite II™) to prevent the interference. The pressure-volume product of the remainder of the gas could then be measured. Then the gas could be passed through a heated copper oxide/cerium catalyst to convert the hydrogen to water, which could be absorbed in magnesium perchlorate. The gas pressure-volume product could be again measured, the difference representing hydrogen.

Since it is not necessary to melt an alloy to extract all of its hydrogen, most vacuum instruments employed temperatures below the melting point of the alloys they were designed to analyze. The Leco and NRC˙ units employed induction heating then dumped the dehydrogenated solid into a collecting reservoir. The Serfass analyzer used a similar approach with a tilting resistance furnace. This latter unit employed a silver/palladium alloy tube to separate hydrogen from other evolved gases by diffusion through the metal. A hot hydrogen determinator that did not melt the sample, but utilized an inert gas/thermal conductivity detection system has also been marketed.˙

2.3.2 Practical Details

Sampling and sample preservation are critical to accurate hydrogen determinations in most matrices. An equally important issue is the matter of reference materials since only the—now rarely used—techniques of pressure-volume product measurement in vacuum fusion or vacuum hot extraction would qualify as definitive methods.

Since hydrogen is mobile in many metal lattice structures the very act of mechanically removing a sample may alter the hydrogen concentration. In

˙ The NRC unit was intended only for titanium alloys.

˙ The Hytest Analyzer (De La Rue Frigistor, Ltd.; Langley, Bucks, England).

Chapter 1 we referred to the once common practice of packing a steel casting with solid carbon dioxide, then using a trepanning tool to drill out a center pin that is broken out with pliers. By whatever means they are taken solid ferritic steel samples for hydrogen (and they must *always* be solids) are usually stored in a liquid nitrogen dewar until they are analyzed. They are commonly identified by securely attaching them to strings, which hang outside the dewar stopper, labeled with tags. In matrices where hydrogen is not especially mobile, such as austenitic stainless steel, or titanium alloys, such special precautions are unnecessary, although care should always be taken to avoid significant heating during sampling and sample preparation.

A related matter—admittedly, somewhat outside of the analytical realm—is how the results from a matrix in which the analyte is very mobile are used. Clearly, if the sampled product is subjected to any further thermal or mechanical processing the hydrogen result is invalidated. In some cases storage for any significant amount of time at room temperature will invalidate the result. Hydrogen results, then, for such alloy systems are most meaningful if they are obtained *and applied* immediately before something will be done to the material lot by either the producer or the consumer. It is fair to say that hydrogen results are most useful as a process control measure during alloy or finished part manufacture. Also, hydrogen analysis is frequently part of an investigation of a cracked casting or forging, or of the failure of a finished part in service.

The issue of reference materials is highly significant for hydrogen since the comparative inert gas fusion technique has become dominant. Unfortunately, very few certified reference materials are available for this application and inconsistencies have been discovered when CRMs from different sources are analyzed by the same technique. In some cases two series of calibration standards from two different sources have been found to each plot linearly, but with two distinctly different slopes.

Besides gas dosing and vacuum fusion/hot extraction there are few applicable alternative methods for hydrogen. There are some nuclear techniques that will be briefly described later in this chapter. Some modern optical emission spectrometers are being represented as an alternative means of hydrogen determination (hydrogen shows lines at 486.1 and 656.3 nm that have been used). Work with spark source mass spectrometry has also been published. None of these procedures represent practical alternatives for the majority of labs at the present time. The apparent bias between molten metal hydrogen determination, such as that described in the previous section, and laboratory-based results is a further complication that has led some companies to rely heavily on one or the other approach for process control. A comparative method that uses levitation melting (and, thus, no crucible) and a thermal conductivity readout has recently been described [24A].

3. X-RAY FLUORESCENCE METHODS

One does not ordinarily think of X-ray fluorescence (or X-ray emission) spectrometry as a trace technique, but it can be a very effective one, both in its time-honored and its state-of-the-art forms. It is unlikely that a laboratory would make the considerable capital investment in conventional XRF equipment exclusively for trace level work. However, where it is already in place its use can be extended into the trace realm. Total reflection X-ray fluorescence spectrometry is a rapidly growing trace analytical technique that utilizes the familiar X-ray sources and detectors in a fundamentally different geometry. For certain applications this new approach has the potential to challenge some of the most sensitive instrumental trace techniques.

3.1 Principles and Instrumentation [25–35]

X-ray emission occurs when an inner shell (nonvalence) electron is ejected from an atom. This event can be brought about by the impact of a charged particle or a high energy X-ray photon. The minimum energy (of a charged particle or of a photon) needed to eject an electron from a specific inner shell is known as the absorption edge energy for that shell. There is one absorption edge for the K shell, three for the L shell, and five for the M shell, corresponding to the number of electrons in the shell.

If protons from an ion accelerator are the excitation source the technique is known as *particle-induced X-ray emission (PIXE)*. An electron beam is utilized for the same purpose in *electron probe microanalysis (EPMA)*. If X-rays (from an X-ray tube or a radioactive material) are the source then the technique is known as *X-ray fluorescence spectrometry*, which is the focus of our attention here. The ejected electron, termed the photoelectron, carries information about the atom it vacated that is studied in the field of *photoelectron spectroscopy* (sometimes known as *ESCA*[*]), a nondestructive surface analysis technique.

The atom responds to the electron ejection by a series of quantum-permitted relaxations in which higher energy electrons drop to lower atomic orbitals, each accompanied by the emission of secondary X-rays of characteristic energies. These emitted X-ray photons constitute the X-ray fluorescence atomic spectrum, which contains a small number of lines characteristic of the element. The X-ray emission lines form a K, L, or M series, depending on whether the ejected electron vacated a K, L, or M shell. Sub-shells are desig-

---

[*] An oddly vague acronym that stands for "Electron Spectroscopy for Chemical Analysis."

nated α and β. Also, the lines may be accompanied by a numeric subscript to designate their relative intensity (1 is the highest, 2 is lower, etc.)

In some cases, particularly with low atomic number elements, a secondary electron, known as an Auger ("o-zhay") electron is emitted instead of secondary X-rays. These events reduce the *fluorescent yield*, making XRF unsuitable for the determination of low Z elements. The Auger electrons carry information about the element, as well, and their spectra are studied in the field of *Auger spectroscopy*, another surface analysis technique.

Since nearly always XRF measurements are made of alloyed (and frequently highly alloyed) material the analyst must become aware of the profound interelement effects that can occur when secondary X-rays from a matrix element have sufficient energy to eject an inner shell electron from the analyte element. Much rarer are so-called third-element effects, in which element I in this manner "enhances" element II, which then "enhances" element III. Self-enhancement, however, does not occur; in other words, secondary X-rays from element Y cannot photoeject an electron from another atom of element Y.

Another phenomenon occurs when X-rays interact with matter. Some of the X-ray photons will be scattered by the outer shell valence electrons of the sample. *Coherent or Rayleigh scattering* results from elastic collisions. No energy is lost in this process. On the other hand *incoherent or Compton scattering* results from inelastic collisions. This produces emission lines in the X-ray spectrum that can, in principle, be used to correct for matrix effects since they are affected by the matrix in a similar manner to the analyte.

Since Planck's relationship tells us that wavelength and energy are proportional:

$$E = hc/\lambda$$

where E is the energy (in joules), λ is the wavelength in cm, h is Planck's constant (6.62620×10^{-34} joule · sec), and c is the speed of light (3×10^{10} centimeters per sec), we can measure the secondary X-rays by dispersing them in space, as with light by a prism or a grating; or by sorting their energies electronically. In the first case the instrument used is a *wavelength dispersive X-ray emission spectrometer*. This device uses a collimator and analyzer crystal to produce a spatially separated array of X-ray lines analogous to an optical emission spectrum produced by a grating. In the second case the instrument used is an *energy dispersive X-ray emission spectrometer*, which frequently uses a semiconductor detector—a high resolution device that produces a signal directly proportional to the energy of the incoming X-ray photon—and pulse processing circuitry to produce an output display with the rates of the pulses arranged according to increasing energy. Except in the low wavelength (high energy) region,[*] wavelength dispersive instruments show a clear advantage in the ability to discriminate neighboring peaks. Energy dispersive instruments are significantly smaller and cheaper than wavelength dispersive instruments. They are extremely useful for identifying unknown or sorting mixed alloys. Some versions are readily portable.

Wavelength dispersive instruments use an X-ray tube source. This is a water-cooled, evacuated envelope in which an electrified filament emits thermal electrons.[†] These flow across a potential gap of between 15 and 100kV, striking a copper anode that is coated with a metal selected for the *characteristic spectrum* of X-ray lines generated (rhodium, platinum, silver, molybdenum, gold, chromium, and tungsten are all frequent choices; although there are others). The primary beam spectrum from the tube also contains a *continuum* that results from the *bremsstrahlung* or "braking" of the thermal electrons as they impinge upon the electron clouds of the atoms of the target metal.

The integrated primary beam continuum intensity is directly proportional to the atomic number of the target element; thus, high Z elements are usually chosen. The target metal must also have a high melting point and good thermal conductivity since over 99% of the electron beam energy must be dissipated as heat, with less than 1% being emitted as X-rays. Thin foils of pure metals are used as filters in the source beam to remove the continuum, leaving the characteristic lines, or, alternatively, to remove a characteristic line that presents an interference. Most manufacturers provide such filters, which can be inserted or removed at the analyst's option.

The primary beam is collimated (i.e., made reasonably parallel) by a series of parallel metal strips (sometimes referred to as a Soller slit) and then it strikes the solid sample surface at a precisely determined angle. Both the characteristic and continuum radiation in the primary beam contribute to the formation of secondary X-rays from the sample. In order to contribute, however, the energy of the characteristic X-rays must be just above the absorption edge of the analyte element. When that situation prevails the characteristic line is a very efficient promoter of secondary X-rays because these photons interact with the sample atoms close to the surface (<1 mm penetration, typically, for metals) where the resultant secondary X-rays are not reabsorbed. However, most sample analytes are excited by the primary beam continuum much deeper below the surface (1–10 mm) and some of the resultant secondary X-rays never escape the specimen.

---

[*] However, even here, the new laminar-structured analyzer crystals seem to have now returned the low wavelength advantage to the wavelength dispersive instruments.

[†] Most modern X-ray tubes are derived from the Coolidge tube, developed in 1913.

The polychromatic secondary X-rays that escape the sample are collimated by a second Soller slit, then strike the analyzer crystal. The different lattice planes of the crystal each diffract the various X-ray wavelengths through distinct angles. After diffraction some waves will be out of phase, interfere destructively, and not be measurable. In-phase waves will interfere constructively and will be observed. These fulfill the relationship known as *Bragg's Law*:

$$n\lambda = 2d \sin \theta$$

where n is the order of the diffraction spectrum (first order: $n=1$, second order: $n=2$, etc.); λ is the wavelength; d is the lattice spacing of the analyzer crystal; and θ is the angle between the incident beam and the diffracted beam. Actually, the value of 2θ is commonly referred to because of the design of the *goniometer*, which is the mechanical arrangement for selecting the angular relationship between the analyzer crystal and the detector. The value of $2d$ is commonly associated with a given type of analyzer crystal, as is a three-digit *Miller index* number that defines the diffracting plane. Table 8.2 lists some common analyzer crystals and their uses. Analyzer crystals are composed of

Table 8.2 Some Common XFS Analyzer Crystals

Common Name	Chemical Name	Miller Index	2d (in angstroms)	Application
Lithium fluoride	Lithium fluoride	220	2.85	
Lithium fluoride	Lithium fluoride	200	4.03	Wavelength <3A
Sodium chloride	Sodium chloride		5.64	
Silicon	Silicon	111	6.27	Reflections from even orders suppressed
Germanium	Germanium	111	6.53	
PET	Pentaerythritol	002	8.74	Best for Al, Si, P, S, and Cl
ADP	Ammonium dihydrogen phosphate	101	10.64	
Mica		002	19.93	Long wavelengths for curved crystal
PHP	Potassium hydrogen phthalate	002	26.3	Best for C, N, O, F, Na, and Mg
Barium stearate	Barium stearate		100.00	Wavelengths >20 angstroms for curved crystal

low Z elements to limit the amount of secondary X-ray emission background from the crystal itself. Crystals are either flat and nonfocusing, or curved and focusing, and the two types are utilized in distinctly different goniometer designs. Modern instruments provide a series of selectable crystals to optimize performance for different needs.

Very old XFS equipment may still employ *geiger tubes* as detectors, but these early devices lacked all ability to distinguish X-ray photon energy. Their slow recovery between counts resulted in serious counting losses at even moderate count rates. *Scintillation counters* consist of a thallium-doped crystal of sodium iodide positioned next to a photomultiplier tube. The crystal emits a flash of 410 nm light when struck by an X-ray photon. The intensity of the light emission is proportional to the energy of the X-ray photon.

Flow proportional counters, like geiger tubes, consist of a tube containing a wire at high potential with respect to the tube housing. They also have a thin polymer window and a continuous flow of P-10 gas (90% argon/10% methane). An X-ray photon entering the window causes argon atoms to ionize. The tube housing neutralizes the argon ions and the wire anode collects the free electrons. The resultant pulse in the circuit is proportional to the X-ray photon's energy, discriminated several times better than possible with a scintillation detector. Both scintillation and flow proportional detectors are utilized in modern wavelength dispersive equipment, and both are completely adequate. Both can recover between counts in about one microsecond.

Energy dispersive instruments may utilize either a low power (10–500 W), often air-cooled, X-ray tube source, or a radioactive material. Tube sources are sometimes coupled with a secondary target to generate specific characteristic secondary X-rays to efficiently excite a specific analyte. Practical radioactive sources include beta emitters coated on, or mixed with a target material (examples are tritiated zirconium or ^{147}Pm on aluminum) and isotopes that emit X-rays as they decay by the K electron capture process (e.g., ^{55}Fe, ^{109}Cd, and ^{153}Gd).

These sources impinge upon the sample surface, generating secondary X-rays that are detected by a semiconductor or a sealed proportional detector a short distance away. These detectors produce an energy-discriminated signal pattern that is processed by a multi-channel analyzer, which counts the pulses, ordering them by increasing energy. A pile-up rejector feature reduces artifacts in the spectrum due to those cases where a pulse arrives before the preceding pulse has been fully processed.

The semiconductor detector usually is a lithium-drifted silicon [Si(Li)] detector, although germanium and mercuric iodide detectors have been used. The Si(Li) detector is composed of a gold-coated p-silicon layer, a lithium-doped layer, and an n-type silicon layer. The p and n silicon layers are maintained under a 600 volt reversed bias. In use an X-ray photon enters the

Table 8.3 Some Solutions to the Problem of X-ray Matrix Effects (Physical)

Solution	Complication
Closely matching reference materials	Availability
Thin film samples are less matrix-sensitive	Sample preparation requires skill and time
Use of internal standards	Added elements may add to matrix effect
Method of additions	Requires skill, time, and knowledge of linear range
Use Compton scatter line as internal standard	Theoretical basis of correction is weak

p-type layer and creates one electron/hole pair for each 3.8 eV of input energy. These devices can discriminate the energy of an X-ray photon with very high resolution (ten times better than a scintillation detector and three times better than a flow proportional detector). The resolution of the Si(Li) detector tends to deteriorate when count rates exceed 50,000 counts/sec, however.

Portable energy dispersive instruments often utilize a radioactive source and a sealed proportional detector. This saves the weight associated with an X-ray tube and eliminates the need for the liquid nitrogen cooling required by the Si(Li) detector and its preamplifier.

Energy dispersive systems are in wide use but their resolution is still no match for wavelength dispersive instruments, except at the highest energies. This does not imply that they are not applicable to trace work, however. All of the methods for extending the detection limits of wavelength dispersive instruments should be applicable to this alternative technology.

The many complex interferences associated with X-ray emission spectrometry—absorbance by the matrix, analyte self-absorbance, and enhancement by the matrix—have yielded to a number of different approaches. Table 8.3 summarizes some of the principal physical solutions to the problem, while Table 8.4 lists some of the important mathematical approaches.

Today, it is well recognized that absorbance effects, including self-absorbance show a hyperbolic relationship with concentration, whereas the enhancement by the matrix has curvature but is *not* hyperbolic. All of the mathematical models shown in Table 8.4 are iterative; however, the fundamental parameters calibration scheme does not require an extensive calibration

For further details concerning the various types of radiation detectors see the discussion in Section 4.1 of this chapter.

Table 8.4 Some Solutions to X-ray Matrix Effects (Mathematical)

Solution	Description
1955—Sherman	One of the first attempts at mathematical correction [*Spectrochim. Acta*, *7*, 283 (1955)].
1964—Lucas-Tooth/Pyne	Intensity correction model [*Advan. X-ray Anal.*, *7*, 523 (1964)].
1966—Traill/LaChance	Improved matrix solution model; influence coefficients [*Can. Spectrosc.*, *11*, 43 (1966)].
1968—Criss/Birks	Fundamental parameters [*Anal. Chem.*, *40*, 1080 (1968)].
1974—Rasberry/Heinrich	Absorption and enhancement treated separately [*Anal. Chem.*, *46*, 81 (1974)].

routine, involving hundreds of samples. In this approach "day-one" values for the coefficients of a complex equation are initially employed for a sample analysis, yielding a first approximation of the results. The equation may contain as many as 30 variables, including terms for fluorescent yields, mass absorption coefficients, and many parameters dealing with the spectrometer, itself. These results are then used to calculate new theoretical intensities with the fundamental parameters equation. The analyte values are then multiplied by the ratio of the first set of theoretical intensities to the second set of theoretical intensities to generate new analyte values. This iterative process repeats until no further change is observed.

Total reflection X-ray fluorescence spectrometry (TRXRF) [33–35] is a trace analysis technique that had its origins in an experiment performed by Arthur Holly Compton in 1930. It was first applied as an analytical technique by Yoneda and Horuchi in 1971. Today, several manufacturers throughout the world produce TRXRF instruments.

The basic concept is quite different from conventional XFS. The primary beam is generated from a line-focus X-ray tube. It is shaped like a narrow paper strip, using carefully aligned slits. This beam is first reflected and filtered by a quartz mirror and a metal foil window to remove the bremsstrahlung continuum. The beam is then reflected from the thin-layer sample at an extremely shallow angle (<0.1°, and typically, 4 min) X-rays are totally reflected at such a grazing angle of incidence. The exact angle where total reflection begins to occur is wavelength dependent. Standing waves that form are detected by a Si(Li) detector immediately above the sample.

The sample is typically a few microliters of solution evaporated onto a highly polished, optically flat substrate. The substrate is frequently quartz, although boron nitride, germanium, glassy carbon, and even plexiglass have been used. The incident and reflected beam both contribute to the measured analyte signal.

Bremsstrahlung radiation must be removed from the primary X-ray beam to reduce the background from both the sample and the substrate. This provides sufficient processing of the primary beam for trace analysis. In the absence of a continuum in the source beam penetration depth is extremely shallow (a few nanometers) and so there is little fluorescence from the substrate. However, TRXFS is also applicable to surface analysis and depth profile analysis, and for these applications an essentially monochromatic primary beam is required. In depth profile work the angle of incidence is varied within the range of total reflection, allowing successive regions to be analyzed down to a depth of about 100 nm. TRXFS also makes possible the direct measurement of the density of surface layers.

At least one major journal has devoted an entire issue* to this technology, which still remains largely unknown among the metals analysis community in the United States.

3.2 Applications [36–54]

The use of conventional XFS for trace analysis almost always involves some form of chemical isolation and/or preconcentration. One obvious alternative is increased counting time, which is inherently limited by analyte sensitivity, background, and instrument stability. The use of an X-ray tube target that emits characteristic lines very close to the analyte's absorption edge, and just on its short wavelength side, will improve analyte sensitivity. Filters to reduce tube continuum near the analyte line will lower the background. There are also other parameters that can be "fine tuned" to increase the analyte signal-to-background. But there are definite limits to the improvements that can be achieved instrumentally.

One approach for reducing background has always been available but seldom employed: the sample is dissolved in acid, diluted to a definite fixed volume and a drop or two is deposited on a filter paper (sometimes in an area delineated by a ring of wax [37–39]. As listed in Table 8.3, such a thin film shows little matrix background.

Chemical manipulations are resisted or not considered by many of today's X-ray spectroscopists, often for no good reason if facilities are available. The fact is that very low limits of detection and of quantification can be achieved with comparatively little time and effort in a chemical laboratory,

---

Spectrochimica Acta, 46B, 1313–1436 (1991).

provided that the individual can work skillfully and safely in such an environment. All such procedures involve dissolution of a weighed test portion, usually in the form of chips or powder. The most common methods rely on the precipitation of the analyte, usually as part of a group, and often with the addition of a co-precipitant/carrier. The solution is then usually filtered through a microporous membrane filter, air-dried, and mounted in some manner.

It is most convenient to use machined nylon disks, cut from rod, as the backing for the membrane. A Mylar film (3–6 mm thick) is stretched over the specimen and secured in place with a rubber band that rests in a groove cut around the edge of the disk. There is a problem with this system, however, especially for light precipitates. A considerable amount of the primary X-radiation is likely to pass completely through the residue and membrane and interact with the backing disk, causing scatter and increased background. For this reason many analysts prefer to mount the membrane sandwiched between two Mylar films over an empty cup designed for use with solutions. The arrangement is secured in place by a retaining ring, similar to the way cloth is held in an embroidery ring. Extra precautions, such as masking, can be taken to prevent scatter from the plastic cup wall. The specimen is always rotated during exposure since the precipitate is unlikely to be uniformly distributed on the membrane.

Numerous analytes can be determined at trace levels using one of these approaches. One of the earliest was the determination of *chlorine in high-purity titanium*, reported by Rudolph and Nadalin. A 3-g test portion was dissolved in hydrofluoric acid with the dropwise addition of nitric acid to oxidize the titanium. Chloride was precipitated with silver nitrate, filtered, washed, dried, mounted, and measured. A calibration curve prepared from pure chloride salt was linear over the range 15–2000 ppm [40].

In 1968 Luke introduced a comprehensive scheme termed "*coprex*" ("coprecipitation X-ray") that utilized 25-mm membrane filters of 0.8- or 5.0-μm pore size. The dried membranes and their precipitates were attached to aluminum masks with stopcock grease. The 1968 paper was comprehensive, but not rigorously quantitative. It suggested precipitation schemes and carriers for trace quantities of most elements, many by groups. Precipitants utilized included sodium diethyldithiocarbamate, ammonium hydroxide, cupferron, hydrogen sulfide, and phenylfluorone for large group separations, as well as tetraphenylboron (for K, Rb, and Cs), dimethylglyoxime (for Pd), α-benzildioxime (for Ni), diammonium hydrogen phosphate (for Ca, Sr, and Ba), sodium carbonate (for Ca and Sr), sodium hydroxide (for Mg), barium chloride (for S), and numerous others. The most unique separation reported was xenon as the barium perxenate, which was measured by XFS without drying due to its extreme volatility [41].

In 1969 Albright, Burke, and Yanak published a method for *selenium in copper-, nickel-, and iron-based alloys* in which the analyte was reduced to elemental form with hydroxylamine hydrochloride, filtered on a 0.22-μm membrane filter and mounted on a phenol fabric disk for X-ray measurement [41A]. Similar work had previously been described for *tellurium* by the same authors [41B].

In 1971 Vassilaros published a quantitative X-ray procedure for *traces of arsenic and selenium in steel* based on his earlier (1968) work with *selenium and tellurium* [42]. This approach utilized arsenic and selenium as carriers for each other when the analytes were precipitated in elemental form with stannous chloride. The precipitates were filtered on 5-μm porosity membranes, washed with 3M hydrochloric acid, and air-dried for 1 min. The dry membranes were mounted on empty solution sample holders and measured. Results were reported for a series of low alloy and stainless steels [43].

In 1976 Vassilaros and Byrnes published an X-ray procedure for *trace niobium, tantalum and tungsten in both ferrous and nonferrous alloys.* It involved the hydrolysis precipitation of the analytes with sulfurous acid. A test portion was selected to contain 0.20 to 1.0 mg of total analytes. Samples of steel, nickel-base alloys, and rare earth metals were dissolved by adding 50 mL of water, then 50 mL of hydrochloric acid. After warming, 10 mL of nitric acid was added to completely dissolve all the sample types. Then 30 mL of 70% perchloric acid was added and the solution was fumed to a volume of 20–25 mL. The cooled salts were dissolved in 200 mL of boiling water; 30 mL of sulfurous acid was added with stirring, and the solution was boiled for 10 min. The solution was allowed to stand for 10 min, then filtered through a 0.22-μm microporous membrane filter. The precipitate was washed with boiling water, air-dried for 1 min, and mounted for X-ray measurement. Results were reported on numerous samples, including several low alloy and stainless steel certified reference materials [44].

Ricci isolated *zirconium and hafnium from nuclear samples* by coprecipitation with iron using ammonium hydroxide. The precipitate was centrifuged and the supernatant solution was drawn off. The residue was washed with hot water, centrifuged, and the supernatant again drawn off. The slurry was then transferred to a Mylar cell, dried on a steam bath, then measured by XRF. Analyte levels up to 200 ppm were measured with detection limits of 0.4 ppm for zirconium and 1.2 ppm for hafnium [45].

Many other precipitation reactions are adaptable to trace level XRF measurements. The rare earths can be brought down as fluorides and zirconium and hafnium as para-bromomandalates, for example [46]. The possibilities are extensive, indeed. Usually a simple calibration with a few points prepared from pure solutions of analyte will suffice.

Another category of chemical preconcentration for XRF measurement involves ion exchange in some form. Conventional column ion exchange chromatography with specific liquid eluents for the analytes of interest is one possibility that is not often used. The eluents can usually be concentrated by evaporation, then measured by XRF in a liquid cell. Another approach is the batch technique, in which the sample solution is agitated with ion exchange resin granules, then filtered. A portion of the filtered granules can then be dried and briquetted for XRF measurement. The most often used approach is probably the multiple filtration of the sample solution through an ion exchange resin-impregnated filter paper, which is then dried and mounted for measurement.

In 1964 Luke used a tiny disk (1/8 in. or 0.3175 cm) punched from a strongly acid cation exchange resin membrane to collect trace and ultratrace levels of numerous analytes and measured them between two sheets of Mylar film with a fully focused curved crystal X-ray milliprobe [47].

Campbell, Spano, and Green evaluated the characteristics of Reeve Angel cation exchange resin-loaded filter papers SA-2 and WA-2 for XRF analysis. Test solutions of known analyte concentrations and correct acidity were filtered through the papers seven times. The same solutions were then filtered seven times through a fresh disk to allow comparison for the completeness of analyte removal. The papers were slowly dried using an infrared lamp, then mounted on a hollow plexiglass holder, held in place with a polyethylene ring-cap. They found that more than 99% of most cations were exchanged after seven filtrations on the first SA-2 paper. They reported limits of detection that ranged from 0.05 μg (Co) to 5.0 μg (Al) with nearly all metallic elements detectable below 1.0 μg [48].

In 1971 Kashuba and Hines used an anion exchange column approach for *cerium, lanthanum, and praseodymium in carbon steels* The dissolved test portion was introduced to the resin in a diluted mixture of nitric acid, methanol, and acetic acid. The resin column was then removed, briquetted, and measured by XRF [49].

Knote and Krivan published a method for the determination of trace levels of *zirconium, molybdenum, hafnium, and tungsten in niobium and tantalum*. One g test portions were dissolved in a hydrofluoric acid/nitric acid mixture. Excess nitric acid was destroyed with urea and the hydrofluoric acid concentration was adjusted. The matrix elements were removed by solvent extraction with diantipyrylmethane in dichloroethane. The aqueous phase was diluted to 100 mL and transferred to a double ion exchange filter paper (Serva SB-2—Heidelberg, Germany). The papers were dried and measured by XFS. Analyte recoveries ranged from 83.1 to 93.9% and detection limits from 0.2 to 0.25 ppm [50].

Applications of total reflection X-ray fluorescence for metals have yet to emerge in great numbers. Chen, Berndt, Klockenkämper, and Tölg published a method for the trace analysis of *high purity iron*. They extracted the dissolved sample with methylisobutylketone to eliminate iron, then measured the bismuth, chromium, copper, lead, manganese, nickel, titanium, and vanadium by TRXFS [51]. *Niobium in steel* has also been determined by means of TRXFS [52]. Laser vapor deposited samples of iron-chromium and copper-zinc binary alloys, a high alloy steel, and $YBa_2Cu_3O_{69}$ high temperature superconducting ceramic were analyzed by TRXFS [52A]. Aluminum samples have been analyzed [53]. The role, if any, that TRXFS may one day be expected to play in the analysis of trace impurities in metals is uncertain. The current requirement to remove the matrix from solution samples may limit its use to highly specialized, niche applications. On the other hand, some breakthrough technology may one day make it a preferred approach and justify the equipment's considerable expense

4. RADIOACTIVATION METHODS

This is a category of trace methods that once enjoyed intense interest. It was the first analytical methodology to promise a nondestructive, multi-element, definitive approach to the trace and ultratrace realm. However, that promise remains largely unfulfilled. Moreover, the activation methods, even in their ascendant phase, never enjoyed widespread use because of cost and safety concerns. Today, these methods remain largely within their provenance—academic and government institutions and only the largest and most well-equipped industrial laboratories.

Despite the fact that there are now better and cheaper ways of reaching the trace realm, activation analysis still has much to offer. It is available as a commercial service, so that even a modest laboratory can avail themselves of the technique—to compare results from other approaches and to access extremely low detection limits for certain elements.

4.1 Principles and Instrumentation [55–61]

The first thing the newcomer to this area must learn is that activation analysis is not one but many methodologies, each with several subcategories. What most people mean by the term, and, in fact, the most widely used form is *thermal neutron activation analysis (TNAA)* and its "instrumental" and "radiochemical" subfields. But this leaves a host of other techniques and their associated acronyms.

What all these methodologies have in common is their central focus on nuclear phenomena. In all the other techniques described in this volume and, indeed, in nearly the entire field of analytical chemistry, only the analyte atom's electron cloud is involved. Here, we are dabbling with the nucleus and its packets of enormous binding energy.

Stable *light* elements have a nucleus containing approximately equal numbers of protons and neutrons, but as Z numbers climb more neutrons are required to provide short-range nuclear strong force to overcome the mutual coulombic repulsion of the proton's charge. One can plot a curve of stable isotopes with the number of protons on the ordinate and the number of neutrons on the abscissa. Nuclei below the curve decay spontaneously by the emission of an electron (β^-); the resultant "daughter" contains one less neutron and one more proton than its "parent." Nuclei above the curve decay spontaneously by the emission of a positron (β^+),* or by the capture of an orbital electron, both resulting in a "daughter" with one less proton and one more neutron. These three *beta decay* processes may be accompanied by the emission of gamma-ray or X-ray photons, as well as by the emission of the massless, or "near massless," neutrino or anti-neutrino. *Alpha decay* is a phenomenon of high Z nuclei, in which a $^4He^{2+}$ nucleus is emitted. It is frequently accompanied by gamma radiation. A chain of such decay processes may occur, traversing the curve of stability back and forth until finally a species is formed that resides on the curve.

The rate at which an unstable nucleus decays is quantified by its *half-life*, which is simply the time required for one-half of a large sample to decay. Alpha particles (basically, helium nuclei) are not very penetrating, being easily stopped by even thin foils. Beta particles (i.e. electrons) are about 500 times more penetrating at the same energy, due to their higher velocity and lower charge. Both types of particles may ionize or excite atoms they encounter by interacting with their orbital electrons. High energy beta particles may also impact the nucleus, generating *bremsstrahlung* radiation. Positrons that have been slowed by interactions may *annihilate* with an electron, producing two 0.511 MeV gamma ray photons.**

While alpha and beta emissions can be monitored the emission of gamma (γ) radiations carries a great deal more information about their source nucleus. By some measures gamma rays are >25 times as penetrating as the same energy beta particles. Gamma rays interact with matter by three processes. The *photoelectric effect* (primarily low energy γ interacting with high Z elements)

---

* The electron and the positron have the same mass but opposite charge. They belong to the class of subatomic particles known as *leptons*. The proton and the neutron belong to the group known as *hadrons*.

** In gamma spectroscopy a peak at 0.511 MeV is the signature of a positron-emitting nucleus.

is a process in which an orbital electron is ejected, carrying away all the gamma photon energy minus its own orbital binding energy. The *Compton effect* results from an elastic collision between a gamma photon and a bound (or free) electron (one γ may interact with multiple e^-). *Pair production* is the process by which a gamma ray of 1.02 MeV or greater interacts with the coulombic field of the nucleus to produce an electron and a positron (in effect, the opposite of the annihilation reaction). Any gamma ray energy in excess of 1.02 MeV is imparted kinetically to the two created particles.

At the core of all forms of activation analysis is some process to artificially produce an unstable nuclide from the analyte so that its concentration can be ascertained by the radiations emitted by the resultant decay mechanisms. The transformation of the analyte is accomplished by bombardment of the nucleus by charged particles, like protons or deuterons, by gamma ray photons, or, most particularly, by neutrons. The interaction of an element's nucleus with one of these beams is defined by a *cross section*, which is an effective area used to express the probability of a nuclear capture reaction. The cross section, σ, is defined by:

$$\sigma = a/bc$$

where a = the number of capture reactions per cm^2 per sec; b = the number of target nuclei per cm^2; and c = the number of incident particles per cm^2 per sec. The *barn* is the common unit for measuring cross sections. One barn is equal to 10^{-24} cm^2 per nucleus.

Charged positive particle bombardment, although an effective analytical tool in PIXE (Section 3.1) and in Rutherford backscattering spectrometry, is not a dominant form of activation analysis. Charged particles must have very high energies to show suitably large capture cross sections. Often, but not always, neutrons are ejected as the analyte is converted to its $Z + 1$ daughter. Thus, for example, protons and deuterons have been used to determine carbon in steel:

$$^{12}C + p \rightarrow {}^{13}N + \gamma$$

or, in shorthand form: $^{12}C(p,\gamma)^{13}N$

and $^{12}C + d \rightarrow {}^{13}N + n$ or, $^{12}C(d,n)^{13}N$

The ^{13}N has a half-life of 10 min, decaying by the emission of a positron and a gamma ray. The penetration depth of charged particles is extremely shallow and the very high energies required cause excessive sample heating.

It should be noted that the cross section is an effective concept also used to indicate the probability of scatter, fission, and certain electron orbital reactions.

Gamma ray photon activation is not used much because the threshold energy is high and cross sections are very small. Some light elements, like beryllium, are feasible analytes because they emit prompt neutrons when activated by moderate gamma energies.

Neutrons, particularly low energy (or thermal) neutrons have been the most popular activation source. The reason for this lies in the fact that the neutron, as a neutral particle (of mass equal to the proton), has no coulombic force to overcome. Moreover, capture cross sections for thermal neutrons are frequently quite large, and even the higher energy neutrons show resonances with useful cross sections.

When a neutron bombards an atomic nucleus three results can occur. It can be *elastically scattered* and in the process lose some of its kinetic energy to the target nucleus. Light nuclei, in particular, will absorb a great deal of its energy. For this reason materials like water and deuterium oxide (D_2O) are used to *thermalize* (or slow down) fast neutrons. The bombarding neutron can also be *inelastically scattered*. Or it can be *captured*.

Thermal neutrons have energies of a few hundredths of an electron volt. Nuclear capture cross sections for these particles range from less than a millibarn (for light elements) to 6.1×10^4 barns (for ^{155}Gd). Nuclear capture of a thermal neutron always produces an atom that is several million electron volts above its ground state because of the contributed strong force binding energy. The system relaxes immediately with the prompt emission of gamma rays. These gamma rays provide useful analytical information.

About 20 elements can be determined by *prompt gamma neutron activation analysis*, including 6 elements that *must* be analyzed in this manner (H, B, C, N, Si, and Gd). In the best experimental arrangement a neutron beam impinges on the sample while the emitted gamma is simultaneously measured by a detector.

Most activated samples are not measured promptly, however. By holding the irradiated samples for various lengths of time before and between measurements it is sometimes possible to determine 30–40 elements. The delays allow interferences to "cool" (or decay away) so that other features of the gamma spectrum can be accessed. For trace and ultratrace multi-element analyses, the irradiation may be the shortest process; measurement, or counting time is next, while holding time may be the lengthiest process. Irradiation time is not usually longer than the half-life of the daughter product measured. Measured activity at shorter irradiation intervals is approximately linear with counting time, but increases more slowly at intervals longer than the halflife.

For all but the lightest elements the contributed binding energy is in the 6–8 MeV range. The neutron's kinetic energy contributes additional energy, as well.

Large automated facilities use computer controlled sample transport and multiple detectors so that multiple samples can be simultaneously irradiated, held, and counted.

Activation with neutrons with energies in excess of 0.5 electron volts is called *epithermal neutron activation (ENAA)*. Usually it is necessary to encase the sample in a thermal neutron absorbing substance (e.g., cadmium or boron) so that lower energy neutrons are screened out. Epithermal neutrons have much smaller cross sections, in general, but about 20 elements show useful, large resonances (e.g., As, Ba, Br, Cs, Ga, In, Mo, Ni, Rb, Sb, Si, Sm, Sr, Zr). The technique is especially useful for elements, like silicon and zirconium, that are difficult by TNAA.

Even higher energies—in *fast neutron activation analysis (FNAA)*—are used for certain elements (Ho, Ta, W, Th, U). A special category is reserved for the 14-MeV neutrons produced by certain accelerators, like the Cockcroft-Walton machine. These interact with nuclei to produce protons, alpha particles, and sometimes, two neutrons. Because cross sections are small, high neutron fluxes are required. It is possible to determine elements that are impossible by TNAA, however (O, Si, P, Zr, Te, and Pb). For example, phosphorus can be determined by $^{31}P(n,\alpha)^{28}Al$. However, corrections must be made for the competing reactions $^{28}Si(n,p)^{28}Al$ and $^{27}Al(n,\gamma)^{28}Al$.

If the sample can be analyzed without any chemical separations using any of these techniques the methodology is referred to as *instrumental neutron activation analysis (INAA)*. If chemical separations are necessary following irradiation to remove matrix interferences the procedure is termed *radiochemical neutron activation analysis (RNAA)*. Interferences take three main forms: 1) nuclear reactions with matrix species produce the same radionuclide as the one sought; 2) nearby gamma ray spectrum peaks cannot be resolved from the analyte's peak; 3) side reactions produce particles that react to form species that interfere.

Chemical separation after a suitable "cooling" period often involves adding a carrier element and then dissolving the sample plus carrier in acids (or molten salts). After some separation the carrier recovery is used to correct the analyte peak activity. Separations usually take the form of precipitations, solvent extractions, or ion exchange. Performing such classical techniques with "hot" samples is not recommended for the uninitiated. Proper training in safe radiochemical manipulations is a must.

It should be noted here that while, in principle, neutron activation is sometimes considered a definitive technique, in practice it is a comparative technique that requires matrix-matched standards. In the case of INAA it is also often necessary that the shape of the standard match that of the sample. The activating neutron flux is seldom homogeneous enough for an accurate flux value to be utilized. The cross section of the reaction is also not known with great accuracy. The sample also has a neutron *absorption* cross section and self-shielding occurs. These phenomena, combined with major and minor interfering side reactions, make the use of reference materials imperative for the most accurate work.

We now can devote the remainder of this section to a brief description of some of the instrumental specifics of this technique. For much greater detail the reader is referred to the exhaustive coverage in References 55–59.

The nuclear reactor is the neutron source in many NAA facilities. All designs produce neutrons in abundance. On average, 99% of the total flux consists of >0.10 MeV neutrons; 66% fall into the 0.5–3.0 MeV range; and above 3.0 MeV the abundance falls off exponentially. The most probable neutron energy is slightly more than 1 MeV and the flux can range from 4×10^7 to 10^{15} neutrons/cm²/sec. The TRIGA design research reactor can operate in a pulsed mode in which 8×10^{15} neutrons/cm²/sec. are produced in a 40 millisecond pulse.* For thermal neutron activation, the sampled beam must be moderated with water, D_2O, or graphite. The energy spectrum of the resultant neutron beam will also depend on the reactor design and the sample irradiation position in the reactor.

Samples are placed in a high-density polyethylene "rabbit," which is transported by gas pressure through a pneumatic tube system to irradiation and counting locations. High-density polyethylene can be manufactured nearly free of all trace elements.

The *Cockcroft-Walton accelerator* is a frequently used alternative neutron source. It produces 14-MeV neutrons by directing a beam of deuterium ions against a tritium target. A typical flux is 10^{10} neutrons/cm²/sec. *Cyclotrons* can be used to produce 10^{12} neutrons/sec by directing high energy deuterium ions against a beryllium target. A similar process works with *electron accelerators*. There are also several types of isotopic neutron sources. The *photoneutron source* utilizes a strong gamma emitter, like ^{124}Sb or ^{226}Ra, mixed with beryllium to generate neutrons. The *alphaneutron source* uses ^{239}Pu, ^{241}Am, or other alpha source mixed with a light element. And the *spontaneous fission source* uses ^{252}Cf (half-life: 2.6 years) or a sub-critical assembly of 3 mg of ^{252}Cf and 1.5 kg of ^{235}U (to generate 3×10^8 neutrons/cm²/sec).

Some of the radiation detectors utilized in gamma, beta, and alpha measurements are recognizable from our discussion of X-ray fluorescence in Section 3.1. They fall into three general categories: *gas ionization detectors, scintillation detectors*, and *solid-state detectors*.

---

* For comparison a 20-kiloton fission bomb produces 4×10^{24} neutrons in 1 microsecond, and a 1 megaton hydrogen bomb produces 2×10^{27} neutrons in 1 microsecond.

The first category includes all devices that have two electrodes at some potential with a nonconducting gas between them. Either a thin film of Mylar (or beryllium or aluminum) separates the electrode chamber from the ionizing radiation, or the radioactive source is placed inside the chamber. An ionizing particle or photon will produce ion pairs in the gas. These are attracted to the cathode, while the released electrons move to the anode. The external electrical circuit causes electrons to flow to neutralize the positive charge on the cathode, resulting in a negative voltage spike. The ion pairs formed are of two types: primary pairs from the direct reaction described above, and secondary pairs, which are generated by the electrons produced by primary pair formation.

When the applied potential between the two electrodes is low (up to about 80 V) only primary ion pairs will form. If the potential is sufficient to prevent recombination all the primary ion pairs will reach the cathode, and the voltage pulse will be a measure of the total energy dissipated in the chamber. *Ionization chambers* are devices into which the radioactive material to be measured is placed. Such chambers operate in this voltage region. The pulses produced are in the nanovolt range and require high gain amplification.

When the electrode potential is in the 80–680 V region secondary ion pairs form, and the pulse that results from the sum of primary and secondary pairs reaching the cathode falls into the easily measurable 1–100 mV range. Proportional counters, both sealed and with flowing gas,[*] operate in this region. Here, the pulse is still proportional to the energy dissipated, and so the pulse carries accurate information about the energy of each incoming particle or photon. The response in this region is also fast enough so that high count rates can be measured accurately (this is referred to as "low dead time").

Above about 680 V proportionality degenerates. In the region in which Geiger-Müller tubes operate (1000–2100 V) the output pulse is high and uniform with sufficient energy to drive a scalar without amplification, but carrying no information about particle or photon energies. Such devices are cheap and portable, but dead time is high and so high count rates are not accurate. The filler gas may be helium with 2–3% of a hydrocarbon or halogen gas.

Scintillation detectors utilize ion pair production and the associated secondary electrons as well. Here the medium is a fluorescent or phosphorescent[†] material, such as thallium-doped sodium iodide crystals or certain organic

---

[*] Often an argon/methane mixture, such as P10: 90% argon / 10% methane.

[†] *Luminescence* is the general term for the non-thermal emission of electromagnetic radiation. *Fluorescence* occurs when the emission ceases within 10^{-8} sec after the stimulus ceases. *Phosphorescence* occurs when the emission persists longer than 10^{-8} sec

compounds that produce a flash of light whose intensity is proportional to the energy of the impinging particle or photon. As discussed in Section 3.1, the fluor or phosphor is very near a photomultiplier tube, which responds to each generated flash of light, producing a pulse whose height is proportional to the intensity of the flash.

Semiconductor detectors were also discussed in Section 3.1. Large surface area Ge(Li) (lithium-drifted germanium) detectors that must be cooled to liquid nitrogen temperature are the type of solid state detector utilized for most neutron activation measurements. The voltage pulses are produced when a gamma ray photon enters the charge-depleted region between the p and n sides of the device. The signal is in the microvolt range and must be preamplified. Such solid state devices are the detectors of choice for high resolution gamma ray spectrometry. Their resolution is about 40 times better than that of scintillation detectors. However, their efficiency (i.e. their sensitivity) is only 10% of that of scintillation devices.

4.2 Applications

Practical published activation methods for the trace analysis of commercial commodities are not abundant in the general analytical literature. Procedures often appear in specialized journals, such as the *Journal of Radioanalytical Chemistry*, *Nucleonics*, and others, as well as in published proceedings of symposia, and in government agency reports. The following is a brief sampling of papers that have appeared in the more generalized analytical journals. The reader is also especially referred to the bibliographic Reference 62.

In 1956 Morrison and Cosgrove published a thermal neutron activation method for trace impurities in *germanium*. Samples and pure element comparison standards were irradiated for three days at a flux of approximately 3.4×10^{12} neutrons/cm^2/sec. Germanium was then distilled off as the tetrachloride, using a lead-shielded apparatus. This removal of the matrix was necessary to reduce the activity that the activated germanium nuclides contributed to the gamma ray spectrum. Trace levels of arsenic, copper, gallium, sodium, and zinc were determined. It was found that the majority of the error was due to an incomplete distillation [63].

Brooksbank, Leddicotte, and Reynolds described a procedure for trace levels of chlorine, copper, manganese, nickel, silicon, tungsten, and vanadium in *titanium and its alloys*. Irradiation times were based on the half-lives of the analyte nuclides; they ranged from 10 min (V) to 25 hrs (W). Except for vanadium, precipitation separations were used to isolate the activated nuclide: chlorine as AgCl; copper as CuS; manganese as MnS; nickel as nickel dimethylglyoxime; silicon as SiO_2; and tungsten as H_2WO_4 [64].

In 1957 Morrison and Cosgrove extended their approach to the trace analysis of *high purity tungsten*. The tungsten background for counting was reduced by adding molybdenum as a carrier, then precipitating both elements with α-benzoinoxime [65].

Notwithstanding earlier efforts in developing a trace procedure for *high purity silicon*, the earliest comprehensive method appears to be that by Thompson, Strause, and Leboeuf in which 29 trace impurities were measured. The majority of the elements were determined by TNAA, but the combined aluminum and magnesium were determined by fast neutron activation. The measured nuclide in this case is ^{24}Na, which forms by two routes: $^{27}Al(n,\alpha)^{24}Na$ and $^{24}Mg(n,p)^{24}Na$. A detailed separation scheme for the thermal neutron activation elements is provided [66].

MacKintosh and Jervis developed two procedures for low concentrations of *hafnium in reactor grade zirconium metal and zirconium alloys*. In an INAA procedure they sandwiched sample turnings between two polyethylene films for a 20 sec irradiation at 6.5×10^{13} neutrons/cm^2/sec. The activity of ^{179}Hf was then measured. In a RNAA method 30 mg of turnings were irradiated at 1.2×10^{13} neutrons/cm^2/sec for 8 days. Then 10 mg of hafnium wire was added and the test portion was dissolved in hydrofluoric acid. Sulfuric acid was added and the solution was heated until free of hydrofluoric acid. Ammonia was used to precipitate hafnium and zirconium; the precipitate was dissolved in hydrochloric acid and taken to dryness. The residue was dissolved in 2M perchloric acid. The solution was separated using an ion exchange scheme, and then hafnium was precipitated as the barium fluorohafniate. The activities of ^{175}Hf and ^{181}Hf were measured. The results obtained by the two approaches agreed within acceptable limits [67].

Aluminum is a good matrix for activation analysis because its principal radioisotope, ^{28}Al, has a half-life of only 2.24 min and is usually gone by measurement time. The only remaining high concentration interferent is usually ^{24}Na, which forms by the reaction $^{27}Al(n,\alpha)^{24}Na$. Brooksbank Jr., Leddicotte, and Dean developed a comprehensive procedure for aluminum-base alloys. The authors determined minor amounts of copper, iron, manganese, nickel, titanium, and zinc by RNAA, and trace levels of antimony, cobalt, silver, and zirconium by INAA [68].

Thompson determined eleven impurities in 80% nickel/20% iron alloys deposited as thin films on glass after a 24-hr irradiation with 2×10^{13} neutrons/cm^2/sec. The films were dissolved off the highly radioactive glass behind a lead shield using 1:1 nitric acid:water. The analytes (Cd, Cr, Cu, Fe, Mo, P, Sn, Ta, W, and Zn) were each separately precipitated and nickel was electroplated [69].

Benson, Holland and Smith determined trace amounts of *iron and uranium in lead foil* by an RNAA method, using irradiation at 1.3 to 3.9×10^{14} neutrons/cm^2/sec for 9 to 19 days. Uranium was determined by measuring the nuclide ^{239}Np, which resulted from the decay of ^{239}U. For iron the ^{55}Fe and ^{59}Fe activities were simultaneously measured [70].

E. E. Wicker describes activation analysis as applied to gases in metals in considerable detail in the book, *Determination of Gaseous Elements in Metals* (edited by Melnick, Lewis, and Holt, John Wiley, NY, 1974). In particular, he describes the use of a Cockcroft-Walton machine for the fast neutron activation analysis of *oxygen* by the $^{16}O(n,p)^{16}N$ reaction. Solid *steel* specimens of 40–50 g weights are irradiated for 15–20 sec at a flux of more than 10^{11} (14-MeV) neutrons/sec, then counted for 20–25 sec. The nuclide ^{16}N has a 7.14 sec half-life and so activity measurement must be quite prompt. The two gamma ray peaks emitted (6.14 and 7.11 MeV) are higher energy than most others, and, therefore, easily distinguished. Fluorine can also produce ^{16}N [by $^{19}F(n,\alpha)^{16}N$], but it is unlikely to be present at significant levels. Very high levels of boron may also interfere [71].

Paul published a review of *hydrogen* determination by prompt gamma ray activation analysis. TNAA is compared to the "cold neutron" (<0.005 eV) activation available at NIST since 1990. The cold neutron flux is 1.5×10^8 neutrons/cm^2/sec. The approach has been applied to *6Al/4V titanium alloy* [72].

Duffey, Balogna, Wiggins, and Elkady demonstrated the usefulness of a ^{252}Cf neutron source and prompt gamma ray measurement for the analysis of commercial *copper ores and related materials.* Low levels of calcium, iron, silicon, and sodium show a significant response. Eight other elements (Au, Co, Cr, Hg, Mn, Ni, Ti, and V) show potential with the technique. The approach is suggested for *in situ* monitoring in exploring copper deposits and for monitoring ore processing operations [73].

Gerard and Pietruszewski reported on the use of a ^{252}Cf subcritical assembly (the first devoted to industrial research) at Eastman Kodak in Rochester, New York. The work involved the determination of halogens in a variety of organic and inorganic materials. Lower concentration limits for the procedure are 100 ppm for chlorine, 20 ppm for bromine and 10 ppm for iodine [74].

Foster and Gaitanis determined trace levels of *phosphorus in aluminum* using an RNAA approach. Several classical precipitation separations were utilized—sulfide, magnesium ammonium phosphate, and phosphomolybdate [75]. MacKintosh determined thirteen impurity elements in *zone-refined aluminum* using RNAA (Co, Cr, Cu, Dy, Fe, Hf, Ga, La, Mn, Sc, Sm, W, and Zn) [76].

Parry *et al.* developed a sensitive RNAA method for *chlorine in stainless steels and high temperature alloys* that are used for reactor components [77]. Wildhagen and Krivan published an RNAA method for determining 26 elements in *high purity titanium.* The post-irradiation separations were accomplished by ion exchange and solvent extraction [78].

5. ELECTROCHEMICAL METHODS

In the 1950s and 1960s electrochemical methods for trace analysis were state-of-the-art, and many metals analysis laboratories adopted the technology to help them measure tramp level impurities for alloy specification requirements. Today, it would be unusual to find a polarograph in regular routine use, although some labs may retain the equipment and the knowledge in the art of its application. Its use today would probably be to verify an important result, such as a certification value, which was obtained by a faster, more "modern" approach (e.g., graphite furnace AA or ICP-MS). Confirmation by an electrochemical approach would lend great confidence to such a result since it represents a truly independent physical measurement.

5.1 Principles and Instrumentation [79–85]

There are numerous electroanalytical methodologies. Some of these are suited for the determination of major quantities of analyte, but the majority are trace and ultratrace techniques, or, at the very least, they excel with minor concentrations.

Voltammetry is a large, important category that includes *polarography*. The general term refers to any method in which the magnitude of the current that passes between two immersed electrodes is plotted against voltage. Polarography is usually reserved for voltammetry using a *dropping mercury electrode (DME)*. Heyrovsky developed the basic apparatus and much of the theoretical basis of polarography in the early 1920s.

As a working electrode, mercury has a useful range, versus the *saturated calomel electrode (SCE)*, of +0.25 to −1.8 V in acidic media and +0.25 to −2.3 V in alkaline media. In most classical dc polarography the DME emits a drop of mercury every 2–6 sec. These drops have masses at detachment that range between 6 and 20 mg. The potential between the DME and the SCE is varied uniformly and the microampere currents that flow are accurately measured.

A *residual current* of about 0.3 μA constitutes a "background" slope upon which the electrochemical redox signal is superimposed. The total polarographic wave is defined by the *limiting current*, which is due mostly to the rate of diffusion of analyte ion to the electrode surface and, thus, is often called the *diffusion current*. In fact, the limiting current is the sum of the diffusion current and several other terms, including a capacitive charging current associated with an *electrical double-layer* of ions (in solution) and electrons (in the mercury) that forms at the solvent/electrode interface. This term, together with the reduction of impurities, accounts for the residual current. Another term is the migration current, which is only significant if the analyte carries a significant amount of the total current. Usually the supporting electrolyte is sufficiently concentrated so that the *transport number* of the analyte is close to zero. There are several other minor terms, as well.

The diffusion current (i_d) is the only part of the limiting current that relates to analyte concentration. The relationship is defined by the Ilkovic equation:

$$i_d = 607nD^{1/2}m^{2/3}t^{1/6}C$$

where n = the number of electrons transferred in the electrode reaction; D = the diffusion coefficient of the analyte, in cm^2/sec; m = the rate of mercury flow through the DME, in mg/sec; t = the drop time, in sec; and C = the concentration of analyte, in mmoles/L. In the above equation i_d is a mean value for the diffusion current and its units are in microamperes.

Unfortunately, as with most of the theoretical treatments we have seen, there are experimental parameters that are beyond exact control (in this case ensuring an undisturbed, diffusion-limited condition), and theoretical parameters that lack certainty. Thus, all polarography is comparative. A calibration curve or the method of additions are generally employed.

The characteristic dc polarogram is a step function with each plateau corresponding to the concentration of an electroactive species. Sometimes, however, maxima occur in place of level plateaus. This phenomenon is due to reducible species streaming tangentally around the mercury droplet. This interfacial effect can often be eliminated by adding a minimum amount of a nonionic surfactant, such as Triton X-100™.* Too much and the additive will produce a distorted analyte response. Dissolved atmospheric oxygen produces two waves that constitute an important interference. At −0.05 V (versus the SCE) oxygen is reduced to hydrogen peroxide; at −0.9 V (versus the SCE) oxygen is reduced to water. These must be removed by purging the solution (usually with nitrogen) prior to inserting the DME capillary. During measurement a gentle flowing sheath of nitrogen is maintained over the solution surface. Polarographic cells have been designed to allow these purges.

There are some additional problems besides maxima and oxygen waves. The growth of a mercury droplet produces a steadily increasing current. And there is the matter of the residual current, due in large part to charging the boundary layer, as described above. Filter circuits and bias currents have been applied with success to compensate for these phenomena. The largest problem, however, is probably the presence of matrix ions that are more easily electrolyzed than the analyte ion. There are two solutions to this dilemma: the

* Often used, but less effective additives are gelatin and methyl red.

analyte can be completely or partially isolated by chemical separation, or the interfering ion can be converted into a complex ion that is more difficult to electrolyze.

Since impurities in the electrolyte may result in an excessive blank some analysts recommend purifying the electrolyte by electrolysis at a potential that is more negative than the half-wave potential of the analyte. The *half-wave potential* is the voltage at which the measured current of a polarographic wave is exactly 50% of the diffusion current. Its value can be used to qualitatively identify a solution component, if the system is free of interfering complexation.

It may be useful here to recount the basic dc polarographic process before proceeding to later modifications and elaborations. The mercury drop-let, which forms by gravity at the end of a glass capillary, is the polarized indicating (or working) electrode. The saturated calomel electrode is the nonpolarized reference electrode. This means that the mercury assumes any potential imposed by the circuitry while the SCE potential remains unchanged. In the case of a metal cation analyte, reduction occurs at the mercury and frequently the reduced metal atom is amalgamated with the mercury. Diffusion of the cations to the mercury drop is the controlling process in the intentionally undisturbed solution.

As a mercury drop begins to form the capacitive charging of the electrical double-layer produces the gently sloping (V vs. A) baseline. As analyte cations begin to be reduced the large Faradaic current begins to be displayed. Increasing potential causes an increased rate of analyte reduction and increased current in the circuit. In time the solution in the immediate vicinity of the mercury becomes depleted of analyte, and analyte cations are reduced the instant that they arrive at the interface. The limiting current then levels off to a plateau, and the system is said to be diffusion limited.

Great pains are taken to ensure that the solution is not stirred or agitated. The vessel is frequently placed in a thermostated bath to prevent thermal convection currents, for example. A refined and reproducible form of stirring *does* occur, however. Each time a mercury droplet falls it creates a "wake" environment for the new droplet that replaces it. This environment is very similar for each droplet (except, of course, the first) and contributes to the steady state diffusion limited equilibrium process.

Because the number of analyte ions reduced and the measured currents are so small polarographic scans can be repeated many times without a decrease

---

Mercury has a high *overvoltage* with respect to hydrogen. This means that most metals are reduced at potentials higher than that at which hydrogen evolution occurs. In contrast platinum evolves hydrogen at only –0.45 V.

in response. Thus, the technique is effectively nondestructive. The diffusion current is approximately proportional to the number of electrons exchanged in the reduction.

Some reductions are fully reversible and the resultant polarographic trace is a vertical "step" wave. Other reductions are irreversible and, typically, the trace is somewhat slanted. This is because for lower potentials the electron transfer mechanism is slower than the diffusion processes. At higher potentials a plateau is finally attained. Both reversible and irreversible electrode processes produce analytically valid polarograms, of course.

For dc polarography a quantification limit of about 5×10^{-5} moles per liter is common. In this concentration region the diffusion current is approximately equal to the residual current. *Normal pulse polarography* is about ten times more sensitive. In this technique the potential applied to the DME is fixed at one value for most of the "life" of the mercury droplet. It is then stepped up to a new value for the final 60 milliseconds before detachment. Current, however, is only measured during the final 17 milliseconds of the droplet's lifetime—a time when its area and electrical characteristics are most stable. Such instruments generally employ a mechanical device for detaching the drop at an exact and reproducible interval.

Differential pulse polarography is 10–100 times as sensitive as normal pulse polarography. In this methodology voltage spikes of a fixed size, separated by a precise time interval, are superimposed on the linearly increasing potential applied to the DME. The timing of the pulses and the action of the mechanical drop separator are arranged so that each drop experiences one pulse during the final stages of its life. The instrument circuitry measures the current just before the application of the pulse, and again during the final milliseconds of the pulse. The difference between the two current readings is plotted as a function of applied voltage. In addition to greatly increased sensitivity the technique also shows improved resolution over conventional dc polarography.

Other modifications of the basic polarographic system have been tried. *Cathode ray polarography*' employs a rapid dc voltage scan while the mercury droplet is most stable, near the end of its growth. The droplets are detached mechanically every 7 sec. A complete polarogram for every droplet that forms is read out as a series of peaks on an image-persistent oscilloscope. The peak-shape response is due to the fact that a steady state diffusion current is not achieved because of the speed of the voltage scan. In this case a *summit potential*, which is proportional to the half-wave potential, is being measured. Sensitivity is somewhat improved over conventional dc polarogra-

---

More generally known as *linear sweep polarography*.

phy, but the resolution of neighboring waves is somewhat poorer. Analytes that are slowly reduced may be a problem, due to the rapid rate of the voltage scan.

Cyclic voltammetry is similar, except that a sawtooth waveform is applied. Here, the voltage is reversed after the analyte reduction potential is exceeded. The analyte then is re-oxidized, producing a closed figure on the oscilloscope trace. This technique is used to study reaction mechanisms. For example, it reveals the extent of reversibility of the electrode reaction. Frequently, other types of indicator electrodes—glassy carbon, pyrolytic carbon, or carbon paste—are utilized; and so now we are nominally no longer in the realm of polarography.

Alternating current polarography is a technique in which a 10 – 100 mV ac component is applied to the linear dc ramp voltage imposed on the DME. Only the ac current that flows is measured. This screens out much of the capacitive charging current encountered in conventional dc polarography and results in improved sensitivity and resolution. An important variant employs a square wave ac signal and is known, predictably, as *square wave polarography*.

The technique with at least the potential for the highest sensitivity is *anodic stripping voltammetry*. This often employs a hanging mercury drop electrode maintained for an extended period at a potential somewhat in excess of the half-wave potential of the analyte. The analyte accumulates in the mercury droplet as an amalgam. After a sufficient interval the potential is reversed and the accumulated analyte is re-oxidized rapidly and the resultant current is measured. A practical estimate of the quantification limit for this technique is approximately 10^{-9} moles/liter. The limitations to this approach include the penetration of the analyte species into the hanging drop capillary where it cannot be re-oxidized. Deposition times longer than 30 min are impractical because of droplet instability. Also, adsorption of the analyte onto the components and vessel walls are an expected problem.

There are many other electrochemical trace techniques—stirred systems, rotating and vibrating electrodes in voltammetric approaches, for example. Also, coulometry—an extremely accurate, precise, and definitive electrochemical method—can sometimes be successfully extended to the trace realm. It will not repay us very much to proceed any further, however.

5.2 Applications

Most of the literature on the application of electrochemical techniques to the determination of trace constituents in metals is quite old. Nevertheless, many of the techniques are still relevant as cross checks on more modern approaches. Because of the limited interest, however, we will be brief.

Applications to ferrous metals were abundant in the literature of the 1950s and '60s. Ferrett and Milner, for example, illustrated the use of the square wave polarograph by determining 0.002% *copper* and 0.0005% *tin* in a *ferrous alloy*, directly, without removal of the iron matrix. The solution they utilized contained a concentration of iron (III) ion 5000 times in excess of the copper and 20,000 times in excess of the tin. This determination is not possible with conventional dc polarography [85,86].

R.C. Rooney determined trace amounts of *lead and bismuth in cast iron* by cathode ray polarography. Removal of the matrix iron is accomplished by extraction with isobutyl acetate from concentrated hydrochloric acid solution. Lead and bismuth were then isolated as the diethyldithiocarbamate complexes by extraction with chloroform from an ammoniacal tartrate/cyanide medium. The solvent phase was evaporated to dryness and the organic residue was destroyed with nitric and perchloric acids. The excess perchloric acid was fumed off and the cooled salts were dissolved in nitric acid and sodium tartrate solution. The solution was then de-oxygenated for 20 min, then measured on a cathode ray polarograph. It was possible to measure lead down to 1 ppm and bismuth down to 0.5 ppm without special precautions. Below these levels it was necessary to specially purify the reagents used [87].

The same author published a method for the determination of trace levels of *aluminum in cast iron*. Most of the matrix components were removed from a dissolved 1 g test portion by chloroform extraction of the diethyldithiocarbamate complexes. Cupferron was added to the aqueous phase and the aluminum cupferrate was extracted with chloroform. The chloroform extract was evaporated to dryness, and then nitric acid and perchloric acid were used to destroy the residual organics. Any chromium color was removed by the cautious dropwise addition of hydrochloric acid to the fuming solution. The solution was heated to dryness and the residue dissolved in dilute perchloric acid. Sodium acetate and Solochrome Violet RS were added and the solution was diluted to a volume in a volumetric flask. The flask was warmed for 5 min, cooled, and cathode ray polarography was used to measure the acid-soluble aluminum. Acid insoluble aluminum was determined on a separate 10 g test portion by filtering off the acid insolubles, igniting the residue at 700–800°C, volatilizing silica with hydrofluoric and sulfuric acids, then fusing the residue with sodium carbonate. The cooled melt was leached with acidified water and evaporated to dryness. The salts were dissolved in hydrochloric acid; acetic acid and sodium acetate were added, and then the solution was extracted with diethyldithiocarbamate and with cupferron, as for the acid-soluble portion [88].

P.H. Scholes determined trace *copper* and alloying levels of lead in *steels* using a cathode ray polarographic approach. The (approx.) 1 g test portion was dissolved in hydrochloric acid, then oxidized with a saturated solution of sodium chlorate. To the cooled solution hydrazine hydrochloride, sodium formate, and a starch solution were added and the solution was reheated but not boiled. The cooled solution was diluted to 20 mL in a calibrated flask and measured polarographically after a nitrogen sparge. Part-per-million levels of copper could be measured if the iron matrix was removed by extraction with isobutyl acetate from 10:1.5 hydrochloric acid:nitric acid [89].

The same author determined *tin in mild and low alloy steel* by first isolating it from the matrix by coprecipitation with manganese dioxide. Cathode ray polarographic measurement was performed in an electrolyte containing peptone˙ and ascorbic acid in a 5M hydrochloric acid medium [90].

Kallmann, Liu, and Oberthin used a steam distillation from a hydrobromic acid medium to isolate *tin* from a variety of dissolved matrices, including *open hearth steel, Cr-W-V steel, ferroniobium, ferroniobium/tantalum, tantalum and tungsten ores*, as well as *brasses, bronzes, lead-base bearing metal, titanium alloy, and zircalloy*. Conventional dc polarography was employed [91].

Higher, alloying levels of tin in zirconium alloys were determined by a direct method by J.T. Porter. Samples were either dissolved under nitrogen or reduced with iron powder. Peptone was added and the analyte was measured in a fixed volume of de-aerated water by dc polarography [92].

L. Meites described the determination of traces of *nickel and zinc in copper metal*. The sample was dissolved in dilute nitric acid; ammonium hydroxide, ammonium chloride, and hydrazine hydrochloride were added and the solution was allowed to stand for several min. A controlled mercury cathode electrolysis was then used to remove nearly all the copper. Triton X-100™ was added to an aliquot as a maximum suppressor and the de-aerated solution was measured by dc polarography [93].

Traces of *sodium in high purity aluminum* have been determined. The test portion was dissolved in hydrochloric acid and the aluminum matrix was precipitated with ammonia gas. The filtrate was evaporated to dryness and dissolved with tetramethylammonium hydroxide, then measured polarographically [94].

Athavale, *et al.* determined *antimony in refined lead*. The sample was dissolved in nitric acid, lead was precipitated as the sulfate, and an aliquot of the supernatant solution evaporated, then dissolved with dilute sulfuric acid and measured [95].

Jennings determined *bismuth, copper, indium, and cadmium in gallium arsenide* at ultratrace levels using square wave polarography. The test portion was dissolved in nitric and hydrochloric acids, evaporated to dryness, and the

---

˙A maximum suppressor.

residue dissolved in dilute hydrochloric acid and potassium bromate [to oxidize As (III) to As (V)]. The analytes were then determined directly in the undiluted solution by square wave polarography [96].

6. OTHER METHODOLOGIES

There remain a few more techniques of generally minor importance, or of unrealized potential, to briefly mention. It continues as our avowed purpose here to focus on the practical. This necessitates shunning the many academic experiments that are unproven for commodities analysis. Our selections are a judgement call, admittedly, and it is very likely that something inchoate and important will be missed. For this we offer apologies and the reader's indulgence. But at least we promise to be brief.

At the outset here it is important to point out that there are a great many very important and useful methodologies that can provide bulk compositional data about trace constituents in alloys, but which have quite a different main purpose. Examples include techniques designed for surface studies (Auger spectroscopy, ion scattering spectroscopy); techniques designed for microscale spatial resolution (electron and ion microprobes); and techniques designed to provide valence state (Mossbauer) or physical property (internal friction) data. We must consider these methodologies outside the scope of this volume since gross compositional analysis is not their *raison d'etre*.

In Chapters 4 and 5 we touched upon the subjects of *molecular* [97] *and atomic fluorescence* [98]. Neither phenomena has been widely utilized in trace metal analysis. Instrument manufacturers have not strongly promoted these technologies, probably because the range of application is narrow. Matrix effects take the form of *quenching* (any collisional process or chemical complexation that lowers the quantum yield of the fluorescence) and *light scatter*. Under ideal conditions the sensitivity of these techniques is quite high, however.

Molecular fluorescence usually involves forming a metal complex that exhibits the phenomenon; however, many rare earths fluoresce as free ions in solution (native fluorescence). Atomic fluorescence requires better optics and electronics than atomic absorption (with their associated cost). The spectrum is, however, considerably simpler than an emission spectrum. One manufacturer marketed for a time an ICP-based system that utilized hollow cathode lamps as the excitation source.

One group of investigators utilized a glow discharge source to determine ppm levels of magnesium and copper in steel, copper in aluminum, and copper and silver in gold by atomic fluorescence spectroscopy [99]. Laser excitation has been utilized in combination with electrothermal atomization for mercury in silver metal [100]. This technique is known by the acronym, LEAFS. The same approach has been used for tin in nickel-base alloys [101]. Tin in steel was also determined by atomic fluorescence, using a flow injection hydride approach [102].

Many of the papers on molecular fluorescence concern the determination of the rare earths [103], which vary in sensitivity in native form. Gadolinium and neodymium are among the least sensitive, while samarium, europium, and praseodymium are among the most sensitive [97]. The cerium complex with Rhodamine 6G has been utilized to enhance its response [104].

There was a period when *gas chromatography* was being actively pursued as a means of inorganic analysis involving the separation and measurement of volatile metal compounds. Some of the detection limits that were achieved were impressive and rival some of the best results from more recent approaches. One still occasionally sees new work in this area, which may one day again become a major field.

The β-diketone derivatives were a dominant research interest because of their thermal stability. Morie and Sweet separated and measured the aluminum, gallium, and indium derivatives of trifluoroacetylacetone (extracted in benzene) on a silanized column of glass microbeads coated with 0.5% DC-550 silicone oil [105]. They later showed similar success with aluminum and iron [106].

Scribner, Treat, Weis, and Moshier used a chloroform/trifluoroacetylacetone extraction for copper, iron, and aluminum [107]. Ross and Sievers determined ultratrace levels of beryllium using a benzene/ trifluoroacetylacetone extraction and an electron capture detector [108]. The same authors determined chromium in steels by a similar approach [109]. Genty, Houin, Malherbe, and Schott applied the methodology to the determination of aluminum and chromium in uranium [110]. Many halogenated derivatives of acetylacetone (2,4-pentanedione) were tried, as well as other chelating and non-chelating complex formers. Several books described some of these studies [111–113].

Liquid chromatography (LC), or as it is sometimes known, *high performance liquid chromatography (HPLC)*, as a distinct form of chromatography dates from the latter part of the 1960s. High pressure pumps carry the mobile phase (an out-gassed high purity solvent) over one of several types of stationary phase, then past a UV/visible, refractive index, electrochemical, or other type of detector. *Normal phase chromatography (NPC)* utilizes a polar staionary phase and a nonpolar mobile phase. *Reversed phase chromatography (RPC)* uses a nonpolar stationary phase and a polar mobile phase [114].

Reversed phase HPLC has been employed to measure nickel, copper, and cobalt complexes with diethyldithiocarbamic acid, following ether extraction. The method has been applied to alloys. Detection limits reported were 50 ppb nickel, 30 ppb copper, and 5 ppb cobalt [115]. Vanadium was determined by RP-HPLC in steel using a ternary complex with citrate, 4-(2-pyridylazo)resorcinol (PAR), and vanadium (V) [116].

Ion chromatography (IC) is a related technique that utilizes an aqueous-based electrolyte mobile phase and an ion exchange resin. The system can be configured to separate either anions or cations. The resin used has high exchange efficiency, but low site capacity. Sometimes this takes the form of inert beads with a thin surface layer of active exchange sites.* For conductivity detection the high conductivity of the pumped mobile phase must be suppressed. This can be accomplished with a suppressor column or fiber or membrane suppressor systems that convert the eluent to a species of low conductivity. In some systems the conductivity of the mobile phase is suppressed electronically. Other detector types include those based on amperometry and spectrophotometry. ICP-OES has also been applied as a detection system with good results [114,117].

Yamane illustrated the utility of the technique with methods for trace levels of manganese and cobalt in high purity aluminum, and for boron, nitrogen, and carbon in iron and steel [118]. Louw applied ion chromatography to the determination of transition metals in irradiated low alloy steels. The analytes were complexed with PAR and measured with a UV/visible spectrophotometric detector [119]. Seubert, Krabichler, Krismer, and Wilhartitz determined trace impurities in high purity molybdenum and tungsten metal (dissolved in hydrogen peroxide) by ion chromatography coupled to both ICP-OES and ICP-MS detection [120]. The IC-ICP-OES procedure was further elaborated using a simultaneous instrument and ultrasonic nebulization. Twenty-two impurity elements were determined at ultratrace levels [121].

Many other methodologies have been applied to the trace determination of selected analytes. *Capillary zone electrophoresis (CZE)* is one promising area. Colburn, Hinton, Sepaniak, and Starnes demonstrated its utility for the rare earths [122]. *Laser induced photoacoustic spectroscopy (LIPAS)* has been applied to the trace determination of transuranic elements (e.g., neptunium) in solutions from the PUREX uranium/plutonium recovery process [123]. Information on additional technologies can be found in general texts dealing with trace analysis [124–126].

REFERENCES

1. Fricioni, R.B.; Essig, L. "High Temperature Combustion" in Metals Handbook, 9th ed., Vol.10: Materials Characterization, American Society for Metals, Metals Park, OH, 1986, pp. 221-225.

---

*Termed a *pellicular* resin.

2. Annual Book of ASTM Standards, Vol.03.06, American Society for Testing and Materials, West Conshohocken, PA, 1998, designation E1019.
3. Dulski, T.R. A Manual for the Chemical Analysis of Metals, MNL25, American Society for Testing and Materials, West Conshohocken, PA, 1996, pp. 177–178.
4. "Steel—Determination of Low Carbon Contents—Part 4: Coulometric Method After Combustion" ISO/TR4830-4:1978, International Organization for Standardization (ISO) (1978).
5. Janicsek, L. Steelmaking Conference Proceedings, Vol.73, Iron and Steel Society, Warrendale, PA, 1990, pp. 111–113.
6. Butin, G.; Hoffert, F.; Ravaine , D.; Gandar, J.L. Progress of Analytical Chemistry in the Iron and Steel Industry, Commission of the European Communities, EUR14113, 1992, pp. 276–290.
7. Janicsek, L. Progress of Analytical Chemistry in the Iron and Steel Industry, Commission of the European Communities, EUR14113, 1992, pp. 494–498.
8. Snell, F.D.; Snell, C.T.; Snell, C.A. Colorimetric Methods of Analysis, Vol.IIA, D. Van Nostrand, Princeton, NJ, 1959.
9. Fricioni, R.B.; McMahon, M.A. Pittsburgh Conference on Analytical Chemistry and Applied Spectroscopy, March, 1970.
10. Kriege, O.H.; Wolfe, A.L. Talanta, 9, 673 (1962).
11. Theodore, M.L.; Colling, D.A.; Aspden, R.G. Transactions of the Metallurgical Society of AIME, 245, 1823 (1969).
12. Watson, A.; Grallath, E.; Kaiser, G.; Tölg, G. Analytica Chimica Acta, 100, 413 (1978).
13. Kurusu, K.; Yamamoto, T. Analytica Chimica Acta, 244, 59 (1991).
14. Annual Book of ASTM Standards, Vol.03.06, American Society for Testing and Materials, West Conshohocken, PA, 1997, designation E1587.
15. Fricioni, R.B.; Essig, L. "Inert Gas Fusion" in Metals Handbook, 9th ed., Vol.10: Materials Characterization, American Society for Metals, Metals Park, OH, 1986, pp. 226–232.
16. Martin, J.F.; Melnick, L.M. "Vacuum and Inert-Gas Fusion" in Determination of Gaseous Elements in Metals (L.M. Melnick, L.L. Lewis, B.D. Holt, eds.)John Wiley, New York, 1974, pp. 113–219.
17. Dallman, W.E.; Fassel, V.A. Analytical Chemistry, 39, 133R (1967).
18. Van Nostrand's Scientific Encyclopedia, 4th ed., D.Van Nostrand, Princeton, NJ, 1968.
19. Takahari, T.; Abiko, K.; Okochi, H.; Furuya, K. Progress of Analytical Chemistry in the Iron and Steel Industry, Commission of the European Communities, EUR14113, 1992, pp. 355–361.
20. Gruner, W.; John, A. Progress of Analytical Chemistry in the Iron and Steel Industry, Commission of the European Communities, EUR14113, 1992, pp. 557–560.
21. Martin, J.F.; Melnick, L.M. "Hydrogen" in Determination of Gaseous Elements in Metals (L.M. Melnick, L.L. Lewis, B.D. Holt, eds.) John Wiley, New York, 1974, pp. 289–320.
22. Lewis, L.L. "Behavior of Gaseous Elements in Metals" in Determination of Gaseous Elements in Metals(L.M. Melnick, L.L. Lewis, B.D. Holt, eds.) John Wiley, New York, 1974, pp. 1–52.

23. Saeki, M.; Ono, A.; Hayakawa, Y.; Senoo, K. Progress of Analytical Chemistry in the Iron and Steel Industry, Commission of the European Communities, EUR14113, 1992, pp. 247-253.
24. Betka, G.; Lindsay, J.; Cunningham, A. Progress of Analytical Chemistry in the Iron and Steel Industry, Commission of the European Communities, EUR14113, 1992, pp. 254–261.
24A. Nishifuji, M.; Ono, A.; Chiba, K. Analytical Chemistry, 68, 3300 (1996).
25. Pella, P.A. "X-ray Spectrometry" in Instrumental Analysis (G.D. Christian and J.E. O'Reilly, eds.) Allyn and Bacon, Boston, 1986, pp. 412–450.
26. Leyden, D.E. "X-ray Spectrometry" in Metals Handbook, 9th ed., Vol.10: Materials Characterization, American Society for Metals, Metals Park, OH, 1986, pp. 82–101.
27. Cahill, T.A. "Particle-Induced X-ray Emission" in Metals Handbook, 9th ed., Vol.10: Materials Characterization, American Society for Metals, Metals Park, OH, 1986, pp. 102–108.
28. Jenkins, R. Analytical Chemistry, 56, 1099A (1984).
29. Jenkins, R.; Gould, R.W.; Gedcke, D. Quantitative X-ray Spectrometry, Marcel Dekker, New York, 1981.
30. Lachance, G.R. Spectrochimica Acta, 48B, 343 (1993).
31. Yokhin, B.; Tisdale, R.C. American Laboratory, July, 24C (1993).
32. Bertin, E.P. Principles and Practice of X-ray Spectrometric Analysis, 2nd ed., Plenum, New York, 1975, pp. 678–697.
33. Wobrauschek, P.; Aiginger, H. Analytical Chemistry, 47, 852 (1975).
34. Klockenkämper, R.; von Bohlen, A. Journal of Analytical Atomic Spectrometry, 7, 273 (1992).
35. Klockenkämper, R.; Knoth, J.; Prange, A.; Schwenke, H. Analytical Chemistry, 64, 1115A (1992).
36. Bertin, E.P. Principles and Practice of X-ray Spectrometric Analysis, 2nd ed., Plenum, New York, 1975, pp. 788–807.
37. Müller, R.O. Spectrochemical Analysis by X-ray Fluorescence (trans. by K. Keil) Plenum, New York, 1972, pp. 303–307.
38. Pfeiffer, H.G.; Zemany, P.D. Nature, 174, 397 (1954).
39. Johnson, J.L.; Nagel, B.E. Mikrochimica Acta, 3, 525 (1963).
40. Rudolph, J.S.; Nadalin, R.J. Analytical Chemistry, 36, 1815 (1964).
41. Luke, C.L. Analytica Chimica Acta, 41, 237 (1968).
41A. Albright, C.H.; Burke, K.E.; Yanak, M.M. Talanta, 16, 309 (1969).
41B. Burke, K.E.; Yanak, M.M.; Albright, C.H. Analytical Chemistry, 39, 14 (1967).
42. McKaveney, J.P.; Baldwin, H.E.; Vassilaros, G.L. Journal of Metals, 54 (1968).
43. Vassilaros, G.L. Talanta, 18, 1057 (1971).
44. Vassilaros, G.L.; Byrnes, C.J. Talanta, 23, 225 (1976).
45. Ricci, E. Analytical Chemistry, 52, 1708 (1980).
46. Vassilaros, G.L.; Byrnes, C.J., private communications.
47. Luke, C.L. Analytical Chemistry, 36, 318 (1964).
48. Campbell, W.J.; Spano, E.F.; Green, T.E. Analytical Chemistry, 38, 987 (1966).
49. Kashuba, A.T.; Hines, C.R. Analytical Chemistry, 43, 1758 (1971).

50. Knote, H.; Krivan, V. Analytical Chemistry, 54, 1858 (1982).
51. Chen, J.S.; Berndt, H.; Klockenkämper, R.; Tölg, G. Fresenius' Journal of Analytical Chemistry, 338, 891 (1990).
52. Hegedus, F.; Winkler, P. Advances in X-ray Analysis, Vol.34, Plenum, New York, 1991, pp. 239–241.
52A. Bredendiek-Kamper, S.; Von Bohlen, A.; Klockenkämper, R.; Quentmeier, A.; Klockow, D. Journal of Analaytical Atomic Spectrometry, 11, 537 (1996).
53. Burba, P.; Willmer, P-G; Becker, M.; Klockenkämper, R. Spectrochimica Acta, 44B, 525 (1989).
54. Klockenkämper, R. Total-reflection X-ray Fluorescence, John Wiley, New York, 1997.
55. Bunker, M.E.; Minor, M.M.; Garcia, S.R. "Neutron Activation Analysis" in Metals Handbook, 9th ed., Vol.10: Materials Characterization, American Society for Metals, Metals Park, OH, 1986, pp. 233–242.
56. Matlack, G.M. "Radioanalysis" in Metals Handbook, 9th ed., Vol.10: Materials Characterization, American Society for Metals, Metals Park, OH, 1986, pp. 243–250.
57. Ehmann, W.D.; Janghorbani, M. "Radiochemical Methods of Analysis" in Instrumental Analysis (G.D. Christian and J.E. O'Reilly, eds.) Allyn and Bacon, Boston, 1986, pp. 594–636.
58. Bowen, H.J.M.; Gibbons, D. Radioactivation Analysis, Oxford University Press, London, 1963.
59. Lyon Jr.,W.S., ed. Guide to Activation Analysis, D. Van Nostrand, Princeton, NJ, 1964.
60. Concise Encyclopedia of Atomic Energy (F. Gaynor, ed.) Philosophical Library, New York, 1950.
61. Van Nostrand's Scientific Encyclopedia, 4th ed., D. Van Nostrand, Princeton, NJ, 1968.
62. Lutz, G.J.; Boreni, R.J.; Maddock, R.S.; Wing, J. Activation Analysis: A Bibliography Through 1971, NIST Technical Note 467, U.S. Government Printing Office, Washington, D.C., 1972.
63. Morrison, G.H.; Cosgrove, J.F. Analytical Chemistry, 28, 320 (1956).
64. Brooksbank, Jr., W.A.; Leddicotte, G.W.; Reynolds, S.A. Analytical Chemistry, 28, 1033 (1956).
65. Cosgrove, J.F.; Morrison, G.H. Analytical Chemistry, 29, 1017 (1957).
66. Thompson, B.A.; Strause, B.M.; Leboeuf, M.B. Analytical Chemistry, 30, 1023 (1958).
67. MacKintosh, W.D.; Jervis, R.E. Analytical Chemistry, 30, 1180 (1958).
68. Brooksbank, Jr., W.A.; Leddicotte, G.W.; Dean, J.A. Analytical Chemistry, 30, 1785 (1958).
69. Thompson, B.A. Analytical Chemistry, 31, 1492 (1959).
70. Benson, P.A.; Holland, W.D.; Smith, R.H. Analytical Chemistry, 34, 1113 (1962).
71. Wicker, E.E. "Activation Analysis" in Determination of Gaseous Elements in Metals (L.M. Melnick, L.L. Lewis, B.D. Holt, eds.) John Wiley, New York, 1974, pp. 75–111.

72. Paul, R.L. Analyst, 122, 35R (1997).
73. Duffey, D.; Balogna, J.P.; Wiggins, P.F.; Elkady, A.A. Analytica Chimica Acta, 79, 149 (1975).
74. Gerard, J.T.; Pietruszewski, J.L. Analytical Chemistry, 50, 906 (1978).
75. Foster, L.M.; Gaitanis, C.D. Analytical Chemistry, 27, 1342 (1955).
76. MacKintosh, W.D. Analytical Chemistry, 32, 1272 (1960).
77. Parry, S.J.; Bennett, B.A.; Benzing, R.; Redpath, D.; Harrison, J.; Wood, P.; Brown, F.J. Analytical Chemistry, 69, 3049 (1997).
78. Wildhagen, D.; Krivan, V. Analytical Chemistry, 67, 2842 (1995).
79. Ewing, G.W. Instrumental Methods of Chemical Analysis, 4th ed., McGraw-Hill, New York, 1975, pp. 291-324.
80. Taylor, J.K.; Maienthal, E.J.; Marinenko, G. "Electrochemical Methods" in Trace Analysis—Physical Methods (G.H. Morrison, ed.) Interscience, New York, 1965, pp. 377–433.
81. Crow, D.R. "Voltammetry" in Metals Handbook, 9th ed., Vol.10: Materials Characterization, American Society for Metals, Metals Park, OH, 1986, pp. 188–196.
82. Milner, G.W.C. The Principles and Applications of Polarography and Other Electroanalytical Processes John Wiley, New York, 1957.
83. Heyrovski, J.; Zuman, P. Practical Polarography, Academic Press, New York, 1968.
84. Meites, L. Polarographic Techniques Interscience, New York, 1955.
85. Meites, L. "Some Recently Developed Electroanalytical Techniques for Determination of Traces of Metals." Symposium on Extension of Sensitivity for Determining Various Constituents in Metals (Proceedings), ASTM STP308, American Society for Testing and Materials, West Conshohocken, PA, 1962.
86. Ferrett, D.J.; Milner, G.W.C. Analyst, 81, 193 (1956).
87. Rooney, R.C. Analyst, 83, 83 (1958).
88. Rooney, R.C. Analyst, 83, 546 (1958).
89. Scholes, P.H. Analyst, 86, 116 (1961).
90. Scholes, P.H. Analyst, 86, 392 (1961).
91. Kallmann, S.; Liu, R.; Oberthin, H.; Analytical Chemistry, 30, 485 (1958).
92. Porter II, J.T. Analytical Chemistry, 30, 484 (1958).
93. Meites, L. Analytical Chemistry, 27, 977 (1955).
94. Heyrovsky, J.; Zuman, P. Practical Polarography Academic Press, New York, 1968, pp. 134–135.
95. Athavale, V.T.; Dhaneshwar, R.G.; Mehta, M.M.; Sundaresa, M. Analyst, 86, 399 (1961).
96. Jennings, V.J. Analyst, 87, 548 (1962).
97. McGown, L.B. "Molecular Fluorescence Spectroscopy" in Metals Handbook, 9th ed., Vol.10: Materials Characterization, American Society for Metals, Metals Park, OH, 1986, pp. 72–81.
98. Siemer, D.D. "Atomic Absorption Spectrometry" in Metals Handbook, 9th ed., Vol.10: Materials Characterization, American Society for Metals, Metals Park, OH, 1986, pp. 45–46.
99. Human, G.C.; Ferreira, N.P.; Kruger, R.A.; Butler, L.R.P. Analyst, 103, 469 (1978).

100. Baker, C.L.; Bolshov, M.A.; Smith, B.W.; Winefordner, J.D. Spectroscopy Letters, 29, 1497 (1996).
101. Yang, K.X.; Lonardo, R.F.; Liang, Z.; Yuzefovsky, A.I.; Preli, Jr., F.R.; Hou, X.; Michel, R.G. Journal of Analytical Atomic Spectrometry, 12, 369 (1997).
102. Chen, H.; Yao, W.; Wu, D.; Brindle, I.D. Spectrochimica Acta, 51B, 1829 (1996).
103. Ozawa, L.; Hersh, H.N. "Significant Improvement of Accuracy and Precision in the Determination of Trace Rare Earths by Fluorescence Analysis" in Accuracy in Trace Analysis: Sampling, Sample Handling, and Analysis, NIST Special Publication 422, National Institute of Standards and Technology, Gaithersburg, MD, 1975, pp. 1103–1107.
104. Jie, N.; Huang, X.; Si, Z.; Yang, D.; Yang, J.; Zhang, Q. Mikrochimica Acta, 126, 93 (1997).
105. Morie, G.P.; Sweet, T.R. Analytical Chemistry, 37, 1552 (1965).
106. Morie, G.P.; Sweet, T.R. Analytica Chimica Acta, 34, 314 (1966).
107. Scribner, W.G.; Treat, W.J.; Weis, J.D.; Moshier, R.W. Analytical Chemistry, 37, 1136 (1965).
108. Ross, W.D.; Sievers, R.E. Talanta, 15, 87 (1968).
109. Ross, W.D.; Sievers, R.E. Analytical Chemistry, 41, 1109 (1969).
110. Genty, C.; Houin, C.; Malherbe, P.; Schott, R. Analytical Chemistry, 43, 235 (1971).
111. Moshier, R.W.; Sievers, R.E. Gas Chromatography of Metal Chelates, Pergamon, Oxford, 1965.
112. Guiochon, G.; Pommier, C. Gas Chromatography in Inorganics and Organometallics, Ann Arbor Science, Ann Arbor, MI, 1973.
113. Mushak, P. "The Gas-Liquid Chromatography of Metal Ions via Chelation and Non-Chelation Techniques" in Handbook of Derivatives for Chromatography (K. Blaue and G.S. King, eds.), Heyden, Philadelphia, 1978, pp. 433–456.
114. Kelly, M.J. "Liquid Chromatography" in Metals Handbook, 9th ed., Vol.10: Materials Characterization, American Society for Metals, Metals Park, OH, 1986, pp. 649–657.
115. Castro Romero, J.M.; Ferandez Sollis, J.M.; Gonzalez Rodriguez, V.; Perez Iglesias, J.; Seco Lago, H.M. Analytical Letters, 27, 1399 (1994).
116. Jarosz, M.; Oszwaldowski, S. Mikrochimica Acta, 126, 241 (1997).
117. Merrill, R.M. "Ion Chromatography" in Metals Handbook, 9th ed., Vol.10: Materials Characterization, American Society for Metals, Metals Park, OH, 1986, pp. 658–667.
118. Yamane, T. Materials Japan, 33, 341 (1994).
119. Louw, I. Fresenius' Journal of Analytical Chemistry, 354, 432 (1996).
120. Seubert, A.; Krabichler, H.; Krismer, R.; Wilhartitz, P. Mikrochimica Acta, 117, 245 (1995).
121. Wilhartitz, P.; Krismer, R.; Bobleter, O.; Dreer, S. Mikrochimica Acta, 125, 45 (1997).
122. Colburn, B.A.; Hinton, R.; Sepaniak, M.J.; Starnes, S.D. Separation Science and Technology, 30, 1511 (1995).
123. Kihara, T.; Matsui, T.; Sakagami, M.; Fugine, S.; Maeda, M.; Kitamori, T. Solvent Extraction 1990 (Conference Proceedings), Elsevier, Essex, U.K., 1992, pp. 497–502.

124. Determination of Trace Elements (Z.B. Alfassi, ed.), VCH, New York, 1994.
125. Trace Analysis—Physical Methods (G.H. Morrison, ed.), Interscience, New York, 1965.
126. Accuracy in Trace Analysis: Sampling, Sample Handling, and Analysis, NIST Special Publication 422, Vols. I and II, National Institute of Standards and Technology, Gaithersburg, MD, 1975.

9
Inclusion and Phase Isolation

Initially, there were some misgivings on the part of the author about this chapter. The material represents a narrow subfield, which is, for the most part, unrecognized and unheralded by even the fairly insular community of metals analysts. And yet there is a salient point that needs to be made concerning our objective with this volume. Simply put, it says: shun these issues at your peril. Impurities do *not* always dissolve passively with the alloy matrix. That black smudge at the bottom of the beaker *cannot* always be ignored.

Because this material is so far outside the normal ken of most metals analysis laboratories the appropriate place might have been a detailed appendix. But it was felt that locating the subject here, in a penultimate chapter properly acknowledges its central, albeit unrecognized, importance. And there is another reason for placing this topic in a prominent location—there is no other readily available collected source for this information in English.

What we have done here is assemble a selection of the available literature, which often comes from sources outside the purview of the analytical chemist. And while it is unlikely that the majority of the readership will immediately rush to apply this work, it will be available here if and when it is needed. In the meantime it can serve to remind us that a great deal of time and effort has been expended in studying that black smudge.

1. INTRODUCTION

Accurate trace analysis can be critically dependent on the acid insoluble portion of a metal or alloy. That is why so much development effort has been expended on devising means for increasing the effectiveness of the fundamentally

See Ref. 1 for such a source in German.

simple acid dissolution approach (e.g., pressure bombs, microwave ovens, reflux systems), and thus eliminating or minimizing the acid insolubles. But what are these, sometimes analyte-bearing, acid-resistant substances?

The terminology is not always consistently applied, but for *our* purposes let us define an *inclusion* (sometimes termed a *nonmetallic inclusion*) as a compound of a metal with oxygen, nitrogen, carbon, sulfur, or a small number of other nonmetallic elements. It may be either stoichiometric or nonstoichiometric. A *phase* (sometimes termed an *intermetallic phase*) is an association of two or more metals (a stoichiometric or nonstoichiometric compound, or simply an alloy mixture) that exists independently of the main alloy matrix. Inclusions and phases may be major, minor, or trace constituents of alloys, but they are usually referred to as *microconstituents*.

There are two reasons why the metals analyst might need to concern himself with these moieties. First, in the compositional analysis of an alloy it is important (and often critical) that all forms of the analyte element are collectively and uniformly measured. In aqueous solution methodologies this means that all forms of the analyte are dissolved. In thermal evolution methods it means that all forms of the analyte are dissociated. In solids optical emission work it means that all forms of the analyte contribute proportionately to produce a plasma plume representative of the total analyte concentration. Thus, it is always very useful (though, admittedly, not always essential) to know the nature of the inclusions and phases one must deal with.

For example, in solution work an insoluble residue suspected of containing a significant amount of analyte might be filtered, ignited, silica volatilized, and the residue fused with a molten salt flux that is then leached in the filtrate. Sometimes it might be necessary to iterate this procedure until no visible residue remains. In thermal evolution analysis a special flux or accelerator might be added, or furnace power might be increased to decompose analyte-bearing refractory inclusions or phases. In optical emission work a high energy prespark might be employed to produce equilibrium conditions representative of the total analyte concentration.

Methods, such as OES, that rely upon electrical discharges to sample and measure a solid are ordinarily subject to special problems. Undesirable tramp elements in the form of nonmetallic precipitates often accumulate at grain boundaries where ordinary spark discharges tend to dwell, resulting in high readings for some trace elements. An opposite error is associated with the fact that many analyte-bearing inclusions are refractory and thus more difficult to excite than the same analyte dissolved in the metal matrix.

It is a common, but confusing, practice to term *both* nonmetallic inclusion compounds and intermetallic associations as "second phases".

Clearly, in all these methodologies some knowledge about the inclusions and phases involved should help in selecting or devising the best remedial protocols.

The second reason for studying inclusions and phases is to provide metallurgical information about the microstructure and related properties of the alloy. Chemical isolation and measurement can often provide a vital supplement to metallographic examination and the most modern instrumental means of studying alloy microstructure.

The great majority of published work in this area involves studies with ferrous-based alloys and, in particular, low alloy and plain carbon steels. These alloys, comparatively simple in gross composition, often contain a huge array of inclusions and phases, whose presence, particulate size, and distribution can be critical to obtaining desired properties in the finished commodity.

The science of chemically extracting inclusions and phases from steels and other metals for chemical analysis and other studies is a largely unrecognized discipline, which, nevertheless, has a long history.' The earliest references in this field date from the turn of the last century when a variety of chemical media were used in an attempt to recover the oxide compounds from steel. Intensive work (especially with electrolysis and halogen dissolution) was published in the 1930s and early 1940s in Germany and Britain. The postwar years were less productive, but there followed a period of greatly renewed interest from 1949 to the late 1970s with many publications from Germany, the U.K., the U.S., and Japan, among others. This period saw the introduction of bromine/methanol and bromine/methyl acetate dissolution, particle size distribution analysis, and differential thermal analysis/effluent gas analysis as important analytical procedures.

In a sense, this work occupies a midground between metallurgy and analytical chemistry. It is clearly distinct from electron- and ion-probe studies that attempt to identify inclusions and phases *in situ*. And it has the potential to be truly quantitative, as such instrumental approaches rarely are. The essence of this work lies in the *isolation*, but the *measurement* may take a variety of forms. Optical emission, X-ray and electron diffraction, infrared absorption, gas chromatography, classical wet chemistry, and mass spectrometry have all been employed. And the results have been correlated with both metallographic and property data. It is also possible to calculate a mass balance in which the sum of the compound and matrix elemental concentrations are compared to gross analytical results for the entire sample.

Some of the historical perspective has been extracted from the Introduction to *Determination of Nonmetallic Compounds in Steel*. ASTM STP 393 (American Society for Metals, West Conshohocken, PA, 1966) by M.K. Weiss.

Our approach in this chapter will be methodological, once we have completed a brief descriptive account of our subject analytes. The emphasis will be very heavily on steels since most of the published work is that area.

2. INCLUSIONS AND PHASES [1–24]

2.1 Inclusions

One can distinguish two types of nonmetallic inclusions in metals. *Endogenous inclusions* are compounds that form during the melting, solidification, heat treatment, or storage of metals. *Exogenous inclusions* are compounds from refractory furnace linings and similar sources that become entrained in the molten metal during its processing. Chemically, these two categories are usually indistinguishable, but knowledge about the alloy system and its processing allow such identifications to be made.

Even the uninitiated are aware of *slag inclusions*, which frequently take the form of large blemishes on the surface of a freshly cut solid specimen. These may be more in evidence when sampling near the top of a cast ingot, where "pipe" (due to riser shrinkage) and "blow-holes" (due to degassing) may also be present.

In routine gross compositional analysis inclusions are often blamed for poor precision. This claim is frequently true in plain carbon and low alloy steels but rarely true with stainless steels and specialty alloys. It is important to recognize that if analyte-bearing inclusions are not accounted for, their distribution will affect *precision*, while the amount of analyte that they contain will affect *bias*.

Exogenous inclusions may comprise any or all of the refractory materials with which the molten alloy and its slag come into contact. These may include silica (SiO_2) and fire-clay (Al_2O_3/SiO_2) for use in contact with acid slags; alumina (Al_2O_3), sintered magnesite (MgO) or dolomite (CaO/MgO), or synthetic periclase (MgO) for basic slag service; or occasionally materials like silicon carbide (SiC), zirconia (ZrO_2), or chromite ($FeO\text{-}Cr_2O_3$). Common sources of refractory particle pickup in steel-making are the furnace and ladle linings and the tundish nozzle in continuous casting. Occasionally, it is possible to locate the source by correlating inclusion type with areas of excessive refractory wear.

In a sense, entrained slag should also be classed as an exogenous inclusion, since it represents reaction products that separated completely from the molten metal, and were then recombined with it due to turbulence and mixing.

One sometimes refers to *gaseous inclusions*, which are due to outgassing of the cooling metal.

Endogenous inclusions are nonmetallic compounds that form by high temperature reactions during metal production and are not collected in the slag layer. Several hundred species of endogenous inclusions have been identified in ferrous-based alloys. Some of the lowest cost commodities, such as plain carbon and low alloy steels may contain the largest arrays of inclusion types. Very small, evenly dispersed nonmetallic inclusions may have a benign or even beneficial effect on physical properties.

It is generally true, however, that oxide inclusions above a critical size will promote defects, such as laminations in rolled sheet. The critical size is both alloy and application specific. Low strength steel rolled to a thick gauge is much more tolerant of large inclusions than high strength steel rolled very thin. Excessive amounts of oxide inclusions have been shown to have a deteriorating effect on fatigue strength, notch toughness, and transverse ductility.

Intergranular corrosion in stainless steels has been associated with the precipitation of chromium carbide and the accompanying chromium depletion at the grain boundaries. This phenomenon has been prevented by designing the alloy with a titanium or a niobium addition since these elements will precipitate as carbides more readily than chromium. In a similar fashion, vanadium, aluminum, or other stable-nitride formers are added to plain carbon steel to prevent the formation of iron nitride, which is known to lead to strain aging. It can be seen, then, that inclusions can serve important functions in certain steels.

2.1.1 Oxides

Oxides are the most studied nonmetallic inclusions. Endogenous oxides form during the deoxidation and solidification of most metals. And most exogenous inclusions are oxides. Oxides, thus, represent the largest and most complex group of precipitates in steel. Table 9.1 lists some of the oxide compounds that have been positively identified in commercial steels. It is, of course, unlikely that more than about a dozen will occur in any given ferrous alloy.

Several important subcategories can be distinguished, notably the silicates, the spinels, and the garnet-like aluminum silicates. At least five of the eleven known crystal structures of SiO_2 and four crystal structures of Al_2O_3 have been identified in steels. The mineral forms found in steels are chemically the same as their namesakes in rocks and ores. The *spinels* take the form

---

Discontinuity defects that appear as seams parallel to the rolling direction.

Strain aging is the return of a pronounced yield point after storage at room temperature. Interstitial nitrogen forms iron nitride, which locks lattice dislocations in place. The result is an abrupt change from elastic to plastic deformation.

Table 9.1 Some Oxide Inclusions Found in Commercial Steels

α-Al_2O_3	(Fe.Mn)O
β-Al_2O_3	$FeO \cdot Cr_2O_3$ (chromite)
θ-Al_2O_3	$2FeO \cdot SiO_2$ (fayalite)
κ-Al_2O_3	$3FeO \cdot Al_2O_3 \cdot 3SiO_2$ (almandite)
$2Al_2O_3 \cdot CaO$	MnO (manganosite)
$6Al_2O_3 \cdot CaO$	Mn_2O_3
$3Al_2O_3 \cdot 2SiO_2$ (mullite)	Mn_3O_4
BeO	$MnO \cdot Cr_2O_3$
$CaO \cdot Al_2O_3$	$MnSiO_3$
$Ca_2Al_2SiO_7$	$MnAl_2O_4$ (galaxite)
$Ca_4Al_2MgSi_3O_{14}$	$MnO \cdot Fe_2O_3$ (jacobsite)
Cr_2O_3	$3MnO \cdot Al_2O_3 \cdot 3SiO_2$ (spessartite)
FeO (wustite)	$2MnO \cdot SiO_2$ (tephroite)
α-Fe_2O_3 (hematite)	SiO_2
γ-Fe_2O_3 (hematite)	TiO_2 (rutile)
Fe_3O_4 (magnetite)	Ti_2O_3
$FeAl_2O_4$ (hercynite)	Ti_3O_5
$FeO \cdot TiO_2$ (ilmenite)	ZrO_2

$FeO \cdot Fe_2O_3$
$FeO \cdot Al_2O_3$
Quartz, tridymite, α-crystabolite, β-crystabolite, glass

$(M)O \cdot (M')_2O_3$, where (M) = Mg, Zn, Fe, or Mn, and (M') = Al, Fe, Mn, or Cr. Thus, chromite ($FeO \cdot Cr_2O_3$ and $MnO \cdot Cr_2O_3$), hercynite ($FeO \cdot Al_2O_3$), jacobsite ($MnO \cdot Fe_2O_3$), galaxite ($MnO \cdot Al_2O_3$), and magnetite ($FeO \cdot Fe_2O_3$) are spinel minerals found in certain steels. The *olivene minerals* take the form $2(M)O \cdot SiO_2$, and examples found in steel include fayalite ($2FeO \cdot SiO_2$) and tephroite ($2MnO \cdot SiO_2$). The *garnet minerals* are given by $3(M)O \cdot Al_2O_3 \cdot 3SiO_2$. Two of the six major garnets have been found in steels: spessartite ($3MnO \cdot Al_2O_3 \cdot 3SiO_2$) and almandite ($3FeO \cdot Al_2O_3 \cdot 3SiO_2$).

There are stories (perhaps, apocryphal) that near gem-quality garnets have been found in steels, formed by the ferrostatic pressure during metal solidification. All the minerals within each of these classes can partially or completely form solid solutions with each other.

Oxygen is fairly soluble in molten steel, but no stable interstitial phase exists at room temperature. It is much more soluble in molten titanium and zirconium metal, where up to 30 atomic percent concentration can be achieved before forming an oxide phase. Nearly as much is soluble in molten niobium, tantalum, and vanadium. And only in vanadium and tantalum has an interstital oxygen phase been identified at room temperature. In contrast, oxygen is virtually insoluble in gold, iridium, mercury, osmium, and platinum.

BeO is a critical component of beryllium metal and its alloys. MgO in magnesium, CaO in calcium, and ZnO in zinc have all been identified and determined.

2.1.2 Nitrides

Nitrides play an important role in the physical properties of many metals, including steel. We have already mentioned how the precipitation of aluminum and other stable nitrides is used as a strategy to prevent the strain aging phenomenon in plain carbon steels. This eliminates the need for an extra cold rolling step (a temper pass) to re-establish a smooth transition from elastic to plastic deformation. It has also been shown that as little as 0.02% nitrogen influences the growth and morphology of grain structure, and thus the magnetic permeability, of high silicon steel. Both α- and β-silicon nitride (Si_3N_4) have been identified in this alloy. It is also generally accepted that AlN is the most critical parameter in controlling the properties of aluminum-killed low carbon steel. Table 9.2 lists some nitrides that have been positively identified in steel.

Unlike oxygen, nitrogen can exist as an interstitial element, as well as a precipitate, in steel at room temperature. High nitrogen levels stabilize the austenite phase in certain high manganese/low nickel (or even nickel-free) stainless steels, for example. Only limited knowledge about the free energy of solution of nitrogen in high alloy materials is available. However, the thermal stability of precipitated nitrides is well characterized.

Iron nitride (Fe_3N) has the lowest range of stability, this is followed by Cr_2N/CrN < VN < Si_3N_4 < TaN < AlN < TiN < ZrN. Predicting composition thermodynamically is nearly impossible, however, because metal/nitride re-

Table 9.2 Some Nitride Inclusions Found in Commercial Steels

AlN	Mn_4N	α-Si_3N_4
AlO_xN_y	$(Mn,Si)N_2$	β-Si_3N_4
BN	Mo_2N	Si_2ON
Be_3N	NbN	TaN
CrN	Nb_2N	TiN
Cr_2N	$NbO_{0.1}N_{0.9}$	$Ti_5C_xN_y$
$(Cr,Mo)N_x$	$NbO_{0.6}N_{0.3}$	VN
Fe_2N	NbC_xN_y	VC_xN_y
Fe_4N	SiN	W_2N
		ZrN

Actually, $Cr_2N_{0.76}$.
Actually, α-Si_3N_4 = $Si_{11.5}N_{15}O_{0.5}$ and β-Si_3N_4 = Si_6N_8 (see Ref. 17, p. 506).

actions seldom are allowed to proceed to equilibrium in alloy production and processing. Some nitrides (e.g., AlN, CrN, VN, Si_3N_4) are considered "heat treatable"—i.e., they can be dissolved in the metal by thermal processing. Others (e.g., TiN and ZrN) cannot be dissolved and are, thus, "non–heat treatable."

Nitrogen as an interstitial element can be distributed throughout the metal lattice or concentrated at dislocations in the lattice. Interstitial nitrogen, like interstitial carbon, acts as a hardener in simple steels. Nitrogen may also occur as the element dissolved in a carbide or in a metallic phase.

The inclusion compounds commonly referred to as *carbonitrides* (or in early texts as "cyano-nitrides") represent a somewhat nebulous category in which the distinction between carbon dissolved in a nitride and nitrogen dissolved in a carbide becomes blurred and may be assigned by relative concentration. In high carbon tool steels and medium carbon stainless steels extractable species such as Ti(C,N) and $(Cr,Mo,V)_2(C,N)$ are reported to exist. However, it is known that while TiC with dissolved nitrogen is "heat treatable," TiN with dissolved carbon is not.

2.1.3 Carbides

The relationship of carbon with iron-based alloy systems is the most studied and, indeed, the *defining* association for steels. In the classical iron–iron carbide phase diagram, cementite (Fe_3C) (6.7% carbon) is shown to be in complex equilibrium with a number of iron phases. In fact it is known that cementite is in reality a metastable compound. Extreme time and temperature will convert it to graphite and α-iron (i.e., ferrite).

Most other carbides that have been isolated from steel are stable in the room temperature metal lattice, although some are pyrophoric when isolated and exposed to air. Table 9.3 lists some carbides that have been found in steel. In addition to those shown, ternary compounds, such as $Ti_4C_2S_2$, have been

Table 9.3 Some Carbide Inclusions Found in Commercial Steels

Cr_2C	$Fe_2Ce_2C_3$	SiC
Cr_3C_2	$Fe_{21}Mo_2C_6$	Ta_2C
Cr_7C_3	Mn_3C	TiC
$Cr_{23}C_6$	MoC	VC
$(Cr,Fe)_7C_3$	Mo_2C	V_2C
$(Cr,Fe)_{23}C_6$	NbC	V_4C_3
Fe_3C	Nb_2C	WC
$(Fe,Cr)_3C$	Nb_4C_3	W_2C
	(Nb,Mo)C	

isolated. Because there is such a high degree of mutual solubility of the alloying components in binary carbides it is sometimes regarded as more useful to refer to them by generic formulas: MC, M_2C, M_3C, M_4C_3, M_6C_3, M_7C_3, $M_{23}C_6$, and possibly others.

The chemically unstable Fe_3C has been isolated from steels in amounts up to 10 weight percent, while some of the other, more stable, carbides have been isolated in various amounts, down to minute quantities. Graphite is often isolated from some iron-base alloys and should be categorized with the carbides. It must be distinguished from amorphous carbon, which may be present due to the breakdown during isolation of chemically unstable carbides.

Like nitrogen, carbon is an interstitial element in steels and can pin lattice dislocations in an analogous manner. Carbon may also be found dissolved in a nitride inclusion or in an intermetallic phase. The precipitation of carbide compounds within the alloy is one of the principal means by which steel is hardened.

2.1.4 Sulfides

Table 9.4 lists some sulfides that have been isolated from steels. Sulfur has played a traditional role in improving the machinability of steels. Manganese and, to some extent, chromium sulfides have been involved. "Stringer" defects in rolled steel are due to sulfides and can be prevented by the use of alloy inoculants to control sulfide morphology. Additives used to prevent "stringers" by promoting the formation of globular sulfides include the rare earths (especially cerium) and calcium, magnesium, titanium, and zirconium. Low temperature fracture can be prevented by replacing manganese sulfide with titanium sulfide or carbosulfide, or one or more of a group of newly studied sulfides (calcium sulfide, magnesium sulfide, zirconium sulfide or carbosulfide, and rare earth sulfides or oxysulfides).

There are other types of nonmetallic inclusion compounds found in steels, as well, but they must be regarded as either very minor constituents, benign in their effect, or difficult to study with current technology. Some examples are Fe_3P and Fe_2P, CaF (from slags or mold powders), and borides.

Table 9.4 Some Sulfide Inclusions Found in Commercial Steels

Al_2S_3	(Mn,Fe)S
Cr_2S_3	$(RE)_2O_2S$*
FeS (triolite)	γ-Ti_2S
MnS (alabandite)	$Ti_4C_2S_2$

* RE = rare earths

Table 9.5 Some Intermetallic Phases Found in Commercial Steels and Nickel Alloys

α-Fe (ferrite)	$Fe_{36}Cr_{12}Mo_{10}$	$Ni_3(Al.Ti)$
δ-Fe (ferrite)	Fe_2Mo	Ni_2AlTi
γ-Fe (austenite)	FeV	Ni_3Cr
FeCr	NiAl	Ni_2FeMo
$FeCr_{0.5}V_{0.5}$	Ni_3Al	Ni_3Mo
		Ni_3Ti

2.2 Intermetallic Phases

By our definition these are microconstituent metallic, stoichiometric or nonstoichiometric compounds, or alloy mixtures embedded in the base alloy matrix. One of the most studied phases is Ni_3Ti. Table 9.5 lists several more, but, clearly, there are hundreds. It is unlikely that techniques will be soon devised to chemically isolate them all, or that there will be a pressing need to do so. In highly alloyed systems, like superalloys, metallic phases are sometimes left undissolved by commonly employed dissolution protocols. On occasion the silvery residue contains a significant portion of the analyte and, thus, it cannot be ignored. The two standard responses are to: 1) try microwave closed vessel dissolution on another test portion with suitable acids, or 2) filter, wash, ignite, volatilize silica, fuse with molten salt, and leach in the filtrate.

Metallic phases are the key to the remarkable physical properties of many superalloys, including iron-, nickel-, and cobalt-based high temperature alloys. Many of these alloy systems derive their important characteristics from intermetallic phase transitions at elevated temperatures. Among light metals magnesium alloys tend to form many stoichiometric intermetallic phases.

3. ELECTROLYSIS [4,25–33]

The use of electrolytic dissolution to remove the metallic matrix, releasing nonmetallic compounds in filterable form, is quite an old technique. In 1931 G.R. Fitterer described what was claimed to be the first published procedure, but he added that electrolytic extraction at that time was being routinely practiced by 20 to 30 steel companies [25]. The Fitterer paper began a long debate that has never been definitively resolved concerning the proper electrolyte for this application.

In 1937 Walter Koch began a series of publications in this field, first with P. Klinger and later with a number of other colleagues, including H. Sundermann [26–29]. Fitterer's apparatus and the early Klinger and Koch design were beset by a number of problems: matrix elements precipitated out of solution contaminating the inclusion isolates; high current densities led to metallic particles in the residue; and the solid specimen did not dissolve uniformly, leaving doubt that the isolated inclusions truly represented the distribution of nonmetallic compounds in the alloy.

Fitterer had originally used a collodion bag to retain the isolated particles while permitting the passage of ions. Other workers have used other semi-permeable materials, including sausage skins. However, the most effective barrier appears to be a cylindrical porous porcelain diaphragm. The test piece, ideally, should be a cylinder, hand-filed with a clean file, degreased, carefully weighed, then attached to a wire by an alligator clip or other means. The entire assembly is then coated with paraffin or a peelable vinyl paint and the desired area stripped off for exposure to the electrolyte.

Fitterer's [25] electrolyte was a solution of 3% (w/v) ferrous sulfate ($FeSO_4 \cdot 7H_2O$)/1% (w/v) sodium chloride (NaCl) in water. His procedure was recommended only for plain carbon steels. Klinger and Koch utilized the so-called "Z-B" electrolyte [5–15% (w/v) sodium citrate/1.2% (w/v) potassium bromide/0.6% (w/v) potassium iodide with pH adjusted to between 6 and 9] for low alloy and medium alloy steels up to about 7% chromium. Their system produced a nonuniform erosion of the test specimen due to the asymmetrical positioning of the two electrodes. It was concluded that this situation sometimes produced nonrepresentative isolates. And so Koch and Sundermann [29] designed a new apparatus in which the electrodes were arranged so as to favor a symmetrical etch.

The principle features of their design were: 1) a central anode chamber for the sample specimen, 2) a surrounding cylinder of porous ceramic, 3) a concentric outer chamber containing a three-piece, nearly cylindrical platinum gauze cathode, 4) an inclusion collection receptacle, immediately below the sample specimen, 5) an inert gas inlet to allow purging the electrolyte of atmospheric oxygen, 6) a funnel and reservoir of fresh electrolyte below the electrolysis cell. In operation, dense, spent electrolyte, full of matrix cations, sank through the funnel to the bottom of the reservoir, pushing fresh electrolyte continuously up into the electrolysis cell.

Walz and Bloom [30] described a design employed at Timken Roller Bearing Company. Their cell (as in the original Klinger and Koch [27] design) used a continuous gravity flow of fresh electrolyte. In their design the electrolyte was maintained at a constant level in the anode cell by a siphon and float system, in which a solenoid valve was controlled by the float interrupting a light beam/photocell arrangement. Many other aspects of the design are similar to that by Koch and Sundermann [29].

Walz and Bloom used a 5% (w/v) sodium citrate/2% (w/v) potassium bromide aqueous solution as the electrolyte in both chambers, but added hydrochloric acid (4 mL per ampere-hr) to the outer (cathode) chamber to prevent the formation of a high pH value in the anode (specimen) chamber.

McKaveney, Raber, Vassilaros, and Snook [4] described a comparatively simple electrolysis cell that utilized a cellophane sheet as the anode/cathode semi-permeable membrane barrier. Their electrolyte was 15% (w/v) ammonium citrate/1.5% (w/v) ammonium thiocyanate/0.02% (w/v) hydrazine sulfate, adjusted, if necessary, to pH 4.68. Electrolyte flow (3L/hr) was controlled in the manner of the Walz and Bloom work. The solution was sparged with nitrogen before and during the electrolysis. The cathode was platinum or titanium, masked, if necessary, so that its electrolye-exposed area was smaller than that of the steel specimen. Like most of the studies that were published since the 1950s they utilized a laboratory ultrasonic transducer to release clinging inclusions and "electrode slime" from the sample surface following completion of the electrolysis. And, as in most early studies their operating aim was to achieve an average current density (in this case 100 milliamperes/cm^2) to dissolve a given amount of metal (50–100 g).

R. Raybeck of J&L Steel utilized a cell design similar to that of Koch and Sundermann but with a gravity-fed electrolyte consisting of 5% (w/v) sodium citrate/2% (w/v) potassium bromide/1% (w/v) potassium thiocyanate in aqueous solution. It yielded a uniform etch with plain carbon steels. Hydrochloric acid was added to the electrolyte in the cathode chamber. A current density of 10 mA/cm^2 was utilized at a current setting of 80 mA. Electrolysis time was 18–20 h to collect isolate for optical emission analysis. Aluminum and silicon results were not considered reliable if the electrolyte was stored in glass. Iron and manganese results were also suspect because of compound instability.

Most published work describes use of a simple constant current voltage supply with conditions adjusted to obtain a desired current density. Since the current resistance could vary it was inaccurate to rely in this way on a fixed current density. However, Koch and Sundermann [29], H. Hughes [31], and Andrews and Hughes [2] described the application of a potentiostat to steels, and Ilschner-Gensch [32] described similar work with nickel-base alloys. This equipment requires the use of a saturated calomel electrode (SCE) connected by a salt bridge to the electrolysis cell, where a platinum cathode serves as an auxiliary electrode in the circuit. With such an apparatus it is possible to generate potentiostatic polarization curves in which the cell potential (versus the SCE) in volts is plotted against current density in milliamperes. For a given electrolyte unique things happen with different alloys. The same basic instrumentation has proven to be of great utility in corrosion studies.

In this application it is useful to distinguish generated curves that correspond to "ferrous active" (Fe^{2+} is produced continuously in direct proportion to the potential), "ferrous partially passive" (the steel is inert until 0–400 mV are attained, then Fe^{2+} begins to be produced—pitting may occur as surface areas are selectively etched), and "ferric active" (the steel is inert up to about 1 V, then Fe^{3+} begins to be produced).

Hughes [31] generated polarization curves for a series of 25% nickel/15% chromium steels, as well as synthesized solid samples that represented an intermetallic phase. Three electrolytes were utilized: 10% (v/v) hydrochloric acid, 20% (v/v) phosphoric acid, and 45% (w/v) ferric chloride ($FeCl_3$). The synthesized phases proved to yield distinctive curves that suggested the potential for their extraction from alloy specimens.

What is isolated from steel by electrolysis procedures depends, at least in part, on the exact experimental conditions and on the nature of the alloy. In many cases the carbides, sulfides, oxides, and nitrides, as well as some intermetallic phases can be retained. This can be a mixed blessing, however, since it is then usually necessary to further separate the extract before useful analytical information can be obtained.

Koch and Sundermann listed the following isolates using their procedure: *oxides*: FeO, Fe_3O_4, MnO · SiO_2, SiO_2, Al_2O_3, TiO_2, Cr_2O_3, spinels, silicates; *nitrides*: AlN, TiN; *carbides*: Fe_3C, TiC, VC, WC, Mo_2C, (Nb,Ta)C, $Cr_{23}C_6$, Cr_7C_3, $M_{23}C_6$, M_6C; *sulfides*: MnS, FeS; *phosphides*: Fe_3P, Fe_2P; *others*: α-Fe, γ-Fe, α-Ti, σ-Cu.

Typically, the combined collected inclusions are suction filtered on a microporous membrane filter of some minute pore size. If they are dried in a stream of inert gas they may catch fire spontaneously when exposed to atmospheric oxygen. They may be scraped off the filter for qualitative work, or the membrane may be dissolved in some solvent, or ignited if the desired analytes are refractory.

Klinger and Koch [27] utilized low temperature chlorination* and high temperature vacuum sublimation to isolate the oxides while removing all interfering components. Walz and Bloom [30] adopted their approach. Koch and Sundermann [29] advocated a number of other secondary separation approaches, such as "sink/float" density separation of the particles using diverse solvent media.

Magnetic separation was their primary interest, however. These authors devised an apparatus with a stirrer and an oscillating magnetic field that collected nonmagnetic particles at the bottom of a vial while magnetic particles were held by an electromagnet to the wall of an inner cylinder. The apparatus was marketed commercially for a short time. They illustrated the technique with the separation of Fe_3P (magnetic) from Fe_2P (nonmagnetic). Other

* For more information about chlorination see Section 4.0.

isolates that are magnetic include Fe_3C (cementite). α-ion. FeO. Fe_3O_4. and $Fe_{21}Mo_6C_6$.

Andrews and Hughes [2] applied a device known as a threshold centrifuge to the separation of carbide and intermetallic phases. This contains a spiral channel with equally spaced restrictions to separate particles of different volumes or densities. Suspended isolates were introduced into the center of the rotor.

McKaveney and coworkers [4] described problems with the chlorination approach (and also with a cupric bromide treatment to destroy carbides that had also been suggested by Klinger and Koch [23]). Instead. they utilized a bromine/methanol treatment of the electrolysis residue to isolate the oxides.

There are numerous other electrolysis approaches and other possible postelectrolysis treatments. For example. we have not discussed the nonaqueous electrolytes that have been occasionally applied to stainless steels. And we will reserve a description of oxygen plasma dry ashing for a later section. It is probably appropriate to stop here and ask. what does anyone do with these isolates?

The measurement categories can be fairly equally divided between qualitative and quantitative approaches. X-ray and electron diffraction are common. while infrared absorption and petrography are somewhat rare in the identification of specific compounds. Differential thermal analysis. especially combined with effluent gas analysis. has been valuable in identifying inclusions. The elemental composition of isolated oxides can be determined by optical emission and the mass balance checked by assigning formulas and summing oxygen concentration. then comparing the sum with the result from a total oxygen determination. Particle size distribution analysis can be performed. as can methods for carbon. sulfur. oxygen. nitrogen. and other specific elements. In the subsequent sections. we will first examine some alternatives to isolation by electrolysis. and then describe some of the chemically-oriented approaches to measurement.

4. HALOGENATION [9,10,34–42]

Halogens have been utilized for a long time to remove metal matrix from insoluble nonmetallic inclusions. As anhydrous gases at elevated temperatures. chlorine. bromine. and iodine react with iron and steel. as well as with most other metals. As aqueous solutions. bromine and iodine have been used to dissolve metals. but it is as anhydrous solutions in alcohols and esters that they have found their greatest utility. Metal halide salts are readily soluble in such solvents and. thus. do not coat the test portion to slow the reaction. We will look at developments in this area in roughly historical order.

4.1 Iodine

In 1927 Willems [34] first used an alcoholic solution of iodine to dissolve steel and collect the inclusions. He utilized absolute methanol and a purge of flowing nitrogen and water cooling to eliminate the hydrolysis of iron, which heavily coats the residues from aqueous iodine dissolution. In 1935 Rooney and Stapleton [35] elaborated on the work, taking special pains to ensure that water and oxygen were excluded from the solution and that the reaction temperature was controlled. At this point it was recognized that iodine/alcohol was a faster and simpler approach than electrolysis, and that it carried the advantage that carbides in plain carbon steels were destroyed, simplifying the isolate.

Speight [36] in 1943 drew attention to a problem: with high carbon/high phosphorus alloys the isolated residues from alcoholic iodine treatment were contaminated with adsorbed phosphorus and iron. Refluxing the residue with 10% (w/v) ammonium tartrate was found to significantly reduce the iron level, and slightly reduce the phosphorus level.

In 1957 Garside and Rooney [37] summarized the historical developments in this area. It was recognized relatively early on that solid pieces, especially thin disks, yielded better results than the most carefully prepared millings. Preliminary nitrogen purging of the iodine/alcohol solution also proved to be important in preventing water hydrolysis of sensitive compounds, such as aluminum nitride. The methanol employed, of course, must be strictly water-free. Another potential shortcoming is that some acid and water may form from the reaction of alcohol and iodine, which would dissolve some of the isolated compounds.

In plain carbon steel the upper carbon limit for the technique appears to be in the 0.4–0.6% range; otherwise many carbides contaminate the residue. High alloy carbides, of course, are insoluble and will be in the residue at any carbon level. When aluminum nitride is present in great amount it will contaminate the residue, invalidating the conversion of aluminum values to Al_2O_3 concentration. Similar effects are expected from silicon, zirconium, and other stable nitride formers. Manganese oxide (MnO) and manganese sulfide (MnS) dissolve at 65°C, but remain intact if the reaction is conducted below 30°C.

Iodine/alcohol procedures are generally considered the least aggressive of the commonly accepted halogen procedures. Dissolution of very slow-reacting samples can often be facilitated by ultrasonically agitating the solution. The general procedure can be employed for plain carbon and low alloy steels, for nickel steels with nickel concentrations up to 30%, and for aluminum (for Al_2O_3 isolation).

Another possible explanation is that iron phosphides and iron carbides are not completely decomposed, as expected (see Ref. 37).

4.2 Chlorination [10,38,39]

The use of anhydrous chlorine gas at elevated temperatures to extract inclusions was described in detail by Short, Roberts, and Croall [38]. It is a slow procedure, requiring specialized apparatus and considerable attention to detail due to the danger of working with chlorine gas. It is essential that the chlorine gas used be free of oxygen and moisture. There are two conceptual approaches—an atmospheric pressure tube furnace system and a vacuum chlorination device. Both can be utilized for solid metal test portions or residues previously isolated by another technique.

The most commonly used temperatures for both schemes are in the range 300°–350°C. The tube furnace version requires a nitrogen-purged furnace with a "hot zone" for the sample boat and a preliminary cooler zone that is used to preheat the chlorine gas. A wide-mouth exit tube allows the collection of the sublimed ferric chloride crystals in two large flasks connected in series. The exiting gas is finally bubbled through concentrated sulfuric acid. Infrared lamps can be used to heat the ferric chloride collection flasks to prevent clogging with crystals. And sodium hydroxide solution scrubbers can replace the sulfuric acid moisture trap at the exit to remove some chlorine before the effluent gas is vented into the stack of an efficient hood. Needless to say, the entire apparatus should be installed in an efficient fume hood. At a flow rate of 20L/hr of chlorine gas with the "hot zone" set at 400°C the chlorination of a 10–13 g solid sample should be complete in 3 h.

Vacuum chlorination, such as that practiced on electrolysis residues by Klinger and Koch [33], and Walz and Bloom [30], requires a more elaborate apparatus, but it allows chlorination to proceed at a somewhat lower temperature. Walz and Bloom transfer their alcohol-moistened residue filtrate to the vacuum chamber and dry it at room temperature under a vacuum. Then the temperature is increased as chlorine is introduced by a slow leak. The chlorine source is shut off and the ferric chloride sublimed away under vacuum. Oxygen can be introduced at this stage to "combust" away carbon in the residue. If a small amount of metal remains after chlorination of a solid test portion it is weighed and subtracted from the original weight.

The chlorination approach yields a reliable isolate of the oxides of silicon and aluminum if the carbon content of the plain carbon steel sample is <0.50%. However, rim grade steel samples yield total oxygen values from the extracted oxides that are too low compared to total oxygen obtained on the metal by inert gas or vacuum fusion. These "unkilled" steels apparently contain oxides that react to some degree with chlorine (e.g., FeO and MnO). Chromium-bearing medium alloy steels yield residues that are contaminated with chromium carbides. However, if the chlorination is conducted at 400°C evidently the chromium carbides are broken down and chromium is evolved as chromyl chloride (CrO_2Cl_2). At the normal range of 300°–350°C, molybdenum, titanium, and chromium carbides are present in the isolate and cause the values for calculated oxide total oxygen to be erroneous.

For plain carbon steels, chlorination yields oxides plus graphite (which can be burned off in air). Differences between values from elemental determinations of chlorination and bromine/ester isolates yield values for the nitrides present.

Chlorination has also been used to isolate alumina (Al_2O_3) from aluminum and titanium oxide (TiO_2) from titanium. All chlorination reactions are exothermic, so careful temperature control must be maintained at all times. Upon completion the furnace and the chlorine are shut off and the sample is allowed to cool in a stream of flowing nitrogen. The isolate is usually quenched in water and filtered and washed with water, 3% (w/v) sodium carbonate, and 5% (w/v) hydrochloric acid, alternately, to remove hydrolysis products, such as silicon or iron compounds from the residue. The isolate is then ignited, and analyzed (often by optical emission techniques).

Sometimes it is useful, and certainly interesting, to examine the residue from a chlorination under a microscope before any further processing. In many cases the oxides from a solid steel specimen retain their spatial orientation, held in place by their own lattice-like framework, just as they occurred in the original test portion.

4.3 Bromine/Alcohol and Bromine/Ester

Bromine and absolute methanol have been used to dissolve the metal matrix from a variety of materials. As pointed out in the 1968 review article by Smerko and Flinchbaugh [9] the use of this reagent combination is a reasoned progression from earlier work with iodine/methanol. It is an exothermic (and sometimes excessively rapid) attack that must be controlled to avoid overheating and physical sample loss. The most common devices are water cooled reflux condensers attached to small beakers or flasks by a ground glass joint. The most efficient means to control the reaction rate is to add the bromine (5mL/g of sample) to the hand-filed monolithic solid test portion (often 1–3 g), then absolute methanol is added at the rate of 5–6 drops/min through a dropping funnel for a final total of 15 mL/g of sample. The reaction begins almost immediately, but under these conditions is unlikely to run out of control.

Bromine/methanol has been used extensively for the isolation of oxides from steel, as well as for isolating Al_2O_3 from aluminum. It has several serious shortcomings, however. Absolute methanol is expensive to purchase and time-consuming to prepare. It readily picks up water in storage. Also, as with the iodine/methanol combination, the potential exists for the halogen to react with the alcohol to form small amounts of a hydrohalic acid, in this case HBr, which would dissolve some of the isolated compounds, and water, which might hydrolyze some of the soluble elements as precipitates. Despite these problems some investigators prefer this mixture because it reacts with most alloys more readily than other halogen/solvent combinations.[*]

The most useful halogen/solvent combination is bromine/methyl acetate, a reaction system that was developed by Hugh F. Beeghly. Methyl acetate is readily available in a water-free form and shows less tendency to absorb atmospheric moisture than does methanol. Bromine/methyl acetate reacts with most metals more rapidly than iodine/methanol and less rapidly than bromine/methanol. Although Beeghly's original work involved water-cooled condenser systems, it is possible to use 300-mL tall-form lipless beakers and a watchglass to contain the reacting test portion (typically, 5 mL of bromine and 15 mL of methyl acetate for each gram of sample). The sample should be a hand-filed solid, between 1 and 6 grams in mass. The prepared test specimen can be placed in the total volume of methyl acetate. The total volume of bromine is then added and the watchglass is immediately placed on top of the beaker. In some cases low heat must be applied to initiate the reaction. If the reaction becomes too violent it is readily slowed by immediately placing the beaker in a shallow ice bath. The tall-form lipless beaker acts in a manner similar to an air condenser with reaction vapors refluxing from the walls and watchglass.

Beeghly's original work with bromine/methyl acetate concerned the characterization of nitrides that were soluble in acid and/or the strong caustic utilized in kjeldahl nitrogen distillations [40–42]. Some of these, notably aluminum nitride (AlN), proved to be insoluble in bromine/methyl acetate and the resultant analyses were used to show that their concentrations were dependent in a sensitive way on the thermal history of the steel sample. Beeghly recognized that the bromine/methyl acetate medium also had the potential to isolate easily destroyed species, such as free ferrous oxide (FeO) and manganous oxide (MnO)[10], although he left the proof and documentation to Raybeck and Pasztor [5]. He also recognized that the sulfides and carbides of iron and manganese were destroyed by bromine/methyl acetate, while the more stable, alloy carbides and nitrides were retained.

Residues from bromine/methyl acetate dissolution are best filtered using either a low porosity paper and gravity filtration or a membrane filter and water aspiration (since it is difficult to protect a mechanical pump from the excess bromine fumes). For nitride retention the solution is filtered on a glass

---

[*] Many combinations have been tried, but, besides those discussed in this section, most are less effective and some have dangerous potential. For example, bromine reacts violently with acetone to produce a potent lachrymator once used as a chemical warfare agent.

fiber pad or other suitable nitrogen-free/asbestos-free medium and washed only with methyl acetate. For oxide work, the paper or membrane is washed alternately with methyl acetate and absolute methanol to remove iron bromide and other soluble bromides, then with a hot aqueous 6% (w/v) sodium carbonate solution to destroy the nitrides. If the residue is to be analyzed by optical emission spectrometry the excess sodium ion is removed by washing with very cold water. The filters are then ignited at 600°C in a platinum crucible or dish.

For nitrogen determination the filter can be digested in hydrochloric acid and fumed in sulfuric acid, and then the solution is distilled using a microkjeldahl approach. For oxide determination the ignited residue can be measured using an optical emission crater electrode technique. Of course many other measurement approaches can be applied to both types of isolate.

Smerko and Flinchbaugh [9] reported that they could greatly reduce the dissolution time using ultrasonic agitation. They were able to completely dissolve a 10 g test portion in 2 h. The same authors described the use of an oxygen plasma dry asher to effectively remove carbon from the bromine/methyl acetate isolate. Carbon from the decomposition of carbides interferes with the X-ray diffraction examination of the residue; however, ignition in a muffle furnace, even at low temperature, destroys the reactive ferrous oxide and manganous oxide in the isolate. A low pressure oxygen plasma can gently remove carbon without oxidizing the lower state oxides. The technique has been proven useful for qualitative identification purposes.

5. DISPLACEMENT

Ionic displacement is the process whereby a metal is dissolved in a neutral solution as another metal's cations plate out of solution. For example, a copper (II) sulfate solution will dissolve iron as metallic copper plates out. In theory the most stable inclusion compounds are left untouched. But the technique suffers from hydrolysis precipitation of dissolved matrix elements. And the metallic form of the displacement agent may coat the test portion before it can dissolve, slowing the reaction. Finally, there is the problem of separating the inclusion isolates from the plated displacement agent. Unstable carbides react with neutral water to form acetylene, methane, and other gaseous hydrocarbons. If these are of interest they can be measured by collecting the gas formed and injecting it into a gas chromatograph using a porous polymer column or another suitable stationary phase. Raybeck [39] suggested that carbides might be characterized in this way

Mercury or copper salts have been used to separate stable oxides and stable carbides, as well as austenite phase from steels. Both salts have been used to isolate Al_2O_3 from aluminum.

6. ACIDS

This is the most frequently used and probably the least effective means of isolating inclusions. Even the most chemically stable compounds are attacked to some extent, especially as minute dispersed particles in the metal matrix.

Acids employed for steel include: 33 (v/v)% hydrochloric acid, 12% sulfuric acid, 20% nitric acid, and 5:1 nitric acid:water, maintained at 2–5°C. Such techniques *do* isolate an interesting array of compounds, but the quantitative aspects of the work must be regarded as empirical and are, therefore, of limited use without an understanding between the laboratory and its clients. The one exception is the case of those alloys in which a refractory material, such as high-fired alumina, is purposely added to impart abrasion existence. Only in such cases is it likely that the analytical result will have absolute significance.

7. PARTICLE SIZE DISTRIBUTION ANALYSIS

In 1967 Wojcik, Raybeck, and Paliwoda [43] published an account of the first application of the Coulter Counter™ to studying the size distribution of oxide inclusions in steel. This important study demonstrated that hot working defects could be related to the size distribution of oxides in small samples. The authors confirmed that the size distribution was a log-normal function that could be used to predict the frequency of large inclusions that result in seam defects. This work utilized dilute acid isolation residues, which were expected to yield results proportional to the actual particle distribution. The particles were filtered onto a membrane filter and then released into the suspension liquid (water/alcohol) with an ultrasonic cleaner.

Flinchbaugh [44] later applied the technique to bromine/methyl acetate and bromine/methanol isolates after oxygen plasma dry ashing to destroy both the carbon from decomposed carbides and the membrane filter used to retain the residue. The same author later modified the Coulter Counter™ aperture to allow recycling of particles for the accurate counting of small numbers of large particles [45].

There are many particle sizing instruments today, utilizing diverse technologies, but at the time this work was undertaken the original Coulter Counter™ device utilized a glass aperture with electrodes on either side. As the suspended particles were made to flow through the aperture they produced a resistance change that resulted in a voltage pulse that was proportional to the particle volume. The pulses were counted and scaled (originally on klystron vacuum counter tubes). The instrument was calibrated with natural pollens and synthetic latex particles. Flinchbaugh's redesign allowed the important large particles of oxide inclusions in plain carbon steels to be accurately measured directly [46].

8. DIFFERENTIAL THERMAL ANALYSIS–EFFLUENT GAS ANALYSIS (DTA–EGA)

W.R. Bandi and coworkers [3,8,47–52] published numerous papers on this technique, which was utilized to enlighten much about the nature and abundance of inclusions in steels. In DTA, two thermocouples are linked in series—one in the test portion (often only a few milligrams of inclusion isolate) and one in a reference material (e.g., alumina). As the temperature of both is raised the difference in output of the two sensors is monitored. An inert or reactive gas (in this work, typically, oxygen) passes through both compartments. Inclusion isolates react or decompose, generating or absorbing heat. The heat content of the sample will vary from that of the reference due to chemical reactions, a phase change, or a change of state (e.g., melting). Both exothermic and endothermic peaks are recorded. The effluent gas is monitored using either a direct thermal conductivity cell or by timed-injection gas chromatography. The combined information from these two sources can be used to identify, and sometimes quantify, the inclusion compounds in isolates from steel.

Bandi and coworkers used a variety of configurations. In one typical approach with oxygen as the reactive gas, the effluent gas from the DTA heating block passed into a catalyst heater to convert CO to CO_2. The effluent gas stream was then sampled at regular intervals by a gas-sampling valve on a cam-driven timer. The gas samples then entered a gas chromatograph with a molecular sieve 5A column and a thermal conductivity detector.

The isolates used for DTA–EGA work were obtained from a variety of dissolution schemes: 1:1 hydrochloric acid:water, 10% (v/v) hydrochloric acid/10% (w/v) tartaric acid, bromine/methyl acetate, iodine/methanol, and electrolysis with a number of different electrolytes. Bandi and coworkers utilized this array of techniques to identify a large number of nitrides, carbides, and intermetallic compounds in high strength low alloy steels and high nickel maraging steels, among other alloy types. In some cases the compounds identified by the DTA–EGA curves could not be resolved by more conventional techniques, such as X-ray and electron diffraction, because of low concentration in the isolate, or interferences from other substances present. For example, cubic niobium carbide, cubic niobium nitride, and cubic niobium carbonitride all have similar lattice parameters, and so their positive identification by X-ray diffraction is difficult. However, DTA–EGA can distinguish these compounds.

9. OTHER GAS-SOLID ANALYTICAL SYSTEMS

There have been a few other attempts to utilize gas-solid reactions to measure microconstituents in metals. Some of these are represented by instruments that amount to temperature-programmable versions of conventional oxygen/nitrogen or carbon/sulfur determinators. For example, a temperature-ramped oxygen/nitrogen determinator using the inert gas/impulse furnace principle was marketed. Sample temperature was monitored as the furnace was ramped upward using an optical sensor, or by very accurate measurement of the current passing through the crucible. With this approach peaks are observed for the nitride and oxide compounds in the metal test portion; although, as with DTA–EGA, compound identification requires knowledge about the alloy [53].

Programmable resistance furnace oxygen combustion for carbon and sulfur has found its greatest application in the determination of surface carbon on metals. However, such instruments are potentially applicable to the determination of carbon and sulfur microconstituents in alloys [54].

Thermal evolution techniques using hydrogen gas require the apparatus to be thoroughly purged with argon, both before and after the introduction of the hydrogen to prevent an explosion. Hot extraction with hydrogen gas has been utilized to evolve hydrogen sulfide gas from FeS in electrolysis residues. The H_2S was then measured by a spectrophotometric procedure [8]. Interstitial nitrogen[*] has been measured in aluminum-deoxidized low carbon steels by passing hydrogen gas over fine millings in a porcelain boat in a tube furnace at 550–600°C. Interstitial nitrogen forms ammonia at the metal surface, which is swept away and absorbed in Nessler's reagent for spectrophotometric measurement. The evolved ammonia can also be titrated with standard acid, or measured directly with an ammonia specific ion electrode.

Oelsen and Sauer [55] developed the concept using a coulometric measurement approach. Dulski and Raybeck [56] identified -80 mesh as an optimum sample particle size range and confirmed that for the steel type studied additional nitrides do not precipitate and common nitrides do not decompose under the experimental conditions.

Fisher and White [57] compared hydrogen hot extraction with the standard procedure of calculating interstitial nitrogen concentration as the difference between total nitrogen and bromine/methyl acetate–insoluble nitrogen. They found the hydrogen hot extraction results to be much more precise. This is consistent with the widely-held belief that some insoluble nitrides, especially AlN, sometimes occur as particles too fine to filter. In fact some of the

---

[*] The term "interstitial nitrogen" is used loosely here to include nitrogen in interstitial lattice sites, nitrogen at lattice defects, and the chemically unstable iron and manganese nitrides.

impetus behind the hydrogen hot extraction work was to verify this contention. Fisher and White also noted that for certain vanadium- and niobium-bearing low alloys nitride precipitation did, in fact, occur. For chromium-bearing steels that lack other, more stable, nitride formers, they also noted the breakdown of some chromium nitride. A commercial instrument utilizing hydrogen hot extraction and an ammonia sensing specific ion electrode was once marketed for the direct determination of mobile nitrogen in steel.

When steels are hydrogen hot extracted, in addition to ammonia, both water (from easily reduced oxides) and methane (from interstitial carbon) are also evolved. The potential exists to exploit these phenomena (using gas chromatography, for example) for analytical purposes.

10. SUMMARY

It is appropriate to close this chapter with a final note to re-emphasize why it has been included in a book about the analysis of trace elemental constituents in metals. Comparatively few commercial alloys are pristine lattice-works of uniform arrays of atoms. Many are dotted and sprinkled with microscopic aggregates and macroscopic blemishes. Some would be best described as a congealed emulsion. It is actually comparatively rare that one simple dissolution step will suffice to dissolve all of the components in a complex alloy. But fortunately, such an ideal solution is also seldom necessary.

The message to the analyst is to learn as much about the sample material as possible. For example, if the analyte is trace boron and an acid insoluble residue remains, one must ask: Is the carbon high? Is the nitrogen high? Only by answering such questions can a reasonable judgement be made about the extra 2 h it will take to filter, ignite, fuse, and recombine the leached fusion with the filtrate.

Steels have been studied for their inclusion and phase contents far more than any other alloy system, but that does not mean that potential problems are absent elsewhere. The prudent metals analyst will study the available literature, not as a materials scientist looking for property data, but as an analytical chemist looking to head off trouble.

REFERENCES

1. Koch, W. Metallkundliche Analysen, Verlag Stahleisen, Dusseldorf, Germany, 1965.
2. Andrews, K.W.; Hughes, H. "The Isolation, Separation, and Identification of Microconstituents in Steels" Determination of Nonmetallic Compounds in Steel.

ASTM STP 393. American Society for Testing and Materials. West Conshohocken. PA. 1966. pp. 3–21.

3. Bandi. W.R.; Straub. W.A.; Karp. H.S.; Melnick. L.M. "Application of Differential Thermal Analysis–Effluent Gas Analysis to the Determination of Nonmetallic Compounds in Steel" Determination of Nonmetallic Compounds in Steel. ASTM STP 393. American Society for Testing and Materials. West Conshohocken. PA. 1966. pp. 22–38.
4. McKaveney. J.P.; Raber. W.J.; Vassilaros. G.L.; Snook. J.M. "Studies on the Determination of Stable Oxides in Low-Alloy Steels Using a Combination of Chemical Separations" Determination of Nonmetallic Compounds in Steel ASTM STP 393. American Society for Testing and Materials. West Conshohocken. PA. 1966. pp. 47–74.
5. Raybeck. R.M.; Pasztor. L.C. "Isolation of Oxide Inclusions from Carbon Steels Using Bromine–Methyl Acetate" Determination of Nonmetallic Compounds in Steel. ASTM STP 393. American Society for Testing and Materials. West Conshohocken. PA. 1966. pp. 75–86.
6. Melnick. L.M.; Lewis. L.L.; Holt. B.D.. eds. Determination of Gaseous Elements in Metals. John Wiley. New York. 1974.
7. Brick. R.M.; Gordon. R.B.; Phillips. A. Structure and Properties of Alloys. 3rd. ed.. McGraw-Hill. New York. 1965.
8. Bandi. W.R. Science. 196. 136 (1977).
9. Smerko. R.G.; Flinchbaugh. D.A. Journal of Metals. 20. 43 (1968).
10. Beeghly. H.F. Analytical Chemistry. 24. 1713 (1952).
11. Baeyertz. M. Nonmetallic Inclusions in Steel. American Society for Metals. Metals Park. OH. 1947. pp. 117–127.
12. Allmand. T.R. Microscopic Identification of Inclusions in Steel. British Iron and Steel Research Association. London. 1962.
13. Van Nostrand's Scientific Encyclopedia. 4th ed.. D. Van Nostrand. Princeton. NJ. 1968.
14. Karp. H.S.; Lewis. L.L.; Melnick. L.M. Journal of the Iron and Steel Institute. 200. 1032 (1962).
15. Langenberg. F.C. Journal of Metals (Transactions of AIME). 206. 1099 (1956).
16. Lewis. L.L. "Behavior of Gaseous Elements in Metals" Determination of Gaseous Elements in Metals (L.M.. Melnick. L.L. Lewis. and B.D. Holt. eds.) John Wiley. New York. 1974. pp. 1-52.
17. Bandi. W.R.; Martin. J.F.; Melnick. L.M. "Iron and Steel" Determination of Gaseous Elements in Metals (L.M. Melnick. L.L. Lewis. and B.D. Holt. eds.). John Wiley. New York. pp. 501-551.
18. Goward. G.W. Analytical Chemistry. 37. 117R (1965).
19. Andrews. K.W.; Hughes. H. Journal of the Iron and Steel Institute. 193. 304 (1959).
20. Andrews. K.W.; Hughes. H.; Dyson. D.J. Journal of the Iron and Steel Institute. 210. 337 (1972).
21. Dyson. D.J.; Andrews. K.W. Journal of the Iron and Steel Institute. 207. 208 (1969).

22. Woodhead, J.H.; Quarrell, A.G. Journal of the Iron and Steel Institute, 203, 605 (1965).
23. Chipman, J. Journal of the Iron and Steel Institute, 180, 97 (1955).
24. Kuo, K. Journal of the Iron and Steel Institute, 173, 363 (1953).
25. Fitterer, G.R. Transactions, American Institute of Mining, Metallurgical and Petroleum Engineers, Iron and Steel Div.,1931, pp. 196–208. Discussion, pp. 208–218.
26. Klinger, P.; Koch, W. Archiv für das Eisenhüttenwesen, 8, 569 (1937).
27. Klinger, P.; Koch, W. Beiträge zur metallkundlichen Analyse, Verlag Stahleisen, Düsseldorf, Germany, 1949.
28. Koch, W.; Sundermann, H. Archiv für das Eisenhüttenwesen, 28, 557 (1957).
29. Koch, W.; Sundermann, H. Journal of the Iron and Steel Institute, 187, 373 (1958).
30. Walz, H.; Bloom, R.A. Journal of Metals, 12, 928 (1960).
31. Hughes, H. Journal of the Iron and Steel Institute, 204, 804 (1966).
32. Ilschner-Gensch, V.C. Archiv für das Eisenhüttenwesen, 31, 97 (1960).
33. Klinger, P.; Koch, W. Archiv für das Eisenhüttenwesen, 11, 569 (1938).
34. Willems, F. Archiv für das Eisenhüttenwesen, 1, 605 (1927).
35. Rooney, T.E.; Stapleton, A.G. Journal of the Iron and Steel Institute, 131, 249 (1935).
36. Speight, G.E. Journal of the Iron and Steel Institute, 148, 257P (1943).
37. Garside, J.E.; Rooney, T.E. Journal of the Iron and Steel Institute, 185, 95 (1957).
38. Short, C.W.; Roberts, R.S.; Croall, G. Journal of the Iron and Steel Institute, 185, 85 (1957).
39. Raybeck, R.M., private communication.
40. Beeghly, H.F. Industrial and Engineering Chemistry, Analytical Edition, 14, 137 (1942).
41. Beeghly, H.F. Analytical Chemistry, 21, 1513 (1949).
42. Beeghly, H.F. Analytical Chemistry, 24, 1095 (1952).
43. Wojcik, W.M.; Raybeck, R.M.; Paliwoda, E.J. Journal of Metals, 19, 36 (1967).
44. Flinchbaugh, D.A. Analytical Chemistry, 41, 2017 (1969).
45. Flinchbaugh, D.A. Analytical Chemistry, 43, 172 (1971).
46. Flinchbaugh, D.A. Analytical Chemistry, 43, 178 (1971).
47. Bandi, W.R.; Karp, H.S.; Straub, W.A.; Melnick, L.M. Talanta, 11, 1327 (1964).
48. Bandi, W.R.; Straub, W.A.; Buyok, E.G.; Melnick, L.M. Analytical Chemistry, 38, 1336 (1966).
49. Bandi, W.R.; Lutz, J.L.; Melnick, L.M. Journal of the Iron and Steel Institute, 207, 348 (1969).
50. Krapf, G.; Buyok, E.G.; Bandi, W.R.; Melnick, L.M. Journal of the Iron and Steel Institute, 211, 353 (1973).
51. Krapf, G.; Bandi, W.R.; Melnick L.M. Journal of the Iron and Steel Institute, 211, 890 (1973).
52. Bandi, W.R.; Krapf, G. Analytical Chemistry, 49, 649 (1977).
53. Fricioni, R.B.; Essig, L. "Inert Gas Fusion" Metals Handbook, Vol. 10: Materials Characterization, American Society for Metals, Metals Park, OH, 1986, pp. 226–232.

54. Fricioni, R.B.; Essig, L. "High Temperature Combustion" in Metals Handbook, Vol. 10: Materials Characterization, American Society for Metals, Metals Park, OH, 1986, pp. 221–225.
55. Oelsen, V.W.; Sauer, K-H Archiv für das Eisenhüttenwesen, 38, 141 (1967).
56. Dulski, T.R.; Raybeck, R.M. Analytical Chemistry, 41, 1025 (1969).
57. Fisher, R.; White, G. "A Critical Evaluation of Methods for the Determination of Free Nitrogen in Carbon and Low Alloy Steels" Open Research Report, Corporate Advanced Process Laboratory, British Steel Corp., Sheffield, England, 1973.

10

Quality in Trace Analysis

1. INTRODUCTION

In this final chapter we will attempt to address some of the confidence issues particular to the trace analysis of metals. A moment's reflection will reveal that the trace and ultratrace realm are beset with particular needs and concerns in regard to analytical data quality. For one thing, appropriate reference materials at these analyte levels are scarce or lacking, and, as we have seen, nearly all trace methods are comparative and, thus, highly dependent upon standards. For another, methods are often stretched to near the noise limit where conventional assurance measures become impractical. Also, sampling and sample homogeneity concerns take on a different perspective in the trace regime.

The reality is that while we, as analytical chemists, would like to believe that we are doing as good a job here as we are in our other work, the state of most arts is not yet at that level. As practical working chemists we must perform our function to the best of our abilities with what is available to us. It has been said that an airplane built to the same safety measures as a railroad bridge would never get off the ground. Similarly, we cannot usually be as certain of our work at 0.1 ppm as we are at 0.1%.

And so we will discuss the compromises that we must make to do our job with the instruments and techniques that are available to us. At the same time we will list some steps we can take to maximize the reliability of our data, as well as some cautionary notes about protocols that can hurt our results. Finally, there are some philosophical issues associated with what we report and how we report it.

Some of what follows has been introduced in a few of the earlier chapters, but most will be new to this volume. The growing literature on this general subject has been sampled, hopefully in a representative manner. We have also included some ideas that are the author's own views about the manner in which trace analytical data should be treated. Certain terms will be applied to this discussion using definitions for which there is currently no compelling consensus. This should not invalidate the underlying arguments. It is hoped that the reader agrees to look beyond the jargon to judge the concepts it represents. As always our approach remains fixed on the practical.

And so we will find ourselves in areas where academic rigor must bow to exigent demands. But this need not be bad science; rather, at its best it should be considered pragmatic science of the very highest order. We must endeavor to, quite literally, make the very best we can from a bad situation.

2. REFERENCE MATERIALS [1–11]

We begin with this subject and will devote considerable attention to it because, with few exceptions, the trace and ultratrace realm are captive to the availability of suitable reference materials of one form or another. Before we can embark on a discussion of their uses and suitability, however, we must delineate some commonly accepted categories. The all-encompassing term "reference materials" includes another, much narrower, use of the phrase.

In this sense *reference materials* refers to materials of known homogeneity with no particular certification of accuracy for the analyte values. The frequently heard acronym, "RM" is often reserved for this meaning. Thus, an "RM" is a metal specimen that is known to be homogeneous with respect to the analyte and can be used to monitor instrumental drift or other analytical performance on a day-to-day basis, or to return analytical performance to "day-one" conditions. In this category are standards used for control chart monitoring, as well as so-called "drift standards," "setup standards," "standardization standards," and a host of related terms.

"RMs" are nearly always metal solids—spetrometric disks or pins for thermal evolution work—and are frequently developed inhouse. They typically have a nominal working value for the analyte concentration, but they are *never* used for the calibration or validation of methods. Their purpose is merely to *verify* that the analytical process was within control limits at the time the test on unknown material was conducted. If it was not, they can sometimes be used to help bring the analytical process back into control. "RMs" are used only with comparative methods and can serve little or no purpose with definitive methods.

In contrast *certified reference materials (CRMs)* are solids or chips that have an associated analyte value attested to by a certifying body. They often also list an associated uncertainty, which represents a range limit for the certified value. CRMs are used for calibration, method validation, and numerous other quality assurance practices, such as analyst- or laboratory-qualification checks, intermethod comparisons, and the settlement of vendor/customer disputes.

Metal CRMs are produced by both government agencies and by private vendors, as well as by individual corporations for their own inhouse use. A worldwide hierarchy of certifying bodies exists, managed by a group known as ISO-REMCO. Among this group's goals is to outline in broad terms traceability paths to SI units by an unbroken chain of comparisons. While this is reassuring for accreditation and audit purposes it carries little meaning for the working analyst who is only concerned about the veracity of the certified values on a given CRM. In the final technical assessment the metrological pedigree of a CRM must take second place to the quality of its numbers.

In the United States the National Institute of Standards and Technology (NIST, formerly the National Bureau of Standards) occupies the highest traceability rung. Its CRM products are known as "Standard Reference Materials" or "SRMs," terms which are registered trademarks of NIST. Many other countries are served by CRM-producing agencies with some degree of governmental status. There are at least an equivalent number of private CRM producers whose main business is the production and sale of metal and metal-related standards. Private and public corporations that produce standards for inhouse use also sometimes market them to outside firms. Some have an aggressive program, while others do not advertise. Table 10.1 is a partial list of governmental and private vendor metal alloy CRM producers and marketers.

When one considers that there are many more alloy producers who sell inhouse produced reference materials (see Table 10.2 for a partial list) one is left with the impression that an abundance of metal standards are available for the analyst's use. Actually, nothing could be further from the truth. Appropriate reference materials are in critically short supply in most metals industries. And nowhere is this more apparent than in the trace analysis realm.

Fortunately, most solution-based comparative techniques can be calibrated using mixtures of pure solutions. These are generally best when prepared inhouse from high purity metals, salts or oxides. These are often distinguished from reference materials by the appellation "chemical standards". Thus, there are *primary chemical standards* of the highest pedigree, *secondary chemical standards*, etc. By preparing calibration solutions from these high purity substances the analyst can assure himself of the preparation scheme, which is *not* a trivial exercise.

For example, consider the following list of cautionary notes:

1. Avoid the use of fine metal powders or "sponge"-forms with large particle surface areas. While they may dissolve easily, surface oxides can be a major problem leading to incorrect weights, especially with high purity transition metals. Sometimes it is possible to remove surface oxides by annealing the powder in a hydrogen atmosphere at 550°C and cooling and storing it under argon. However, it

Table 10.1 Some Sources of Metal Alloy Certified Reference Materials [1-10]

Supplier	Address
(ARMI) Analytical Reference Materials International	P.O. Box 2246 Evergreen, CO 80437-2246
(ASMW) Office for Standardization, Metrology, and Control	Furstenwalder Damm 388, Berlin, Germany
(BAM) Federal Agency for Material Testing	87 D-1000 Berlin–Dahlem 45, Germany
(BAS) Bureau of Analysed Samples, Ltd.	Newham Hall, Newby, Middlesbrough, Cleveland TS89EA, U.K.
(BCR) Community Bureau of Reference	Commission of European Communities 20 rue de la Loi, B-1049 , Brussels, Belgium
(BEL) Breitlander Eichproben Labormaterial, GmbH`	Postfach 8046, D-4700 Hamm 3, Germany
(BS) Brammer Standard Co	14603 Benfer Road , Houston, TX 77069-2895
(CANMET) Canada Centre of Mineral and Energy Technology	Energy, Mines, and Resources Canada 555 Booth St., Ottawa, Ontario K1A 0G1, Canada
(CKD) Research Institute	Na Harfe 7 19002, Praha 9, Czechoslovakia
(CMSI) China Metallurgical Standardization Research Institute	7 District 11, Hepinje, Chaoyangqu, Beijing 100013, People's Republic of China
(CTIF) Foundry Industry Technical Center	44, avenue de la Division Laclere, F-92310 Sevres, France
(IMN) Institute of NonFerrous Metals	ul. Sovinsklego, 5 44-100, Gliwice, Poland
(IMZ) Institute of Ferrous Metals	ul. Karola Miarki 12, Gliwice, Poland
(IPT) Technological Research Institute	Cidade Universitaria Armando de Salles Oliveira 05508, Sao Paulo, SP-Brazil
(IRSID) Research Institute of the Iron and Steel Works Industry	Voie Romaine, F-57210 Maizieres-les-Metz, France
(JK) Swedish Institute for Metal Research	Drottning Kristinas vag 48 S-11428 Stokholm, Sweden
(JSS) Iron and Steel Institute of Japan	Keidanren Kaikan (3rd floor) 9-4 Otemachi 1-chome Chiyoda-ku, Tokyo 100 Japan
(LECO) Leco Corp.	3000 Lakeview Ave., St. Joseph, MI 49085-2396, U.S.A.
(LGC) Laboratory of the Government Chemist	Office of Reference Materials, Queens Road, Teddington, Middlesex TW11 OLY, U.K.
(MBH) MBH Analytical Ltd.	Holland House, Queens Road, Barnet, Herts, EN5 4DJ, U.K.
(NIST) National Institute of Standards and Technology	Standard Reference Materials Program Gaithersburg, MD 20899-0001
(RCM) Royal Canadian Mint	320 Sussex Drive, Ottawa, Ontario K1A OG1, Canada

Suppliers marked with an asterisk stock and sell materials from many other producers.

Table 10.2 Some Metal Companies Who Sell Reference Materials

Alcan International Limited (Canada)
Aluminum Company of America (U.S.)
Aluminum Pechiney (France)
Bethlehem Steel Corporation (U.S.)
Carpenter Technology Corporation (U.S.)
Imperial Metal Industries (U.K.)
Inco Alloys International (U.S.)
Reynolds Metals (U.S.)
Showa Aluminum Industries (Japan)
Sumitomo Chemical Company (Japan)
(SKF) Ovako Steel (Sweden)
(VAW) Vereingte Aluminum Werke (Germany)
Wolverhampton Metal (U.K.)

is very much preferable to begin with a low surface area metal form, such as wire or thin rod that is kept sealed from the atmosphere. Such forms are easily cut with clean cutting pliers retained exclusively for this purpose.

2. Avoid delinquescent salts, such as $ZrOCl_2$, which dissolve readily in water, but which also absorb moisture rapidly from the air. Such compounds are likely to produce a fairly accurate standard solution upon first use when the reagent bottle seal is first broken but an erroneous one months later when the standard solution needs to be remade. If the analyst insists on using such salts, it is always prudent to standardize the prepared solution by means of classical wet chemical techniques.
3. Be cautious of oxide forms since some may be mixtures of analyte oxides and not exclusively the listed stoichiometric compound. Some commercial "high purity" oxides have even been known to be contaminated with carbonates. "High-fired" stable oxides, such as Al_2O_3, are likely to be highly pure, but they are also extremely difficult to dissolve.

Also, the analyst should not avoid inconvenient forms of otherwise suitable high purity materials. For example, chromium is best as a shot-like bead form that is too hard to cut. It also cannot usually be crushed without contaminating it. In this case a "catch weight" of some amount close to the needed weight is perfectly acceptable. Standard solutions of exactly even number concentrations are an anachronism from the days before calculators and personal computers. However, if the analyst insists on an even number concentration, another option is dilution by weight on a top-loading balance.

The analyst must also be keenly aware of chemical compatibilities and incompatibilities in preparing mixed element standards either "from scratch" or by mixing commercial solutions. For a detailed discussion of this topic see Ref. 11.

The busy lab today is, in fact, more likely to purchase commercial elemental standard solutions rather than prepare them. Admittedly, this can save a great deal of time and effort provided that the right source is selected. Good quality assurance practice requires that each new lot of each elemental standard solution be qualified by measuring it against an appropriate metal alloy reference material. Certificates for such solutions are typically provided by the manufacturer, and each lot's certificate should be filed for reference by any auditing or accrediting agencies that wish to check the traceability of a calibration. In the case of "homemade" standard solutions, certificates for each lot of the high purity metals or compounds used to prepare them need to be stored as well.

The analyst needs to pay careful attention to labelling and expiration dates with both "homemade" and commercial elemental standard solutions. For either type, the solution bottle label should list not only the element and its concentration, but also the exact solution matrix, the starting material, the date prepared and the expected shelf life. Some labs also require the initials or name of the preparer. In the case of commercial solutions some labs require the date the solution was received and the date the seal was first broken to be noted on the label.

Good laboratory practice requires that a standard solution be kept tightly sealed and then mixed thoroughly immediately before each use. The time the lid is off the bottle should be kept to a minimum. A standard solution bottle less than half-full should be used with caution and the results it produces should be thoroughly checked with metal alloy CRMs.

Some private vendor reference materials producers have taken to marketing acid dissolved metal alloy CRMs diluted to a stated concentration. While this approach appears on the surface to provide an additional savings of cost, time, and effort, such solutions leave the unsophisticated analyst in peril of serious error. What such solutions do, in effect, is to remove the dissolution sequence from any validation or verification process. As a result, important analyte losses or significant contamination could occur in the sample while the standard continues to yield the correct result. As an example, consider the determination of trace levels of phosphorus, arsenic, and selenium in steel. It is possible that a neophyte analyst might dissolve his test portion in hydrochloric acid, measure it against a series of such dissolved CRM solution standards, and obliviously report "none detected" for all three analytes. Of, perhaps, even greater concern is the possibility that the vendor organization may not have retained and preserved all the analytes in the preparation of the product.

2.1 Uses of Reference Materials

The use of reference materials in trace analysis (and in general) is a very large subject that touches in some way or another nearly every form that the analytical process can take. We have already begun using some terms that require a precise definition for this discussion. And we must include additional terms in the brief glossary that follows in order to chart a course through this topic. Since there is little agreement about semantics in this area, sources for the listed definitions have been given when the words were not the author's own.

Calibration: "The set of operations which establish, under specified conditions, the relationship between values indicated by a measuring instrument or measuring system, and the corresponding standard or known values derived from the standard." (American National Standard ANSI/NCSL Z540-1994)

Certified Reference Material: "Reference material, accompanied by a certificate, one or more of whose property values are certified by a procedure which establishes its traceability to an accurate realization of the unit in which the property values are expressed, and for which each certified value is accompanied by an uncertainty at a stated level of confidence." (ISO Guide 30)

Comparative Analytical Method: "An analytical procedure that requires the use of CRMs, reference materials (RMs), or, in certain instances, primary chemical standards for calibration. Methods vary widely in the number of CRMs required and the degree to which such CRMs must match unknown samples. (ASTM Standard Guide E1724)

Definitive Analytical Method: "An analytical procedure that does not require the use of CRMs, RMs, or primary chemical standards to achieve accurate results. Examples include gravimetry, coulometry, specific titration methods, and isotope dilution mass spectrometry. Each individual laboratory should validate its performance of such methods with CRMs, RMs, or primary chemical standards." (ASTM Standard Guide E1724)

Metrology: the science of measurement.

"Calibration Laboratories and Measuring and Test Equipment—General Requirements" American National Standards Institute, approved July 27, 1994.

"Terms and Definitions Used in Connection with Reference Materials" ISO/REMCO N223.

"Standard Guide for Testing and Certification of Metal and Metal-Related Reference Materials" American Society for Testing and Materials, approved Aug. 15, 1995 (appears in Vol. 03.06 of *Annual Book of ASTM Standards*, 1997)

Reference Material: "A material or substance one or more of whose property values are sufficiently homogeneous and well established to be used for the calibration of an apparatus, the assessment of a measurement method, or for assigning values to materials." (ISO Guide 30)

Primary Standard: "A standard which has the highest metrological quality in a given field." (ISO Guide 30)

Secondary Standard: "A standard whose value is fixed by comparison to a primary standard." (ISO Guide 30)

Standardization: "The process of adjusting instrument output to a previously established calibration." (ASTM Terminology Standard E135)'

Uncertainty (of a certified value): "The estimate attached to a certified value of a quantity which characterizes the range of values within which the "true value" is asserted to lie with a stated level of confidence. (ISO Guide 30)

Traceability: "The property of a result of a measurement whereby it can be related to appropriate measurement standards, generally international or national standards, through an unbroken chain of comparisons." (ISO Guide 30)

Verification: "Evidence by calibration that specified requirements have been met." (American National Standard ANSI/NCSL Z540-1-1994)

Validation (of an analytical method): "Evidence that a method yields accurate results on a test sample because it yields accurate results on a CRM of similar composition which was analyzed at the same time." (ASTM Standard Guide E1724)

We will have occasion to define some additional terms in a later section, but at this point we now have the rudiments of a "language" in which to discuss the use of reference materials. It should first be noted that the highest circles in metrology have agonized over each word in some of these definitions. Some of the basic sources in this field are: the *International Vocabulary of Basic and General Terms in Metrology* (known as "VIM"), the *Vocabulary of Legal Metrology* (VML), and ISO 3534: "Statistics—Vocabulary and Symbols".

Organizations involved in metrology at an international level derive their origin from the Meter Convention (signed in Paris, France by 17 countries, including the U.S. in 1875). Bodies created by this "Treaty of the Meter" include the General Conference on Weights and Measures (CGPM—*Conference Générale des Poids et Mesures*), the International Bureau of Weights

---

'"Standard Terminology Relating to Analytical Chemistry for Metals, Ores, and Related Materials" American Society for Testing and Materials (appears in Vol.03.05 of *Annual Book of ASTM Standards*, 1997).

and Measures (BIPM—*Bureau International des Poids et Mesures*), and the International Committee for Weights and Measures (CIPM—*Comité International des Poids et Mesures*). These organizations maintain responsibility for the worldwide unification of measurement under the International System of Units or SI units (from *Le Systeme International d'Unites*).

The International Organization for Standardization (ISO) in 1975 created a committee on reference materials, known as REMCO. The broad scope of this organization is to harmonize all matters relating to reference materials throughout the world. Issues that are being addressed include definitions, classification of reference materials, uses of reference materials, and accreditation of reference materials producers. The United States participates in this work through its member body, ANSI (the American National Standards Institute).

The American Society for Testing and Materials (ASTM) also addresses the issue of reference materials as an essential feature of its methods writing committees, such as Committee E-1 on Analytical Chemistry for Metals, Ores, and Related Materials.

Reference materials are used to calibrate and validate methods and to verify that the analytical process is in control. Within this "shell" of activity the analyte values and their stated uncertainties for the associated reference materials remain unquestioned whether they are certificate values or merely "day one" readings.

2.1.1 Calibration [12,13]

Analytical procedures vary widely in their calibration requirements, ranging from definitive techniques, such as gravimetry, with no need for reference materials, to such comparative techniques as certain forms of X-ray fluorescence spectrometry, which may require a hundred or more reference materials. To understand this relative dependence it is helpful to further refine the distinction between methodologies.

For purposes of discussion let us term a procedure that is totally independent of all forms of calibration standards, Class I. Let us call a method that can be calibrated with primary or secondary elemental standard solutions, Class II. And finally, we can call a method that must be calibrated with closely matching matrix and analyte levels, Class III. This distinction says nothing about the relative accuracy of these techniques, it merely reminds us of their relative dependence upon reference materials.

One is hard-pressed to name a trace level technique that is Class I. Activation analysis techniques come to mind, but they have been quite definitely shown to fail the requirement in practical application. Coulometry certainly can be Class I, but it is weak for most trace applications. Only isotope dilution mass spectrometry can pass the requirement in any sense, and, as we have seen, it requires the quantitative addition of an isotope of the analyte—in some regards, a reference material, since the accuracy of the result is dependent upon the accuracy of its assay.

The majority of trace techniques are Class II. They utilize elemental standard solutions prepared from high-purity metals, salts, or oxides to calibrate, and the amount of synthetic matrix-matching practiced is kept to a minimum. Examples include the conventional forms of ICP-OES and ICP-MS, most forms of flame and electrothermal atomic absorption, gas dosing thermal evolution methods, most molecular absorption spectrophotometry methods, and some types of voltammetric methods. The key assumptions with Class II trace methodology are that: 1) the matrix of the sample does not adversely affect analyte measurement, or 2) the matrix interference has been handled instrumentally, mathematically, or chemically in so perfect a fashion that the corrective measures do not need to be tracked in the calibration process.

A calibration curve prepared from fixed concentration increments of pure analyte is the signature example of a Class II method. There is, however, an important alternative in which matrix interference is compensated for by the calibration technique itself—namely, the method of additions and its two-point abbreviated version, known in some labs as the "spiking technique." These are technically Class II approaches since only increments of pure analyte solution are required, but they mimic Class III in their ability to directly compensate for very severe and complex matrix effects.

Whether a series of identical dissolved alloy test portions is spiked incrementally (method of additions) or a single dissolved test portion is split and one half is spiked (spiking technique), two things are always required. First the analyst must know or be able to accurately measure the true method blank for the specific sample matrix. Second, the analyst must know the extent of the linear range. These approaches are extremely powerful in all types of trace level methodologies but have been used especially effectively in electrothermal atomic absorption, in inductively coupled plasma-mass spectrometry, and in polarography.

Class III methods are not limited to solids techniques and solids techniques are not limited to Class III methods. They *do* go together to a considerable extent, however. Arc/spark-source OES, glow discharge OES, spark ablation/ICP-OES, glow discharge AA, glow discharge-MS, spark source-MS, and other solids techniques come readily to mind as classic Class III approaches. The mass spectrometry methodologies can all do reasonably well with relative sensitivity factors established from pure element solids—arguably, making them Class II methods. But high reliability work is known to require closely matching solid alloy reference materials.

Whenever an alloy matrix of any complexity is synthesized from mixtures of pure element standard solutions—for example, in ICP-OES work—the method becomes Class III.

We have drawn these distinctions in required calibration approaches in order to illustrate several points. First, the fact that analytical calibration methods vary widely in their requirements for reference materials—both in number and how closely they must match the sample matrix. Second, some techniques are capable of operating in two or more classes as needs and exigencies require. Volumetric titrations, for example, can be based on the normality of primary standard solutions (Class I), on a titer established from a pure element standard (Class II), or on a titer established from a dissolved alloy reference material (Class III). Last is the fact that reference material dependence can sometimes be traded for modifications to the procedure that compensate for matrix effects—chemical separations, instrumental changes, mathematical corrections, or even the mode of calibration, itself.

2.1.2 Validation [12,14,15]

Validation is the process by which we convince ourselves (and others) that a method works, that it works in the hands of a given individual or a given laboratory, and that it was working at the time the unknown samples were analyzed. So we see, initially, that validation means different things in different situations. The concept is not useful without a context.

The fact that a given analytical procedure is called a *standard method* by some consensus organization (or by somebody, somewhere) *implies* that it has been *validated* in some way, somewhere, at some time. This, by no means, should suggest that it requires no further validation in its *application*.

Suppose we are considering a wet chemical multiseparation/gravimetric finish procedure that was developed 40 years ago, and now is in the hands of a trained but inexperienced recent hire. The method was validated many years before by an extensive array of certified reference materials, but will it produce valid results now? Obviously, here the method must be revalidated in the context of its present use, and every time it is used.

There is an interesting inverse relationship between the calibration class of a method (as discussed above) and its validation requirements. Class I methods, which require no calibration reference materials, are generally the procedures that require the highest degree of manipulative skill on the part of the analyst. Since even the most talented individual can have an "off-day," it becomes essential that each data set from such procedures be heavily validated with suitable reference materials. The results from the validation standards, analyzed at the same time and in the same manner as the unknowns, must become part of the official records of the testing. It is normally *inappropriate* to correct any observed bias from the validation results for Class I methods because of the degree of human manipulations involved in such procedures. In such methods random error is much more likely than systematic error. An apparent bias is usually sufficient cause to discard the entire data set.

Class II methods require validation with certified reference materials to only a slightly less extensive degree than Class I methods, but the reason is different. These techniques generally require less manipulative skill, but they are often sensitive to matrix variation outside the tested range. Only by measuring a CRM validation standard that duplicates the matrix and the analyte level of the unknown can we assure ourselves that the method works for the test sample. Here, an observed minor bias from a suite of accompanying appropriate CRMs may justify a bias correction, although its cause should be sought out and the method appropriately modified.

Class III methods have the lowest validation requirements. Once the laboratory and all the operators have been qualified by validation on a standard method, sometimes only periodic checks of the validation are required. Such methods are frequently continuously monitored by control charts, especially in a production environment. These quality checks are usually designed around RMs to conserve valuable CRM stores. This process of monitoring any divergence from "Day-One" conditions has a special name: *verification*.

2.1.3 Verification [16–22]

Verification is the process of determining that a method is still doing what it did when it was last calibrated and validated. It has its most meaningful application in Class III methods that are not frequently calibrated. We normally speak of verifying a *result*, in the sense that we verify that the result on an unknown material is *correct* because the method was *in control* at the time the result was obtained. *In control* is defined for the specific application by the type of control protocol used and the rules associated with it.

Spectroscopists, in particular, tend to use the term "standardization" as the process of using an RM or CRM to bring a drifting instrument back into control. Thus, the "verifier" is an RM used for frequent checks on the control condition, and the "standardant" is an RM or CRM used to return the instrument to "day-one" (or controlled) conditions, basically by shifting the calibration curve in some manner. To complete the litany of jargon, the "calibrants" are the CRMs and RMs used to recalibrate, when and if that becomes necessary.

The most common and useful way to monitor the control state of a method is with a control chart. There are several common types of control charts. The most generally useful are probably the $\bar{x}$ and r charts, originally developed by W.A. Shewhart [21] and still often referred to as *Shewhart control charts*. The $\bar{x}$ *chart* is a sequential record of the average of two or more determinations of the verifier, which is analyzed in replicate on a regular schedule. The upper and lower *control limits* are calculated from $\bar{x} \pm 3s/(n)^{1/2}$, where $\bar{x}$ is the established mean for the verifier (the central line), s is the established standard deviation and n is the number of replicates used to establish them.

When a reading at or beyond these limits is obtained the analytical system is declared to be out of control. Within the area bounded by these lines are two lines at $\bar{x} \pm 2s/(n)^{1/2}$, which are *warning limits* used to initiate corrective measures before control is lost. The values of $\bar{x}$, s, and n must be predetermined before control charting can begin,' but they can also be modified from time to time. Theoretically at least, improved estimates of $\bar{x}$ and s should result from the accumulated data from the operation of the control chart.

The *r chart* is a plot of the range of highest to lowest values obtained on each replicate analysis of the verifier. In this case the established standard deviation is multiplied by factors from a table (see References 17–19) to establish the central line, the upper control limit and the upper warning limit. The lower control and warning limits are normally both zero in analytical applications with their ordinarily small numbers of replicates.

In most cases, both an $\bar{x}$ chart and an r chart are kept in monitoring the control of a routine, frequently used, analytical procedure. What we expect to see on the $\bar{x}$ chart is a random pattern of points on both sides of the central line between the two warning limits. When data points fall between the central line and the warning limits no action is ordinarily called for. But when even one data point falls between the warning limit and the control limit the *Westgard Rules* are often applied. Under these guidelines corrective action is called for if:

1. The next data point is also in excess of the warning limit.

or

2. The corresponding range of the two consecutive measurements both exceed 4s.

or

3. The corresponding ranges of the next three data points all exceed 1s,

or

4. The next nine data points are on the same side of the central line.

Less useful, although, perhaps, more frequently used, is the *control chart for individuals*. This is merely a plot of the result of a single measurement of the verifier with 3s control and 2s warning limits surrounding a central line that is the best estimate of the analyte value. Another approach is to calculate the control limits from the mean moving range (R)—i.e., the average difference between successive measurements. In this case the control limits are $\bar{x} \pm 2.66R$.

In practice a small number of tests (say, 10-20) are used to establish initial estimates.

A "cumulative sum" or *cusum chart* plots the sum of the deviations of all verifier result averages from the central line. This sum includes the present deviation and all previous deviations. Such a chart rises or dips sharply at the exact point where the analytical process went out of control. The *average run length*—i.e., the number of data points needed to positively detect an out of control condition—tends to be lower for a cusum chart than for an equivalent Shewhart chart.

2.1.4 Other Uses and an Alternative

Reference materials can be used to provide evidence of equivalency for alternative methods, laboratories, and analysts. In customer/vendor disputes, sometimes certified reference materials are exchanged between laboratories to provide a uniform basis for calibration or validation. In method development certified reference materials are used to confirm that both systematic and random error are within acceptable limits. Analysts and laboratories are qualified to perform new procedures on the basis of results obtained on CRMs, sometimes in a blind or a double-blind study.

We have seen in Sections 2.1.1 and 2.1.2 that all three classes of methods ordinarily use reference materials for calibration and/or validation. But there is one alternative to the use of reference materials in validating procedures and removing doubt about an analytical result—namely, the use of an alternative analytical technique. Two methods based on different chemical or physical principles that produce the same result on an unknown material are powerful evidence that the result is a close estimate of the truth.

The concept of using two or more truly independent methods to establish an analytical result has been recognized as a premier strategy for nearly a century [23]. However, like all roads to truth, there are obstacles here, as well. Perhaps, the most significant is the simple fact that for many analytes, especially in a particular matrix and at a particular concentration level, there is only one best method at the current state of the technology. Sometimes an inferior alternative method can be modified and extended to approach the performance of the best procedure. But that raises another concern: what is the appropriate way to combine the results from two methods, each with different population statistics? Schiller and Eberhardt have addressed such concerns [24].

The third problem is, at least partially, metaphysical: what constitutes sufficient difference in the analytical methodology for two techniques to be considered truly independent? Of foremost concern is the sampling and sample preparation steps, which may be identical for widely variant measurement techniques. An undetected systematic error in these two initial tasks would transmit seamlessly to both methods.

Concerns like this might have us looking to compare a solutions-based technique with a solids-based technique. At a later stage in the analytical process we may feel better about comparing a technique requiring chemical separations with one which does not. And, finally, there is the measurement process, itself. How different must it be? Does the inductively coupled plasma torch too closely link ICP-OES and ICP-MS? Are electrothermal and flame AA different enough?

Another question arises about the method of calibration. Two, otherwise different, Class II methods, calibrated from solutions prepared from the same stock standard solution (commercial, or "homemade") are subject to possible undetectable systematic error. Some may say that at the very least we should insist on different sources for a commercial stock standard solution. Or, is that overkill if the calibrations are verified and the methods are validated?

Clearly, only the analyst, the laboratory management, and the lab's clients can answer such questions to reach a mutually acceptable comfort level when critical work is involved. The point is moot for many common situations where only one method is ever employed for a given task. In trace and ultratrace work the "second method" strategy can lend credence to results when CRMs are lacking. Few laboratories are willing or able to commit to the time and effort that such a program requires, however. Table 10.3 is a fanciful comparison of two laboratory reports on the same unknown sample. Who does the reader believe?

2.2 Suitability of Reference Materials

Once again at the outset of an extensive subject it becomes appropriate to define some terms. As in the previous listing, sources are provided when we

Table 10.3 Two Hypothetical Laboratory Reports

Laboratory A	Laboratory B
Report on Sample 7-L	Report on Sample 7-L
GFAA: 7.49 ppm Tl (Av. of 10 replicates)	GFAA: 8.84 ppm Tl (Av. of 3 replicates)
No certified reference material available	ICP–MS: 8.59 ppm Tl (Av. of 3 replicates)
	ASV: 8.91 ppm Tl (Av. of 3 replicates)
	Grand Average: 8.78 ppm Tl
	(No certified reference materials available)

GFAA: Graphite furnace atomic absorption
ICP–MS: Inductively coupled plasma/mass spectrometry
ASV: Anodic stripping voltammetry

have adopted wording. Some of what follows represents ground previously covered, although, hopefully we can apply it as part of a new synthesis.

Accuracy: "The degree of agreement of a measured value with the true or expected value of a quantity of concern." (John K. Taylor, *Handbook for SRM Users*, NIST Special Publication 260-100, National Institute of Standards and Technology, Gaithersburg, MD, 1985)

Bias: Systematic error (see).

Detection Limit: The lowest concentration of analyte that can be confidently distinguished from zero using a given method.

Error: "Difference between the true or expected value and the measured value of a quantity or parameter." (Taylor, *Handbook for SRM Users*)

Outlier: "A measurement that, for a specific degree of confidence, is not part of the population." (ASTM Standard Practice E876*)

Precision: The degree of agreement among a set of measurements.

Quantification Limit: The lowest concentration of analyte that can be confidently assigned a value using a given method (also called: *quatitation limit*).

Random Error: "A component of the error of a measurement which, in the course of a number of measurements of the same measurand, varies in an unpredictable way. (*International Vocabulary of Basic and General Terms in Metrology*—VIM)

Repeatability: "The closeness of agreement between the results of successive measurements of the same quantity carried out by the same method, by the same observer, with the same measuring instruments, in the same laboratory, at quite short intervals of time." (*Vocabulary of Legal Metrology*—VML)

Reproducibility: "The closeness of the agreement between the results of measurements of the same quantity, where the individual measurements are made by different methods, with different measuring instruments, by different observers, in different laboratories, after intervals of time quite long compared with the duration of a single measurement, under different normal conditions of use of the instruments employed." (VML).

Systematic Error: "A component of the error of a measurement which, in the course of a number of measurements of the same measurand, remains constant or varies in a predictable way." (VIM)

Tolerance Interval: "That range of values, calculated from an estimate of the mean and the standard deviation, within which a specified percentage

---

*"Standard Practice for Use of Statistics in the Evaluation of Spectrometric Data" *Annual Book of ASTM Standards*, Vol.03.06, American Society for Testing and Materials, West Conshohocken, PA 1998.

of individual values of population (measurements or sample) are expected to lie with a stated level of confidence (Taylor, *Handbook for SRM Users*)

Some further clarification is needed here before we proceed with the topic of reference material suitability. Two of the preceding short list of terms, "accuracy" and "precision" are qualities or attributes (of methods, data, etc.) and, as such, have no quantitative statistical definitions. The remaining terms are statistical quantities, and as such can be defined by mathematical equalities.

We will not spend a great deal of time discussing specific statistical data treatments, but it should be pointed out that "reproducibility," as defined here, is being used in its broadest metrological sense. In contrast, in the narrow context of the interlaboratory testing of a specific method, one requires and expects the same exact procedure to be tested in all laboratories by all analysts.

Also, we have chosen to include the definition of "tolerance interval" here to contrast it with "uncertainty" as defined in Section 2.1, because many analysts appear to confuse these quite distinct concepts. An uncertainty range is a feature of a value, such as an analyte value on a reference material certificate. A tolerance interval is a feature of a method which describes its expected performance.

2.2.1 Selecting Reference Materials

It is rare, indeed, that some compromise is not involved in the selection of standards for calibration and verification. The considerations nearly always include the matrix, the analyte concentration and its associated uncertainty, and the number and variety of standards needed. There are also intangibles involved, such as the degree of confidence the analyst places in the certifying agency, and in the homogeneity of the reference material.

Both Class II and Class III methods vary widely in the extent to which they require matrix-matched calibration. The difference between them is, of course, that for Class II methods an appropriate matrix can usually be relatively "painlessly" synthesized from pure element solutions, while Class III methods are usually at the mercy of available solid reference materials. Some laboratories have taken to the highly questionable practice of melting and casting their solid metal RMs in a laboratory arc or induction button furnace. Such a practice is surely born out of desperation due to the unavailability of suitable matrix-matched RMs, but it cannot be condoned. Elemental recoveries from such an operation are not exhaustively characterized, and without a supporting classical wet chemistry program, the resultant solids are relatively worthless as reference materials.

The "graded series" of solid standards has replaced, for the most part, the binary and ternary standard alloy sets that many spectroscopists at one time felt they needed. These graded series standards are a general matrix type, such as 300 series stainless steel, with each alloy incremented in the certified concentration of a number of analytes. These sets are often designed so that while the concentrations of a number of elemental components are going up, the concentrations of others are going down. They can be expensive and difficult to produce but have become the province of certain private reference material vendors.

In X-ray fluorescence spectrometry the "type standard" has become an essential addendum to "global calibrations." Thus, in this field today it is, for example, common to see a nickel-base alloy calibration, perhaps derived from a hundred standards. Results are then "fine tuned" with, say, an Inco 718™ type standard. Type standardization is also applied when calibrations are based on fundamental parameters.

The analyte content of a solid standard is not usually a quantity that the analyst can pick and choose at will. He or she must be satisfied with what is available, and, as we have mentioned, especially in the trace realm, world stocks are slim. A related factor is the number of significant digits included in the certified value. A standard certified at 0.001%, for example, is of little use, even as the high point on a calibration curve, if the analyst must report a lead result to a tenth of a part-per-million.

The number of standards needed to calibrate or validate a method is always to some degree a judgement call on the part of the analyst, unless it is specified in the standard operating procedure. One must be reminded always of the maxim, "extrapolation is dangerous and interpolation is safe." Thus, we should always have some (and at least one) reference point at a concentration just higher than our highest unknown sample. If we are operating near the lower noise limit of the method we should have at least one standard that allows us to unequivocally define the quantification limit of the procedure in this specific application. If we report actual test results down to 0.1 ppm we should have evidence in hand that at the time the test was performed, at least, we could confidently measure 0.1 ppm (and not merely distinguish it from zero).*

It is an unwritten rule that as calibration curves depart from linearity, the need for incremental data points increases. Another way of saying this is that a curvilinear relationship requires better empirical definition than one that is known to be rectilinear. Analysts have been getting around this complication for decades with a number of "hat tricks". For example, narrowly bracketting the unknown concentration by a line segment between two stan-

---

* Changing detection and quantification limits are a reality. They should be dutifully checked and reported. The results may, however, require an educational program for the lab's clients.

dards works for all but the most severely curved relationships. Or, one can assume a polynomial relationship of one kind or another, and by digital or analog techniques draw a smooth curve through a sparse array of data points. In an ideal world, of course, there would always be a sufficient number of reference materials to positively define any sort of nonlinear calibration curve.

For most work a calibration curve can be correctly regarded as linear from the origin to some "point of highest linearity," whereupon the slope begins to change as curvature sets in with higher and higher concentrations. The majority of analytical curves tend toward a line parallel with the concentration axis. A few may even "roll over"* and begin to show diminishing response with increasing analyte concentration.

Another area of special concern in trace analysis is the region near the origin, where strange nonlinear effects can sometimes be observed. If we choose to regard only the linear portion of the calibration the uncertainty envelope surrounding the line begins to flare at its upper and lower ends, indicating the need for a good array of reference materials for both regions in order to define the linearity limits.

This brings up another consideration in the selection and suitability of reference materials—one large enough for its own section.

2.2.2 Uncertainty [25–33]

One does not find the term "uncertainty" mathematically defined in elementary statistics texts. In metrology circles in recent years it has come to be widely used to refer to the measure of confidence that the user can place in a certified value. Recent dissatisfaction with the ISO Guide 30 definition (see Section 2.1) has led to the following suggested change:

Uncertainty of measurement: A parameter associated with the result of a measurement that characterizes the dispersion of the values that could reasonably be attributed to the measurand." [25–26]

This revision avoids reference to an unknowable "true value," which some objected to, while still, presumably conceding its existence. Mathematical treatments to arrive at an uncertainty are all based on a summation of the errors introduced at each stage in the certification process—the so-called *uncertainty budget* for each certified analyte. However, it is fair to say that despite some ponderous tomes that speak with great authority, there is not yet agreement on how best to accomplish this task for chemical analysis measurements [30].

---

* Zeeman-corrected GFAA can show this effect. Another example is self-absorption in optical emission spectrometry.

Many metal alloy certified reference materials do include uncertainty values on the accompanying certificate. All NIST SRMs produced after January, 1993 incorporate uncertainty statements, for example [32]. The question we need to consider is what to make of these numbers.

The most common error in regard to published uncertainties is the assumption that an analytical result obtained for the CRM must fall within the published uncertainty range. In fact, all that is required to validate a procedure with a given CRM is for the analytical method precision (its tolerance interval, as defined, for example, by its within-laboratory repeatability) to overlap the uncertainty range of the CRM.

If one is in doubt about the precision of the test procedure the best course is to run the CRM several times, then calculate a mean and standard deviation. One then develops a confidence interval for the mean from:

$$U = \pm ts/(n)^{1/2}$$

where t is the value from a two-sided student's t-test table for the appropriate degrees of freedom [defined by $(n - 1)$, where n is the number of replicates] at the desired confidence interval (typically 95%); and s is the standard deviation $\{\pm [\Sigma(\bar{x} - x)^2/(n - 1)]^{1/2}\}$. If the value of U is larger than the difference between the certified value and the mean the method has been validated with respect to that CRM. If the value of U is larger than the difference between the largest divergence between the certified uncertainty range of the CRM and the mean the method has also been validated.

Certified reference materials with published uncertainties are *not* more accurate than CRMs that lack uncertainties. In fact, there are those who would argue that the recent deterioration of the infrastructure for producing metal alloy CRMs has resulted in materials that are significantly *less* useful than those of an earlier era, when certificates almost universally lacked uncertainties or sported something with that appellation, but which was, in fact, a judgement call based on little more than a feel for "analytical windage."

The analyst must never abrogate his responsibility to his assigned task by blind acceptance. There are occasions when a certified value does not seem to fit—situations when that cast-in-stone number with its narrow range of uncertainty seems to contradict all the other available evidence. Then the analyst must scrutinize every aspect of what he is doing; redo the calibrations and rethink the protocols. Only when all this has been exhaustively pursued and no errors are found—only then is it time to question the unquestionable.

2.3 Traceability [34–37]

Traceability is an issue that has been creeping into the consciousness of the practical working analyst for more than a decade. It arises principally from the audits and accreditations that are the way commercial business is being conducted these days. In its essence, traceability is an appeal to a higher authority to imbue truth on the work we have just completed. If we look at the definition in Section 2.1 we see a linkage to "stated references, usually national or international standards." The language is broad enough to allow linkage to a reference method, to a reference material, or to one of the seven *Systéme International d'Unites* (SI) base units. For chemical measurements these would most typically be the mole or the kilogram.

On the surface it appears that classical gravimetry and titrations based on normality, which are unquestionably definitive techniques of Class I, are also directly traceable to the kilogram and mole, respectively. In fact they are not, in any easily understood sense. Gravimetry results are typically expressed as a *ratio* of masses derived from weight *differences*. They would be identical tomorrow even if every atom in the universe suddenly lost half of its mass. The mole relates atom, ion, or molecule aggregates to macroscopic masses and is useful to prepare even-quantity concentrations of titrant, but again, it is the *ratio* of the reactants that is important in obtaining an analytical result. The other method categories, Class II and Class III, are even further removed from traceabilty to SI units. It appears that the analytical chemist is limited in his work to traceability to a reference method or a reference material.

A reference method can take two forms: an empirical or semi-empirical standard method that everyone agrees to follow rigorously, or an umpire method that is too long or costly to run routinely, or that requires specialized equipment or expertise not normally accessible in an industrial working laboratory. In these two distinct cases we can say that a result is traceable to a reference method—in the first instance because nearly everybody finds the same result on the same sample, and in the second because the umpire method finds the same result as the "routine" method.

The great majority of traceability data, however, involves reference materials, and, most particularly, certified reference materials. The simplest and most frequently used documentation of traceability are the results from validation CRMs analyzed at the same time and in the same manner as the sample. If the CRMs have a high metrological pedigree, a recorded result that closely matches the certified value goes a long way toward establishing traceability to at least the national laboratory level.

Class II methods can be traceable to *both* the pure element used to prepare the calibration solutions and to CRMs used to validate the method.

Length—meter: mass—kilogram: time—second: electric current—ampere: temperature—kelvin: amount of substance—mole: luminosity—candela.

But why do we care about traceability? The answer, at its heart, is to show each other that we are all playing on the same field and, thus, our results are *comparable*. At a deeper level we want to demonstrate that our results reflect the *truth* at humanity's current level of understanding of the universe.

2.4 The Certification Process [13,14,38–42]

Clearly, it is beyond the scope of this book to describe in minute detail the step-by-step process involved in the making of a metal alloy CRM. However, some metal producers (not always the largest) have a program in place for the development of inhouse reference materials. Others may be considering such a program. So it is worthwhile to at least touch upon this rather involved subject.

The first step in the process, once a need has been identified and it is confirmed that no applicable commercial CRM already exists, should be some type of cost/benefit study, considering alternatives. Since reference material production is costly, one must contrast the costs associated with *not* developing the material. Should a special "heat" of metal be melted or can a production lot be diverted for this purpose? Can all the work be performed inhouse, or is it better to outsource some of the production steps? Since it is not cost-effective to produce small lots of reference materials that are soon expended one needs to consider how large a lot can be practically handled.

Once the material has been procured it must undergo one or more processing steps to convert it into its final form—generally, spectrometric solids, chips, or pins. This may involve heat treating, forging, centerless grinding, swaging, or drawing. More than 90% of the solid reference materials in the world are wrought forms, which are the product of some form of mechanical hot or cold working.

But some spectrometric techniques—X-ray fluorescence, for example—are known to be sensitive to cast/wrought structure differences for certain types of alloys. Examples among ferrous alloys are cast irons and high speed steels. Cast-structure standards are much more difficult to produce in large lots than are wrought-structure standards. They *are* needed in some circles, however. Hot isostatically pressed atomized metal powders have been used to produce solid spectrometric standards for certain types of alloys, but they are not the answer for all applications.

When pieces are cut out of larger forms it is important to maintain their original position identity by assigning location codes. This allows for a rational sampling plan when homogeneity testing is initiated, and also to identify areas of detected inhomogeneity so that they can be discarded.

Homogeneity testing may involve XRF, OES, and/or thermal evolution methods. Special types of standard development may require the use of additional techniques. For small lots of, say, less than 50 pieces 100% testing is called for. Larger lots must be sampled in a representative manner—for example, the ends and middle of cut bars, or a helical pattern of sample positions down the length of a billet.

Each element to be certified must be homogeneity tested since it is entirely possible that otherwise homogeneous material may be inhomogeneous with respect to one or a few analytes. The statistical treatment of the data involves the testing of a null hypothesis that the analytical variability does not exceed the variability due to sample location.

If the material is judged to be homogeneous, the next step involves the production of drillings or millings from an appropriate number of randomly or representatively selected pieces. Special procedures are often employed to obtain a uniform size of chips. The chips are degreased with a low residue solvent, dried, and blended.

Wet chemical analysis may be performed by one or a network of laboratories. In either case, unambiguous testing protocols must be provided to the analysts. These would include the number of days on which the test is to be repeated, the number of replicates per test, the number of significant figures in the results to be reported (at least one more than is planned for the certified result). The method to be used may be specified, or the laboratories may be exhorted to use their own best method.

As referred to above, the results of such testing gains weight from a consensus of diverse methodologies. Definitive methods should be encouraged where feasible. Validation CRMs may be specified or left to the analyst's or the laboratory's discretion. But the validation results must always accompany the data. Comments on the work should also be requested from each analyst.

The testing coordinator compiles the data and calculates the statistical results. This may include the arithmetic mean, the within laboratory variability (repeatability), between laboratory variability (reproducibility), tests for bias, and the calculation of an uncertainty for each certified value. The testing coordinator then prepares the certificate, which may include, in addition to analyte values and associated uncertainties, a list of the measurement techniques employed, instructions for the intended use of the reference material (possibly even including a minimum sample size), statements about stability and storage, details of the standard preparation protocols, and other information.

Certificates occasionally require revision for reasons ranging from discovered "typos" to revised values or added analytes. The revision should include the original issue date and the revision history of the certificate. In addition, the responsible agency must keep a detailed data file for each certificate, including data from all revisions.

3. SAMPLING AND SAMPLE PREPARATION FOR TRACE ANALYSIS

It is not our intention here to reprise all the topics covered in Chapters 1 and 2, which treated these two areas in considerable detail. It *is* appropriate, however, to extract and re-emphasize some key issues that pertain to quality at the trace level.

Recall Pierre Gy's distinction between constitution heterogeneity and distribution heterogeneity. The former describes the analyte constitution of individual particle types. The latter describes the analyte concentration effect of the mixture of particle types in the sample [43,44]. This is *not* a remote subject for some types of trace work with metals.

The process of machining chips from various alloys is an inherently complex and incompletely understood mechanism. Some stainless steels, for example, machine like chewing gum, forming deformed or partially extruded chips. Other metals undergo a kind of partial brittle fracture that releases fines as well as chips. If machined chips from certain alloy grades were to be sieved and each fraction analyzed, quite different compositions would be found. Fines are likely to be enriched in elements like silicon and carbon when those elements are present in the base metal at high concentrations. Fines may also contain elevated levels of tramp impurities, which are often associated with hard and brittle grain boundary precipitates.

We must ensure that the removal of a test portion is a *probalistic* process—that is, a process in which all particle sizes have *some* probability of being selected. This is clearly not the case if fines are removed or if the sample is not properly mixed. Ideally, we must strive to improve the probability that the test portion will be selected in a way that accurately represents the distribution of analyte in the original solid. This can be as simple as shaking a bottle or envelope before weighing out a portion. Or it can be as elaborate as sieving a large weighed sample of chips at several mesh sizes, weighing each mesh fraction, then preparing a test portion by weighing proportionately-sized weights from each mesh fraction.

The size of the test portion is a critical concern in trace analysis. Unless the material is known to be extremely homogeneous with respect to the analyte, it is always best to give precedence to sampling considerations over analytical considerations. In other words, take a suitably large test portion, dissolve it, and take an appropriate aliquot. Do not take a small test portion weight just because your method is sensitive enough to allow it. Except in unusual cases, no one should be attempting solution-based trace analysis with a metal alloy test portion smaller than 500 mg. In Chapter 1 we spent some time on the subject of test portion size. It was there that we noted that a very small test portion that yields very precise results is probably producing an erroneous average.

In some ways a properly designed sampling plan for trace analysis affords the analyst more latitude since minor weighing and dilution errors do not show up significantly in what is, in most cases, a single significant figure result. That is not to say that a metals lab can lower its vigilance about analytical balance calibration, or the design of a pipetting scheme that minimizes the propagation of errors. Such fundamentals are essential in the preparation of calibration solutions for Class II methods, for example, as well as for all work at the "macro" level [45–47].

We must also, of course, remain critically mindful of those essential features of trace analysis that make it difficult. Most trace methods require a carefully measured blank—even after the application of strategies to reduce its size. Blanks in trace methods can sometimes represent a significant part of the total analytical signal. Contamination from reagents, equipment, the lab environment, and people, as well as positive sample matrix interferences, all contribute to the blank. Something can usually be done to reduce some or all of these, but some fraction of the total effect will usually remain. If only one point from this volume can produce a change, the author believes it should be the widespread use of representative *replicate* blanks.

In trace analysis the absence of an accurate blank puts the analyst out of business in a hurry. A replicate average is always a wise choice; and the blank solution should represent the matrix to the extent needed. But representation is not always easy to accomplish when matrix interference effects are complex or incompletely understood, as is often the case in ICP–MS or GFAA work, for example. Sometimes it is possible to use a "high-purity" version of the alloy matrix as a blank. If such a material is found it should be thoroughly characterized as "analyte-free," preferably by two or more independent techniques. The use of a proper array of validation standards provides the best check on the validity of the blank employed.

The opposite effect—diminished signal due to negative interference effects and loss of analyte—is also extremely important in trace analysis. Not only can low concentrations of analyte be lost by volatilization or precipitation, but chemisorption and physical adsorption effects on vessel walls can result in a dramatic drop in measured analyte concentration. And so selection and cleaning of lab equipment becomes critical in any well-thought-out dissolution and separation scheme. Negative interference effects from matrix elements are usually chemical (e.g., coprecipitation) and must be addressed where feasible.

Finally, there is the matter of replicate sample determinations. No one *wants* to do replicate analyses if they are convinced that a single determination will suffice. And such is the conventional wisdom in trace metals analysis; but nothing could be further from the truth. Veteran tramp analysts who have worked in the field of some narrow range of alloys have sometimes convinced themselves that spikes in the data do not (and by inference, cannot) occur. And so, from this fatuous and erroneous perspective, replicates become a waste of time.

In fact, it may take many replicates to properly represent some samples when unavoidably low test portion sizes are employed. The effect is especially important in solids instrumental work, where test portion sizes may range from milligrams to monolayers. In trace work with solid samples quadrants of a disk and several repolished surfaces may have to be examined before the analyst can convincingly demonstrate that his grand average is a reasonable estimate of the truth.

4. DETECTION AND QUANTIFICATION LIMITS [48–51]

Here, again, we briefly return to a subject that has been treated elsewhere in this volume—in this case in Chapters 5 and 6. Both terms in this section's title have been defined in Section 2.2 above.

The detection limit is an overworked concept that, in the final analysis, has not proven to be especially useful. There are several reasons for this. First, it is not common for a laboratory's clients to be uninterested in a quantitative result once they are informed that a component of interest has *been detected*. The next question is likely to be: what is your detection limit? If you respond with your dutifully calculated detection limit value, you have, in effect, presented your client with a quantitative result, despite whatever qualifying addenda or protests you include. This is especially regrettable because of the second concern about detection limits—they are based on very loosely specified statistics.

In Chapter 5 (Section 7) and Chapter 6 (Section 4.2.2) we used an equation of the general form:

$$C_{DL} = (\Delta C/\Delta x)sk$$

where $(\Delta C/\Delta x)$ is the reciprocal slope of the calibration curve (i.e., change in analyte concentration, ΔC, with change in analyte signal, Δx); k is a statistical factor $[2 \times (2)^{1/2}$, or approximately, 3]; and s is an estimate of the standard deviation of a result from a very low analyte concentration material, or from the blank.

The problem arises because the exact analyte concentration used and the number of replicates taken to achieve a value for s are not specified. And so it is conceivable that a broad range of values for C_{DL} might be obtained by a variety of analysts and labs conducting the same test. Labs that utilized a reference material with a higher concentration of analyte, and that analyzed it with many replicates are likely to show much lower detection limits than labs that used a reference material with an extremely low analyte concentration, or a blank, and that analyzed it only a few times. If we were to standardize our calculation by basing in on a standard deviation determined at a point well into the optimum concentration range for the method, we are likely to obtain a consistent, reproducible value for the detection limit. But it may not reflect very accurately the point at which we can first unequivocally distinguish the presence of analyte.

Substituting the Background Equivalent Concentration (BEC) for $(\Delta C/\Delta x)$ and the relative standard deviation of the background (RSD_B) for s does not simplify matters. In Chapter 6 we saw a frequently used expression in optical emission work:

$$C_{DL} = 0.01k(RSD_B)(BEC)$$

where, again, k is approximately 3. Here, the problem of accurately determining (RSD_B) is identical in nature to that described above. Pure instrumental methodologies with few manipulative steps greatly simplify the process of obtaining a reliable estimate for (RSD_B), however.

Accurate determination of the precision of a measurement at some level just above zero concentration requires an unconventional approach to data collection for many instrumental techniques. In this situation it becomes necessary to collect all readings—positive, negative, and zero—in the statistical data gathering phase. It may help the analyst to regard these numbers as response measurements, even though they may be labeled as concentration. The statistics remain viable so long as no filter, human or electronic, is imposed to remove the negative values.

Michael Thompson has argued convincingly that detection limits, indeed, quantification limits, as well, are unnecessary. He suggests plotting a curve of uncertainty versus analyte concentration, and on the same scale and axes also plotting the specific variability tolerance for the analysis versus analyte concentration. So long as the second curve remains above the first, the method is fit for the application and results can be reported [48]. This "fitness for purpose" has been a theme, albeit not yet a dominant one, in discussions in the current literature.

The quantification (or quantitation) limit has traditionally, and somewhat arbitrarily, been defined as $10s(\Delta C/\Delta x)$—that is, ten times the standard deviation times the reciprocal slope of the calibration curve. Another definition is simply $x_0 \pm 10s_0$, where x_0 is the average blank measurement and s_0 is the standard deviation of the blank. The graphical treatment from the book by Welz [50], described in Chapter 5, is, perhaps, the most convincing treatment of the concept.

If one adopts such concepts then it is appropriate to report analytical values above the quantification limit, to report "less than" values between the quantification and detection limit, and to report "none detected" below the detection limit.

Gibbons, Coleman, and Maddalone have developed a new approach to the concept of the quantification limit that they term the Alternative Minimum Level (AML). Their treatment incorporates the relationship between variability and concentration, the uncertainty in the calibration, and the uncertainty in the standard deviation of the blank [51].

One key feature of detection and quantification limits or what is used to replace them is that they change—with time, with instrumental and environmental conditions, and between instruments, analysts, and laboratories. One must *not* assume otherwise. When such limits must be declared in official documents (e.g., in a Standard Operating Procedure), or to the laboratory's clients, it is prudent to be conservative. Such limits should be frequently checked, in some cases each time the method is run.

In the final analysis, what emerges from a reflection about these terms is a need for consistency, so that the laboratory's clients become familiar with its "reporting style" in this regard. Some labs, for example, choose never to report "none detected" if a quantitative result is requested; instead, they report "less than" the quantification limit, whether any analyte signal has been detected or not. There is more to this reporting style concept, which we will address in the next section.

5. SPECIFICATIONS AND REPORTING CONVENTIONS [52,53]

Specifications are agreed upon limits for some attribute of a commercial commodity. The metals analyst sees these as concentration limits for an elemental component. They are typically expressed by an absolute concentration—a minimum or a maximum—or as an absolute range of concentrations. The analyst is expected to match his results against these criteria for the alloy product, which are typically set by economic, production-related, or physical property considerations. But unlike the specification limits, the analyst's data are never absolute numbers—they are estimates of the truth with an associated range of uncertainties. And therein lies an essential problem.

If any part of the analytical error bar overlaps a specification limit, the material is, theoretically, rejectable as out-of-specification. This point is nearly always difficult to make with the lab's clients, but it is correct in all respects. The answer appears to lie in a dialogue between the analytical chemist and the specification writers. Chemists must argue for wider specification limits where that is feasible, and for analytical control limits within the specification limits so that the statistics for a high (or low) range result still fall short of crossing a specification line. For the same reason melting metallurgists should be exhorted to aim for the center and not the upper or lower edges of a specification range. Finally, the chemist should campaign against technically and legally insupportable language in specifications, such as "all other impurities," which still persists for certain metal commodities.

There are also a number of key issues concerning the manner in which data are treated that directly relate to the interface with specifications. A prudent laboratory manager will establish a Standard Operating Procedure that outlines the reporting conventions that the laboratory will adhere to. This official consistency, if strictly followed, will produce a body of work with a uniformity that may reveal data trends that would not be apparent otherwise. At the very least, such a document will give the analyst the guidance to treat his own result sets consistently so that related samples are reported on the same basis.

Most analysts know that rounding is always the last step in a calculation, and that when the last digit of an average is a 5, the value is rounded to the *even* number, and also that zero is an even number. But it doesn't hurt to reiterate these rules in such a policy document. More subtle are some situations in trace analysis where statistical texts offer no guidance. For example, consider the situation in which six replicate results are obtained very near the noise limit of a method. Let us assume that the sample material is known to be homogeneous with respect to the analyte and that the detection limit and the quantification limit of the method have been recently and correctly characterized. Table 10.4 shows the model data set.

Table 10.4 A Model Trace Level Determination (Se in Ni-base Alloy—Method A)

3-Sigma Detection Limit: 0.00002% Se
10-Sigma Quantification Limit: 0.00006% Se
Reporting Requirement: 5 decimal places
Specification: 0.00008% Se (Max.)

% Se Found	Reporting Style	Result Reported, % Se
0.00002	Statistical	<0.00012
0.00009	Literal	0.00005
0.00000	Conservative	<0.00020
0.00005	Specification-driven	<0.00008
−0.00003		
0.00016		

Uncensored Statistical Mean: 0.000048
Censored Statistical Mean: 0.000053

It is not a pretty picture. But it is also not very far removed from actual situations that a metals analyst may have to face from time to time. To further pressurize the setting let us assume that there is no time to do any additional work—a value must be reported today.

An examination of the data reveals a method stretched to its very lowest limits, the analyte signal nearly lost in noise. By any criterion based on the precision of the measurements, no result should be reported. But a result must be reported.

Four of the six results are below the specification limit maximum, and two are above it. The readings stretch from -0.00003 to .00016%. Two readings are below the detection limit and one is at the detection limit. Two readings are above the quantification limit and one is between the detection limit and the quantification limit. Censoring negative values (as most analysts and many instruments do)—that is, equating them to "none detected" or zero in *this particular case* has little effect on the average. (In other examples such a practice could have a profound effect on the average.) Here, both the uncensored and the censored averages show the material to be within specification. But is it?

Perhaps, a better question for the reader who has followed this exercise so far is: has the analytical work allowed us to say something about the selenium level in this hypothetical sample of nickel-base alloy? At first glance one might suspect that the 0.00016% value is a rejectable statistical outlier, although in our example it has no assignable cause.* There are certain tests that can be applied in these cases, but their use is discouraged for small data sets like this one. It could also be argued that nearly all analytical chemistry data sets are "small" in the statistical sense. Nevertheless, if we apply the Dixon test in this instance (see any good statistics text, like Reference 20, for a description) the questionable 0.00016% value is *not* rejectable.

In this example, we are faced with two data points that show the test material to be out of specification, and four data points that show it to be in specification. While our average shows the material to be in specification,it has an associated uncertainty (at the 95% confidence level) of ± 0.72 ppm. In other words, we know with 95% certainty that the true value is less than 0.00012%

At the bottom of Table 10.4 are listed some results that might be reported in this situation using four different reporting conventions (or reporting philosophies, if you will). The "Statistical" style dutifully goes through

---

* It can be convincingly argued that no data point should ever be rejected without a known error having been observed. It is always legitimate, however, to reject a complete data set and repeat the entire test protocol.

the calculations and reports the value obtained. The "Literal" style merely takes the average and reports it. The "Conservative" style stems from a concern that the most damaging data might be the only correct data. After all, at least one of the six results was above the 0.00012% level, and there is a 5% chance of the true value being above 0.00012%). To add an extra measure of assurance, this approach raises the "less than" level to the next higher four-place digit. The "Conservative" style will also fail a material if *any* of a set of replicates shows it to fail. Finally, the "Specification-driven" style adopts a literal approach, but uses the specification value as its reporting platform. In this case only the technically questionable "Literal" and "Specification-driven" styles are likely to make the laboratory's client happy.

No doubt any analyst worth his mettle would like to see such an unruly data set repeated after some refinements to the method have improved its signal-to-noise ratio. Table 10.5 shows a hypothetical repeat set of six values for the same sample material from an improved version of the procedure.

This data set appears somewhat more behaved, but it, also, provides an illustration of reporting style differences. In this case, the mean (there is no censored mean) has an associated uncertainty (at 95% confidence) of ±0.27 ppm. The Statistical style would report "less than or equal to" the specification limit. The Conservative style would be concerned about the one value at the specification limit, and the one at 0.00009%, and so it would report <0.00010%. Once again the lab's client is only likely to be happy with the problematic Literal and Specification-driven results.

Table 10.5 A Model Repeat Trace Level Determination (Se in Ni-base alloy: Method A-1)

3-Sigma Detection Limit: 0.00001% Se
10-Sigma Quantification Limit: 0.00004% Se
Reporting Requirement: 5 decimal places
Specification: 0.00008% Se (Max.)

% Se Found	Reporting Style	Result Reported, % Se
0.00004	Statistical	≤0.00008
0.00008	Literal	0.00006
0.00002	Conservative	<0.00010
0.00009	Specification-driven	<0.00008
0.00006		
0.00005		

Statistical Mean: 0.000057

And upon this point there is just a little more that needs to be addressed, whereupon we can bring this chapter and this book to a conclusion.

6. CONCLUSION

The analysis of metals and alloys for trace level elemental components is pursued because materials scientists and metallurgists have concluded that such compositional information is vital to understanding and predicting the performance of the product. The analytical chemist, for his part, seeks to reconcile an envelope of uncertainty with inflexible limits, often in concentration regions that tax the capabilities of his tools and talents. The examples of the previous section are not reflective of the norm, but neither are they a fantasy. Such situations *do* occur. And the reason appears to be that no one consulted the tester before they ordered the test.

It would be extremely useful if a dialogue could be opened between the metals analyst and those who establish specifications and design new alloys. The metals analyst would be happy to oblige his client's requests with trace and ultratrace methods for which the needed lower working limit shows a relative percent error significantly below 50%. In some instances this is not the case, either because the technology to do so does not exist, or is so expensive or so arcane as to place it beyond the reach of all but the largest research facilities. A dialogue between the specification writers and the analysts would not solve this problem, but it could lead to more understanding about what the lab results mean. It may even lead to specifications that account for analytical uncertainty.

At the very least, it might give some future specification writer pause to stop and ask: "Can the lab analyze this stuff?"

REFERENCES

1. 1997–1998 Brammer Catalog, Brammer Standard Company, Houston, TX, pp. vi–vii.
2. Roelandts, I. Spectrochimica Acta, 45B, 815 (1990).
3. Roelandts, I. Spectrochimica Acta, 45B, 1275 (1990).
4. Roelandts, I. Spectrochimica Acta, 46B, 1101 (1991).
5. Roelandts, I Spectrochimica Acta, 47B, 749 (1992).
6. Roelandts, I. Spectrochimica Acta, 48B, 461 (1993).
7. Roelandts, I. Spectrochimica Acta, 49B, 1039 (1994).
8. Roelandts, I. Spectrochimica Acta, 49B, 1097 (1994).
9. Roelandts, I. Spectrochimica Acta, 49B, 1103 (1994).
10. Roelandts, I. Spectrochimica Acta, 50B, 205, (1995).

11. Dulski, T.R. A Manual for the Chemical Analysis of Metals, ASTM MNL 25, American Society for Testing and Materials, West Conshohocken, PA, 1996, pp. 189–192.
12. Dulski, T.R. A Manual for the Chemical Analysis of Metals, ASTM MNL 25, American Society for Testing and Materials, West Conshohocken, PA, 1996, pp. 189–192.
13. ISO Guide 32 "Calibration of Chemical Analyses and Use of Certified Reference Materials" International Organization for Standardization, Geneva, Switzerland.
14. ISO Guide 33 "Uses of Certified Reference Materials" International Organization for Standardization, Geneva, Switzerland.
15. Taylor, J.K. Analytical Chemistry, 55, 600A (1983).
16. Annual Book of ASTM Standards, Vol.03.06, American Society for Testing and Materials, West Conshohocken, PA, 1998, Designation E882.
17. Annual Book of ASTM Standards, Vol.03.06, American Society for Testing and Materials, West Conshohocken, PA, 1998, Designation E1329.
18. Manual on Presentation of Data and Control Chart Analysis, ASTM MNL 7, 6th ed., American Society for Testing and Materials, West Conshohocken, PA, 1990.
19. Natrella, M.G. Experimental Statistics, NIST Handbook 91, National Institute of Standards and Technology, Gaithersburg, MD, 1996, pp. 18.1–18.4.
20. Miller, J.C.; Miller, J.N. Statistics for Analytical Chemistry, 2nd ed., John Wiley, New York, 1988, pp. 92–98.
21. Shewhart, W.A. Economic Control of Quality of Manufactured Product, Van Nostrand, New York, 1931.
22. Harmonized Guidelines for Internal Quality Control in Analytical Chemistry Laboratories, ISO/REMCO N271, Revision, International Organization for Standardization, Geneva, Switzerland, November, 1994.
23. Epstein, M.S. Spectrochimica Acta, 46B, 1583 (1991).
24. Schiller, S.B.; Eberhardt, K.R. Spectrochimica Acta, 46B, 1607 (1991).
25. International Vocabulary of Basic and General Terms in Metrology, 2nd ed., International Organization for Standardization, Geneva, Switzerland, 1993.
26. Guide to the Expression of Uncertainty in Measurement, International Organization for Standardization, Geneva, Switzerland, 1995.
27. Quantifying Uncertainty in Analytical Measurement, Eurachem, Laboratory of the Government Chemist, Middlesex, UK, 1995.
28. Guidelines for Evaluating and Expressing the Uncertainty of NIST Measurement Results, NIST Technical Note 1297, National Institute of Standards and Technology, Gaithersburg, MD, 1994.
29. Analytical Methods Committee, The Royal Society of Chemistry, Analyst, 120, 2303 (1995).
30. Horwitz, W.; Albert, R. Analyst, 122, 615 (1997).
31. Ellison, S.L.R.; Gregory, S.; Hardcastle, W.A. Analyst, 123, 1155 (1998).
32. NIST Standard Reference Materials Catalog (N.M. Trahey, ed.) NIST Special Publication 260, 1998.
33. Taylor, J.K. Standard Reference Materials: Handbook for SRM Users NIST Special Publication 260-100, National Institute of Standards and Technology, Gaithersburg, MD, 1985.

34. Taylor, B.N. Guide for the Use of the International System of Units, NIST Special Publication 811, National Institute of Standards and Materials, Gaithersburg, MD, 1995.
35. Valcarcel, M.; Rios, A. Analyst, 120, 2291 (1995).
36. King, B. Analyst, 122, 197 (1997).
37. Thompson, M. Analyst, 122, 1201 (1997).
38. An.ual Book of ASTM Standards, Vol.03.06, American Society for Testing and Materials, West Conshohocken, PA, 1998, Designation E1724.
39. Annual Book of ASTM Standards, Vol.03.06, American Society for Testing and Materials, West Conshohocken, PA, 1998, Designation E1831.
40. ISO Guide 30 "Terms and Definitions Used in Connection with Reference Materials" International Organization for Standardization, Geneva, Switzerland.
41. ISO Guide 31 "Contents of Certificates, Certification Reports, and Labels of Reference Materials" International Organization for Standardization, Geneva, Switzerland.
42. ISO Guide 34 "Quality System Guidelines for the Production of Reference Materials" International Organization for Standardization, Geneva, Switzerland.
43. Gy, P.M. Sampling of Heterogeneous and Dynamic Material Systems, Elsevier, Amsterdam, 1992.
44. Pitard, F.F. Pierre Gy's Sampling Theory and Sampling Practice, Vols. 1 and 2, CRC Press, Boca Raton, FL, 1989.
45. Harris, G. ASTM Standardization News, April, 44 (1993).
46. Zehr, B.D.; Maryott, M.A. Spectrochimica Acta, 48B, 1275 (1993).
47. Youden, W.J. Experimentation and Measurement, NIST Special Publication 672, National Institute of Standards and Technology, Gaithersburg, MD, 1991.
48. Thompson, M. Analyst, 123, 405 (1998).
49. Julicher, B.; Gowik, P.; Uhlig, S. Analyst, 123, 173 (1998).
50. Welz, B. Atomic Absorption Spectrometry (trans., C. Skegg), 2nd ed., VCH, Weinheim, Germany, 1985.
51. Gibbons, R.D.; Coleman, D.E.; Maddalone, R.F. Environmental Science and Technology, 31, 2071 (1997).
52. Annual Book of ASTM Standards, Vol.03.06, American Society for Testing and Materials, West Conshohocken, PA, 1998, Designation E1282.
53. Dulski, T.R. A Manual for the Chemical Analysis of Metals, ASTM MNL 25, American Society for Testing and Materials, West Conshohocken, PA, 1996, pp. 196–202.

Glossary

Terms included here either have not been fully defined in the text or are used so widely that their definition bears repeating for the reader's convenience.

Absorbance: n. the diminution of electromagnetic radiation by its passing through a substance; minus the $\log_{10}$ of the transmittance.

Absorption edge energy: n. the minimum energy (of electromagnetic radiation or of an impinging charged particle) needed to eject an electron from an atom's specific inner electron shell.

Absorptivity: n. absorbance per [unit pathlength times the analyte concentration]. Pathlength is in centimeters and analyte concentration is in kg/m^3 (see also: *molar absorptivity*). Also known as "specific absorption coefficient."

Abundance sensitivity: n. a parameter that describes the extent of tailing on both sides of a mass-to-charge peak in mass spectrometry.

Aliquant: n. (obsolete). a quantitative portion of a solution that represents an irrational fraction of the whole (e.g., 3 mL from 100 mL).

Aliquot: n. 1. any quantitative portion of a solution. 2. (obsolete) a quantitative portion of a solution that represents a rational fraction of the whole (e.g., 2 mL from 100 mL).

Analyte: n. the sample component that is determined.

Annihilation reaction: n. a nuclear reaction in which a particle and its corresponding antiparticle (e.g., an electron and a positron) collide and disappear, producing gamma ray photons whose combined energy is equal to the mass and kinetic energy of the two annihilated particles.

Astigmatism: n. an optical aberration in which the image of a point off the optical axis is a pair of lines.

Argide: n. a polyatomic ion incorporating argon.

Auger spectroscopy: n. a surface analysis technique that utilizes electrons emitted from the relaxation of atoms that have lost an inner shell electron.

Austenite: n. a face-centered cubic, nonmagnetic, metallic iron phase in certain stainless steels and other alloys.

Barn: n. a unit for the measurement of capture cross sections; 10^{-24} cm^2/nucleus.

BEC: abbrev. Background Equivalent Concentration. The concentration of analyte that would produce the same response as the background at the location of the analyte line in emission spectrometry.

Bragg's Law: n. an expression for the diffraction of electromagnetic radiation by a crystal; $n\lambda = 2d(\sin \theta)$, where n is the order of reflection, λ is the wavelength, d is the spacing between crystal planes, and θ is the angle between the incident beam and the lattice planes.

Bremsstrahlung radiation: n. the continuum radiation produced by the deceleration of incident electrons as they interact with the atom clouds of the target.

Cathode fall: n. in glow discharge sources, a nonluminous region immediately above the sample surface with a high discharge potential across it; also known as the "cathode dark space."

Characteristic concentration: n. the concentration of analyte that produces 1% absorbance (0.0044 absorbance units) in flame atomic absorption methods.

Characteristic mass: n. the mass of analyte that produces 1% absorbance (0.0044 absorbance units) in electrothermal atomic absorption methods.

Chromatic aberration: n. the effect in which a lens shows slightly different focal lengths for different colors (due to the fact that glass refracts light of different wavelengths by different amounts).

Collimator: n. a lens or other arrangement for producing parallel rays.

Collisional broadening: n. the broadening of an atomic absorption resonance line due to the collision of atoms in the atom cloud; also called "Lorenz broadening."

Comminution: n. crushing, grinding, or pulverizing to diminish average particle size.

Compton effect: n. the decrease in frequency (increase in wavelength) of gamma or X-rays that results from interaction with a bound or free electron.

Compton scattering: n. inelastic scattering of X-ray photons by the outer shell valence electrons of the target.

Concentration ratio: n. the ratio of the analyte response to that of the internal standard, which may be the main matrix element or an inoculant.

Coprecipitate: v.t. to precipitate under the influence of another element or species that forms an insoluble compound under the experimental conditions.

Cross section: n. an effective area used to express the probability of a capture, scatter, fission, or other nuclear or chemical reaction.

Dalton: n. the mass of one nucleon (proton or neutron).

Dark current: n. a background current in photon detectors generated in the absence of a photon flux by the temperature dependent emission of thermal electrons; especially used in reference to photomultiplier detectors.

Dark noise: n. a synonym for *dark current* often used in reference to solid state photon detectors; also sometimes called "dark charge."

Detection limit: n. the lowest concentration of analyte that can be reliably distinguished from zero.

Diffusion current: n. that portion of the total polarographic wave that is due to diffusion of analyte ion to the surface of the electrode.

Doppler width: n. the broadening of an atomic absorption resonance line that is due to the thermal motion of atoms in the atom cloud.

Dry asher: n. a device that utilizes a low pressure oxygen plasma to destroy organic material.

Dry filter: v.t. to filter a solution using a dry funnel, filter paper, and collection vessel so that no dilution occurs.

Eluate: n. the effluent from a chromatographic column containing some species of interest.

Eluent: n. the solution that is employed to desorb some species of interest from a chromatographic column.

Endogenous inclusions: n. nonmetallic compounds in a metal or alloy that have formed by chemical reactions during the melting or processing of the product.

Epitaxy: n. growth of a crystalline material on a crystalline substrate with a fixed spatial orientation between the two lattice structures.

Equivalent noise input: n. the light flux that produces a photomultiplier signal equal to the root mean square current from all noise sources; a measure of photon detection threshold.

Exogenous inclusions: n. nonmetallic compounds in a metal or alloy that have been incorporated from refractory containment contact by physical processes.

Faradaic current: n. that portion of the total polarographic wave due to the reduction of analyte cations.

Ferrite: n. a body-centered cubic, magnetic iron phase that occurs in many iron-base alloys.

Fixation line pair: n. in optical emission spectrometry, a pair of lines that respond differently in the event of changing source (or other spectrometric) conditions. Their ratio is used to monitor the stability of the excitation source.

Fluorescent yield: n. in X-ray fluorescence spectrometry, the portion of excited analyte atoms that emit secondary X-rays rather than Auger electrons.

Fourier transform absorption spectrometry: n. an arrangement in which a moving mirror that forms one leg of a Michaelson interferometer is employed with a mathematical transform that converts mirror displacement and detector response to an absorption spectrum. Now common for infrared instruments.

Fume: v.t. to heat a solution containing sulfuric, perchloric, or phosphoric acids to evolve the acid anhydride vapor.

FWHM: abbrev. Full Width at Half Maximum. the width of a peak at half of its height. Sometimes used to define Δm in the ratio ($m/\Delta m$), used as a measure of resolution in mass spectrometry

Gettering: n. the absorption of gases by a metallic film. A potential source of error in vacuum fusion analysis; also known as "contact gettering."

Global calibration: n. the use of a single calibration curve for analyte determination in a wide range of alloys.

Grab sample: n. a nonprobalistic sample resulting from an opportunistic single use of a sampling device.

Grotian diagram: n. an energy level diagram for electronic transitional states in atoms.

Hadron: n. a subatomic particle that interacts by means of the strong nuclear force (e.g., protons and neutrons).

Half-life: n. the time required for one-half of a large sample of unstable nuclides to decay.

Half-wave potential: n. the voltage at which the measured current of a polarographic wave is exactly 50% of the diffusion current.

HEPS: (abbrev.) High Energy Pre-Spark. a high current discharge prior to the data measurement phase in optical emission spectrometry. It is designed to rapidly heat, melt, and homogenize a region on the solid sample surface.

Homologous: adj. 1. in chemistry, pertaining to a series of related compounds that differ from one another by small fixed structural increments. 2. in spectroscopy, pertaining to a line pair that behave similarly under changing conditions (thus, the analyte line and the internal standard line must be a homologous line pair).

Hydrolysis: n. 1. a reaction involving water. 2. a reaction with water to form a precipitate of hydrous oxide.

Hysteresis: n. in general, the condition of a reversible system whose exact present state depends on past influences. Such systems may return to an initial condition by a different path from that by which they initially changed. The term has been applied to describe magnetic, electrical, and elastic systems.

Ignite: v.t. to heat at a high temperature.

Ilkovic equation: n. in polarography, a theoretical expression that relates analyte concentration to the diffusion current.

Intensity ratio: n. the ratio of the analyte response to the response from an added or indigenous internal standard element.

Intermetallic phase: n. a stoichiometric or nonstoichiometric metallic compound or an association of metallic elements dispersed in the matrix of an alloy.

IRZ: (abbrev.) Initial Radiation Zone. the region just below the Normal Observation Zone (NOZ) (see) in an operating inductively coupled plasma torch.

Isobestic wavelength: n. the wavelength at which two (or more) molecular species that are in equilibrium with one another exhibit the same molar absorptivity constant.

Isoelectronic sequence: n. a series of elements of increasing Z number that are increasingly ionized [e.g., a Z excited atom (zero valence), a (Z+1) monovalent ion, and a (Z+2) divalent ion]. Such sequences show similar types of optical emission spectra.

Isopiestic distillation: n. room temperature purification of volatile mineral acids by gas phase diffusion into high purity water in a closed chamber.

Laboratory sample: n. the sample submitted to the laboratory.

lattice dislocations: n. imperfections in the regular pattern of atoms in a crystal or metal that can move under ordinary conditions.

Leach: v.t. 1. to dissolve a solidified molten salt fusion in water or some aqueous reagent. 2. to selectively dissolve an analyte by treatment of a particulate solid with acid or other reagent; generally, regarded as an empirical technique.

Lepton: n. a subatomic particle of small mass that interacts by means of the weak nuclear force (e.g., the electron, the positron, and the various types of muons and neutrinos).

Liberation size: n. the largest particle size at which the analyte can be regarded as fully released from association with a specific particle type.

Limiting current: n. in polarography, the sum of the diffusion current, the capacitive charging current at the electrode surface, and the migration current; in effect, the limiting current defines the observed polarographic wave.

Load coil: n. the induction coil in an induction furnace or a plasma torch.

Local thermal equilibrium: n. a stable thermal environment, especially during spectrometric measurement; sometimes referred to by its abbreviation: "LTE."

Maraging steels: n. 17–20% nickel steels with major alloying amounts of cobalt and molybdenum. Such steels are precipitation hardening while "aged" at 480° C in the martensitic condition (see *martensite*).

Martensite: n. a hard and brittle, body-centered tetragonal phase of iron/carbon.

Matrix: n. 1. all components of the sample except the analyte. 2. the principal component of the sample (e.g., "an iron-matrix").

Measurand: n. that which is measured.

Metrology: n. the science of measurement.

Migration current: n. in polarography, a term for the current carried by the analyte species (usually a negligibly small part of the limiting current).

Miller index number: n. in crystallography, a system for designating the plane family to which a crystal belongs; generally, a 3-digit number, except for hexagonal crystals which require 4 digits.

Molar absorptivity: n. absorbance divided by the product of the path length and the analyte concentration in moles/liter. Also known as the "molar absorption coefficient."

MOS–FET: (abbrev.) Metal Oxide Substrate–Field Effect Transistor. An insulated gate transistor that can be utilized in various modes of operation.

Negative glow: n. the nearly field-free region where ionization occurs in a glow discharge source.

Nonprobalistic sampling: n. an incorrect sampling procedure in which some forms of the analyte have a zero probability of being included.

NOZ: (abbrev.) Normal Observation Zone. The optimal region for measuring the analyte emission spectrum in an inductively coupled plasma plume (typically, 15–20 mm above the top of the load coil).

Nucleons: n. protons and neutrons.

Nuclide: n. an atom characterized by the specific structure of its nucleus; thus isotopes of the same element are each separate nuclides.

Occlude: v.t. to absorb and retain one substance in another substance; often used to denote an undesirable condition (e.g., analyte loss).

Optic axis: n. a reference line through the vertices and foci of lenses and mirrors in the graphical construction of optical systems.

Oscillator strength: n. the number of classical electron oscillators (i.e., the average number of electrons) per atom that are involved in the transition associated with the absorption of a resonant wavelength of light by ground state analyte atoms. A measure of the sensitivity of an atomic absorption resonance line that is useful in comparing resonance lines for the same element.

Overvoltage: n. in electrolysis, the difference between an observed potential at which a reaction occurs and the potential predicted by the reversible half-cell reaction. E.g., mercury has a high overvoltage with respect to hydrogen evolution.

Pellicular: adj. a term applied to a form of low capacity ion exchange resin used in ion chromatography, consisting of inert beads coated with a thin layer of exchange sites.

Peltier effect: n. at the junction of two dissimilar metals an absorption or a release of heat caused by the application of an electrical current. Heat flow direction and magnitude are proportional to current direction and magnitude.

Penning ionization: n. ionization by reaction with excited neutral species (e.g., $M^{o} + Ar^{\cdot} \rightarrow M^{+} + Ar^{o} + e^{-}$, where $Ar^{\cdot}$ represents an excited state of the argon atom).

Peptization: n. 1. dispersion into colloidal form. 2. (obsolete) dissolution.

Petrography: n. 1. the systematic description of rocks and minerals. 2. the application of a petrographic microscope to the identification of minerals.

Photoelectron spectroscopy: n. a surface analysis technique that measures information from the photoejected inner shell electrons when the sample is bombarded with X-rays; sometimes known as "ESCA": "electron spectroscopy for chemical analysis."

PIXE: (abbrev.) Particle Induced X-ray Emission. A methodology that utilizes ion beams (typically 2–5 MeV proton beams) to produce characteristic X-rays from the sample for analytical measurement.

Pixel: n. a contraction of "picture element;" the smallest portion of a digital image.

Police: v.t. to remove traces of a precipitate from a vessel using a rubber paddle-like device (policeman) attached to a rod.

Prills: n. 1. any spherical particles about the size of buckshot. 2. in fire assay, the doré beads that result from the cupellation of lead-gold or lead-silver buttons. 3. small spheres that result from the spray solidification of molten salt.

Probalistic sampling: n. sampling in which all forms of the analyte have some probability of being selected by one operation of a sampling device. Probalistic sampling makes possible but does not guarantee correct sampling.

Profiling: n. in emission spectroscopy, the process by which the slit image is aligned with the dispersion element so that correct wavelengths are measured. The term is applied most often for wavelength calibration of direct reader instruments. Typically, a stable line source, such as a mercury vapor lamp, is utilized.

Quantification limit: n. the lowest concentration of analyte that can be measured with acceptable accuracy; also called "quantitation limit."

Quantum efficiency: n. for photon detection devices, the efficiency with which incident photons are converted to a measurable signal.

Quenching: n. any collisional or chemical process that lowers the quantum yield (i.e., sensitivity to analyte) in atomic or molecular fluorescence spectrometry.

Rayleigh scattering: n. coherent scattering that results from elastic collisions in which the intensity of a scattered wavelength at a given angle, θ, is directly proportional to $(1 + \cos^2\theta)$ and inversely proportional to the wavelength.

Ray tracing: n. the geometric construction of the path of a light ray from a point, through an optical system, to an endpoint (in a spectrometer the endpoint is the detector).

Reciprocal linear dispersion: n. the number of nanometers (or angstroms) separating two spectral lines that are 1 mm apart on the focal plane; a measure of the dispersing power of a grating.

Reciprocity failure: n. the failure of a photographic emulsion to respond *exclusively* to light exposure.

Relative aperture: n. the ratio of the focal length to the slit area in an emission spectrometer; a measure of light throughput. Also called the "f number."

Relative sensitivity factors: n. in mass spectrometry, (analyte concentration/measurement response) ratios that are sometimes applied to a wide range of matrices.

Rutherford backscattering: n. an instrumental surface analysis technique in which magnetically filtered 2 MeV helium ions bombard the solid sample surface and are backscattered and measured. The electronics sorts the results by the energy of the backscattered ion and plots the results versus the corresponding number of counted pulses.

Saha equation: n. an expression that describes ionization as an equilibrium process.

Sample: n. a part of a lot intended to represent the lot.

Sampler cone: n. the outer cone of an ICP–MS interface; it is usually nickel or platinum with a 1.0–1.2 mm orifice.

Saturated calomel electrode: n. a common type of reference electrode used in various forms of electrochemical measurements. It consists of a saturated potassium chloride solution, solid calomel (mercurous chloride), and metallic mercury.

Shot noise: n. a statistically distributed background emission of photoelectrons in photomultiplier tubes.

Skimmer cone: n. the inner cone of an ICP/MS interface; it is usually nickel or platinum with an orifice size that ranges from 0.04 to 1.0 mm, depending on system design.

Soller slit: n. an X-ray collimator consisting of a series of parallel metal strips.

Space charge: n. a region in vacuum occupied by ions where their electrostatic charges are not balanced by opposite charges.

Space charge effects: n. in mass spectrometry, a sensitivity bias toward higher mass analytes due to the disproportionate coulombic repulsion of lower mass analytes.

Sparge: v.t. to bubble a gas through a liquid to either cause or prevent a reaction.

Spectrography: n. the science and art of photographing and interpreting photographs of spectra.

Spectrometry: n. the measurement of spectra.

Spectrophotometry: n. a term reserved for the measurement of absorption spectra.

Spectroscopy: n. the scientific discipline that covers the theory, measurement, and application of spectra.

Summit potential: n. in cathode ray (linear sweep) polarography, the maxima of the readout peaks; the summit potential is proportional to the steady-state *half-wave potential:* (see).

Tare: v.t. to weigh an empty vessel so that its weight can be subtracted from the result of a subsequent (or a previous) weighing.

Test portion: n. the mass (or volume) of test sample used for a single analytical determination.

Test sample: n. that sample, prepared from the laboratory sample, from which test portions are removed.

Thermalize: v.t. to slow down fast neutrons by elastic scattering from collisions with light nuclei. Water and deuterium oxide are commonly used to thermalize neutrons.

Third element effect: n. in X-ray fluorescence spectrometry, a rarely encountered enhancement effect: matrix element A emits secondary X-rays that photoeject an inner shell electron from matrix element B, which then emits secondary X-rays that photoeject an inner shell electron from the analyte element.

Transmittance: n. the ratio of the transmitted radiant power to the incident radiant power as photons pass through a substance.

Transport number: n. in electrochemistry, the fraction of the total current carried by a given ion.

Triturate: v.t. to grind to a powder (e.g., by means of a mortar and pestle).

Unified Numbering System: n. a classification system for all metal alloys developed jointly by the Society of Automotive Engineers (SAE), and the American Society for Testing and Materials (ASTM). Abbreviation: "UNS."

Work function: n. the energy required to release an electron from a metal and move it to a point (infinity) where the metal exerts no influence on it.

Bibliography

The following are some significant texts in selected subjects covered in this volume. The reader should be aware that the lists are not exhaustive in any of the categories and that some of the volumes included are currently out of print.

SAMPLING AND SAMPLE PREPARATION

Bock, R.A. A Handbook of Decomposition Methods in Analytical Chemistry, John Wiley, New York, 1979.

Dolezal, J.; Povondra, P.; Sulcek, Z. Decomposition Techniques in Inorganic Analysis, Elsevier, New York, 1966.

Gy, P.M. Sampling of Heterogeneous and Dynamic Material Systems, Elsevier, Amsterdam, 1992.

Gy, P.M. Sampling of Particulate Materials, Theory and Practice, Elsevier, Amsterdam, 1982.

Kingston, H.M.; Jassie, L.B. (eds.) Introduction to Microwave Sample Preparation, American Chemical Society, Washington, D.C., 1988.

Pitard, F.F. Pierre Gy's Sampling Theory and Sampling Practice, Vols. I and II, CRC Press, Boca Raton, FL, 1989.

Sulcek, Z.; Povondra, P. Methods of Decomposition in Inorganic Analysis, CRC Press, Boca Raton, FL, 1989.

Zief, M.; Mitchell, J.W. Contamination Control in Trace Element Analysis, John Wiley, New York, 1976.

Zief, M.; Speights, R.M. (eds.) Ultrapurity: Methods and Techniques, Marcel Dekker, New York, 1972.

SEPARATION SCIENCE

Cheng, K.L.; Ueno, K.; Imamura, T. CRC Handbook of Organic Analytical Reagents, CRC Press, Boca Raton, FL, 1982.

Erdey, L. Gravimetric Analysis, Parts I–III (G. Svehla, trans.; I. Buzas, ed.), Pergamon, New York, 1965.
Hillebrand, W.F.; Lundell, G.E.F.; Bright, H.A.; Hoffman, J.I. Applied Inorganic Analysis, 2nd ed., John Wiley, New York, 1953.
Johnson, W.C. Organic Reagents for Metals, Chemical Publishing, New York, 1955.
Kodama, K. Methods of Quantitative Inorganic Analysis, Interscience, New York, 1963.
Lundell, G.E.F.; Hoffman, J.I. Outlines of Methods of Chemical Analysis, John Wiley, New York, 1938.
Minczewski, J.; Chwastowska, J.; Dybczynski, R. Separation and Preconcentration Methods in Inorganic Trace Analysis (M.R. Mason, trans. and ed.) Ellis Horwood/John Wiley, New York, 1982.
Morrison, G.H.; Freiser, H. Solvent Extraction in Analytical Chemistry, John Wiley, New York, 1957.
Rieman III, W.; Walton, H.F. Ion Exchange in Analytical Chemistry, Pergamon, New York, 1970.

MOLECULAR ABSORPTION SPECTROMETRY

Boltz, D.F. (ed.) Colorimetric Determination of Nonmetals, Interscience, New York, 1958.
Sandell, E.B. Colorimetric Determination of Traces of Metals, Interscience, New York, 1959.
Snell, F.D.; Snell, C.T. Colorimetric Methods of Analysis, D. Van Nostrand, New York, 1948.
Snell, F.D.; Snell, C.T.; Snell, C.A. Colorimetric Methods of Analysis, D. Van Nostrand, Princeton, NJ, 1959.

ATOMIC ABSORPTION SPECTROMETRY

Dean, J.A.; Rains, T.C. (eds.) Flame Emission and Atomic Absorption, Vols. 1–3, Marcel Dekker, New York, 1975.
Fuller, C.W. Electrothermal Atomization for Atomic Absorption Spectrometry, The Chemical Society, London, 1977.
Price, W.J. Spectrochemical Analysis by Atomic Absorption, Heyden, Philadelphia, 1979.
Robinson, J.W. Atomic Absorption Spectroscopy, Marcel Dekker, New York, 1966.
Slavin, W. Atomic Absorption Spectroscopy, John Wiley, New York, 1978.

Welz, B. Atomic Absorption Spectrometry (C. Skegg, trans.), 2nd ed., VCH, Weinheim, Germany, 1985.

ATOMIC EMISSION SPECTROMETRY

Boumans, P.W.J.M. Line Coincidence Tables for Inductively Coupled Plasma Emission Spectrometry, Vols. I and II, Pergamon, New York, 1980.
Boumans, P.W.J.M. (ed.) Inductively Coupled Plasma Emission Spectroscopy, John Wiley, New York, 1987.
Massachusetts Institute of Technology Wavelength Tables, MIT Press, Cambridge, MA, 1969.
Montaser, A.; Golightly, D.W. (eds.) Inductively Coupled Plasmas in Analytical Atomic Spectrometry, 2nd ed., VCH, New York, 1992.
Parsons, M.L.; Forster, A.; Anderson, D. An Atlas of Spectral Interferences in ICP Spectroscopy, Plenum, New York, 1980.
Slickers, K. Automatic Emission Spectroscopy, Bruhlsche Universitats-druckerei, Geissen, Germany, 1980.
Thomsen, V.B.E. Modern Spectrochemical Analysis of Metals, ASM International, Materials Park, OH, 1996.
Winge, R.K.; Fassel, V.A.; Peterson, V.J.; Floyd, A. Inductively Coupled Plasma Emission Spectroscopy, Elsevier, New York, 1985.

INORGANIC MASS SPECTROMETRY

Adams, F.; Gijbels, R.; Van Grieken, R. (eds.) Inorganic Mass Spectrometry, John Wiley, New York, 1988.
Evans, E.H.; Giglio, J.J.; Castillano, T.M.; Caruso, J.A. Inductively Coupled and Microwave Induced Plasma Sources for Mass Spectrometry, The Royal Society of Chemistry, Cambridge, U.K., 1995.
Holland, G.; Eaton, A.N. (eds.) Applications of Plasma Source Mass Spectrometry, The Royal Society of Chemistry, Cambridge, U.K., 1991.
Watson, J.T. Introduction to Mass Spectrometry, 3rd ed., Lippincott-Raven, Philadelphia, 1977.

THERMAL EVOLUTION

Melnick, L.M.; Lewis, L.L., Holt, B.D.(eds.) Determination of Gaseous Elements in Metals, John Wiley, New York, 1974.

X-RAY FLUORESCENCE SPECTROMETRY

Bertin, E.P. Principles and Practice of X-ray Spectrometric Analysis, 2nd ed., Plenum, New York, 1975.
Jenkins, R.; Gould, R.W.; Gedcke, D. Quantitative X-ray Spectrometry, Marcel Dekker, New York, 1981.
Klockenkämper, R. Total-reflection X-ray Fluorescence, John Wiley, New York, 1997.
Müller, R.O. Spectrochemical Analysis by X-ray Fluorescence (K. Keil, trans.), Plenum, New York, 1972.

ACTIVATION ANALYSIS

Bowen, H.J.M.; Gibbons, D. Radioactivation Analysis, Oxford University Press, London, 1963.
Lyon Jr., W.S. (ed.) Guide to Activation Analysis, D. Van Nostrand, Princeton, NJ, 1964.

ELECTROCHEMICAL METHODS

Heyrovski, J.; Zuman, P. Practical Polarography, Academic Press, New York, 1968.
Meites, L. Polarographic Techniques, Interscience, New York, 1955.
Milner, G.W.C. The Principles and Applications of Polarography and Other Electrochemical Processes, John Wiley, New York, 1957.

GAS CHROMATOGRAPHY OF METAL CHELATES

Guiochon, G.; Pommier, C. Gas Chromatography in Inorganics and Organometallics, Ann Arbor Science, Ann Arbor, MI, 1973.
Moshier, R.W.; Sievers, R.E. Gas Chromatography of Metal Chelates, Pergamon, Oxford, 1965.

TRACE INSTRUMENTAL METHODS (COMPREHENSIVE TEXTS)

Alfassi, Z.B. (ed.) Determination of Trace Elements, VCH, New York, 1994.
Christian, G.D., O'Reilly, J.E. (eds.) Instrumental Analysis, Allyn and Bacon, Boston, 1986.

Ewing, G.W. Instrumental Methods of Chemical Analysis, 4th ed., McGraw-Hill, New York, 1975.
Metals Handbook, 9th ed., Vol.10: Materials Characterization, American Society for Metals, Metals Park, OH, 1986.
Morrison, G.H. (ed.) Trace Analysis—Physical Methods, Interscience, New York, 1965.
Skoog, D.A.; West, D.M. Principles of Instrumental Analysis, 2nd ed., Saunders, Philadelphia, 1980.
Sibilia, J.P. (ed.) A Guide to Materials Characterization and Chemical Analysis, VCH, New York, 1996.
Winefordner, J.D. (ed.) Trace Analysis: Spectroscopic Methods for Elements, John Wiley, New York, 1976.
Vandecasteele, C.; Block, C.B. Modern Methods for Trace Element Determination, John Wiley, Chichester, U.K., 1997.

CHEMICAL ANALYSIS OF METALS

ASTM Chemical and Spectrometric Test Methods for Steel, American Society for Testing and Materials, West Conshohocken, PA, 1992.
Beamish, F.E.; Van Loon, J.C. Recent Advances in the Analytical Chemistry of the Noble Metals, Pergamon, New York, 1972.
Beamish, F.E. The Analytical Chemistry of the Noble Metals, Pergamon, New York, 1966.
Busev, A.I. Analytical Chemistry of Molybdenum (J. Schmorak, trans.) Ann Arbor-Humphrey, Ann Arbor, MI, 1969.
Codell, M. Analytical Chemistry of Titanium Metals and Compounds, Interscience, New York, 1959.
Donaldson, E.M. Some Instrumental Methods for the Determination of Minor and Trace Elements in Iron, Steel, and Nonferrous Metals, Monograph 884, Energy, Mines, and Resources, Canada, 1982.
Dulski, T.R. A Manual for the Chemical Analysis of Metals, ASTM MNL 25, American Society for Testing and Materials, West Conshohocken, PA, 1996.
Elinson, S.I.; Petrov, K.I. Analytical Chemistry of Zirconium and Hafnium (N. Kaner, trans.) Ann Arbor-Humphrey, Ann Arbor, MI, 1965.
Elwell, W.T.; Scholes, I.R. Analysis of Copper and Its Alloys, Pergamon, New York, 1967.
Elwell, W.T.; Wood, D.F. Analysis of the New Metals: Titanium, Zirconium, Hafnium, Niobium, Tantalum, Tungsten, and Their Alloys, Pergamon, New York, 1966.
Elwell, W.T.; Wood, D.F. Analytical Chemistry of Molybdenum and Tungsten, Pergamon, New York, 1971.

Gibalo, I.M. Analytical Chemistry of Niobium and Tantalum, Ann Arbor-Humphrey, Ann Arbor, MI, 1970.

Gregory, E.; Stevenson, W.W. Chemical Analysis of Metals and Alloys, 2nd ed., Blackie & Son, London, 1942.

Lewis, C.L.; Ott, W.L. Analytical Chemistry of Nickel, Pergamon, New York, 1970.

Lewis, C.L.; Ott, W.L.; Sine, N.M. The Analysis of Nickel, Pergamon, New York, 1966.

Lundell, G.E.F.; Hoffman, J.I.; Bright, H.A. Chemical Analysis of Iron and Steel, John Wiley, New York, 1931.

Moshier, R.W. Analytical Chemistry of Niobium and Tantalum, MacMillan-Pergamon, New York, 1964.

Peshkova, V.M.; Savostina, V.M. Analytical Chemistry of Nickel (J. Schmorak, trans.) Ann Arbor-Humphrey, Ann Arbor, MI 1969.

Pyatnitskii, I.V. Analytical Chemistry of Cobalt (N. Kaner, trans.) Ann Arbor-Humphrey, Ann Arbor, MI, 1969.

Rodden, C.J. (ed.) Analytical Chemistry of the Manhatten Project, McGraw-Hill, New York, 1950.

Ryabchikov, D.I.; Ryabukhin, V.A. Analytical Chemistry of Yttrium and the Lanthanide Elements (A. Aladjim, trans.) Ann Arbor-Humphrey, Ann Arbor, MI, 1970.

Van Loon, J.C.; Barefoot, R.R. Determination of Precious Metals, Selected Instrumental Methods, John Wiley, New York, 1991.

Vickery, R.E. Analytical Chemistry of the Rare Earths, Pergamon, New York, 1961.

Young, R.S. The Analytical Chemistry of Cobalt, Pergamon, New York, 1966.

QUALITY

Miller, J.C.; Miller, J.N. Statistics for Analytical Chemistry, 2nd ed., John Wiley, New York, 1988.

Natrella, M.G. Experimental Statistics, NIST Handbook 91, National Institute of Standards and Technology, Gaithersburg, MD, 1996.

Sargent, M.; MacKay, G. Guidelines for Achieving Quality in Trace Analysis, CRC Press, Boca Raton, FL, 1995.

Taylor, J.K. Standard Reference Materials: Handbook for SRM Users, NIST Special Publication 260-100, National Institute of Standards and Technology, Gaithersburg, MD 1985.

Youden, W.J. Experimentation and Measurement, NIST Special Publication 672, National Institute of Standards and Technology, Gaithersburg, MD, 1991.

Index